Hans-Joachim Hübschmann

Handbook of GC/MS

Fundamentals and Applications

Weinheim · New York · Chichester · Brisbane · Singapore · Toronto

Dr. Hans-Joachim Hübschmann
OmegaTech GmbH
Mikroforum Ring 2
D-55234 Wendelsheim
Germany

> This book was carefully produced. Nevertheless, author and publisher do not warrant the information contained therein to be free of errors. Readers are advised to keep in mind that statements, data, illustrations, procedural details or other items may inadvertently be inaccurate.

Die Deutsche Bibliothek – CIP-Cataloguing-in-Publication Data
A catalogue record for this publication is available from
Die Deutsche Bibliothek
  ISBN 3-527-30170-4

© WILEY-VCH Verlag GmbH, D-69469 Weinheim (Federal Republic of Germany), 2001

Printed on acid-free paper

All rights reserved (including those of translation in other languages). No part of this book may be reproduced in any form – by photoprinting, microfilm, or any other means – nor transmitted or translated into a machine language without written permission from the publishers. Registered names, trademarks, etc. used in this book, even when not specifically marked as such, are not to be considered unprotected by law.

Composition: ProSatz Unger, D-69469 Weinheim
Printing: betz-druck GmbH, D-64291 Darmstadt
Bookbinding: Wilh. Osswald & Co., D-67433 Neustadt/Weinstr.

Printed in the Federal Republic of Germany

Foreword

The coupling of gas chromatography with mass spectrometry is probably the oldest hybrid technique in instrumental analysis. The number of articles alone which were written on each individual technical aspect during the development of the method is enormous, not to mention articles on its applications. It is therefore to be expected that there would be a sufficient number of comprehensive and instructive books on the subject. However, this is not the case. Since the classical text of Bill McFadden (*Techniques of Combined Gas Chromatography/Mass Spectrometry* written in 1973), which is, of course, completely out of date but should still be read for historical interest, there is nothing which really meets current demands.

This deficit is all the more serious as GC/MS has been regarded as a single method and has been developed as such for some time. Benchtop instruments, which have been available from many suppliers for several years, can be used in almost all routine operations and in mobile form even outdoors in the field far away from a laboratory. A large majority of analysts will probably work with such an instrument at some point. A comprehensive account of the methods is therefore not only very important, but also necessary. In addition the number of analyses prescribed by legislation, which must be carried out using GC/MS, is constantly increasing.

Now at last a scientist experienced in the field has taken up the challenge of filling the gap with the present book. It cannot be easy to find one's way through the large quantity of technically relevant information, the complex theoretical aspects and the large number of important applications in widely varying areas of analytical chemistry. The presentation needs to tread the difficult path between not being removed from the practice of the art, while not concentrating so intensively on everyday problems that it carries within it the seed of its own decay. Another matter of concern here is that such a text might give rise to prejudice and perhaps also to incorrect practices. Here also the greatest care is therefore necessary in the choice of words and examples.

However, with the present treatise the author has succeeded in overcoming these problems. He has prepared a book which is more than just a handbook. It has not only been written for the practical chemist who needs to consult it in the laboratory, but also for the analyst who wishes to familiarise himself with any or all the aspects of GC/MS.

The instrumental aspects of sample preparation, gas chromatographic separation and mass spectrometry are described carefully and comprehensively, including the most recent developments. A comparison with other common methods of detection is also included. The particular features of GC/MS coupling are then discussed and finally a large number of selected examples are given, showing how far this method has penetrated the field of analysis in environmental chemistry and medicine.

After the chapter on sample preparation, in which the versatile possibilities of various extraction methods are described, injection techniques are discussed. The author then deals with the important aspects of separation by gas chromatography and gives the relevant theoretical background and parameters for effecting and optimising it. The chromatography section ends with a short comparison of different GC detectors.

The book then turns to mass spectrometry, explaining the most important terms with great clarity and concentrating on the most important aspects. After giving the strengths and weaknesses of the types of interface still in use today, the role of data systems in the production, determination and evaluation of data is given in appropriate detail. Finally the important classes of substance are listed and their analysis described with all the important practical aspects from separation to the choice of the correct internal standard. Special emphasis is placed on quantitative determination, showing that the idea that GC/MS coupling is difficult to use for precise quantitative measurements is no longer justifiable. The opposite is, in fact, the case. Comprehensive quality assurance is more possible with GC/MS because of the higher information content of the results, compared with classical detectors. Inaccuracies and errors are easier to find and to avoid than with nonspectrometric GC detectors.

The final chapter contains examples of applications, a large number of which come from analysts in industry, research institutes and universities. Through its extensive and precise presentation of the most important parameters and results, this chapter gives the practical chemist some useful ideas which will make his everyday work more reliable and successful. Even if a particular application is no longer being used at some future time, the analyses presented here are so instructive that they will still be of value for a long period. Finally it should be mentioned that the references have been set out in such a way that both experienced personnel and beginners can find further details in the most important primary literature.

This book, therefore, fills a long existing gap. It should enable the practical chemist to compare his procedure for the detection of a substance with others and perhaps to optimise it, to get stimulation or simply to recognise that the procedure which he is looking for has been well worked out in a similar form. The beginner will find preliminary advice on working out his own procedures and clear information on problems and special cases, besides all the important basic concepts. The book can definitely be used to teach the use of GC/MS, even though the scope, intention and price of this book do not directly suggest it as a textbook. It certainly contains much important material of the required quality.

Prof. Michael Linscheid
Humboldt University Berlin

Preface to the English Edition

More than three years have elapsed since the original German publication of the Handbook of GC/MS. GC/MS instrument performance has significantly improved in these recent years. GC/MS methodology has found its sound place in many "classical" areas of application of which many application notes are reported as examples in this handbook. Today the use of mostly automated GC/MS instrumentation is standard. Furthermore GC/MS as a mature analytical technology with a broad range of robust instruments increasingly enters additional analytical areas and displaces the "classical" instrumentation.

The very positive reception of the original German print and the wide distribution of the handbook into different fields of application has shown that comprehensive information about functional basics as well as the discussion about the practical use for different applications is important for many users for efficient method development and optimization.

Without the support from interested users and the GC/MS community concerned, the advancement and actualisation of this handbook would not be possible. My special thanks go to the active readers for their contribution to valuable discussions and details. Many of the applications notes have been updated or replaced by the latest methodology.

I would like to express my personal thanks to Dr. Brody Guggenberger (ThermoQuest Corp., Austin TX), Joachim Gummersbach (ThermoQuest GmbH, Egelsbach), Gert-Peter Jahnke (ThermoQuest APG GmbH, Bremen), Prof. Dr. Ulrich Melchert (Robert-Koch-Institut, Berlin), Dr. Jens P. Weller (Institut für Rechtsmedizin der Medizinischen Hochschule, Direktor Prof. Dr. med. H. D. Tröger, Hannover), and Dr. John Ragsdale jr. (ThermoQuest Corp., Austin TX) for their valuable discussions and contributions with application documentation and data.

My sincere thanks to Dr. Elisabeth Grayson for the careful text translation.

I wish all users of this handbook an interesting and informative read. Comments and suggestions concerning further improvement of the handbook are very much appreciated.

Sprockhövel, August 2000   Hans J. Hübschmann

Preface to the German Edition

The great extension of the use of mass spectrometers has been accelerated considerably through their coupling with gas chromatography. From the middle of the 1980s integrated GC/MS systems have been increasingly used in what were previously purely gas chromatography laboratories. At first the detector philosophy predominated, as expressed in the trade names 'mass specific detector' or 'ion trap detector'. Now, more than ten years later, highly developed powerful mass spectrometers have replaced mass spectrometric detectors. As far as GC/MS coupling compatibilities are concerned, these benchtop machines are just as powerful as the early large instruments for organic analysis.

Gas chromatography frequently provides the only inlet in GC/MS benchtop instruments. The construction of the instruments, their operation and the software for data evaluation are specially orientated towards total integration (hyphenated technique). In this aspect GC/MS systems differ significantly from large mass spectrometers (which can also be connected to a gas chromatograph). In the operation and in the evaluation software there are correspondences or direct connections with chromatography data systems.

In this book GC/MS is treated as a developed, stable and flexible coupling technique. The mutual dependence, effects and increase in power achieved by the coupling are clearly covered in the consideration of the basic functions and in the examples of applications. In the individual areas of gas chromatography and mass spectrometry extensive and up to date books have been published. However, the analytical potential of the coupling of the two processes with consideration of the special instrumental technique and its applications has, up till now, only been covered on the periphery.

In this handbook both the basis of the GC/MS coupling technique and the practical aspects of the evaluation of analyses are described. This is followed by detailed coverage of current applications. The discussion of typical sample preparation procedures, which are coupled with GC/MS either on- or off-line, takes into account the emerging trend of integrated sample preparation in GC/MS systems. The basis, coupling techniques and parameters for selected instrumental methods, which are frequently used in direct association with GC/MS analysis as well as classical procedures, are also given.

The practical aspects of sample injection into capillary columns and the choice of columns for gas chromatography are described in detail. The possibilities for optimising a particular separation are outlined. For GC/MS applications, as with classical detectors, the quality of the chromatography is the limiting factor for the whole system.

In mass spectrometry emphasis is placed on practice of GC/MS applications. The analytical background and the use of the ionisation techniques EI and CI are dealt with in detail. The interested reader can consult the extensive current literature available on the basics of the production of mass spectra and their interpretation. The widely used techniques for obtaining full scan mass spectra and the recording of individual masses (SIM) are supplemented by the GC/MS techniques currently available even with benchtop instruments.

The interpretation of GC/MS chromatograms for qualitative and quantitative analysis has been given special consideration and explored in detail for everyday applications. The comparison of mass spectra with those already published can initially be used for identification of substances. Modern computer techniques give rapid results and can also tempt the user into reaching rash positive or negative conclusions. A scheme for the interpretation and details of the decoding of mass spectra can form a starting point for further work in the explanation of unknown signals. For quantitative determinations, calibration procedures are given, in particular the use of labelled internal standards.

As a compendium for practical work there is also extensive coverage of applications from different areas. For selected groups of substances methods of resolving problems from recognised laboratories are presented. The aim of this section is to give information and stimulus for practical situations and to promote the exchange of experiences between users.

My grateful thanks extend to all those who have made writing this handbook possible through their contributions, support and critical discussions: P. Bachhausen (Landesumweltamt NRW, Düsseldorf), H. Bittdorf (Kernforschungszentrum, Karlsruhe), P. Brand (Axel Semrau GmbH, Sprockhövel), Dr. U. Demme (Institut für Rechtsmedizin, Klinikum der Friedrich-Schiller-Universität, Jena), Th. Egert (Institut Fresenius, Taunusstein), H.-D. Eschke (Ruhrverband, Essen), Dr. P. Fürst (Chemisches Landesuntersuchungsamt Nordrhein-Westfalen, Münster), W. Gatzemann (Dortmunder Stadtwerke AG, Schwerte), B. Grass (Institut für Spektrochemie und angewandte Spektroskopie, Dortmund), Dr. K. Grob (Kantonales Labor, Zurich, Switzerland), F. Heimlich (Institut für Spektrochemie und angewandte Spektroskopie, Dortmund), Dr. D. Henneberg (Max-Planck-Institut für Kohleforschung, Mülheim/Ruhr), H. Korpien (Stadtwerke Frankfurt am Main), Dr. J. Kurz (Institut Fresenius, Taunusstein), M. Landrock (Institut Fresenius Sachsen, Dresden), P. Lönz (Terrachem Essen GmbH, Essen), U. Matthiesen (Heinrich-Heine-Universität, Düsseldorf), L. Matter (Dinslaken), H. Merten (Institut Fresenius Sachsen, Dresden), M. & M. Mierse (Berlin), A. Mülle (Axel Semrau GmbH, Sprockhövel), J. Niebel (Axel Semrau GmbH, Augsburg), Dr. J. Nolte (Institut für Spektrochemie und angewandte Spektroskopie, Dortmund), Dr. W. Ockels (SpectralService GmbH, Cologne), H.-G. Ostrop (AGR mbH, Gelsenkirchen), Dr. Th. Paschke (Gesamtuniversität Siegen), H. Richter (Institut Fresenius Sachsen, Dresden), Dr. H. Sachs (Institut für Rechtsmedizin, Munich), Dr. C. Schlett (Gelsenwasser AG, Gelsenkirchen), Prof. Dr. H.-F. Schöler (Institut für Sedimentforschung, Ruprecht-Karls-Universität, Heidelberg), A. Semrau (Axel Semrau GmbH, Sprockhövel), Prof. Dr. H.-J. Stan (Institut für Lebensmittelchemie, Technische Universität, Berlin), M. Tschickard (Landesamt für Umweltschutz und Gewerbeaufsicht Rheinland-Pfalz, Mainz), Dr. H. Vorholz (Ingenieurgemeinschaft Technischer Umweltschutz, Potsdam), Dr. M. Wislicenus (Staatliche Milchwirtschaftliche Lehr- und Forschungsanstalt, Kempten).

I would like to thank the following companies for their generous provision of product documentation and application descriptions: Baker Chemikalien (Gross-Gerau), Balzers Hochvakuum GmbH (Wiesbaden-Nordenstedt/Asslar), Chrompack GmbH

(Frankfurt), Finnigan MAT GmbH (Bremen/San Jose CA, USA), Fischer GmbH (Meckenheim), HNU/Nordion (Helsinki, Finland), ICT GmbH (Bad Homburg v. d. H.), ISCO Corporation (Lincoln NE, USA), Palisade Corporation (Newfield NY, USA), Perkin Elmer GmbH (Überlingen), Pyrol AB (Lund, Sweden), Restek Corporation (Bellafonte PA, USA), Fa. Seekamp (Achim), SGE GmbH (Weiterstadt), Supelco Deutschland GmbH (Bad Homburg v. d. H.), Suprex Corporation (Pittsburgh, PA, USA), Tekmar Company (Cincinatti, OH, USA).

I would also like to express my gratitude to the VCH editors Mrs Cornelia Reinemuth and Dr. Steffen Pauly for their generous support and Mrs Claudia Grössl for her untiring assistance in the compilation of this book.

For much stimulation and valuable help in drawing up the manuscript, I particularly thank my wife Gudrun. I am also grateful to our children for their limitless patience and the many coloured pictures they drew on my manuscripts!

Sprockhövel, November 1996   Hans-Joachim Hübschmann

Contents

Foreword V
Preface to the English Edition VII
Preface to the German Edition IX

1	**Introduction**	1
2	**Basics**	7
2.1	Sample Preparation	7
2.1.1	Sample Preparation by Solid Phase Extraction	10
2.1.2	Sample Preparation by SFE	14
2.1.3	Headspace Techniques	26
2.1.3.1	Static Headspace Technique	27
2.1.3.2	Dynamic Headspace Technique (Purge & Trap)	35
2.1.3.3	Headspace versus Purge & Trap	45
2.1.4	Adsorptive Enrichment and Thermodesorption	51
2.1.5	Pyrolysis	59
2.2	Gas Chromatography	67
2.2.1	GC/MS Sample Inlet Systems	67
2.2.1.1	Hot Sample Injection	68
2.2.1.2	Cold Injection Systems	72
2.2.1.3	Injection Volumes	81
2.2.1.4	On-column Injection	83
2.2.1.5	Cryofocusing	86
2.2.2	Capillary Columns	88
2.2.2.1	Sample Capacity	89
2.2.2.2	Internal Diameter	89
2.2.2.3	Film Thickness	97
2.2.2.4	Column Length	98
2.2.2.5	Adjusting the Carrier Gas Flow	99
2.2.2.6	Properties of Stationary Phases	100
2.2.3	Chromatography Parameters	103
2.2.3.1	The Chromatogram and its Meaning	103
2.2.3.2	Capacity Factor k'	106
2.2.3.3	Chromatographic Resolution	107
2.2.3.4	Factors Affecting the Resolution	110
2.2.3.5	Maximum Sample Capacity	112

2.2.3.6	Peak Symmetry	113
2.2.3.7	Optimisation of Flow	113
2.2.4	Classical Detectors for GC/MS Systems	117
2.2.4.1	FID	117
2.2.4.2	NPD	119
2.2.4.3	ECD	121
2.2.4.4	PID	123
2.2.4.5	ELCD (Electrolytical Conductivity Detector)	125
2.2.4.6	FPD (Flamephotometric Detector)	127
2.2.4.7	Connection of Classical Detectors Parallel to the Mass Spectrometer	128
2.3	Mass Spectrometry	130
2.3.1	Resolution in Mass Spectrometry	131
2.3.1.1	High Resolution	131
2.3.1.2	Unit Mass Resolution	133
2.3.1.3	High and Low Resolution in the Case of Dioxin Analysis	138
2.3.2	Ionisation Procedures	140
2.3.2.1	Electron Impact Ionisation	140
2.3.2.2	Chemical Ionisation	144
2.3.3	Measuring Techniques in GC/MS	164
2.3.3.1	Detection of the Complete Spectrum (Full Scan)	164
2.3.3.2	Recording Individual Masses (SIM/MID)	166
2.3.3.3	MS/MS – Tandem Mass Spectrometry	177
2.3.4	Mass Calibration	185
2.4	Special Aspects of GC/MS Coupling	192
2.4.1	Vacuum Systems	192
2.4.2	GC/MS Interface Solutions	196
2.4.2.1	Open Split Coupling	196
2.4.2.2	Direct Coupling	198
2.4.2.3	Separator Techniques	199
	References for Chapter 2	200
3	**Evaluation of GC/MS Analyses**	213
3.1	Display of Chromatograms	213
3.1.1	Total Ion Current Chromatograms	214
3.1.2	Mass Chromatograms	215
3.2	Substance Identification	218
3.2.1	Extraction of Mass Spectra	218
3.2.2	The Retention Index	225
3.2.3	Libraries of Mass Spectra	230
3.2.3.1	Universal Libraries of Mass Spectra	231
3.2.3.2	Application-orientated Libraries of Mass Spectra	233
3.2.4	Library Search Procedures	236
3.2.4.1	The INCOS/NIST Search Procedure	236
3.2.4.2	The PBM Search Procedure	243

3.2.4.3	The SISCOM Procedure	247
3.2.5	Interpretation of Mass Spectra	251
3.2.5.1	Isotope Patterns	253
3.2.5.2	Fragmentation and Rearrangement Reactions	259
3.2.6	Mass Spectroscopic Features of Selected Substance Classes	266
3.2.6.1	Volatile Halogenated Hydrocarbons	266
3.2.6.2	Benzene/Toluene/Ethylbenzene/Xylenes (BTEX, Alkylaromatics)	266
3.2.6.3	Polynuclear Aromatics	273
3.2.6.4	Phenols	273
3.2.6.5	Plant Protection Agents	279
3.2.6.6	Polychlorinated Biphenyls (PCBs)	293
3.2.6.7	Polychlorinated Dioxins/Furans (PCDDs/PCDFs)	297
3.2.6.8	Drugs	298
3.2.6.9	Explosives	301
3.2.6.10	Chemical Warfare Agents	306
3.2.6.11	Flameproofing Agents	309
3.3	Quantitation	310
3.3.1	Decision Limit	311
3.3.2	Limit of Detection	312
3.3.3	Limit of Quantitation	312
3.3.4	Sensitivity	314
3.3.5	The Calibration Function	314
3.3.6	Quantitation and Standardisation	316
3.3.7	The Standard Addition Procedure	320
3.4	Frequently Occurring Impurities	322
	References for Chapter 3	329
4	**Applications**	333
4.1	Air Analysis According to EPA Method TO-14	333
4.2	BTEX using Headspace GC/MS	341
4.3	Simultaneous Determination of Volatile Halogenated Hydrocarbons and BTEX	345
4.4	Static Headspace Analysis of Volatile Priority Pollutants	349
4.5	MAGIC 60 – Analysis of Volatile Organic Compounds	355
4.6	Vinyl Chloride in Drinking Water	363
4.7	Chloral Hydrate in Surface Water	366
4.8	Field Analysis of Soil Air	369
4.9	Field Analysis of Solvent Contamination of Soil	373
4.10	Residual Monomers and Polymerisation Additives	377
4.11	Field Analysis of Odour Emissions	380
4.12	Geosmin and Methylisoborneol in Drinking Water	385
4.13	Substituted Phenols in Drinking Water	389
4.14	Pesticides in Tea	394
4.15	GC/MS/MS Target Compound Analysis of Pesticide Residues in Difficult Matrices	403

4.16	Triazine Herbicides in Drinking and Untreated Water	415
4.17	Nitrophenol Herbicides in Water	421
4.18	Dinitrophenol Herbicides in Water	424
4.19	Hydroxybenzonitrile Herbicides in Drinking Water	430
4.20	Determination of Phenoxyalkylcarboxylic Acids in Water Samples	437
4.21	Pentachlorophenol in Leather Goods	442
4.22	Polycyclic Aromatic Hydrocarbons	444
4.23	Polyaromatic Hydrocarbons in City Air	454
4.24	Routine Analysis of 24 PAHs in Water and Soil	459
4.25	Polychlorinated Biphenyls in Milk and Milk Products	463
4.26	Polychlorinated Biphenyls in Indoor Air	470
4.27	Screening for Dioxins and Furans	473
4.28	Analysis of Military Waste	483
4.29	SFE Extraction and Determination of Nitroaromatics	493
4.30	Detection of Drugs in Hair	497
4.31	Detection of Morphine Derivatives	500
4.32	Detection of Cannabis Consumption	504
4.33	Determination of Phencyclidine	507
4.34	Analysis of Steroid Hormones using MS/MS	509
4.35	Determination of Prostaglandins using MS/MS	513
4.36	Amphetamines – Differentiation by CI	520
4.37	Identification and Quantitation of Barbiturates	526
4.38	Detection of Clenbuterol by CI	532
4.39	Systematic Toxicological-chemical Analysis	536
4.40	Clofibric Acid in Aquatic Systems	541
4.41	Polycyclic Musks in Waste Water	545
	References for Chapter 4	551
5	**Glossary**	559
Subject Index		575
Index of Chemical Substances		584

1 Introduction

Detailed knowledge of the chemical processes in plants and animals and in our environment has only been made possible through the power of modern instrumental analysis. In an increasingly short time span more and more data are being collected. The absolute detection limits for organic substances lie in the attomole region and counting individual molecules per unit time has already become a reality. We are making measurements in the area of background contamination (whatever that is meant to be). Most samples subjected to chemical analysis are now mixtures, as are even blank samples. With the demand for decreasing detection limits, in the future effective sample preparation and separation procedures in association with highly selective detection techniques will be of critical importance for analysis. In addition the number of substances requiring detection is increasing and with the broadening possibilities for analysis, so is the number of samples. The increase in analytical sensitivity is exemplified in the case of 2,3,7,8-TCDD.

Year	Instrumental technique	Limit of detection [pg]
1967	GC/FID (packed column)	500
1973	GC/MS (quadrupole, packed column)	300
1976	GC/MS-SIM (magnetic instrument, capillary column)	200
1977	GC/MS (magnetic sector instrument)	5
1983	GC/HRMS (double focusing magnetic sector instrument)	0.15
1984	GC/MSD-SIM (quadrupole benchtop instrument)	2
1986	GC/HRMS (double focusing magnetic sector instrument)	0.025
1989	GC/HRMS (double focusing magnetic sector instrument)	0.010
1992	GC/HRMS (double focusing magnetic sector instrument)	0.005

Capillary gas chromatography is today the most important process in organic chemical analysis for the determination of individual substances in complex mixtures. Mass spectrometry as the detection procedure gives the most meaningful data, arising from the direct determination of the substance molecule or of fragments. The results of mass spectrometry are therefore used as a reference for other indirect processes and finally for confirmation of the facts. The complete integration of mass spectrometry and gas chromatography into a single GC/MS system has shown itself to be synergistic in every respect. While at the beginning of the 1980s mass spectrometry was considered to be expensive, complicated and time-consuming or personnel-intensive, there is now hardly a GC laboratory which is not equipped with a GC/MS system. At the beginning of the 1990s mass spectrometry became more widely recognised and furthermore an indispensable detection procedure for gas chromatography. The simple construction, clear function and an operating procedure, which has become easy because of modern computer systems, have resulted in the fact that GC/MS is widely used alongside traditional spectroscopic methods. The universal detection technique together with high specificity and very high sensitivity have made GC/MS important for a broad spectrum of applications. The number of benchtop GC/MS instruments installed worldwide will exceed 20 000 in the coming years and in future will be determined by the number of gas chromatographs required. Out of a promising process for the expensive explanation of spectacular individual cases, a universally used routine process has developed within a few years. The serious reservations of experienced spectroscopists wanting to keep mass spectrometry within the spectroscopic domain, have been found to be without substance because of the broad success of the coupling procedure. The control of the chromatographic procedure still contributes significantly to the exploitation of the analytical performance of the GC/MS system (or according to Konrad Grob: chromatography takes place in the column!). The analytical prediction capabilities of a GC/MS system are, however, dependent upon mastering the spectrometry. The evaluation and assessment of the data is leading to increasingly greater challenges with decreasing detection limits and the increasing number of compounds sought or found. At this point the circle goes back to the earlier reservations of renowned spectroscopists.

The high performance of gas chromatography lies in separation of the substance mixtures. With the introduction of fused silica columns GC has become the most important and powerful method of analysing complex mixtures of products. GC/MS accommodates the current trend towards multimethods or multicomponent analyses (e.g. of pesticides, solvents etc) in an ideal way. Even isomeric compounds, which are present, for example in terpene mixtures, in PCBs and in dioxins, are separated by GC, while in many cases their mass spectra are almost indistinguishable. The high efficiency as a routine process is achieved through the high speed of analysis and the short turn-round time and thus guarantees a high sample throughput. Adaptation and optimisation for different tasks only requires a quick change of column. In many cases, however, and here one is relying on the predicting capabilities of the mass spectrometer, one type of column can be used for different applications by adapting the sample injection technique and modifying the method parameters.

The area of application of GC and GC/MS is limited to substances which are volatile enough to be analysed by gas chromatography. The further development of column technology in recent years has been very important for application to the analysis of high-boiling compounds. Temperature-stable phases now allow elution temperatures of up to 500 °C. A pyrolyser in the form of a stand-alone sample injection system extends the area of application to involatile substances by separation and detection of thermal decomposition products. A typical example of current interest for GC/MS analysis of high-boiling compounds is the determination of polyaromatic hydrocarbons, which has become a routine process using the most modern column material. It is incomprehensible that, in spite of an obvious detection problem, HPLC is still frequently used in parallel to GC/MS to determine polyaromatic hydrocarbons in the same sample.

The coupling of gas chromatography with mass spectrometry using fused silica capillaries has played an important role in achieving a high level of chemical analysis. In particular in the areas of environmental analysis, analysis of residues and forensic science the high information content of GC/MS analyses has brought chemical analysis into focus through sometimes sensational results. For example, it has been used for the determination of anabolic steroids in cough mixture and the accumulation of pesticides in the food chain. With the current state of knowledge GC/MS is an important process for monitoring the introduction, the location and fate of man-made substances in the environment, foodstuffs, chemical processes and biochemical processes in the human body. GC/MS has also made its contribution in areas such as the ozone problem, the safeguarding of quality standards in foodstuffs production, in the study of the metabolism of pharmaceuticals or plant protection agents or in the investigation of polychlorinated dioxins and furans produced in certain chemical processes, to name but a few.

The technical realisation of GC/MS coupling occupies a very special position in instrumental analysis. Fused silica columns are easy to handle, can be changed rapidly and are available in many high quality forms. The optimised carrier gas streams show good compatibility with mass spectrometers. Coupling can therefore take place easily by directly connecting the GC column to the ion source of the mass spectrometer. The operation of the GC/MS instrument can be realised because of the low carrier gas flow in the widely used benchtop instruments even with a low pumping capacity. Only small instruments are therefore necessary, and these also accommodate a low pumping capacity. A general knowledge of the construction and stable operating conditions forms the basis of smooth and easily learned service and maintenance. Compared with GC/MS coupling, LC/MS coupling, for example, is still much more difficult to control.

The limits of GC and GC/MS lie where actual samples contain involatile components (matrix). In this case the sample must be processed before the analysis proper. The clean-up is generally associated with enrichment of trace components. In many methods there is a trend towards integrating sample preparation and enrichment in a single instrument. Even today the headspace and purge and trap techniques, thermodesorption, SPME (solid phase microextraction) or SFE (supercritical fluid extraction) are coupled on-line with GC/MS and are integrated stepwise into the data system for smooth control. Development will continue in this area in future, and as a result will move the focus from the previously expensive mass spectrometer to the highest possible sample throughput and will convert positive substance detection in the mass spectrometer into an automatically performed evaluation.

The high information content of GC/MS analyses requires powerful computers with intelligent programs to evaluate them. The evaluation of GC/MS analyses based on data systems is therefore a necessary integral component of modern GC/MS systems. Only when the evaluation of mass spectrometric and chromatographic data can be processed together can the performance of the coupling process be exploited to a maximum by the data systems. In spite of the state of the art computer systems, the performance level of many GC/MS data systems has remained at the state it was 20 years ago and only offers the user a coloured data print-out. The possibilities for information processing have remained neglected on the part of the manufacturers and require the use of external programs (e.g. the characterisation of specimen samples, analysis of mixtures, suppressing noise etc).

Nonetheless development of computer systems has had a considerable effect on the expansion of GC/MS systems. The manual evaluation of GC/MS analyses has become practically impossible because of the enormous quantity of data. A 60-minute analysis with two spectra per second over a mass range of 500 mass units gives 3.65 million pairs of numbers! The use of good value but powerful PCs allows the systems to be controlled but gives rapid processing of the relevant data and thus makes the use of GC/MS systems economically viable.

The Historical Development of the GC/MS Technique

The GC/MS technique is a recent process. The foundation work in both GC and MS which led to the current realisation was only published between the middle and the end of the 1950s. At the end of the 1970s and the beginning of the 1980s a rapid increase in the use of GC/MS in all areas of organic analysis began. The instrumental technique has now achieved the required level for the once specialised process to become an indispensable routine procedure.

1910	The physicist J.J. Thompson developed the first mass spectrometer and proved for the first time the existence of isotopes (^{20}Ne and ^{22}Ne). He wrote in his book 'Rays of Positive Electricity and their Application to Chemical Analysis': *'I have described at some length the application of positive rays to chemical analysis: one of the main reasons for writing this book was the hope that it might induce others, and especially chemists, to try this method of analysis. I feel sure that there are many problems in chemistry which could be solved with far greater ease by this than any other method'*. Cambridge 1913.
1910	In the same year M.S. Tswett published his book in Warsaw on 'Chromophores in the Plant and Animal World'. With this he may be considered to be the discoverer of chromatography.
1918	Dempster used electron impact ionisation for the first time.
1920	Aston continued the work of Thompson with his own mass spectrometer equipped with a photoplate as detector. The results verified the existence of isotopes of stable elements (e.g. ^{35}Cl and ^{37}Cl) and confirmed the results of Thompson.
1929	Bartky and Dempster developed the theory for a double-focusing mass spectrometer with electrostat and magnetic sector.
1934	Mattauch and Herzog published the calculations for an ion optics system with perfect focusing over the whole length of a photoplate.
1935	Dempster published the latest elements to be measured by MS, Pt and Ir. Aston thus regarded MS to have come to the end of its development.
1936	Bainbridge and Jordan determined the mass of nuclides to six significant figures.
1937	Smith determined the ionisation potential of methane (as the first organic molecule).
1938	Hustrulid published the first spectrum of benzene.
1941	Martin and Synge published a paper on the principle of gas liquid chromatography, GLC.

1946	Stephens proposed a time of flight (TOF) mass spectrometer: velocitron.
1947	The US National Bureau of standards (NBS) began the collection of mass spectra as a result of the use of MS in the petroleum industry.
1948	Hipple described the ion cyclotron principle, known as the 'Omegatron' which now forms the basis of the current ICR instruments.
1950	Gohlke published for the first time the coupling of a gas chromatograph (packed column) with a mass spectrometer (Bendix TOF, time of flight).
1950	The Nobel Prize for chemistry wass awarded to Martin and Synge for their work on gas liquid chromatography (1941).
From 1950	McLafferty, Biemann and Beynon applied MS to organic substances (natural products) and transferred the principles of organic chemical reactions to the formation of mass spectra.
1952	Cremer and coworkers presented an experimental gas chromatograph to the ACHEMA in Frankfurt; parallel work was carried out by Janák in Czechoslovakia.
1952	Martin and James published the first applications of gas liquid chromatography.
1953	Johnson and Nier published an ion optic with a 90° electric and 60° magnetic sector, which, because of the outstanding focusing properties, was to become the basis for many high resolution organic mass spectrometers (Nier/Johnson analyser).
1954	Paul published his fundamental work on the quadrupole analyser.
1955	Wiley and McLaren developed a prototype of the present time of flight (TOF) mass spectrometer.
1955	Desty presented the first GC of the present construction type with a syringe injector and thermal conductivity detector. The first commercial instruments were supplied by Burrell Corp., Perkin Elmer, and Podbielniak Corp.
1956	A German patent was granted for the QUISTOR (quadrupole ion storage device) together with the quadrupole mass spectrometer.
1958	Paul published information on the quadrupole mass filter as – a filter for individual ions, – a scanning device for the production of mass spectra, – a filter for the exclusion of individual ions.
1958	Ken Shoulders manufactured the first 12 quadrupole mass spectrometers at Stanford Research Institute, California.
1958	Golay reported for the first time on the use of open tubular columns for gas chromatography.
1958	Lovelock developed the argon ionisation detector as a forerunner of the electron capture detector (ECD, Lovelock and Lipsky).
1964	The first commercial quadrupole mass spectrometers were developed as residual gas analysers (Quad 200 RGA) by Bob Finnigan and P.M. Uthe at EAI (Electronic Associates Inc., Paolo Alto, California).
1966	Munson and Field published the principle of chemical ionisation.
1968	The first commercial quadrupole GC/MS system for organic analysis was supplied by Finnigan Instruments Corporation to the Stanford Medical School Genetics Department.

1978	Dandenau and Zerenner introduced the technique of fused silica capillary columns.
1978	Yost and Enke introduced the triple-quadrupole technique.
1982	Finnigan obtained the first patents on ion trap technology for the mode of selective mass instability and presented the ion trap detector as the first universal MS detector with a PC data system (IBM XT).
1989	Prof. Wolfgang Paul, Bonn University received the Nobel Prize for physics for work on ion traps, together with Prof. Hans G. Dehmelt, University of Washington in Seattle, and Prof. Norman F. Ramsay, Harvard University.

2 Basics

2.1 Sample Preparation

The preparation of analysis samples is today already an integral part of practical GC/MS analysis. The trend is clearly going in the direction of automated instrumental techniques and limits manual work to the essential. The concentration processes in this development are of particular importance for coupling with capillary GC/MS, as in trace analysis the limited sample capacity of capillary columns must be compensated for. It is therefore necessary both that overloading of the stationary phase by the matrix is avoided and that the limits of mass spectrometric detection are taken into consideration in the development of a process. To optimise separation on a capillary column, strongly interfering components of the matrix must be removed before applying an extract. The primarily universal character of the mass spectrometer poses conditions on the preparation of a sample which are to some extent more demanding than those of an element-specific detector, such as ECD or NPD. Concentration, which forms part of sample preparation, must therefore in principle always be regarded as a preparative step for GC/MS analysis. The differences in the concentration ranges between various samples, differences between the volatility of the analytes and that of the matrix and the varying chemical nature of the substances are important for the choice of a suitable sample preparation procedure.

Off-line techniques (as opposed to on-line coupling or hyphenated techniques) have the particular advantage that samples can be processed in parallel and the extracts can be subjected to other analytical processes besides GC/MS. On-line techniques have the special advantage of sequential processing of the samples without intermediate manual steps. Here there is an optimal time overlap which gives the sample preparation the same amount of time as the analysis of the preceding sample. This permits maximum use of the instrument and automatic operation.

On-line processes generally offer potential for higher analytical quality through lower contamination from the laboratory environment and, for smaller sample sizes, lower detection limits with lower material losses. Frequently total sample transfer is possible without taking aliquots or diluting. Volatility differences between the sample and the matrix allow, for example, the use of extraction techniques such as the static or dynamic (purge and trap) headspace techniques as typical GC/MS coupling techniques. These are already used as on-line techniques in many laboratories. Where the volatility of the analytes is insufficient, other extraction procedures are being increasingly used on-line. Solid phase extraction in the form of microextraction, LC/GC coupling, or extraction with supercritical fluids show high analytical potential here.

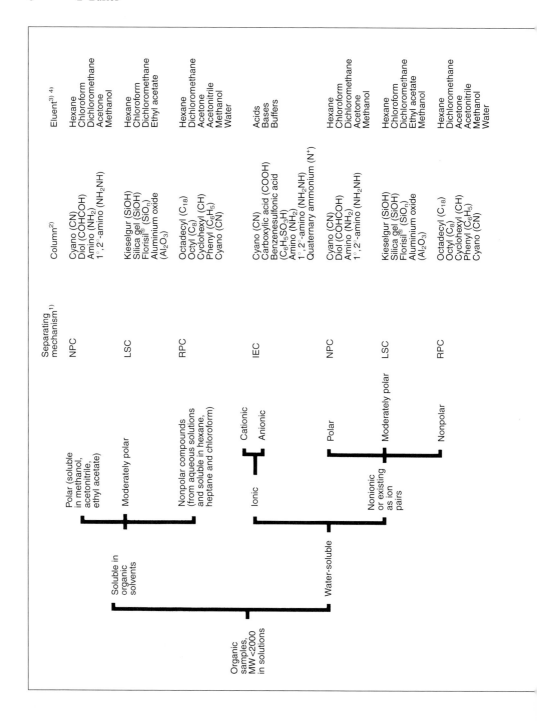

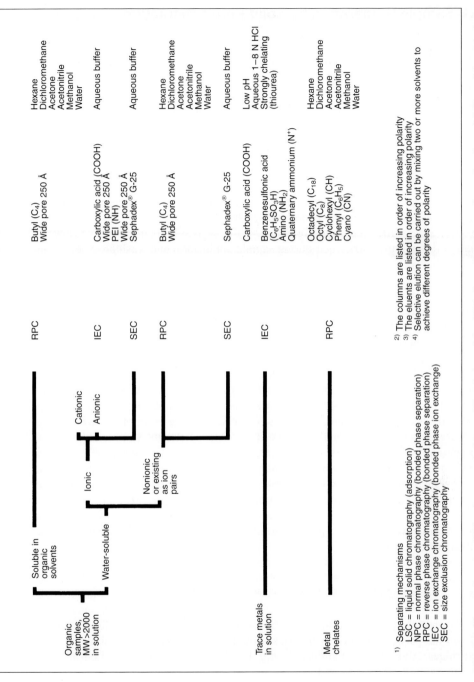

Fig. 2-1 Key to choosing SPE columns and eluents. The choice of the SPE phase depends on the molecular solubility of the sample in a particular medium and on its polarity. The sample matrix is not considered (Baker Chemikalien)

2.1.1 Sample Preparation by Solid Phase Extraction

From the middle of the 1980s solid phase extraction (SPE) began to revolutionise the enrichment, extraction and clean-up of analytical samples. Following the motto 'The separating funnel is a museum piece', the time-consuming and arduous liquid/liquid extraction has increasingly been displaced from the analytical laboratory. Today the euphoria of the rapid and simple preparation with disposable columns has lessened as a result of a realistic consideration of their performance levels and limitations. A particular advantage over the classical liquid/liquid partition is the low consumption of expensive and sometimes harmful solvents. The amount of apparatus and space required is low for SPE. Parallel processing of several samples is therefore quite possible. Besides a good clean-up, the necessary concentration of the analyte frequently required for GC/MS is achieved by solid phase extraction.

In solid phase extraction strong retention of the analyte is required, which prevents migration through the carrier bed during sample application and washing. Specific interactions between the substances being analysed and the chosen adsorption material are exploited to achieve retention of the analytes and removal of the matrix. An extract which is ready for analysis is obtained by changing the eluents. The extract can then be used directly for GC and GC/MS in most cases. The choice of column materials permits the exploitation of the separating mechanisms of adsorption chromatography, normal-phase and reversed-phase chromatography, and also ion exchange and size exclusion chromatography (Fig. 2-1).

The physical extraction process, which takes place between the liquid phase (the liquid sample containing the dissolved analytes) and the solid phase (the adsorption material) is common to all solid phase extractions. The analytes are usually extracted successfully because the interactions between them and the solid phase are stronger than those with the solvent or the matrix components. After the sample solution has been applied to the solid phase bed, the analytes become enriched on the surface of the SPE material. All other sample components pass unhindered through the bed and can be washed out. The maximum sample volume that can be applied is limited by the breakthrough volume of the analyte. Elution is achieved by changing the solvent. For this there must be a stronger interaction between the elution solvent and the analyte than between the latter and the solid phase. The elution volume should be as small as possible.

In analytical practice two solid phase extraction processes have become established. Cartridges are mostly preferred for liquid samples (Figs. 2-2 and 2-3). If the GC/MS analysis reveals high contents of plasticisers, the plastic material of the packed columns must first be considered and in special cases a change to glass columns must be made. For sample preparation using slurries or turbid water, which rapidly lead to deposits on the packed columns, SPE disks should be used. Their use is similar to that of cartridges. Additional contamination, e.g. by plasticisers, can be ruled out for residue analysis in this case (Fig. 2-4).

A large number of different interactions are exploited for solid phase extraction (see Fig. 2-2). Selective extractions can be achieved by a suitable choice of adsorption materials. If the eluate is used for GC/MS the detection characteristics of the mass spectrometer in particular must be taken into account. Unlike an electron capture de-

2.1 Sample Preparation 11

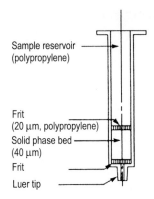

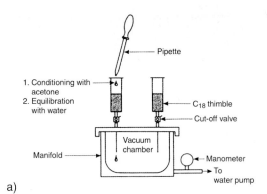

Fig. 2-2 Construction of a packed column for solid phase extraction (Baker Chemikalien)

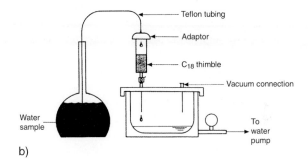

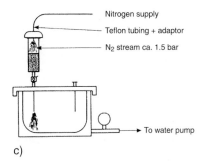

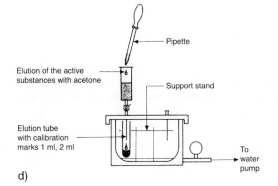

Fig. 2-3 Enrichment of a water sample on solid phase cartridges with C_{18}-material (Hein/Kunze 1994):
(a) conditioning
(b) loading (extraction)/washing
(c) drying
(d) elution of active substances

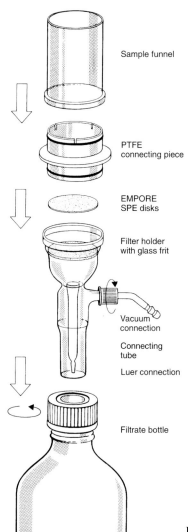

Fig. 2-4 Apparatus for solid phase extraction with SPE disks (Baker Chemikalien)

tector (ECD), which is currently widely used as a selective detector for substances with a high halogen content in environmental analysis, a GC/MS system can be used to detect a wide range of nonhalogenated substances. The use of mass selectivity therefore requires better purification of the extracts obtained by SPE. In the case of the processing of PCBs from waste oil, for example, a silica gel column charged with sulfuric acid is also necessary. Extensive oxidation of the nonspecific hydrocarbon matrix is thus achieved. The quality and reproducibility of SPE depends on criteria comparable to those which apply to column materials used in HPLC.

The solvent-free extraction technique *solid phase microextraction* (SPME) recently developed by Prof. J. Pawliszyn (University of Waterloo, Ontario, Canada) is an important step towards the instrumentation and automation of the solid phase extraction technique for on-line sample preparation and introduction to GC/MS. It involves exposing a fused silica fibre coated with a liquid polymeric material to a sample containing the analyte. The typical dimensions of a fibre are 1 cm × 100 μm. The analyte diffuses from gaseous or liquid samples into the fibre surface and partitions into the coating according to the first partition coefficient. The agitated sample can be heated during the sampling process if desired to achieve maximum recovery and precision for quantitative assays. After equilibrium is established, the fibre with the analyte is withdrawn from the sample and transferred into a GC injector system manually or via an autosampler. The analyte is desorbed thermally from the coating.

SPME offers several advantages for sample preparation including reduced time per sample, less sample manipulation resulting in an increased sample throughput and, in addition, the elimination of organic solvents and reduced analyte loss.

An SPME unit consists of a length of fused silica fibre coated with a phase similar to those used in chromatography columns. The phase can be mixed with solid adsorbents, e.g. divinylbenzene polymers, templated resins or porous carbons. The fibre is attached to a stainless steel plunger in a protective holder used manually or in a specially prepared autosampler. The plunger on the syringe retracts the fibre for storage and piercing septa, and exposes the fibre for extraction and desorption of the sample. A spring in the assembly keeps the fibre retracted, reducing the chance of it being damaged (see Fig. 2-5).

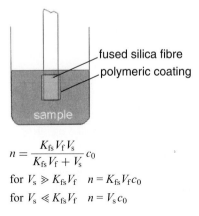

with

K_{fs} partition coefficient fibre/sample
V_f volume of fibre
V_s volume of sample
c_0 concentration of analyte in sample

$$n = \frac{K_{fs} V_f V_s}{K_{fs} V_f + V_s} c_0$$

for $V_s \gg K_{fs} V_f$ $n = K_{fs} V_f c_0$
for $V_s \ll K_{fs} V_f$ $n = V_s c_0$

Fig. 2-5 Equipment for solid phase microextraction

The design of a portable, disposable SPME holder with a sealing mechanism gives flexibility and ease of use for on-site sampling. The sample can be extracted and stored by placing the tip of the fibre needle in a septum. The lightweight disposable holders can be used for the lifetime of the fibre. SPME could also be used as an indoor air sampling device for GC and GC/MS analysis. For this technique to be successful, the sample must be stable to storage.

A major shortcoming of SPME is the lack of fibres that are polar enough to extract very polar or ionic species from aqueous solutions without first changing the nature of the species through prior derivatisation. Ionic, polar and involatile species have to be derivatised to GC-amenable species before SPME extraction.

Since the introduction of SPME in 1993 a variety of applications have been established, including the analysis of volatile analytes and gases consisting of small molecules. However, SPME was limited in its ability to retain and concentrate these small molecules in the fibre coating. Equilibria were obtained rapidly and distribution constants were low, which resulted in high minimum detection limits. The thickness of the phase is important in capturing small molecules, but a stronger adsorbing mechanism is also needed. The ability to coat porous carbons on a fibre has enabled SPME to be used for the analysis of small molecules at trace levels.

The amount of analyte extracted is dependent upon the distribution constant. The higher the distribution constant, the higher is the quantity of analyte extracted. Generally a thicker film is required to retain small molecules and a thinner film is used for larger molecules with high distribution constants. The polarity of the fibre and the type of coating can also increase the distribution constant.

Types of SPME fibres currently available:

Nonpolar fibres	100, 30, 7 µm polydimethylsiloxane (PDMS)
Polar fibres	85 µm polyacrylate, 65 µm Carbowax/divinylbenzene (CW-DVB)
Bipolar fibres	65 µm PDMS-DVB 75 µm Carboxen-PDMS
Blended fibre coatings	These fibre coatings contain porous materials, such as divinylbenzene and Carboxen, blended in either PDMS or Carbowax.

Summary of method performance for several analyses

Parameter	Formaldehyde	Triton-X-100	Phenylurea pesticides	Amphetamines
LOD	2–40 ppb	1.57 µg/l	< 5 µg/l	1.5 ng/ml
Precision	2–7%	2–15%	1.6–5.6%	9–15%
Linear range	50–3250 ppb	0.1–100 mg/l	10–10000 µg/l	5–2000 ng/ml
Linearity (r_2)	0.9995	0.990	0.990	0.998
Sampling time	10–300 s	50 min	8 min	15 min

2.1.2 Sample Preparation by SFE

Supercritical fluid extraction (SFE) has the potential of replacing conventional methods of sample preparation involving liquid/liquid and liquid/solid extractions (the current soxhlet extraction steps) in many areas of application. The soxhlet extraction (Fig. 2-6) has, up to now, been the process of choice for extracting involatile organic compounds from solid matrices, such as soil, sewage sludge or other materials.

The prerequisite for the soxhlet extraction with organic solvents is a sample which is as anhydrous as possible. Time-consuming drying of the sample (freeze drying) is

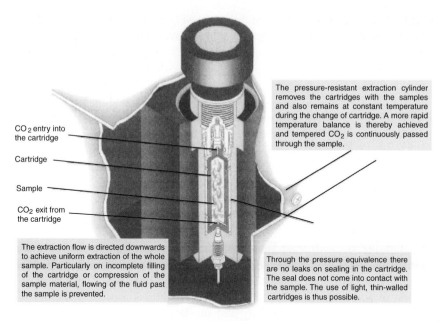

Fig. 2-6 SFE extraction system employing the principle of pressure equivalence (the cartridge with the sample is in a pressure-resistant cylinder, the internal and external pressures on the cartridge are kept equal, ISCO)

therefore necessary. The time required for a soxhlet extraction is, however, ca. 15–30 h for many samples in environmental analysis. This leads to the running of many extraction columns in parallel. The requirement for pure solvents (several hundred ml) is therefore high.

While the development of chromatographic separation methods has made enormous advances in recent years, equivalent developments in sample preparation, which keep up with the possibilities of modern GC and GC/MS systems, have not been made. The technical process for extraction of natural products (e.g. plants/perfume oils, coffee/caffeine) with supercritical carbon dioxide has already been used for a long time and is particularly well known in the area of perfumery for its pure and high-value extracts.

Only in recent the years has the process achieved increased importance for instrumental analysis. SFE is a rapid and economical extraction process with high percentage recovery which gives sample extracts which can be used for residue analysis usually without further concentration or clean-up. SFE works within the cycle time of typical GC and GC/MS analyses, can be automated, and avoids the production of waste solvent in the laboratory. The EPA (US Environmental Protection Agency) is working on the replacement of extraction processes with dichloromethane and freons used up till now, with particular emphasis on environmental analysis. The first EPA method published was the determination of the total hydrocarbon content of soil using SFE sample preparation (EPA # 3560). The EPA processes for the determination of polyaromatic hydrocarbons (EPA # 3561) and of organochlorine pesticides and PCBs (EPA # 3562) using SFE extraction and detection with GC/MS systems soon followed.

In the fundamental work of Lehotay and Eller on SFE pesticide extraction, carbamates, triazines, phosphoric acid esters and pyrethroids were all analysed using a GC/MS multimethod.

Table 2-1 Comparison of the physical properties of different aggregate states (in orders of magnitude)

Phase	Density [g · cm^3]	Diffusion [cm^2 · s^{-1}]	Viscosity [g · cm^{-1} · s^{-1}]
Gas	10^{-3}	10^{-1}	10^{-4}
Supercritical fluid	0.1–1.0	10^{-3}–10^{-4}	10^{-3}–10^{-4}
Liquid	1	<10^{-5}	10^{-2}

Supercritical fluids are extraction agents which are above their critical pressures and temperatures during the extraction phase. The fluids used for SFE (usually carbon dioxide) provide particularly favourable conditions for extraction (Fig. 2-7). The particular properties of a supercritical fluid arise from a combination of gaseous sample penetration, liquid-like solubilising capacity and substance transport (Table 2-1). The solubilising capacity of supercritical fluids reaches that of the liquid solvents when their density is raised. The maximum solubility of an organic compound can, nevertheless, be higher in a liquid solvent than in a supercritical phase. The solubility plays an important role in the efficiency of a technical process. However, in residue analysis this parameter is of no practical importance because of the extremely low concentration of the analytes.

The rate of extraction with a supercritical fluid is determined by the substance transport limits. Since supercritical fluids have diffusion coefficients an order of mag-

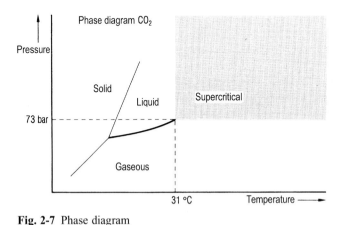

Fig. 2-7 Phase diagram
Physical data for CO_2:
Mp   −78.5 °C
Bp   −56.6 °C
Vapour pressure 57.3 bar at 20 °C

Critical data:
Pressure   73.8 bar
Temperature 31.1 °C
Density   0.464 g/l

nitude higher and viscosities an order of magnitude lower than those of liquid extraction agents, SFE extractions can be carried out in a much shorter time than, for example, the classical soxhlet extraction. Quantitative SFE extractions are typically finished in ca. 10–60 min, while normal solvent extractions for the same quantity of sample take several hours, often even overnight.

The solubilising capacity in an SFE step can easily be controlled via the density of the supercritical fluid, whereas the solubilising capacity of a liquid extraction agent is essentially constant. The density of the supercritical fluid is determined by the choice of pressure and temperature during the extraction. Raising the extraction temperature to ca. 60–80 °C also has a favourable effect on the swelling properties of the sample matrix. High-boiling polyaromatic hydrocarbons are now even extracted from real life samples at temperatures above 150 °C. This makes it necessary to use pressures of above 500 bar to achieve maximum extraction.

At constant temperature low pressures favour the extraction of less polar analytes. The continuation of the extraction at higher pressures then elutes more polar and higher molecular weight substances. This allows the optimisation of the extraction for a certain class of compound by programming the change in pressure. In this way extracts are produced by SFE which are ready for analysis and generally require no further clean-up or concentration. The SFE extracts are often purer than those obtained by classical solvent extraction. Selective extractions and programmed optimisation are possible through the collection of extract fractions.

Carbon dioxide, which is currently used as the standard extraction agent in SFE, is gaseous under normal conditions and evaporates from the extracts after depressurising the supercritical fluid. Various techniques are used to obtain the extract. An alternative to simply passing the extract into open, unheated vessels involves freezing out the extracts in a cold trap. Here the restrictor is heated to ca. 150 °C and the extract is frozen out in an empty tube or one filled with adsorbent. The necessary cooling is achieved using liquid CO_2 on the outside of the trap. For many volatile substances this type of receiver leads to abnormally low values for quantitative analyses because of aerosol formation. The direct collection of the extract by adsorption on solid material (e.g. C_{18}) or passing the extract by means of a heated restrictor into a cooled, pressurised solvent has proved successful in many areas of application. The extraction of samples containing fats can also be carried out reliably using this method. Where adsorption is used, elution with a suitable solvent is carried out after the extraction as with solid phase extraction (SPE). If the extract is passed directly into a cooled solvent, the contents of the receiver can be further processed immediately. The choice of the appropriate collection technique is critical when modifiers are used. Even the addition of 10% of the modifier results in the capacity of the cold trap being exceeded. Where the extract is collected on C_{18} cartridges increasing quantities of modifier can lead to premature desorption, because, in general, good solvents for the analyte are used as modifiers.

The extracts thus obtained can be analysed directly using GC, GC/MS or HPLC. Only in the processing of samples containing fats is the removal of the fats necessary before the subsequent analysis. This can be readily achieved using SPE. On the other hand, extracts from liquid extractions must be subjected to a clean-up before analysis and also concentrated because of their high dilution. The steps require additional time and are often the cause of low percentage recoveries.

Carbon dioxide has further advantages as a supercritical fluid. It is nontoxic, inert, can be obtained in high purity, and is clearly more economical with regard to the costs of obtaining it and disposing of it than liquid halogenated extraction agents. Because of the critical temperature of CO_2 (31 °C) SFE can be carried out at low temperatures so that thermolabile compounds can be extracted.

The mode of function of the analytical SFE unit (Fig. 2-8) has clear parallels with that of an HPLC unit. An analytical SFE unit consists of a pump, the thermostat (oven) in which the sample is situated in an extraction thimble, and a collector for the extract.

Carbon dioxide of the highest purity ('for SFE/SFC', 'SFE grade', total purity <99.9998 vol. %) is used as the extraction agent and is currently available from stock from all major gas suppliers. The pumping unit consists of a syringe or piston pump. The liquid CO_2 is pumped out of the storage bottle, via a riser and compressed. Piston pumps require CO_2 bottles with an excess helium pressure (helium headspace ca. 120 bar when full). On the high pressure side the liquid modifier is passed into the compressed CO_2 with a suitable pump.

The extraction thimbles for SFE are either situated in pressure-resistant cells (pressure equivalent extraction) or are the pressurised cells themselves. Nonextractable plastics or aluminium with a low thermal mass are used as the construction materials. Pressure-resistant vessels are made exclusively from stainless steel and are specified to resist up to 500 or 680 bar. The extraction thimbles should be filled as completely as possible with the sample to avoid empty spaces. They should be extracted in a vertical position to prevent the CO_2 from flowing past the sample. Carbon dioxide flow from above to below has been shown to be most favourable, as even with compression homogeneous penetration of the sample on the base of the extraction chamber is guaranteed. For analytical purposes extraction thimbles with volumes of 0.5 ml to 50 ml are usual. The cap of the extraction thimble consists of a stainless steel frit with a pore size of 2 µm. In the case of pressure-resistant stainless steel vessels frits made of PEEK material are used as seals for the screw-on caps.

The CO_2 converted into a critical state through compression and variation of the temperature of the oven is passed continuously through the sample. In a static extraction step the extraction thimble is first filled with the supercritical fluid. At the selected oven temperature and through the effects of a selected modifier, in this step of ca. 15–30 min the sample matrix is digested and partition of the analytes between the matrix and the fluid is achieved. The swelling of the matrix has been shown to be of critical importance for real life samples as it leads to higher and more stable the yields of extract.

The dynamic extraction step is introduced by using a 6-way valve (Fig. 2-8). In this step supercritical CO_2 is passed continuously through the equilibrated sample and under the chosen conditions extractable analytes are transported out of the extraction chamber.

The pressure ratios in the extraction chamber and in the tube carrying the extract are maintained by a restrictor at the end of the carrier tube. The dynamic extraction is quantitative and thus purifies the system for the next sample. The time required depends on the sample volume chosen.

At the restrictor the supercritical CO_2 is depressurised and the sample extract collected. In recent years a wide variety of restrictor constructions have been used with varying degrees of success. The choice of restrictor significantly affects the sample throughput of the SFE unit, the reproducibility of the process, and the breadth of application of SFE with regard to volatility of the analytes and the use of predried or moist

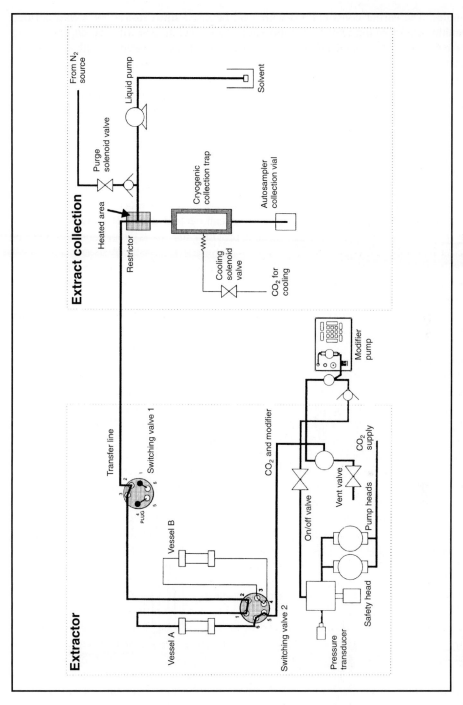

Fig. 2-8 Flow diagram of an SFE unit with cold trap and fraction collection, off-line coupling to GC (after Suprex)

samples. The earlier constructions with steel or fused silica restrictors with a defined opening size (e.g. integral restrictors) have not proved successful. The depressurising of the CO_2 usually took place in small glass vessels. These were filled with solvent into which the restrictor was dipped. A portion of the volatile substances was lost and the necessary heat of evaporation for depressurising was not supplied adequately. All samples which were not carefully freeze-dried led here to a sudden freezing out of the extracted water and thus to frequent blocking of the restrictor. Consequently at this stage of the development the results of the process had low reproducibility.

A significant advance in restrictor technology came with the introduction of variable restrictors (Fig. 2-9). The width of the opening of the variable restrictor is controlled by the flow rate and has a fine adjustment mechanism in order to achieve a constant flow of the supercritical fluid. Diminutions in flow with simultaneous deposition of the matrix or particles at the restriction are thus counteracted. In this construction the restriction is situated in a zone which can be heated to 200 °C, which completely prevents water from freezing out and provides the necessary energy for the depressurising of the CO_2.

Polar solvents, such as water, methanol, acetone, ethyl acetate or even toluene are predominantly used as modifiers in SFE (Table 2-2). The modifiers are added to swell the matrix and improve its surface activity and to increase the polarity of the CO_2. In many cases the use of modifiers leads to an increase in the yield of the extraction and

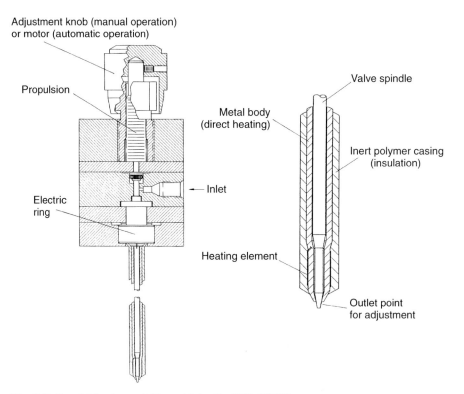

Fig. 2-9 Coaxial heater, variable restrictor for SFE (ISCO)

Table 2-2 Development of an SFE method for the extraction of PCDDs from fly ash (after Onuschka)

Experiment no.	1	2	3	4	5	6	7	8	9	10	11	12	13	14
Pressure	350	400	400	350	400	400	400	400	350	350	400	350	350	400
Replicate number	2	1	1	2	1	2	1	3	2	2	3	2	2	2
Supercritical fluid	N_2O	CO_2	N_2O	N_2O	CO_2	N_2O	N_2O	N_2O	CO_2	N_2O	N_2O	N_2O	N_2O	N_2O
Modifier identity	M	–	T	M	T	T	T	M	M	M	M	M	M	M
Pretreatment	–	F	F	–	F	F	F	HCl	T	T	T	E	–	–
Time-static [min]	–	64	–	120	64	64	32	64	40	40	60	50	40	64
Time-dynamic [min]	60	16	60	30	16	16	8	16	10	10	12	16	10	16
Program no.	B	A	B	C	A	A	A	A	C	C	C	D	A	A

Remarks:

Codes:
- M   methanol 2%
- T   toluene 5%
- F   formic acid
- E   pre-extraction with ethylene
- HCl hydrochloric acid treatment

Program A 2 min static equilibration repeated 8 times followed by 30 seconds of leaching and recharging with a fresh supercritical fluid. This program, which takes 20 min to complete the 15 step run, was usually repeated 2 to 4 times.

Program B dynamic leaching only

Program C 20 steps static extraction; 20 min static equilibration time + 3 min purging, usually repeated 2 to 6 times. 20 step program consisting of 14 min static extraction followed by 30 seconds of leaching supplied 13 times. The program requires a total of 20 min and was repeated 2 to 6 times.

Program D 12 steps static extraction; 15 min equilibration time + 3 min purging. 12 steps static extraction; 6 static steps, the first one lasting 5 min, the rest are set for 2 min duration. Leaching steps are set for 30 seconds. This program was repeated up to 7 times.

Table 2-3 Extraction yields from the SFE of PCBs for selected modifiers (after Langenfeld, Hawthorne et al.)

PCB isomer no.	1	2	3	4
18–2,2',5	3.46 (2)	71 (12)	61 (1)	56 (4)
28–2,4,4'	2.21 (5)	63 (3)	72 (1)	65 (5)
52–2,2',5,5'	4.48 (1)	71 (4)	74 (1)	70 (8)
44–2,2',3,5'	1.07 (11)	94 (14)	101 (1)	88 (10)
66–2,3',4,4'	0.93 (1)	68 (4)	73 (1)	66 (1)
101–2,2',4,5,5'	0.82 (1)	84 (3)	87 (1)	81 (4)
118–2,3',4,4',5	0.51 (2)	87 (4)	92 (2)	84 (5)
138–2,2',3,4,4',5'	0.57 (2)	93 (2)	95 (4)	79 (7)
180–2,2',3,4,4',5,5'	0.16 (6)	106 (1)	109 (2)	99 (8)

(1) Certified contents in µg/g of a sediment sample according to NIST, determined by two sequential Soxhlet extractions of 16 h.

% recovery (% RSD) for the addition of the modifiers
(2) 10% methanol
(3) 10% methanol/toluene 1:1
(4) 1% acetic acid

SFE conditions: ISCO model 260D and SFX 2–10, 400 bar, 80 °C, 0.5 g sediment, 2.5 ml extraction cell, extraction agent CO_2, addition of modifier to the extraction cell (10% addition corresponds to 2.5 µl modifier), 5 min static and 10 min dynamic extraction.

Table 2-4 Extraction yields from the SFE of polyaromatic hydrocarbons for selected modifiers (after Langenfeld, Hawthorne et al.)

Polyaromatic hydrocarbon	1	2	3	4	5
Phenanthrene	4.5 (7)	99 (5)	100 (2)	238 (7)	265 (3)
Fluoranthene	7.1 (7)	84 (7)	96 (2)	175 (5)	195 (3)
Pyrene	7.2 (7)	69 (7)	81 (2)	138 (3)	151 (2)
Benzo[a]anthracene	2.6 (12)	63 (3)	94 (4)	97 (6)	89 (1)
Chrysene + triphenylene	5.3 (5)	67 (2)	90 (4)	56 (5)	67 (2)
Benzo[b,k]fluoranthene	8.2 (5)	55 (3)	108 (8)	66 (1)	67 (1)
Benzo[a]pyrene	2.9 (17)	20 (8)	60 (8)	32 (1)	34 (1)
Benzo[ghi]perylene	4.5 (24)	7 (5)	26 (11)	16 (4)	13 (11)
Indeno[1,2,3-cd]pyrene	3.3 (15)	14 (7)	57 (13)	35 (10)	28 (9)

(1) Certified contents in µg/g of a dust sample (urban air particulate matter) according to NIST, determined by soxhlet extraction over 48 h.

% Recovery (% RSD) for the addition of modifiers
(2) 10% methanol
(3) 10% toluene
(4) 10% methanol/toluene 1:1
(5) 10% acetic acid

SFE conditions: ISCO model 260D and SFX 2-10, 400 bar, 80 °C, 0.3 g dust, 2.5 ml extraction cell, extraction agent CO_2, addition of modifier to the extraction cell (10% addition corresponds to 2.5 µl modifier), 5 min static and 10 min dynamic extraction.

Note: the high recoveries for some polyaromatic hydrocarbons are real and were verified using blank values and carry-over experiments.

even ionic compounds can be quantitatively extracted. The choice of modifier and the adjustment of its concentration are usually carried out empirically. Experience shows that modifiers which are themselves good solvents for the analytes can also be used successfully in SFE (Tables 2-3 and 2-4). Changes in the critical parameters of the fluid on addition of the modifier should be taken into account. Suitable software programs for calculating the quantities are available (e.g. SFE-Solver, ISCO Corp., Lincoln (NE)). While methanol is currently the most widely used modifier, ethyl acetate, for example, has been shown to be particularly effective in the extraction of illegal drugs from hair (see Section 4.30).

The modifier can be added in two different ways. The simplest involves the direct addition of the modifier to the sample in the extraction thimble. As many samples, in particular most foodstuffs, already have a high water content, in such cases lyophilising the sample is not necessary. Here a suitable restrictor is necessary. The water contained in the sample can be used as the modifier for the extraction and the addition of other modifiers is unnecessary. In other cases a few ml of the modifier are added to the sample. The direct addition procedure is suitable for all extractions which employ the static extraction step.

By using an additional modifier pump (HPLC pump, syringe pump) a preselected quantity of modifier can be mixed continuously with the supercritical carbon dioxide on the high pressure side. This elegant procedure which is generally controlled by the system's software in modern instruments allows both the static step and also the continuous dynamic extraction in the presence of the modifier.

In SFE the sample quantity used has a pronounced effect on the concentrations of the analytes in the extract and on the length of the extraction. For residue analysis using SFE, sample volumes of up to 10 ml are generally used. Larger quantities of sample increase the time required for the quantitative extraction and also increase the CO_2 consumption, which is ca. three to ten times the empty volume of the extraction thimble. Further shortening of the extraction time and lowering of the sample volume is possible using on-line extraction.

On coupling with GC and GC/MS the possible water content of the extracts must be taken into consideration. The GC column must be chosen so that water can be chromatographed as a peak and does not affect the determination of the analytes. All stationary phases of medium and high polarity are suitable for this purpose. Generally it must be assumed that if there is a constant water background the response of the analytes in the mass spectrometer will be strongly impaired. An accumulation with decreasing response factors can be observed in the course of a day with ion sources with unheated lens systems.

On-line coupling with GC/MS can be easily realised with a cold injection system (e.g. PTV, programmable temperature vaporiser, Fig. 2-10, see Section 2.2.1.2). Here a fused silica column as a restriction and the transfer line from the 6-way valve of the SFE unit are connected to the injector (see Fig. 2-8). The injector can be filled with a small quantity of an adsorption material, e.g. Tenax, depending on the task required. During the extraction the injector is kept at a low temperature with the split open, to ensure the expansion of the supercritical CO_2 and the trapping of the analyte. At the end of the extraction the 6-way valve of the SFE unit switches the carrier gas into the transfer line to prevent further passage of CO_2 on to the GC column. Injection requires controlled heating of the injector with the split closed and the start-up of the analysis program (Fig. 2-11).

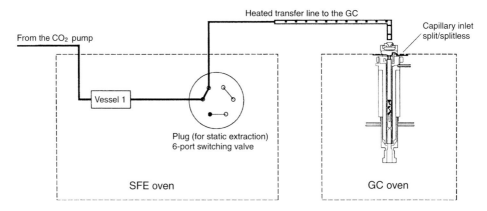

Fig. 2-10 On-line coupling of SFE with GC and GC/MS via a heated transfer line (connection directly at the 6-port switching valve) and PTV cold injection system

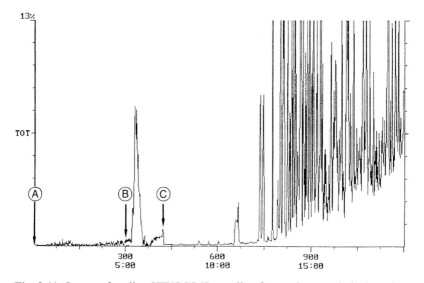

Fig. 2-11 Course of on-line SFE/GC/MS coupling for a polyaromatic hydrocarbon analysis
(A) Start of the dynamic extraction and data recording
  Injector cold, split open
(B) End of the dynamic extraction and start of the injection
  Injector heats up, split shut
(C) Start of the GC temperature program
  Injector hot, split open

The principal advantages of on-line SFE are that the manual steps between extraction and analysis are omitted and high sensitivity is achieved. All the extract is transferred quantitatively to the chromatography column. The quantity of sample with a particular expected analyte content applied must be modified according to the capacity of the GC column. As far as the mass spectrometer is concerned, at least for ion trap instruments, no special measures need to be taken on account of the on-line coupling. The quality of the analysis which can be achieved corresponds without limitations to the requirements of residue analysis.

The example of the extraction of polychlorinated dibenzodioxins (PCDDs) from fly ash clearly shows the comparison between the conventional soxhlet extraction and the way in which the current method is progressing. For the extraction 25 mg portions of homogenised fly ash were weighed out. The determination of the PCDDs was carried out with labelled internal standards using high resolution GC/MS. In the experiments presented here the variables sample pretreatment, fluid, modifier, pressure, and extraction time and program were systematically varied (Fig. 2-12 and 2-13).

Besides the analytical aspects, the ecological and economic aspects of SFE are also important. An environmental analysis technique which produces large quantities of

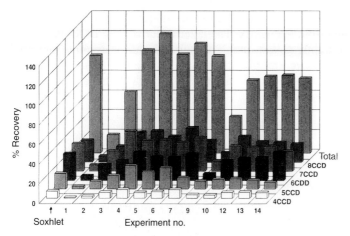

Fig. 2-12 Recoveries in SFE for PCDD extraction from fly ash compared with Soxhlet extraction (after Onuschka)

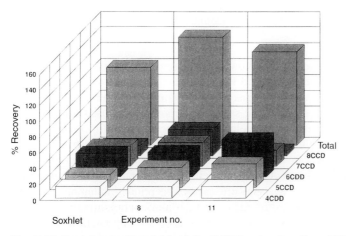

Fig. 2-13 Experiments 8 and 11 of the PCBD extraction from Table 2-4 compared with the Soxhlet extraction (after Onuschka)

potentially environmentally hazardous waste is paradoxical and not acceptable in the long term. At the same time the laboratory personnel have to handle smaller quantities of harmful solvents. The critical analysis of the costs during one year shows that the SFE procedure for sample preparation in the routine laboratory reduces the cost to two-thirds that of the conventional soxhlet extraction.

2.1.3 Headspace Techniques

One of the most elegant possibilities for instrumental sample preparation and sample transfer for GC/MS systems is the use of the headspace technique (Fig. 2-14). Here all the frequently expensive steps, such as extraction of the sample, clean-up and concentration are dispensed with. Using the headspace technique the volatile substances in the sample are separated from the matrix. The latter is not volatile under the conditions of the analysis. The tightly closed sample vessels, which, for example, are used for the static headspace procedure, can frequently even be filled at the sampling location. The danger of false results (loss of analyte) as a result of transportation and further processing is thus reduced.

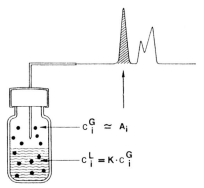

Fig. 2-14 Principle of headspace analysis (after Kolb)
A_i area of the GC signal of the ith component
c_{Gi}, c_{Li} concentrations in the gas and liquid phases respectively
k' partition coefficient

The extraction of the analytes is based on the partition of the very and moderately volatile substances between the matrix and the gas phase above the sample. After the partition equilibrium has been set up the gas phase contains a qualitatively and quantitatively representative cross-section of the sample and is therefore used for analysing the components to be determined. All involatile components remain in the headspace vial and are not analysed. For this reason the coupling of headspace instruments and GC/MS systems is particularly favourable. Since the interfering organic matrix is not involved, a longer duty cycle of the instrument and outstanding sensitivity are achieved. Furthermore headspace analyses are easily and reliably automated for this reason and achieve a higher sample throughput in a 24-hour operating period.

There are limitations to the coupling of headspace analysis with GC/MS systems if moisture has to be driven out of the sample. In certain cases water can impair the focusing of volatile components at the beginning of the GC column. Impairment of the

GC resolution can be counteracted by choosing suitable polar and thick film GC columns, by removing the water before injection of the sample, or by using a simple column connection.

It is also known that water affects the stability of the ion source of the mass spectrometer detector, which nowadays is becoming ever smaller. In the case of repeated analyses the effects are manifested by a marked response loss and poor reproducibility. In such cases special precautions must be taken, in particular in the choice of ion source parameters.

The headspace technique is very flexible and can be applied to the most widely differing sample qualities. Liquid or solid sample matrices are generally used, but gaseous samples can also be analysed readily and precisely using this method. Both qualitative and, in particular, quantitative determinations are now carried out coupled to GC/MS systems.

There are two methods of analysing the headspace which have very different requirements concerning the instrumentation: the static and dynamic (purge and trap) headspace techniques. The areas of use overlap partially but the strengths of the two methods are demonstrated in the different types of applications.

2.1.3.1 Static Headspace Technique

The term headspace analysis was coined in the early 1960s when the analysis of substances with odours and aromas in the headspace of tins of food was developed. The equilibrium of volatile substances between a sample matrix and the gas phase above it in a closed static system is the basis for static headspace chromatography (HSGC). Currently the term headspace is now used without the word static, e.g. headspace analyses, headspace sampler etc.

An equilibrium is set up in the distribution of the substances being analysed between the sample and the gas phase. The concentration of the substances in the gas phase then remains constant. An aliquot is taken from the gas phase and is fed into the GC/MS system via the transfer line (Fig. 2-15).

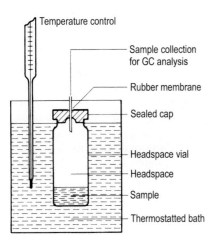

Fig. 2-15 Sample handling in the static headspace method (after Hachenberg)

The static headspace method is therefore an indirect analysis procedure, requiring special care in performing quantitative determinations. The position of the equilibrium depends on the analysis parameters (e.g. temperature) and also on the sample matrix itself. The matrix-dependence of the procedure can be counteracted in various ways. The matrix can be standardised, e.g. by addition of Na_2SO_4 or Na_2CO_3. Other possibilities include the addition method, internal standardisation, or the multiple headspace extraction procedure (MHE) published by Kolb in 1981 (Fig. 2-16).

> **Static Headspace Analysis**
>
> In static headspace analysis the samples are taken from a closed static system (closed headspace bottle) after the thermodynamic equilibrium (partition) between the liquid/solid matrix and the headspace above it has been established.

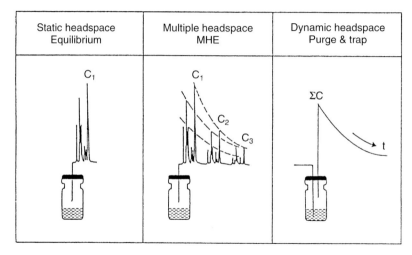

Fig. 2-16 Gas extraction techniques (after Kolb)

For coupling of HSGC with mass spectrometry the internal standard procedure has proved particularly successful for quantitative analyses. Besides the headspace-specific effects, possible variations in the MS detection are also compensated for. The best possible precision is thus achieved for the whole procedure. The MHE procedure can be used in the same way.

In static headspace analysis the partition coefficient of the analytes is used to assess and plan the method. For the partition coefficient k of a volatile compound the following equation is valid:

$$k = \frac{c_S}{c_G} = \frac{\text{concentration in the sample}}{\text{concentration in the gas phase}} \tag{1}$$

The partition coefficient depends on the temperature at equilibrium. This must therefore be kept constant for all measurements.

Rearranging the equation gives:

$$c_S = k \cdot c_G \tag{2}$$

As the peak area A determined in the GC/MS analysis is proportional to the concentration of the substance in the gas phase c_G, the following is valid:

$$A \approx c_G = \frac{1}{k} \cdot c_S \tag{3}$$

To be able to calculate back to the concentration of a substance c_0 in the original sample, the mass equilibrium must be referred to:

$$M_0 = M_S + M_G \tag{4}$$

The quantity M_0 of the volatile substance in the original sample has been divided at equilibrium into a portion in the gas phase M_G and another in the sample matrix M_S. Replacing M by the product of the concentration c and volume V gives:

$$c_0 \cdot V_0 = c_S \cdot V_S + c_G \cdot V_G \tag{5}$$

By using the definition of the partition coefficient given in equation (2), the unknown parameter c_S can be replaced by $k \cdot c_G$:

$$c_0 \cdot V_0 = k \cdot c_G \cdot V_S + c_G \cdot V_G \tag{6a}$$

$$c_0 = c_G \cdot \frac{V_S}{V_0} \cdot \left(k + \frac{V_G}{V_S}\right) \tag{6b}$$

The starting concentration c_0 of the sample, assuming $V_0 = V_S$, is given by:

$$c_0 = c_G \cdot \left(k + \frac{V_G}{V_S}\right) \tag{7}$$

As the peak area determined is proportional to the concentration of the volatile substance in the gas phase c_G, the following equation is valid, corresponding to the proportionality between the gas phase and the peak area:

$$c_0 \approx A \cdot (k + \beta) \tag{8}$$

whereby β is the phase ratio V_G/V_S (see equation (7)) and therefore describes the degree of filling of the headspace vessel (Ettre, Kolb 1991).

The effects on the sensitivity of a static headspace analysis can easily be derived from the ratio given in equation (8):

$$A \approx c_0 \cdot \frac{1}{k + \beta} \quad \text{with } \beta = V_G/V_S \tag{9}$$

The peak area determined from a given concentration c_0 of a component depends on the partition coefficient k and the sample volume V_S (through the phase ratio β).

For substances with high partition coefficients (e.g. ethanol, isopropanol, dioxan in water) the sample volume has no effect on the peak area determined. However, for substances with small partition coefficients (e.g. cyclohexane, trichloroethylene, xylene in water), the phase ratio β determines the headspace sensitivity. Doubling the quantity of sample leads in this case to a doubling of the peak area (Fig. 2-17). For all quantitative determinations of compounds with small partition coefficients, an exact filling volume must be maintained.

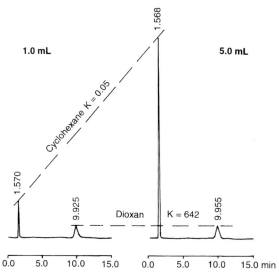

Fig. 2-17 Sample volume and sensitivity in the static headspace for cyclohexane ($K = 0.05$) and dioxan ($K = 642$) with sample volumes of 1 ml and 5 ml respectively (after Kolb)

Table 2-5 Partition coefficients of selected compounds in water (after Kolb)

Substance	40 °C	60 °C	80 °C
Tetrachloroethane	1.5	1.3	0.9
1,1,1-Trichloroethane	1.6	1.5	1.2
Toluene	2.8	1.6	1.3
o-Xylene	2.4	1.3	1.0
Cyclohexane	0.07	0.05	0.02
n-Hexane	0.14	0.04	<0.01
Ethyl acetate	62.4	29.3	17.5
n-Butyl acetate	31.4	13.6	7.6
Isopropanol	825	286	117
Methyl isobutyl ketone	54.3	22.8	11.8
Dioxan	1618	642	288
n-Butanol	647	238	99

The sensitivity of detection in static headspace analysis can also be increased through the lowering of the capacity factor k by setting up the equilibrium at a higher temperature. Substances with high partition coefficients profit particularly from higher equilibration temperatures (Table 2-5, e.g. alcohols in water). Raising the temperature is, however, limited by the danger of bursting the sample vessel or the seal.

Changes in the sample matrix can also affect the partition coefficient. For samples with high water contents an electrolyte can be added to effect a salting out process (Fig. 2-18). For example, for the determination of ethanol in water the addition of NH_4Cl gives a twofold and addition of K_2CO_3 an eightfold increase in the sensitivity of detection. The headspace sensitivity can also be raised by the addition of nonelectrolytes. In the determination of residual monomers in polystyrene dissolved in DMF, adding increasing quantities of water, e.g. in the determination of styrene and butanol, leads to an increase in peak area for styrene by a factor of 160 and for n-butanol by a factor of only 25 (Fig. 2-19).

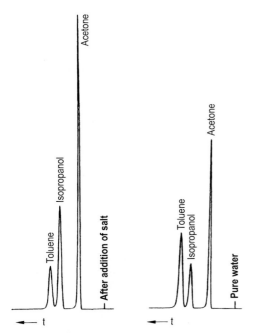

Fig. 2-18 Matrix effects in static headspace. The effect of salting out on polar substances (after Kolb)

To shorten the equilibration time the most up-to-date headspace samplers have devices available for mixing the samples. For liquid samples vigorous mixing strongly increases the phase boundary area and guarantees rapid delivery of analyte-rich sample material to the phase boundary.

Teflon-coated stirring rods have not proved successful because of losses through adsorption and cross-contamination. However, shaking devices which mix the sample

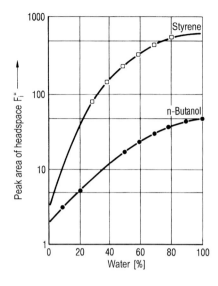

Fig. 2-19 Increasing water concentrations in the determination of styrene as a residual monomer lead to a sharp increase in response (after Hachenberg)

in the headspace bottle through vertical or rotational movement have proved effective (Table 2-6). A particularly refined shaker uses changing excitation frequencies of 2–10 Hz to achieve optimal mixing of the contents at each degree of filling (Fig. 2-20).

The resulting equilibration times are in the region of 10 min on using shaking devices compared with 45–60 min without mixing.

In addition quantitative measurements are more reliable. The use of shaking devices in static headspace techniques lowers the relative standard deviations to less than 2%

Table 2-6 Comparison of the analyses with/without shaking of the headspace sample (volatile halogenated hydrocarbons), average values in ppb, standard deviations (after Tekmar)

Substance	Without shaking			With shaking		
Ethylbenzene	353	18	(5.2%)	472	8	(1.7%)
Toluene	336	20	(5.9%)	411	4	(1.0%)
o-Xylene	324	13	(4.1%)	400	7	(1.8%)
Benzene	326	18	(5.4%)	372	5	(1.3%)
1,3-Dichlorobenzene	225	13	(5.6%)	255	5	(2.1%)
1,2,4-Trichlorobenzene	207	9	(4.2%)	225	6	(2.5%)
Bromobenzene	213	11	(5.2%)	220	5	(2.1%)

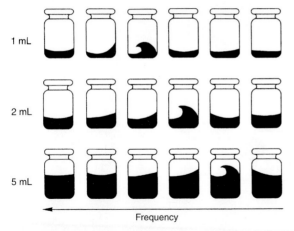

1 mL	
2 mL	
5 mL	

←──────────────────────
              Frequency

Fig. 2-20 Effect of variable shaking frequencies (2–10 Hz) on different depths of filling in headspace vessels (Perkin Elmer)

Static Headspace Injection Techniques

- *Pressure Balanced Injection* (Fig. 2-21):
  Variable quantity injected: can be controlled (programmed) by the length of the injection process, injection volume = injection time x flow rate of the GC column, no drop in pressure to atmospheric pressure, depressurising to initial pressure of column.
  Change in pressure in the headspace bottle: reproducible pressure build-up with carrier gas, mixing, initial pressure in column maintained during sample injection.
  Sample losses: none known.

- *Syringe Injection* (Fig. 2-22):
  Variable injection volume through pump action: easily controlled, drop in pressure to atmospheric on transferring the syringe to the injector.
  Pressure change in the headspace bottle: varies with the action of the syringe plunger (volume removed) on sample removal from the sealed bottle, compensation by injection of carrier gas necessary.
  Sample losses: possible through condensation on depressurising to atmospheric pressure, losses through evaporation from the syringe after it has been removed from the septum cap of the bottle, in comparison the largest surface contact with the sample.

- *Sample Loop* (Fig. 2-23):
  Quantity injected on the instrument side: changes in the volume injected by disconnection of the sample loop, drop in pressure to atmospheric on filling the sample loop.
  Pressure change in the headspace bottle: complete depressurising of the pressure built up in the bottle through tempering, the expansion volume must be a multiple of the sample loop to be able to fill it reproducibly.
  Sample losses: for larger sample loops attention must be paid to dilution with carrier gas during the pressure build-up phase, possible condensation on depressurising to atmospheric pressure.

34   2 Basics

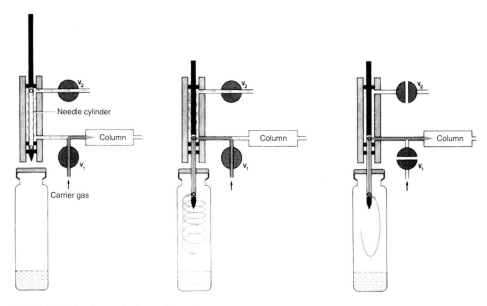

Fig. 2-21 Injection techniques for static headspace: the principle of pressure balanced injection (Perkin Elmer)

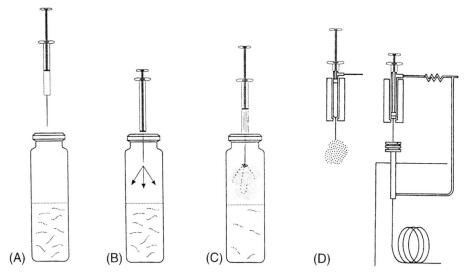

Fig. 2-22 Injection techniques for static headspace: the principle of transfer with a gastight syringe; (A) Sample heating, (B) Pressure, (C) Sampling, (D) Inject

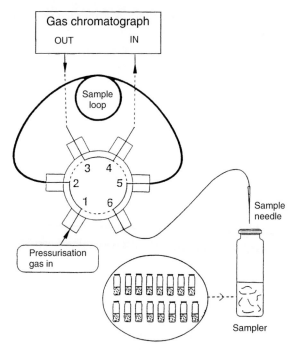

Fig. 2-23 Injection techniques for static headspace: the principle of application with a sample loop (Tekmar).
Sequence: 1 Heat sample at a precise temperature over a set period of time
2 Pressurise with carrier gas
3 Fill the sample loop
4 Inject by switching the 6-port switching valve

2.1.3.2 Dynamic Headspace Technique (Purge & Trap)

The trace analysis of volatile organic compounds (VOCs, volatile organic carbons) is of continual interest, e.g. in environmental monitoring, in outgasing studies on packaging material or in the analysis of flavours and fragrances. The dynamic headspace technique, known as purge and trap (PAT), is a process in which highly and moderately volatile organic compounds are continuously extracted from a matrix and concentrated in an adsorption trap. The substances driven out of the trap by thermal desorption reach the gas chromatograph as a concentrated sample where they are separated and detected.

Early purge and trap applications in the 1960s already involved the analysis of body fluids. In the 1970s purge and trap became known because of the increasing requirement for the testing of drinking water for volatile halogenated hydrocarbons. The analysis of drinking water for the determination of a large number of these compounds in ppq quantities (concentrations of less than 1 µg/l) only became possible through the concentration process which was part of purge and trap gas chromatography (PAT-GC). This technique is now frequently used to detect residues of volatile organic compounds in the environment and in the analysis of liquid and solid foodstuffs.

In particular, the coupling with GC/MS systems allows its use as a multicomponent process for the automatic analysis of large series of samples.

An analytically interesting variant of the procedure is the fine dispersion of liquid samples in a carrier gas stream. This so-called spray and trap process uses the high surface area of the sample droplets for an effective gas extraction. The procedure also works very well with foaming samples. A detergent content of up to 0.1% does not affect the extraction result. The spray and trap process is therefore particularly suitable for use with mobile GC/MS analyses.

Modes of Operation of Purge and Trap Systems

A purge and trap analysis procedure consists of three main steps:

1. Purge phase with simultaneous concentration
2. Desorption phase
3. Baking out phase

1. The Purge Phase

During the purge phase the volatile organic components are driven out of the matrix. The purge gas (He or N_2) is finely divided in the case of liquid samples (drinking water, waste water) by passing through a special frit in the base of a U-tube (frit sparger, Fig. 2-24A). The surface of the liquid can be greatly increased by the presence of very small gas bubbles and the contact between the liquid sample and the purge gas maximised.

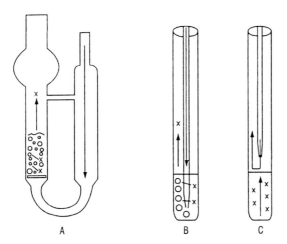

Fig. 2-24 Possibilities for sample introduction in purge and trap
(A) U tube with/without frits (*fritless/frit sparger*) for water samples
(B) Sample vessel (*needle sparger*) for water and soil samples (solids)
(C) Sample vessel (*needle sparger*) for foaming samples for determination of the headspace sweep

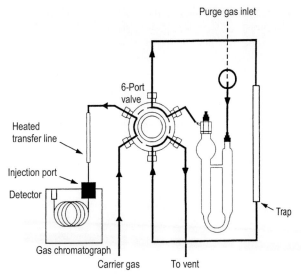

Fig. 2-25 Gas flow schematics of purge & trap-GC coupling, switching of phases at the 6-port switching valve:
Bold line: purge and baking out phase
Dotted line: desorption phase

The purge gas extracts the analytes from the sample and transports them to the trap. The analytes are retained in this trap and concentrated while the purge gas passes out through the vent. The desorption gas, which comes from the carrier gas provision of the GC, enters this phase via the 6-port switching valve in the gas chromatograph and maintains the constant gas flow for the column (Fig. 2-25).

Solid samples are analysed in special vessels (needle sparger, see Fig. 2-24 B, C) into which a needle with side openings is dipped. For foaming samples the headspace sweep technique can be used.

The total quantity of volatile organic compounds removed from the sample depends on the purge volume. The purge volume is the product of the purge flow rate and the purge time. Many environmental samples are analysed at a purge volume of 440 ml. This value is achieved using a flow rate of 40 ml/min and a purge time of 11 min. A purge flow rate of 40 ml/min gives optimal purge efficiency. Changes in the purge volume should consequently only be made after adjusting the purge time. Although a purge volume of 440 ml is optimal in most cases, some samples may require larger purge volumes for adequate sensitivity to be reached (Fig. 2-26).

The purge efficiency is defined as that quantity of the analytes which is purged from a sample with a defined quantity of gas. It depends upon various factors. Among them are: purge volume, sample temperature, and nature of the sparger (needle or frit), the nature of the substances to be analysed and that of the matrix. The purge efficiency has a direct effect on the percentage recovery (the quantity of analyte reaching the detector).

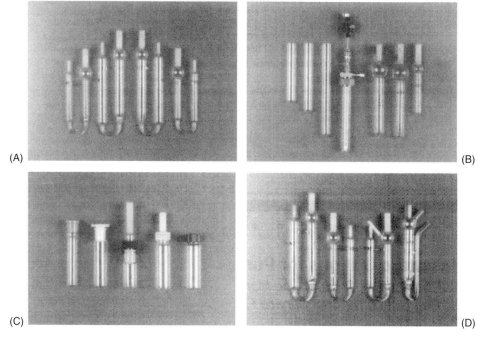

Fig. 2-26 Glass apparatus for the purge and trap technique (Tekmar)
(A) U-tube with/without frit (5 ml and 25 ml sizes)
(B) Needle sparger: left: single use vessels, middle: glass needle with frits, right: vessels with foam retention, 5 ml, 20 ml and 25 ml volumes
(C) Special glass vessels: 25 ml flat-bottomed flask, 40 ml flask with seal, 20 ml glass (two parts) with connector, 40 ml glass with flange, 40 ml screw top glass
(D) U-tube for connection to automatic sample dispensers: left: 25 ml and 5 ml vessel with side inlet, right: 5 ml vessel with upper inlet for sample heating, 25 ml special model with side inlet

Control of Purge Gas Pressure during the Purge Phase

The adsorption and chromatographic separation of volatile halogenated hydrocarbons is improved significantly by regulating the pressure of the purge gas during the purge phase. By additional back pressure control during this phase, a very sharp adsorption band is formed in the trap, from which, in particular, the highly volatile components profit, since a broader distribution does not occur. The danger of the analytes passing through the trap is almost completely excluded under the given conditions.

During the desorption phase the narrow adsorption band determines the quality of the sample transfer to the capillary column. The result is clearly improved peak symmetry, and thus better GC resolution and an improvement in the sensitivity of the whole procedure.

The Dry Purge Phase

To remove water from a hydrophobic adsorption trap (e.g. with a Tenax filling) a dry purge phase is introduced. During this step most of the water condensed in the trap is blown out by dry carrier gas. Purge times of ca. 6 min are typical.

2. The Desorption Phase

During the desorption phase the trap is heated and subjected to a backflush with carrier gas. The reversal of the direction of the gas flow is important in order to desorb the analytes in the opposite direction to the concentration by the trap. In this way narrow peak bands are obtained.

The time and temperature of the desorption phase affect the chromatography of the substances to be analysed. The desorption time should be as short as possible but sufficient to transfer the components quantitatively on to the GC column. Most of the analytes are transferred to the GC column during the first minute of the desorption step. The desorption time is generally 4 min.

The temperature of the desorption step depends on the type of adsorbent in the trap (Table 2-7). The most widely used adsorbent, Tenax, desorbs very efficiently at 180 °C without forming decomposition products. The peak shape of compounds eluting early can be improved by inserting a desorb-preheat step. Here the trap is preheated to a temperature near the desorption temperature before the valve is switched for desorption and before the gas flows freely through the trap. Gas is not passed through the trap during the preheating step, but the analytes are nevertheless desorbed from the carrier material. When the gas stream is passed through the trap after switching the valve, it purges the substances from the trap in a concentrated carrier gas cloud. Highly volatile compounds which are not focused at the beginning of the column thus give rise to a narrower peak shape. A preheating temperature of 5 °C below the desorption temperature has been found to be favourable.

VOCARB material, which is used in all current applications involving volatile halogenated hydrocarbons, can be desorbed at higher temperatures than Tenax (up to 290 °C). At higher desorption temperatures, however, the possibility of catalytic decomposition of some substances must be taken into account (see also Section 2.1.4).

Moisture Removal by the Moisture Control System (MCS)

Water is driven out of aqueous or moist samples, most of which is disposed of by the dry purge step, particularly in Tenax adsorption traps. Residual moisture would be transferred to the GC column during the desorption step. As the resolution of highly volatile substances on capillary columns would be impaired and the detection by the mass spectrometer would be affected, additional devices are used to remove the water. A moisture control system (MCS) guarantees maximum efficiency and retention of the residual moisture during the desorption process (Fig. 2-27).

Table 2-7 List of traps used in the purge and trap procedure with details of applications and recommended analysis parameters

Trap #	Adsorbent	Application	Drying possible?	Drying time [min]	Desorb pre-heat [°C]	Desorb temperature [°C]	Bake-out temperature [°C]	Bake-out time [min]	Conditioning temperature [°C] time [min]	Remarks
1	Tenax	All substances down to CH_2Cl_2	Yes	2–6	220	225	230	7–10	230 / 10	Low response with brominated substances, high back pressure, background with benzene, toluene, ethylbenzene
2	Tenax Silica gel	All substances except freons	No	–	220	225	230	10–12	230 / 10	Low response with brominated substances, high back pressure, background with benzene, toluene, ethylbenzene
3	Tenax Silica gel Activated charcoal	All substances including freons	No	–	220	225	230	10–12	230 / 10	Low response with brominated substances, high back pressure, background with benzene, toluene, ethylbenzene
4	Tenax Activated charcoal	All substances down to CH_2Cl_2 and gases	No	–	220	225	230	7–10	230 / 10	Low response with brominated substances, high back pressure, background with benzene, toluene, ethylbenzene
5	OV-1 Tenax Silica gel Activated charcoal	All substances including freons	No	–	220	225	230	10–12	230 / 10	Low response with brominated substances, high back pressure, background with benzene, toluene, ethylbenzene
6	OV-1 Tenax Silica gel	All substances except freons	No	–	220	225	230	10–12	230 / 10	Low response with brominated substances, high back pressure, background with benzene, toluene, ethylbenzene
7	OV-1 Tenax	All substances down to CH_2Cl_2	Yes	2–6	220	225	230	7–10	230 / 10	Low response with brominated substances, high back pressure, background with benzene, toluene, ethylbenzene

Table 2-7 (continued)

Trap #	Adsorbent	Application	Drying possible?	Drying time [min]	Desorb pre-heat [°C]	Desorb temperature [°C]	Bake-out temperature [°C]	Bake-out time [min]	Conditioning temperature [°C] time [min]	Remarks
8	Carbopak B Carbosieve SIII	All substances including freons	Yes	11	245	250	260	4–10	260 20–30	Losses of CCl_4
9	VOCARB 3000	All substances including freons	Yes	1–3	245	250	260	4	290 4 h	Response factors see Table 2-8
10	VOCARB 4000	All substances including freons	Yes	1–3	245	250	260	4	270 4 h	Low response with chlorinated compounds, high back pressure, quantitative losses of chloroethyl vinyl ethers
11	BTEX Carbopak B Carbopak C	All substances including freons	Yes	1–3	245	250	260	4	270	–
12	Tenax GR Graphpac-D	All substances including freons	Yes	1–4	245	250	260	12	–	–

Trap nos. 1–8: Tekmar, trap nos. 9–11: SUPELCO, trap no. 12: Alltech

Table 2-8 Response factors and standard deviations for components with EPA method 624 using a VOCARB 3000 trap and 5 ml of sample (250 °C desorption temperature, reference bromochloromethane/difluorobenzene, concentrations in ppb, Supelco)

Compound	20	50	100	150	200	Average	Standard deviation	% Relative standard deviation
Methyl chloride	0.933	0.962	0.984	0.906	1.287	1.014	0.139	13.7
Vinyl chloride	1.037	1.167	1.466	1.536	1.333	1.308	0.185	14.1
Methyl bromide	1.018	1.257	1.079	1.077	1.290	1.144	0.108	9.5
Ethyl chloride	0.419	0.582	0.503	0.557	0.536	0.519	0.056	10.9
Trichlorofluoromethane	0.972	1.533	1.266	1.491	1.507	1.354	0.213	15.8
1,1-Dichloroethylene	0.873	1.207	1.203	1.264	1.180	1.145	0.139	12.1
Dichloromethane	1.550	1.134	1.231	1.325	1.11	1.270	0.159	12.5
1,2-Dichloroethylene	0.973	0.991	0.944	1.124	1.007	1.008	0.062	6.1
1,1-Dichloroethane	2.064	2.093	2.013	2.186	2.175	2.106	0.066	3.1
Chloroform	2.513	2.368	2.193	2.419	2.373	2.373	0.104	4.4
Tetrachloroethane	1.205	1.472	1.302	1.325	1.476	1.356	0.105	7.7
Carbon tetrachloride	1.067	1.426	1.350	1.240	1.335	1.284	0.124	9.6
Benzene	0.947	0.949	0.906	0.973	1.024	0.960	0.039	4.0
1,2-Dichloroethane	0.046	0.045	0.047	0.050	0.049	0.047	0.002	4.2
Trichloroethylene	0.412	0.495	0.485	0.504	0.540	0.487	0.042	8.6
1,2-Dichloropropane	0.535	0.530	0.513	0.537	0.549	0.533	0.012	2.2
Bromodichloromethane	0.082	0.080	0.084	0.088	0.087	0.084	0.003	3.4
2-Chloroethyl vinyl ether	0.043	0.039	0.043	0.047	0.041	0.043	0.003	6.2
cis-1,3-Dichloropropene	0.898	0.977	1.001	1.086	1.082	1.009	0.070	6.9
Toluene	0.861	1.208	1.141	1.250	1.283	1.149	0.151	13.2
trans-1,3-Dichloropropene	0.312	0.300	0.280	0.299	0.289	0.296	0.011	3.7
1,1,2-Trichloroethane	0.593	0.486	0.488	0.526	0.507	0.520	0.039	7.5
Tetrachloroethylene	0.364	0.488	0.492	0.497	0.558	0.480	0.063	13.2
Dibromochloromethane	0.818	0.712	0.731	0.787	0.785	0.767	0.039	5.1
Chlorobenzene	1.135	1.116	1.117	1.183	1.242	1.159	0.048	4.2
Ethylbenzene	0.561	0.508	0.476	0.504	0.543	0.518	0.030	5.8
Bromoform	1.070	0.882	0.920	1.013	0.918	0.961	0.070	7.3
1,1,2,2-Tetrachloroethane	0.082	0.069	0.070	0.078	0.065	0.073	0.006	8.4
1,3-Dichlorobenzene	1.277	1.215	1.237	1.338	1.451	1.304	0.085	6.5
1,4-Dichlorobenzene	1.397	1.279	1.291	1.391	1.509	1.374	0.084	6.1
1,2-Dichlorobenzene	1.351	1.233	1.188	1.284	1.350	1.281	0.064	5.0

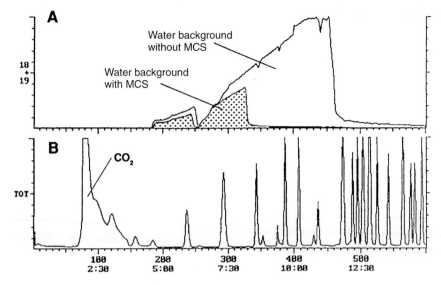

Fig. 2-27 GC/MS analysis using purge & trap
(A) Mass chromatogram for water (m/z 18+19) with and without removal (without MCS) of the moisture driven out.
(B) Total ion chromatogram with volatile substances (volatile halogenated hydrocarbons) after removal of the water (with MCS)

If the dew point for water is not reached in the MCS, a stationary water phase is formed which, for example, for BTEX and volatile halogenated hydrocarbons, the analytes can pass through unaffected. Polar components can also pass through the MCS system at moderate temperatures with a short retardation. When the desorption phase is finished, the tubing of the MCS is dried by baking out in the countercurrent (Fig. 2-28).

In particular, where the purge and trap technique is used in capillary gas chromatography and when using ECD or mass spectrometers as detectors, reliable removal of the moisture driven out is necessary.

3. Baking Out Phase

When required, the trap can be subjected to a baking out phase (trap-bake mode) after the desorption of the analytes. During desorption involatile organic compounds are released from the trap at elevated temperatures and driven off. Baking out conditions the trap material for the next analysis.

Coupling of Purge and Trap with GC/MS Systems

There are two fundamentally different possibilities for the installation of a purge and trap instrument coupled to a GC/MS system, which depend on the areas of use and the number of purge and trap samples to be processed per day.

In many laboratories the most flexible arrangement involves connecting a purge and trap system to the gas supply of the GC injector. The purge and trap concentrator is connected in such a way that either manual or automatic syringe injection can still be carried out. The carrier gas is passed from the GC carrier gas regulation to the central 6-port valve of the purge and trap system (see Fig. 2-25). From here the carrier gas flows back through the continuously heated transfer line to the injector of the gas chromatograph. A particular advantage of this type of installation lies in the fact that the injector can still be used for liquid samples. This allows the manual injection of control samples or the operation of liquid autosamplers. In addition it guarantees that the whole additional tubing system remains free from contamination even when the purge and trap instrument is on standby.

When the sample is transferred to the GC column it should be ensured that adequate focusing of the analytes is achieved. For this purpose GC columns are available with film thicknesses of ca. 1.8 µm or more, which have already been used for the analysis of volatile halogenated hydrocarbons using GC/MS systems (types 502, 624 or volatiles). The injector system is operated using a split ratio which can be selected depending on the detection limit required.

A complete splitless injection of the analytes on to the GC column can only be achieved with cryofocusing (see also Section 2.2.1.5). Here the column is connected to the transfer line or the central 6-port valve of the purge and trap apparatus in such a way that the beginning of the column passes through a zone ca. 10 cm long which can be cooled by an external cooling agent, such as liquid CO_2 or N_2, to $-30\,°C$ to $-150\,°C$. All components which reach the GC column after desorption from the trap can be frozen out in a narrow band at the beginning of the column. This enables the highest sensitivities to be achieved, in particular with mass spectrometers. As large quantities of water are also concentrated together with the analytes in the case of moist or aqueous samples, removal of moisture in the desorption phase is particularly important with this type of coupling. This can, for example, be effected by means of a moisture control system (MCS) (Fig. 2-28). Insufficient removal of water leads to the deposition of ice and blockage of the capillaries. The consequences are poor focusing or complete failure of the instrument.

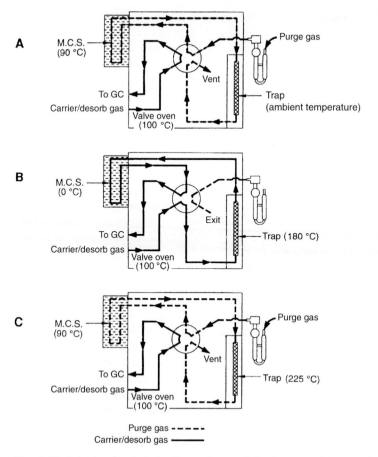

Fig. 2-28 Scheme showing the three phases of the purge and trap cycle with water removal (MCS, Tekmar).
(A) Purge phase
(B) Desorption phase
(C) Baking out phase

2.1.3.3 Headspace versus Purge & Trap

Both instrumental extraction techniques have specific advantages and disadvantages when coupled to GC and GC/MS. This should be taken into consideration when choosing an analysis procedure. In particular, the nature of the sample material, the concentration range for the measurement and the work required to automate the analyses for large numbers of samples play a significant role. The recovery and the partition coefficient, and thus the sensitivity which can be achieved, are relevant to the analytical assessment of the procedure. For both procedures it must be possible to vaporise the substances being analysed below 150 °C and then to partition them in the gas phase. The vapour pressure and solubility of the analytes, as well as the extraction temperature, affect both procedures (Fig. 2-29).

How then do the techniques differ? For this the terms recovery and sensitivity must be defined. For both methods, the recovery depends on the vapour pressure, the solubility and the temperature. The effects of temperature can be dealt with because it is easy to increase the vapour pressure of a compound by raising the temperature. With the purge and trap technique the term percentage recovery is used. This is the amount of a compound which reaches the gas chromatograph for analysis relative to the amount which was originally present in the sample. If a sample contains 100 pg benzene and 90 pg reach the GC column, the percentage recovery is 90%. In the static headspace technique, a simple expression like this cannot be used because it is possible to use a large number of types of vial, injection techniques and injection volumes.

The commonly used term connected with the static headspace method is the partition coefficient, as mentioned above. The partition coefficient is defined as the quantity of a compound in the sample divided by the quantity in the vapour phase. Therefore the smaller the partition coefficient is, the higher the sensitivity. It should be noted that a partition coefficient is only valid for the analysis parameters at which it has been determined. These include the temperature, the size of the sample vial, the quantity of sample (weight and volume), the nature of the matrix and the size of the headspace. After the partition coefficient, the quantity injected is the next parameter affecting headspace sensitivity. The quantity injected is limited by a range of factors. For example, only a limited quantity can be removed from the headspace of a closed vessel. Attempts to remove a larger quantity of sample vapour would lead to a partial vacuum in the sample vessel. This is extraordinarily difficult to reproduce. Furthermore, only a limited quantity can be injected on to a GC column without causing peak broadening. For larger quantities of sample cryofocusing is necessary. Capillary columns require cold trapping at injection quantities of more than ca. 200 µl.

An alternative injection system involves pressure balanced injection. Here a needle is passed through the septum into the headspace in order to create pressure in the headspace vial which is at least as high as the column pre-pressure. After equilibrium has been established this pressure is reduced during injection on to the GC column over a short programmable time interval. This allows larger quantities to be injected. It is impossible to measure the exact quantity injected; however, the reproducibility of this method is extremely high.

First example: volatile halogenated hydrocarbons

A comparison with actual concentration values makes the differences between the static headspace and purge and trap techniques very clear. The percentage recovery for the purge and trap technique is, for example, for an environmental sample of volatile halogenated hydrocarbons: 95% for chloroform, 92% for bromodichloromethane, 87% for chlorodibromomethane and 71% for bromoform. For a sample of 5 ml, which contains 1 ppb of each substance (i.e. a total quantity of 5 ng of each compound), 4.75 ng, 4.6 ng, 4.35 ng and 3.55 ng are recovered. In a typical static headspace system with a sample vessel of 21 ml containing 15 ml of sample at 70 °C the partition coefficients for the corresponding volatile halogenated hydrocarbons are: 0.3, 0.9, 1.5 and 3.0. This means that the quantities in the 5 ml of headspace are: 11.5 ng, 7.9 ng, 6.0 ng and 3.8 ng. On injection of 20 µl of the headspace gas mixture

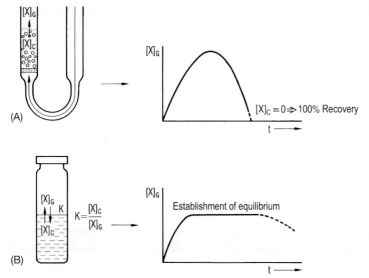

Fig. 2-29 Comparison of the purge & trap and static headspace techniques
(A) Purge & trap
(B) Static headspace

Static headspace	Dynamic headspace
Headspace	Purge & trap
Extraction: waiting for establishment of equilibrium	Continuous disturbance of the equilibrium by the purge gas
Intermediate steps: closed vessel at constant temperature	Enrichment of the substances driven off in a trap
Sample injection: Removal of a preselected volume from the headspace	Thermal desorption from the trap

on to a standard capillary column, the quantities injected are: 0.05 ng, 0.03 ng, 0.02 ng and 0.015 ng. For a larger injection (0.5 ml) using cryofocusing, the quantities injected are: 1.2 ng, 0.8 ng, 0.6 ng, and 0.4 ng. The purge and trap technique is therefore more sensitive than the static headspace procedure for these volatile halogenated hydrocarbons by factors of 4.1, 5.8, 7.2 and 9.3 (Table 2-9).

Table 2-9 Lower application limits [μg/l] for headspace and purge & trap techniques in the analysis of water from the river Rhine (after Willemsen, Gerke, Krabbe)

Substance	Lower application limits [μg/l] for headspace	Lower application limits [μg/l] for purge and trap
Dichloromethane	1	0.5
Chloroform	0.1	0.05
1,1,1-Trichloroethane	0.1	0.02
1,2-Dichloroethane	5	0.5
Carbon tetrachloride	0.1	0.02
Trichloroethylene	0.1	0.05
Bromodichloromethane	0.1	0.02
1,1,2-Trichloroethane	0.5	0.5
Dibromochloromethane	0.1	0.05
Tetrachloroethylene	0.1	0.02
1,1,2,2-Tetrachloroethane	0.1	0.02
Bromoform	0.1	0.2
1,1,2,2-Tetrachloroethane	0.1	0.02
Benzene	5	0.1
Toluene	5	0.05
Chlorobenzene	5	0.02
Ethylbenzene	5	0.1
m-Xylene	5	0.05
p-Xylene	5	0.05
o-Xylene	5	0.1
Triethylamine	–	0.5
Tetrahydrofuran	–	0.5
1,3,5-Trioxan	–	0.5

Second example: cooking oils
For compounds with poorer recoveries, e.g. in the analysis of free aldehydes in cooking oils, the difference is even clearer. At 150 °C the purge and trap analysis gives recoveries of 47%, 59% and 55% for butanal, 2-hexenal and nonan. For a sample of 0.5 ml, containing 100 ppb of the compounds, 24 ng, 30 ng and 28 ng are recovered. In the static headspace analysis, the partition coefficients for these compounds at 200 °C are all higher than 200. Assuming the value of 200, the quantity of each of these compounds in 5 ml of headspace is 0.7 ng. For an injection of 0.5 ml the quantity injected is therefore only 0.07 ng. The differences in sensitivity favouring purge and trap analysis are therefore 343, 428 and 400!

Third example: residual solvents in plastic sheeting
Comparable ratios are obtained in the analysis of a solid sample, for example, the analysis of residual solvents in a technical product. A run using the purge and trap technique and 10 ml of sample at 150 °C gave a recovery of 63% for toluene. The sample contained 1.6 ppm, which corresponds to a quantity of 101 ng. The partition coefficient in the static headspace technique at 150 °C (for a sample of 1 g) is 95. The quantity of residual solvent in 19 ml of headspace is therefore 17 ng. For an injection of

0.5 ml, 0.4 ng are injected. The quantity injected is therefore smaller by a factor of 250 than that in the purge and trap analysis. Furthermore, the reproducibility of this analysis was 7% for the purge and trap technique and 32% for the static headspace analysis (relative standard deviation).

The theoretically achievable or effectively necessary sensitivity is not the only factor deciding the choice of procedure. The specific interactions between the analytes and the matrix, the performance of the detectors available, and the legally required detection limits play a more important role.

Besides the sensitivity, there are other aspects which must be taken into account when comparing the purge and trap and static headspace techniques.

The static headspace technique is very simple and relatively quick. The procedure is well documented in the literature, and for many applications the sensitivity is more than adequate, so that its use is usually favoured over that of the purge and trap technique. There are areas of application where good results are obtained with the static headspace technique which cannot be improved upon by the purge and trap method. These include: the determination of alcohol in blood, of free fatty acids in cell cultures, of ethanol in fermentation units or drinks, and residual water in polymers. This also applies to studies on the determination of ionisation constants of acids and bases and the investigation of gas phase equilibria.

However, for many other samples specific problems besides sensitivity arise on use of the static headspace technique, which can be overcome using the purge and trap procedure. In every static headspace system all the compounds present in the headspace are injected, not only the organic analytes. This means that an air peak is also obtained. All air contaminants (a widely occurring problem in many laboratories) are visible and oxygen can impair the service life of the capillary column. In addition a larger quantity of water is injected than in the purge and trap method. For drinks containing carbonic acid CO_2 can lead to the build-up of excess pressure. Even at room temperature an undesirably high pressure can build up in the sample vials, which leads to flooding of the GC column during the analysis. In addition, a very large quantity of CO_2 is injected. Heating the sample enhances these effects. For safety reasons sample vials with safety caps to guard against excess pressure have to be used in these cases. Dust particles lead to further problems in the case of powder samples. To achieve the necessary sensitivity for an analysis, the sample often has to be heated to a temperature which is higher than that necessary in the purge and trap technique. This can lead to thermal decomposition, which is frequently observed with foodstuffs. In addition, oxygen cannot be eliminated from the sample before heating in a static headspace system. It is therefore impossible to prevent oxidation of the sample contents. During the thermostatting phase additional problems arise as a result of the septum used. Substances emitted from the septum (e.g. CS_2 from butyl rubber septums) falsify the chromatogram. The permeability of the septum to oxygen presents a further hazard.

Advantages of the Static Headspace Technique

– The static headspace can be easily automated. All headspace samplers now operate automatically for 20, 40 or 100 samples. For manual qualitative preliminary samples a gas-tight syringe is satisfactory.

- Samples, which tend to foam or contain unexpectedly high concentrations of analyte, do not generally lead to faults or cross-contamination.
- All sample matrices (solid, liquid or gaseous) can be used directly usually without expensive sample preparation.
- Headspace samplers are often readily portable and, when required, can be rapidly connected to different types of GC instruments.
- The sample vessels (special headspace vials with caps) are only intended for single use. There is no additional workload of cleaning glass equipment and hence cross-contamination does not occur.
- Headspace vials can be filled and sealed at the sampling point outside the laboratory in certain cases. This dispenses with transfer of the sample material and eliminates the possibility of loss of sample. Moreover, the danger of inclusion of contamination from the laboratory environment is reduced.
- Because of the high degree of automation, the cost of analysing an individual sample is kept low.

Disadvantages of the Static Headspace Technique
- On filling headspace vials a corresponding quantity of air is enclosed in the vial unless filling is carried out under an inert gas atmosphere. However, this is very expensive. During thermostatting undesirable side reactions can occur as a result of atmospheric oxygen in the sample.
- During injection from headspace vials air regularly gets into the GC system and can damage sensitive column materials.
- In the case of moist or aqueous samples a considerable quantity of water vapour gets on to the column. This requires special measures to ensure the integrity of the early eluting peaks.
- On use of mass spectrometers special attention must be paid to the stability of ion sources. On insufficient heating the surface of the source and the lens system become increasingly coated with moisture in the course of the work (caused by the injection of water vapour). As a result the focusing of the mass spectrometer can change and this can impair quantitative work.
- Quantitation is matrix-dependent. Standardisation measures are necessary, such as addition of carbonate, internal standards and MHE procedures.
- The ability of substances to be analysed is limited by the maximum possible filling of the headspace vials and by the partition coefficients. For small partition coefficients larger quantities of sample do not lead to an increase in sensitivity. If the maximum possible equilibration temperature is being used, it is almost impossible to increase the sensitivity further.
- The quantity injected is limited and where the sensitivity is insufficient, multiple extraction with cryofocusing is necessary.
- Undesired blank values can be obtained through contaminated air (laboratory air) which gets into the headspace vial or through bleeding of the septum caps.
- Excess pressure can cause headspace vials to burst (e.g. with drinks containing carbonic acid or high equilibration temperatures). This always puts instruments out of operation for long periods and results in considerable clean-up costs. Only the use of special vial caps (with spring rings) together with special sealed vials which can release excess pressure into the atmosphere prevents bursting.

Advantages of the Purge and Trap Technique

- The sample quantity (maximum 25–50 ml) can easily be adapted to give the required sensitivity.
- A pre-purge step can remove atmospheric oxygen from the sample, even at room temperature if required.
- No septum or other permeable material is placed in the way of the sample.
- Substances with high partition coefficients in the static headspace can be determined with good yields.
- There are no excess pressure problems. The entire gas stream is passed pressure-free through the trap to the vent.
- Water vapour can be kept completely out of the GC/MS system by means of a dry purge step and the moisture control system (MCS).
- In automatic operations mixing in an internal standard without manual measures is quite straightforward.

Disadvantages of the Purge and Trap Technique

- Foaming samples require special treatment, the headspace sweep technique.
- The purge vessels (made of glass) must be cleaned carefully. Economical single-use vessels are currently only available in polymer materials.
- Larger quantities of sample require longer purge times.
- For highly contaminated samples the breakthrough volume of the trap must be taken into consideration. If the baking out step is inadequate, the danger of a carry-over exists.
- Coupling with capillary GC necessitates the use of small split ratios or cryofocusing.

2.1.4 Adsorptive Enrichment and Thermodesorption

To determine volatile organic compounds, GC/MS coupling is the method of choice. In the analysis of air or gas samples an extraordinarily large number of components of the most widely differing classes of compounds and over a very wide concentration range have to be considered. Usually, the concentration of the substances of interest is too low for the direct measurement of an air sample, and therefore enrichment on suitable adsorbents is necessary. The concentration on solid adsorption material allows the accumulation of organic components from large volumes of gas. Typical areas of use include soil air, workplace monitoring, gases from landfill sites, air pollution, and air inside buildings.

Besides a high storage capacity to achieve high breakthrough volumes (BTV), the adsorption material is expected to have a low affinity for water (air moisture). This is not only important for GC/MS coupling. Neither water nor CO_2 should have a negative effect on the breakthrough volume of the organic components. The surrounding air generally has a high moisture content. However, particular precautions must be taken in the case of combustion gases. The desorption of the enriched components from the carrier materials used should be complete and without thermal changes.

The adsorption material used must have a high thermal stability and must not contribute to the background of the analysis. The expected adsorption and stability properties should not change for a large number of analyses, even on repeated use.

In air analysis adsorption materials, such as Tenax, Carbotrap and XAD resins, are generally used (Fig. 2-30, Table 2-10). Pure activated charcoal is indeed an outstanding adsorbent for all organic compounds; however, these can only be sufficiently desorbed when displaced using liquid solvents. Complete thermal desorption requires extremely high temperatures (>600 °C). This can lead to pyrolytic decomposition of the organic compounds which are then no longer detected in the residue analysis.

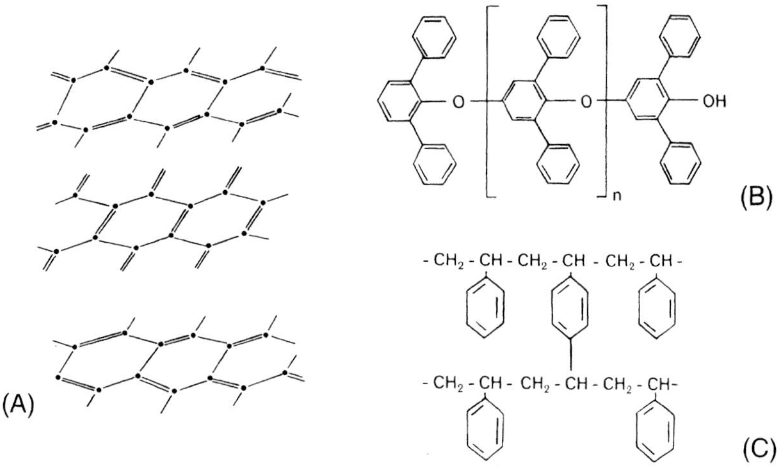

Fig. 2-30 Surface model for common adsorbents (Supelco)
(A) Carbotrap, surface area ca. 100 m^2/g, uniform charge distribution over all carbon atom centres
(B) Tenax (2,6-diphenyl-p-phenylene oxide), surface area ca. 24 m^2/g, nonuniform charge distribution, the charge is essentially localised on the oxygen atoms. (Tenax-TA has replaced Tenax-GC as a new material of higher purity; Tenax-GR is a graphitised modification)
(C) Amberlite XAD-2, surface area ca. 300 m^2/g, nonuniform charge distribution, less polar than Tenax (XAD-4, ca. 800 m^2/g)

A recent development for retaining the high adsorptivity of activated charcoal with simultaneously favourable desorption properties involves the use of graphitised carbon black (Carbotrap, Carboxen). Its surface consists of graphite crystals. The associated hydrophobic properties and the exploitation of nonspecific interactions have led to wide use of these carrier materials.

Like Tenax, graphitised carbon blacks also have a low affinity for water. These adsorption materials can be dried with a dry gas stream, e.g. the GC carrier gas (in the direction of adsorption!), without significant loss of material (dry purge).

VOCARB traps are special combinations of the adsorption materials Carbopack (graphitised carbon black, GCB) and Carboxen (carbon molecular sieve). These combinations have been optimised in the form of VOCARB 3000 and VOCARB 4000 for volatile and involatile compounds respectively, corresponding to the EPA methods 624/1624 and 542.2. VOCARB 4000 exhibits higher adsorptivity for less volatile

Table 2-10 Adsorption materials and frequently described areas of use

XAD–4	C_1/C_2-chlorinated hydrocarbons R11 halogenated narcotics vinyl chloride ethylene oxide styrene
Tenax TA 35-60 mesh	for boiling points 80–200 °C C_1/C_2-chlorinated hydrocarbons general solvents, volatile halogenated hydrocarbons BTEX phenols
Porapak	high-boiling, nonpolar substances halogenated narcotics ethylene oxide
Carbotrap/VOCARB	for boiling points –15 to –120 °C C_1/C_2-chlorinated hydrocarbons volatile halogenated hydrocarbons BTEX styrene
Molecular sieve 5 Å	N_2O

components, such as naphthalenes and trichlorobenzenes. However, it shows catalytic activity towards 2-chloroethyl vinyl ether (complete degradation!), 2,2-dichloropropane, bromoform and methyl bromide.

Volatile polar components are enriched on highly polar carrier materials. The combination of Tenax TA and silica gel has proved particularly successful for the enrichment of polar compounds.

To cover a wider range of molecular sizes various adsorption materials are combined with one another. For example, the Carbotrap multibed adsorption tube (Fig. 2-31) consists of three materials: Carbotrap C with its low surface area of 12 m²/g is used to enrich high molecular weight components, such as alkylbenzenes, polyaromatic hydrocarbons or PCBs, directly at the inlet. All the more volatile substances pass through to the sub-

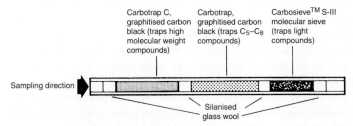

Fig. 2-31 Multibed adsorption/desorption tube Carbotrap 300 (Supelco)
Carbotrap, Carbotrap C: graphitised carbon black, GCB, surface area ca. 12 m²/g
Carbosieve S-III: carbon molecular sieve, surface area ca. 800 m²/g, pore size 15–40 Å

sequent layers. A layer of Carbotrap particles (Carbotrap B) separated by glass wool is characterised by the higher particle size of 20/40 mesh. Volatile organic substances, in particular, are excellently adsorbed on Carbotrap and thermally desorbed with high recoveries. To adsorb C_2-hydrocarbons Carbosieve S-III material with a particularly high surface area of 800 m^2/g and a pore size of 15–40 Å is placed at the end of the adsorption tube (patent, BASF AG, Ludwigshafen) (Table 2-11). The hydrophobic properties of the Carbosieve material allow its use in atmospheres with high moisture contents.

Table 2-11 Breakthrough volumes [l] for Carbotrap 300 adsorption/desorption tubes (Supelco)

Substance	Carbosieve™ S-III	Carbotrap	Carbotrap C
Vinyl chloride	158		
Chloroform		1.1*	
1,2-Dichloroethane		0.4	
1,1,1-Trichloroethane		2.7*	
Carbon tetrachloride		4.7*	
1,2-Dichloropropane		6.8	
Trichloroethylene		2.5	
Bromoform		1.7	
Tetrachloroethylene		2.2	
Chlorobenzene		316	
n-Heptane		262	
1-Heptene		284*	
Benzene		2.3	
Toluene		130	
Ethylbenzene			12.9
p-Xylene			11.2
m-Xylene			11.0*
o-Xylene			11.0*
Cumene			27.8*

* theoretical value

Sample Collection

Sample collection can be either passive or active (Fig. 2-32). For passive collection diffusion tubes with special dimensions are used. The content of substances in the surrounding air is integrated, taking the collection time into account.

Active collection devices require a calibrated pump with which a predetermined volume is drawn through the adsorption tube. Having estimated the expected concentrations, for example for indoor air 100 ml/min and for outdoor air 1000 ml/min, the air is drawn through the prepared adsorption tube over a period of 4 h. After sample collection the adsorption tube must be closed tightly to exclude additional uncontrolled contamination.

Brown and Purnell carried out thorough investigations on the determination of breakthrough volumes. The latter generally vary widely with the collection rate. On use of Tenax as the adsorbent the ideal collection rate is 50 ml/min, in any case, however, between 5 and 600 ml/min. Moisture does not affect the breakthrough volumes

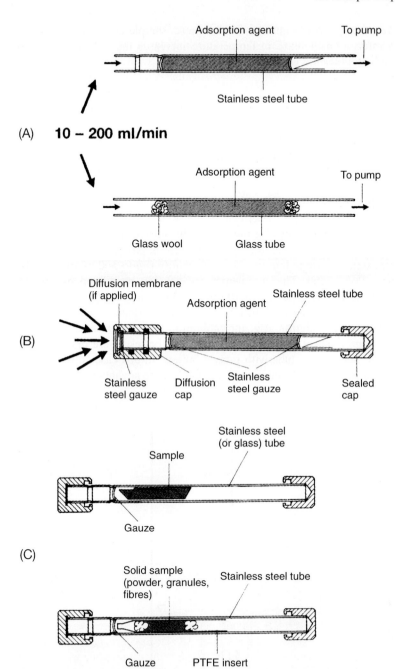

Fig. 2-32 Sample collection with thermodesorption tubes (Perkin-Elmer)
(A) Active sample collection with pump (e.g. personal air sampler)
(B) Passive sample collection by diffusion
(C) Direct introduction of solid samples

with Tenax (unlike other porous materials). Furthermore, sample collection is greatly affected by temperature. An increase in temperature increases the breakthrough volume (ca. every 10 K doubles the volume).

Calibration

In the calibration of thermodesorption tubes the same conditions should predominate as in sample collection. Methods such as the liquid application of a calibration solution to the adsorption materials or the comparison with direct injections have been shown to be unsatisfactory.

The process for the preparation of standard atmospheres by continuous injection into a regulated air stream has been described in the VDI guidelines (Fig 2-33). In this process an individual component or a mixture of the substances to be determined is continuously charged to an injector through which air is passed (complementary gas), using a thermostatted syringe burette. The air quantity (up to 500 ml/min) is adjusted using a mass flow meter. The complementary gas can be diluted by mixing with a second air stream (dilution gas up to 10 l/min). Moistening the gas can be carried out inside or outside the apparatus. In this way concentrations in the ppm range can be generated. For further dilution, e.g. for calibration of pollution measurements, a separate dilution stage is necessary. The gas samples to be tested are drawn out of the calibration station into a glass tube with several outlets. In this case active or passive sample collection is possible. Continuous injection has the advantage that the preparation of mixtures is very flexible (Tables 2-12 and 2-13).

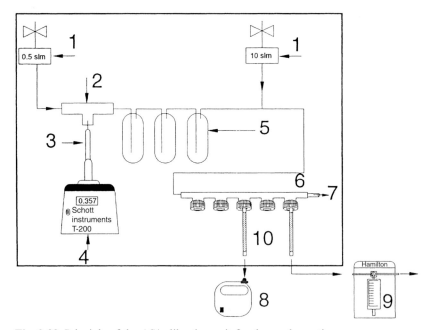

Fig. 2-33 Principle of the AS/calibration unit for thermodesorption

Table 2-12 Evaluation of test tubes which were prepared using the AS/calibration unit by continuous injection ($n = 10$)

Component	Mean value	Standard deviation	Relative standard deviation [%]
1,1,1-Trichlorethane	562.6	4.95	0.88
Dichloromethane	538.6	7.93	1.47
Benzene	753.3	9.48	1.25
Trichloroethylene	627.3	16.7	2.67
Chloroform	626.6	12.1	1.94
Tetrachloroethylene	698.2	6.11	0.88
Toluene	1074	6.88	0.64
Ethylbenzene	358.1	2.91	0.81
p-Xylene	736.3	4.85	0.66
m-Xylene	731.6	4.77	0.65
Styrene	755.8	4.60	0.61
o-Xylene	389.5	3.49	0.89

Table 2-13 Analysis results of BTEX determination of certified samples after calibration with the AS/calibration unit

Mass [µg]	Benzene	Toluene	m-Xylene
Measured value	1.071	1.136	1.042
Required value (certified)	1.053	1.125	1.043
Standard deviation (certified)	0.014	0.015	0.015

Desorption

The elution of the organic compounds collected involves extraction by a solvent (displacement) or thermal desorption. Pentane, CS_2 or benzyl alcohol are generally used as extraction solvents. CS_2 is very suitable for activated charcoal, but cannot be used with polymeric materials, such as Tenax or Amberlite XAD, because decomposition occurs. As a result of displacement with solvents the sample is extensively diluted, which can lead to problems with the detection limits on mass spectrometric detection. With solvents additional contamination can occur. The extracts are usually applied as solutions. The readily automated static headspace technique can also be used for sample injection. This procedure has also proved to be effective for desorption using polar solvents, such as benzyl alcohol or ethylene glycol monophenyl ether (1% solution in water after Krebs).

In thermal desorption, the concentrated volatile components are liberated by rapid heating of the adsorption tube and after preliminary focusing, usually within the instrument, are injected into the GC/MS system for analysis (Table 2-14). Automated thermodesorption gives better sensitivity, precision and accuracy in the analysis. The number of manual steps in sample processing is reduced. Through frequent re-use of

Table 2-14 Desorption temperatures for common adsorption materials and possible interfering components which can be detected by GC/MS

Adsorbent	Desorption	Maximum temperature	Interfering components
Carbotrap	Up to 330 °C	>400 °C	Not determined
Tenax	150–250 °C	375 °C	Benzene, toluene, trichloroethylene
Molecular sieve	250 °C	350 °C	Not determined
Porapak	200 °C	250 °C	Benzene, xylene, styrene
VOCARB 3000	250 °C	>400 °C	Not determined
VOCARB 4000	250 °C	>400 °C	Not determined
XAD-2/4	150 °C	230 °C	Benzene, xylene, styrene

the adsorption tube and complete elimination of solvents from the analysis procedure, a significant lowering of cost per sample is achieved.

Thermodesorption has now become a routine procedure because of program-controlled samplers. The individual steps are prescribed by the user in the control program and monitored internally by the instrument. For the sequential processing of a large number of samples, autosamplers with capacities of up to 50 adsorption tubes are used.

The adsorption tubes are fitted with temporary Teflon caps or septums. The Teflon caps are removed inside the instrument before measurement and the adsorption tube is inserted into the desorption oven. Before the measurement is carried out the tightness of the seal is tested by monitoring an appropriate carrier gas pressure for a short time. Carrier gas is passed through the adsorption tubes in the desorption oven at temperatures of up to 400 °C (in the reverse direction to the adsorption!). The components liberated are stored in a cold trap inside the apparatus. The sample is transferred to the GC column by rapidly heating the cold trap. This two-stage desorption and the use of the multiple split technique enable the measuring range to be adapted to a wide range of substance concentrations. The sample quantity can be adapted to the capacity of the capillary column used through suitable split ratios both before and after the internal cold trap (Fig. 2-34).

Thermodesorption is now usually carried out automatically. The high performance achieved does not mean that it has no disadvantages. A gas sample in an adsorption tube is always unique, so a single measurement cannot be repeated. Possible and, for certain carrier materials, already known decomposition reactions have already been mentioned. For this reason another method is favoured by the EPA for air analysis. Passivated, nickel-coated canisters (ca. 2 l, maximum up to 15 l) are evacuated for sample collection. The samples collected on opening the canister can be measured several times in the laboratory. Suitable samplers are used, which are connected on-line with GC/MS. Cryofocusing is used to concentrate the analytes from the volumes collected. If required, the sample can be dried with a semipermeable membrane (Nafion drier) or by condensation of the water (moisture control system, see also Section 2.1.3.2). Adsorption materials are not used in these processes.

Step 1: primary (tube) desorption

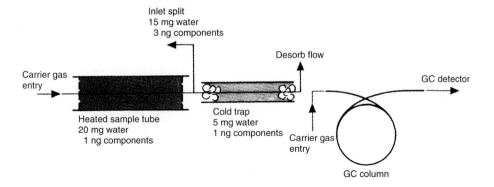

Step 2: secondary (cold trap) desorption

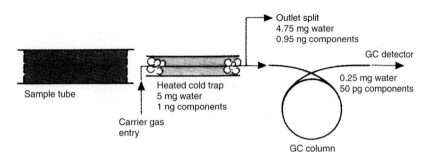

Fig. 2-34 Multiple split technique for thermodesorption (Perkin-Elmer)

2.1.5 Pyrolysis

The use of pyrolysis extends the area of use of GC/MS coupling to samples which cannot be separated by GC because they cannot be desorbed from a matrix or evaporated without decomposition. In analytical pyrolysis a large quantity of energy is passed into a sample so that fragments which can be gas chromatographed are formed reproducibly. The pyrolysis reaction initially involves thermal cleavage of C-C bonds, e.g. in the case of polymers. Thermally induced chemical reactions within the pyrolysis product are undesired side reactions and can be prevented by reaching the pyrolysis temperature as rapidly as possible. The reactions initiated by the pyrolysis are temperature-dependent. To produce a reproducible and quantifiable mixture of pyrolysis products, the heating rates and pyrolysis temperatures, in particular, should be kept constant. The sample and its contents can be characterised using the chromatographic sample trace (pyrogram) or by the mass spectroscopic identification of individual pyrolysis products.

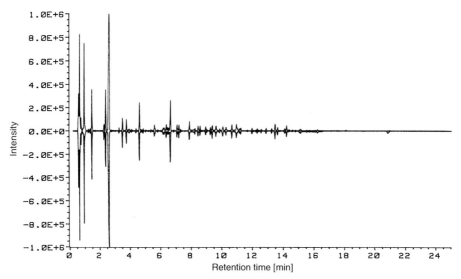

Fig. 2-35 Reproducibility of the pyrolysis of an automotive paint on two consecutive days using a foil pyrolyser coupled to GC/MS, mirrored representation (Steger, Audi AG)

The use of pyrolysis apparatus with GC/MS systems imposes particular requirements on them. The sample quantity applied must correspond to the capacity of commercial fused silica capillary columns. It is usually in the μg range or less. By selecting a suitable split ratio peak equivalents for the pyrolysis products can be achieved in the middle ng range. These lie completely within the range of modern GC/MS systems. The small sample quantities are also favourable for analytical pyrolysis in another aspect. The reproducibility of the procedure increases as the sample quantity is lowered as more rapid heat transport through the sample is possible (Fig. 2-35). Side reactions in the sample itself and as a result of reactive pyrolysis products are increasingly eliminated.

Because of the small sample quantities and the high reproducibility of the results, analytical pyrolysis has experienced a renaissance in recent years. Both in the analysis of polymers with regard to quality, composition and stability, and in the areas of environmental analysis, foodstuffs analysis and forensic science, pyrolysis has become an important analytical tool, the significance of which has been increased immensely by coupling with GC/MS.

Analytical pyrolysis is currently dominated by two different processes: high frequency pyrolysis (Curie point pyrolysis) and foil pyrolysis. The processes differ principally through the different means of energy input and the different temperature rise times (TRT). Both pyrolysis processes can now easily be connected to GC and GC/MS systems. The reactors currently used are constructed so they can be placed on top of GC injectors (split operation) and can be installed for short term use only if required.

Analytical pyrolysis procedures		
Procedure	Foil	High frequency
Carrier	Pt foil	Fe/Ni alloys
Pyrolysis temperature	Can be freely selected up to 1400 °C	Fixed Curie temperatures
TRT	<8 ms	ca. 30 ≈ 100 ms

Foil Pyrolysis

In foil pyrolysis the sample is applied to a thin platinum foil (Fig. 2-36). The thermal mass of this device is extremely low. After application of a heating current any desired temperature up to ca. 1400 °C can be achieved within milliseconds. The extremely high heating rate results in high reproducibility. The temperature of the Pt foil can be controlled by its resistance. However, the temperature can be measured and controlled more precisely and rapidly from the radiation emitted by the Pt foil. An exact calibration of the pyrolysis temperature can be carried out and the course of the pyrolysis is recorded by this feedback alone. Besides endothermic pyrolyses, exothermic processes can also be detected and recorded.

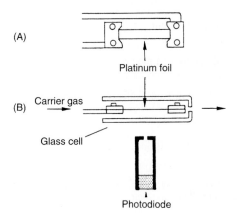

Fig. 2-36 Scheme of a Pt foil pyrolyser with temperature control by means of fibre optic cable (Pyrola)
(A) View of the Pt foil with carrier gas and current inlets
(B) Side view of the pyrolysis cell with the glass cell (ca. 2 ml volume, pyrex) and photodiode under the Pt foil for calibration and monitoring of the pyrolysis temperature

Pyrolysis Nomenclature

- Pyrolysis:
  A chemical degradation reaction initiated by thermal energy alone.
- Oxidative pyrolysis:
  A pyrolysis which is carried out in an oxidative atmosphere (e.g. O_2).
- Pyrolysate:
  The total products of a pyrolysis.
- Analytical pyrolysis:
  The characterisation of materials or of a chemical process by instrumental analysis of the pyrolysate.
- Applied pyrolysis:
  The production of commercially usable materials by pyrolysis.
- Temperature/time profile (TTP):
  The graph of temperature against time for an individual pyrolysis experiment.
- Temperature rise time (TRT):
  The time required by a pyrolyser to reach the pyrolysis temperature from the start time.
- Flash pyrolysis:
  A pyrolysis which is carried out with a short temperature rise time to achieve a constant final temperature.
- Continuous pyrolyser:
  A pyrolyser where the sample is placed in a preheated reactor.
- Pulse pyrolyser:
  A pyrolyser where the sample is placed in a cold reactor and then rapidly heated.
- Foil pyrolyser:
  A pyrolyser where the sample is applied to metal foil or band which is directly heated as a result of its resistance.
- Curie point pyrolyser:
  A pyrolyser with a ferromagnetic sample carrier which is heated inductively to its Curie point.
- Temperature-programmed pyrolysis:
  A pyrolysis where the sample is heated at a controlled rate over a range of temperatures at which pyrolysis occurs.
- Sequential pyrolysis:
  Pyrolysis where the sample is repeatedly pyrolysed for a short time under identical conditions (kinetic studies).
- Fractionated pyrolysis:
  A pyrolysis where the sample is pyrolysed under different conditions in order to investigate different sample fractions.
- Pyrogram:
  The chromatogram (GC, GC/MS) or spectrum (MS) of a pyrolysate.

Curie Point Pyrolysis

High frequency pyrolysis uses the known property of ferromagnetic alloys of losing their spontaneous magnetism above the Curie temperature (Curie point). At this temperature a large number of properties change, such as the electrical resistance or the specific heat. Above the Curie temperature ferromagnetic substances exhibit paramagnetic properties. The possibility of reaching a defined and constant temperature using the Curie point was first realised by W. Simon in 1965. In a high frequency field a ferromagnetic alloy does not absorb any more energy above its Curie point and remains at this temperature. As the Curie temperature is substance-dependent, another Curie temperature can be used by changing the material. If a sample is applied to a ferromagnetic material and is heated in an energy-rich high frequency field (Fig. 2-37), the pyrolysis takes place at the temperature determined by the choice of alloy and thus its Curie point. The temperature rise curves for various metals and alloys are shown in Fig. 2-38.

In practice sample carriers in the form of loops, coils or simple wires made of different alloys at fixed temperature intervals are used. A disadvantage of Curie point pyrolysis is the longer temperature rise time of ca. 30–100 ms required to reach the Curie point compared with foil pyrolysis. There are also effects due to the not completely inert surface of the sample carrier. They manifest themselves in the inadequate reproducibility of the pyrolysis process for analytical purposes. Copolymers with thermally reactive functional groups (e. g. free OH, NH_2 or COOH groups) cannot be analysed by Curie point pyrolysis.

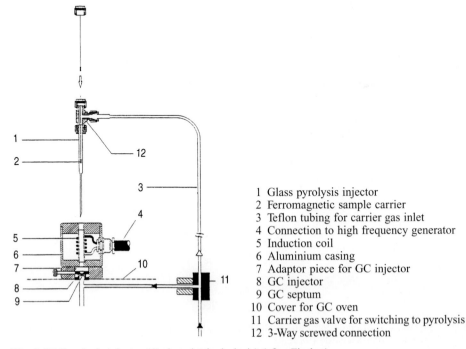

1 Glass pyrolysis injector
2 Ferromagnetic sample carrier
3 Teflon tubing for carrier gas inlet
4 Connection to high frequency generator
5 Induction coil
6 Aluminium casing
7 Adaptor piece for GC injector
8 GC injector
9 GC septum
10 Cover for GC oven
11 Carrier gas valve for switching to pyrolysis
12 3-Way screwed connection

Fig. 2-37 Pyrolysis injector (Curie point hydrolysis) (after Fischer)

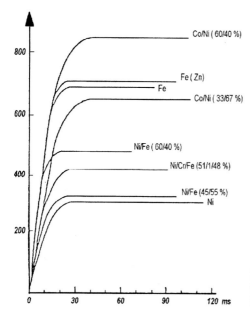

Fig. 2-38 Curie temperatures and temperature/time profiles of various ferromagnetic materials (after Simon)

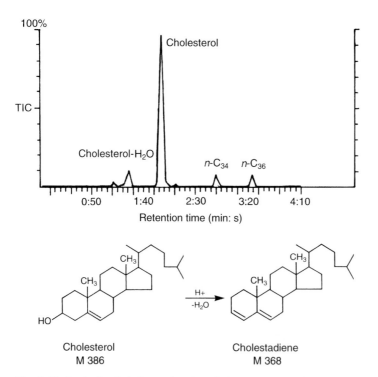

Fig. 2-39 Analysis of cholesterol and n-alkanes (C_{34}, C_{36}) for testing the pyrolysis coupling to the GC injector (after Richards 1988).
Conditions: GC column 4 m × 0.22 mm ID × 0.25 µm CP-Sil5, 200–320 °C, 30 °C/min, GC-ITD

For the analytical assessment of the coupling of pyrolysers with GC and GC/MS systems, a high-boiling mixture of cholesterol with n-alkanes has been proposed (Gassiot-Matas). Fig. 2-39 shows the evaporation of cholesterol (500 ng) with the C_{34}- and C_{36}-n-alkanes (50 ng of each). Before the intact cholesterol is detected, its dehydration product appears. The intensity of this peak increases with small sample quantities and increasing temperatures in the region of substance transfer in the GC injector. The high peak symmetries of the signals of the alkanes, the cholesterol and its dehydration product indicate that the coupling is functioning well.

There are various possibilities for the evaluation of pyrograms obtained with a GC/MS system. With classical FID detection the pyrogram pattern is only compared with known standards. However, with GC/MS systems the mass spectra of the individual pyrolysis products can be evaluated. By using libraries of spectra, substances and substance groups can be identified. GC/MS pyrograms can be selectively investigated for trace components even in complex separating situations by using the characteristic mass fragments of minor components. The comparison of sample patterns becomes meaningful through the choice of mass chromatograms of substance-specific fragment ions.

In the library search for the spectra of pyrolysis products special care must be taken. Commercial spectral libraries consist of spectra of particular substances which have been described. Depending on the sample material, however, an extremely large number of reaction products are formed on pyrolysis, which cannot be completely separated even under the best GC conditions. If the detection is sensitive enough this situation is shown clearly by the mass chromatogram. Also, for polymers there is a typical appearance of homologous fragments, which must also be taken into account. These series can also be shown easily using mass chromatograms through the choice of suitable fragment ions.

Quantitative determinations using pyrolysis benefit particularly from the selectivity of GC/MS detection. The precision is comparable to that of liquid injections.

Thermal Extraction

Besides the processes of foil and Curie point pyrolysis described, thermal extraction (Fig. 2-40) covers the area between pyrolysis and thermodesorption. In thermal extraction the sample (manual or automatic) is placed in a quartz oven, which can be operated at a constant temperature of up to 750 °C. This allows both the thermal extraction of volatile components from solid samples as well as their pyrolysis.

The difference between thermal extraction and the analytical pyrolysis systems described above lies in the consideration of the sample quantity. The typical capacity of a thermal extractor is between 0.1 mg and ca. 500 mg sample material. This allows weighing out and reweighing. Even inhomogeneous materials can be investigated in this way.

The wide temperature range allows a broad spectrum of applications. The stripping of volatile compounds, e. g. solvents, volatile halogenated hydrocarbons and BTEX, is carried out typically below 80 °C (Fig 2-41). At temperatures of up to 350 °C involatile compounds are liberated, for example from plastics or composite materials. Pyrolytic cleavage reactions begin at temperatures above 350 °C. Because of the large sample quantities and the indirect heating of the sample in an oven, the temperature in the sample increases relatively slowly. The pyrograms obtained can be compared with pyrolysis patterns which take precedence.

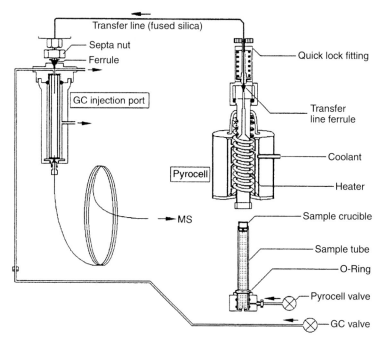

Fig. 2-40 Thermal extractor, cross section and coupling to a GC injector (after Ruska)

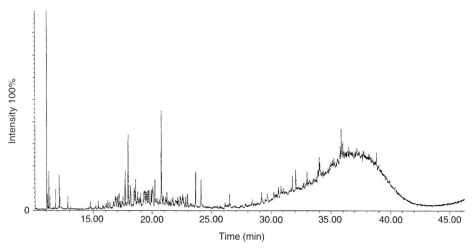

Fig. 2-41 GC/MS analysis of volatile components of a floor covering by thermal extraction at 80 °C (after Ruska)

2.2 Gas Chromatography

2.2.1 GC/MS Sample Inlet Systems

Much less attention was paid to this area at the time when packed columns were used. On-column injection was the state of the art and was in no way a limiting factor for the quality of the chromatographic separation. With the introduction of capillary techniques in the form of glass or fused silica capillaries, high resolution gas chromatography (HRGC) maintained its presence in laboratories and GC and GC/MS made a great technological advance. Many well known names in chromatography are associated with important contributions to sample injection: Desty, Ettre, Grob, Halasz, Poy, Schomburg and Vogt among others. The exploitation of the high separating capacity of capillary columns now requires perfect control of a problem-orientated sample injection technique.

According to Schomburg, sample injection should satisfy the following requirements:

- achieving the optimal efficiency of the column,
- achieving a high signal/noise ratio through peaks which are as steep as possible in order to be certain of the detection and quantitative determination of trace components at sufficient resolution (no band broadening),
- avoidance of any change in the quantitative composition of the original sample (systematic errors, accuracy),
- avoidance of statistical errors which are too high for the absolute and relative peak areas (precision),
- avoidance of thermal and/or catalytic decomposition or chemical reaction of sample components,
- sample components which cannot be evaporated must not reach the column or must be removed easily (guard column). Involatile sample components lead to decreases in separating capacity through peak broadening, and shortening of the service life of the capillary column.
- In the area of trace analysis it is necessary to transfer the substances to be analysed to the separating system with as little loss as possible. Here the injection of larger sample volumes (up to >100 µl) is desirable.
- Simple handling, service and maintenance of the sample injection system play an important role in routine applications.
- The possibility of automation of the injection is important, not only for large numbers of samples, but also as automatic injection is superior to manual injection for achieving a low standard deviation.

Careless injection of a sample extract frequently overlooks the outstanding possibilities of the capillary technique. In all modern GC and GC/MS systems, sample injection is of fundamental importance for the quality of the chromatographic analysis. Poor injection cannot be compensated for even by the choice of the best column material. This also applies to the choice of detectors. The use of a mass spectrometer as a detector can be much more powerful if the chromatography is of the best quality. In fact, more than 10 years experience in the use of GC/MS systems shows that as soon as the GC is coupled to an MS system, the chromatography is rapidly degraded to an inlet route. Effort is put too quickly into the optimisation of the parameters of the MS detector without exploiting the much wider possibilities of the GC. Each GC/MS system is only as good as the chromatography allows!

The starting point for the discussion of sample inlet systems is the target of creating a sample zone at the top of the capillary column for the start of the chromatography which is as narrow as possible. This narrow sample band principally determines the quality of the chromatography as the peak shape at the end of the separation cannot be better (narrower, more symmetrical etc.) than at the beginning. As an explanatory model, chromatography can be described as a chain of distillation plates. The number of separation steps (number of plates) of a column is used by many column producers as a measure of the separating capacity of a column. In this sense sample injection means application to the first plate of the column. In capillary GC the volume of such a plate is less than 0.01 µl. The sample extracts used in trace analysis are generally very dilute, making larger quantities of solvent (>1 µl) necessary. This shows the importance of sample injection techniques. In this connection Pretorius and Bertsch should be cited:

"When the column is described as the heart of chromatography, sample injection can be identified as the Achilles heel. It is the least understood and the most confusing aspect of modern gas chromatography."

In practice, different types of injectors are used (see Table 2-17). The sample injection systems are classified as hot or cold according to their function. A separate section is dedicated to direct on-column injection.

2.2.1.1 Hot Sample Injection

Classical injection techniques involve applying the sample solution in constantly heated injectors. Both the solvent and the dissolved sample evaporate in an evaporation tube specially fitted for the purpose (insert) and mix with the carrier gas. Temperatures of ca. 200 °C to above 300 °C are used for evaporation. The operating procedures of split injection and total sample transfer (splitless) differ according to whether there is partial or complete transfer of the solvent/sample on to the column.

The problem of discrimination on injection into hot injectors arises with the question of what is the best injection technique. Figure 2-42 shows the effects of various injection techniques on the discrimination between various alkanes. While in the chain length range of up to ca. C_{16} hardly any differences are observed, discrimination can be avoided for higher-boiling, long-chain compounds by a suitable choice of injection technique. For example, filling the injection syringe with a solvent/derivatising agent plug before drawing up the sample and observing the needle temperature during a short period after insertion into the injector.

Solvent Flush/Hot Needle: This is the injection procedure of choice for hot injectors and gives favourable discrimination properties. Ca. 0.5–1 µl solvent or derivatising agent are first taken up in the syringe, then 0.5 µl air and finally the sample extract

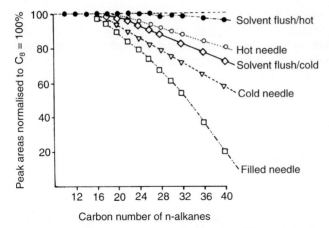

Fig. 2-42 Discrimination among alkanes using different injection techniques (after Karasek 1988). The peak areas are normalised to $C_8 = 100\%$

(sandwich technique). Before the injection the liquid plug is drawn up into the body of the syringe. The volumes can thus be read better on the scale. The injection involves first inserting the needle into the injector, waiting for the needle to warm up (ca. 2 s) and then rapidly injecting the sample. The solvent flush/hot needle procedure can be carried out manually or using a programmable autosampler.

Hot Needle: This technique is definitely the most frequently used variant. Only the sample extract is drawn up into the syringe; the plug is thus drawn up into the body of the syringe so that the volume can be read off on the scale. After inserting the needle, there is a pause for warming up and then the sample is rapidly injected.

Solvent Flush/Cold Needle: Here the sandwich technique described above is used to fill the syringe. The injection is, however, carried out very quickly (usually using an autosampler) without waiting for the injection needle to warm up.

Cold Needle: Only the sample extract is drawn up into the syringe; the plug is held in the body of the syringe in such a way that the volume on the scale can be read. The injection is carried out very rapidly (usually with an autosampler) without waiting for the needle to warm up.

Filled Needle: This injection procedure is no longer up-to-date. It is associated with certain types of syringe which, on measuring out the sample extracts, can only allow the liquid plug into the injection needle. Warning: certain automatic liquid sample injectors use this procedure.

Split Injection

After evaporation in the insert, with the split technique the sample/carrier gas stream is divided. The larger, variable portion leaves the injector via the split exit and the smaller portion passes on to the column. The split ratios can be adapted within wide limits to the sample concentration, the sensitivity of the detector and the capacity of the capillary column used. Typical split ratios mostly lie in the range $1:10$ to $1:100$ or more. The start values for the temperature program of a GC oven are independent of the injection procedure using the split technique.

For concentrated samples the variation of the split ratio and the volume applied represents the simplest method of matching the quantity of substance to the column load and to the linearity of the detector. Even in residue analysis the split technique is not unimportant. By increasing the split stream the carrier gas velocity in the injector is increased and this allows the highly accelerated transport of the sample cloud past the orifice of the column. This permits a very narrow sample zone to be applied to the column. To optimise the process the possibility of a split injection at a split ratio of less than $1:10$ should also be considered. In particular, on coupling with static headspace or purge and trap techniques better peak profiles and shorter analysis times are achieved. The smallest split ratio which can be used depends on the internal volume of the insert. Inserts with small volumes (smaller internal diameters) are preferred for the split technique.

A disadvantage of the split injection technique is the uncontrollable discrimination with regard to the sample composition. This applies particularly to samples with a wide boiling point range. Quantitation with an external standard is particularly badly affected by this. Because of the deviation of the effective split ratio from that set up, this value should not be used in the calculation. Quantitation with an internal standard or alternatively the standard addition procedure must be used.

Total Sample Transfer (splitless injection)

With the total sample transfer technique the sample is injected into the hot injector with the split valve closed (Fig. 2-43). The volume of the injector insert must be able to hold the solvent/sample vapour cloud completely. Because of this special insert liners (vaporiser) are recommended for splitless operations. Depending on the insert used and the solvent, there is a maximum injection volume which allows the vapour to be held in the insert. Inserts which are too small lead, on explosive evaporation of the solvent, to expansion of the sample vapour beyond the inserts into the cold regions of the injector, such as the septum area or the split tubing, causing considerable contamination problems. Inserts which are too wide lead to significant dilution of the sample cloud and thus to prolonged transfer times and losses through diffusion. A recent compromise made by many manufacturers involves insert volumes of 1 ml for injection volumes of about $1-2$ µl. The septum flush should not be completely closed even with splitless injection. On correct choice of insert the sample cloud does not reach the septum so that the low purge flow does not have any effect on the injection itself.

The carrier gas flushes the sample cloud continuously from the injector on to the column. This process generally takes ca. $30-90$ s for complete transfer. There is an

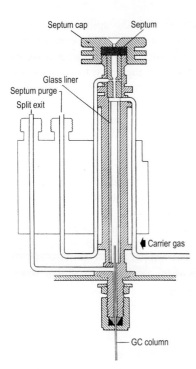

Fig. 2-43 Hot split/splitless injector (Carlo Erba)

exponential decrease in concentration caused by mixing and dilution with the carrier gas. Longer transfer times are generally not advisable because the sample band becomes broadened on the column. Ideal transfer times allow three to five times the volume of the insert to be transferred to the column. This process is favoured by high carrier gas flow rates. Hydrogen is preferred to helium as the carrier gas with respect to sample injection. For the same reason a column diameter of 0.32 mm is preferred for the splitless technique compared to narrower diameters, in order to be able to use optimal flow rates in the injector. In the same way a pressure surge effected using electronic pressure regulation favours sample transfer.

As the transfer times are long compared with the split injection and lead to a considerable distribution of the sample cloud on the column, the resulting band broadening must be counteracted by a suitable temperature of the column oven. Only at start temperatures below the boiling point of the solvent is sufficient refocusing of the sample achieved because of the solvent effect. A rule of thumb is that the solvent effect operates best at an oven temperature of 10–15 °C below the boiling point of the solvent. The solvent condensing on the column walls acts temporarily as an auxiliary phase, accelerates the transfer of the sample cloud on to the column, holds the sample components and focuses them at the beginning of the column with increasing evaporation of solvent into the carrier gas stream (solvent peak). Only after the sample transfer is complete is the split valve opened until the end of the analysis to prevent the further entry of contaminants on to the column. The splitless technique therefore requires working with temperature programs. Because of the almost complete transfer of the sample on to the column, total sample transfer is the method of choice for residue analysis.

Because of the longer residence times in the injector, with the splitless technique there is an increased risk of thermal or catalytic decomposition of labile components. There are losses through adsorption on the surface of the insert, which can usually be counteracted by suitable deactivation. Much more frequently there is adsorption at involatile sample residues in the insert or septum particles collect. This makes it necessary to regularly check and clean the insert.

2.2.1.2 Cold Injection Systems

The cold sample injection technique involves injecting the liquid 'cold' sample directly on to the column (on-column, see Section 2.2.1.4) or into a specially constructed vaporiser. The sample extract is ejected from the syringe needle in liquid form into the insert at temperatures which are well below the boiling point of the solvent. Heating only begins after the syringe needle has been removed from the injection zone. In the area of residue analysis programmed temperature vaporiser (PTV) cold injection systems for split and splitless injection are becoming more widely used, as are on-column systems for exclusively splitless injection.

The cold injection of a liquid sample eliminates the selective evaporation from the syringe needle in all systems, which, in the case of hot injection procedures, leads to discrimination against high-boiling components. Discrimination is also avoided as a result of explosive evaporation of the solvent in hot injectors. Colder regions of the injection system, such as the septum area or the tubing leading to the split valve, can serve as expansion areas if the permitted injection volume is exceeded and can retain individual sample components through adsorption. The individual concepts of cold injection differ in the transfer of the sample to the column and in the possibility for using the split exit. There are definite advantages and limitations for the operation of injectors in practice and for the areas of use envisaged. These exist both between different cold injection techniques and in comparison with hot injection techniques.

The PTV Cold Injection System

The temperature-programmed evaporation with split or splitless operations using the currently available injectors is based on the systematic work of Poy (1981) and Schomburg (1981) (Figs. 2-44 and 2-45). In particular, emphasis was placed on the precise and accurate execution of quantitative analyses of complex mixtures with a wide boiling point range. Particularly at the beginning of the experiments, absence of discrimination for substances up to above C_{60} was documented and its suitability for involatile substances, such as polyaromatic hydrocarbons, was stressed. Examples of the analysis of triglycerides demonstrate the injection of samples up to above C_{50} and the analysis of crude oil fractions up to C_{90}! The PTV process combines the advantages of the hot and on-column injection techniques.

2.2 Gas Chromatography

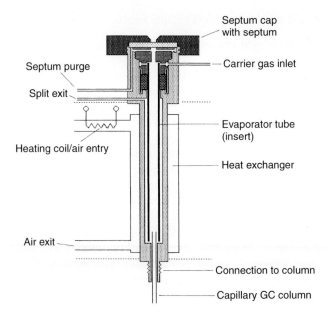

Fig. 2-44 PTV with air as the heating medium from the design by Poy (AS)

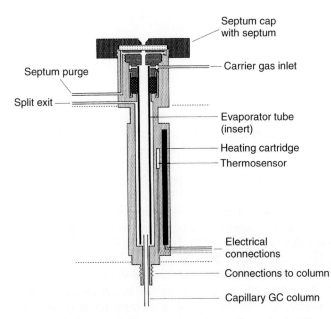

Fig. 2-45 PTV with electrical heating, air or CO_2 as the cooling medium (AS)

There are many advantages of cold split and splitless sample injection:

- Discrimination as a result of fractionated evaporation effects from the syringe needle does not occur.
- A defined volume of liquid can be injected reproducibly.
- The sample components evaporate as a result of controlled heating of the injection area in the order of their boiling points. The solvent evaporates first and leaves the sample components in the injection area without causing a distribution of the analytes in the injector as a result of explosive evaporation.
- Aerosol and droplet formation is avoided through fractional evaporation.
- As the evaporator does not have to take up the complete expansion volume of the injected sample solution, smaller inserts with smaller internal volumes can be used. The consequently more rapid transfer to the column lowers band broadening of the peaks and thus improves the signal/noise ratio.
- If the boiling points of the sample components and the solvent differ by more than 100 °C larger sample volumes (up to more than 100 µl) can be injected (solvent split).
- Impurities and residues which cannot be evaporated do not get on to the column.
- With concentrated samples the possibility can be used of adapting the injection to the capacity of the column by selecting the split ratio.

Currently the PTV cold injection technique is mainly used for residue analysis in the areas of plant protection agents, pharmaceuticals, polyaromatic hydrocarbons and PCBs. However, the enrichment of volatile halogenated hydrocarbons, the direct analysis of water and the formation of derivatives in the injector also demonstrate the broad versatility of the PTV type injector.

The PTV Injection Procedure

PTV Total Sample Transfer

The splitless injection for the injection of a maximum sample equivalent on to the column is the standard requirement for residue analysis. The sample is taken up in a suitable solvent and injected at low temperature. The split valve is closed. The PTV temperature at injection should correspond to the boiling point of the solvent at 1013 mbar. During the injection phase the oven temperature is kept below the PTV temperature. It should certainly be below the boiling point of the solvent in order to exploit the necessary solvent effect according to Grob for focusing the substances at the beginning of the column (Fig. 2-46). If the focusing of the substances is unsuccessful or insufficient, the peaks of the components eluting early are broad and are detected with a low signal/noise ratio.

The injector is heated a few seconds after the injection when the solvent has already evaporated and has reached the column (Fig. 2-47). Typically this time interval is between 5 and 30 s. For high-boiling substances in particular longer residence times have been found to be favourable. The heating rate should be moderate in order to achieve a smooth evaporation of the sample components required for transfer to the

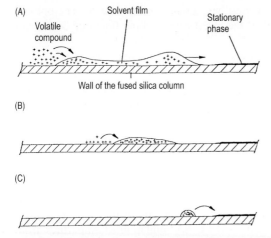

Fig. 2-46 Refocusing by means of the solvent effect in splitless injection (after Grob)
(A) Condensation of solvent at the beginning of the capillary column by lowering the oven temperature below the boiling point. The condensed solvent acts as a stationary phase and dissolves the analytes. At the same time this front migrates and evaporates in the carrier gas flow.
(B) The continuous evaporation of the solvent film concentrates the analytes on to a narrow ring (band) in the column. For this process no stationary phase at the beginning of the column is necessary.
(C) The substances concentrated in a narrow band meet the stationary phase. The separation begins with a sharp band.

PTV Total Sample Transfer

– Split valve closed
– PTV at the boiling point of the solvent
– Oven temperature below the boiling point of the solvent
– Start of PTV heating ca. 10 s after injection
– Start of the GC temperature program ca. 30–120 s after injection
– PTV remains hot until the end of the analysis

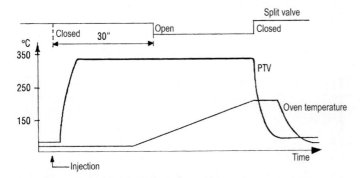

Fig. 2-47 PTV total sample transfer (splitless injection)

column. Heating rates of ca. 200–300 °C/min have proved to be suitable. The optimal heating rate depends on the dimensions of the insert liner and the flow rate of the carrier gas in the insert.

Filling the insert with silanised glass wool has proved effective for the absorption of the sample liquid and for rapid heat exchange. However, the glass wool clearly contributes to enlarging the active surface area of the injector and should therefore only be used in the analysis of noncritical compounds (alkanes, chlorinated hydrocarbons etc.). For the injection of polar or basic components an empty deactivated insert is recommended.

After the transfer of the substances to be analysed to the column, which is complete after ca. 30–120 s, the temperature program of the GC oven is started and the injector purged of any remaining residues by opening the split valve. If required, a second PTV heating ramp can follow to bake out the insert at elevated temperatures. The high injection temperature of the PTV is retained until the end of the analysis to keep the injector free from possible adsorptions and the accumulation of impurities from the carrier gas inlet tubing. The cooling of the PTV is adjusted so that both the PTV and the oven are ready for the next analysis at the same time.

A special form of the PTV cold injection system consists of the column and insert connected via a press-fit attachment. Total sample transfer is possible in the same way as the classical PTV injection, but no split is present to allow flushing of the injection area after evaporation. Injectors of this type can only be used for total sample transfer (e.g. SPI injector, Fig. 2-48).

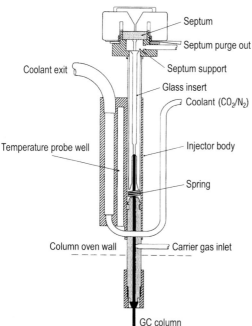

Fig. 2-48 Septum-equipped programmable injector, SPI (Finnigan/Varian)

PTV Split Injection

In this mode of operation the split valve is open throughout (Fig. 2-49). In classical sample application this mode of injection is suitable for concentrated solutions, whereby the column loading can be adapted to its capacity and the nature of the detector by regulating the split flow. Cold injection systems are characterised by the particular dimensions of the inserts. This allows a high carrier gas flow rate at the split point. Compared to hot split injectors, smaller split ratios of ca. 1 : 5 are possible. In an individual case it is possible to work in the split mode and thus achieve better signal/noise ratios and shorter analysis times, compared with total sample transfer.

All other PTV adjustments of time and temperature are completely unchanged compared with total sample transfer! Of particular importance for the split injection using the PTV cold injection system is the fact that the sample is injected into the cold injector (see also Section 2.2.1.3). The choice of start temperature of the GC oven is now no longer coupled to the boiling point of the solvent because of the solvent effect and is chosen according to the retention conditions.

PTV Split Injection

– Split valve open
– PTV below the boiling point of the solvent
– Start of PTV heating ca. 10 s after injection
– Oven temperature to be chosen freely
– PTV remains hot until the end of the analysis

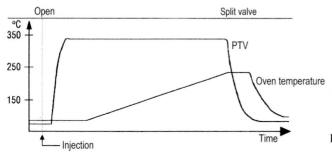

Fig. 2-49 PTV split injection

PTV Solvent Split Injection

This technique is particularly suitable for residue analysis for the injection of solutions with low concentrations, which would undergo loss of analyte on further concentration (Fig. 2-50) or to avoid time-consuming pre-concentration steps. With suitable parameter settings very large solvent quantities, limited only by practical considerations, can be applied.

To inject larger sample volumes (from ca. 2 µl to well over 100 µl) it is advantageous for the insert to be filled at least with silanised glass wool or packing of Tenax

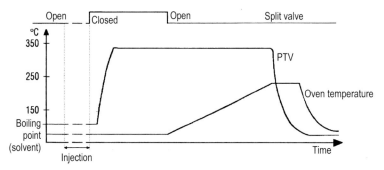

Fig. 2-50 PTV solvent split injection

Supelcoport or another inert carrier material ca. 0.5–1 cm wide. The split valve is open during the injection phase. The PTV is kept at a temperature which corresponds to the boiling point of the solvent. The oven temperature is below the PTV temperature and thus below the boiling point of the solvent. The maximum injection rate depends upon how much solvent per unit time can be evaporated in the insert and carried out through the split tubing by the carrier gas. During injection a high split flow

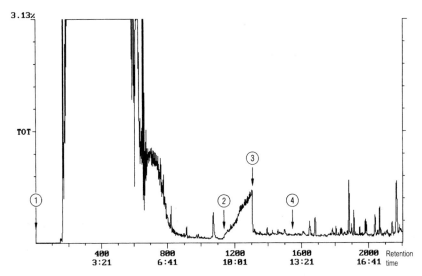

Fig. 2-51 The course of an analysis with large quantities of solvent for the PTV cold injection system: 100 µl PCB solution (200 pg/µl). After the dead time a broad solvent peak is registered. After the solvent peak has subsided, the split valve is closed and the PTV heated up. A smaller roughly triangular solvent peak is produced from adsorbed material. After injection the split valve is opened again and the temperature program started. The peaks eluting show good resolution and are free from tailing.
1 Start of the injection: split open
2 Start of PTV heating: split closed
3 Baking out the PTV: split open
4 Start of the GC temperature program: split open

of >100 ml/min is recommended to focus the analytes on the packing material by local consumption of evaporation heat. At the beginning of the injection the data recording of the GC or GC/MS system is started so that the course of the solvent peak can be followed (Fig. 2-51). For GC/MS systems the compatibility of the ion source with the solvent quantity, which can be considerable and have long-lasting effects, must first be tested. In the case of ion trap mass spectrometers, only by the repeated injection of 100 µl of sample volume can it be proved that the performance of the instrument is not impaired (Fig. 2-52).

When quantities of more than 10 µl are applied, the split valve is only closed when the end of the solvent peak begins to show on the display, meaning that the injector is free of most of the solvent. After the split valve has been closed, the PTV injector can be heated up as usual. After ca. 30–120 s the transfer of the enriched sample from the insert to the column is complete and the GC temperature program can be started. The split valve then opens and the PTV is held at the injection temperature until the end of the analysis.

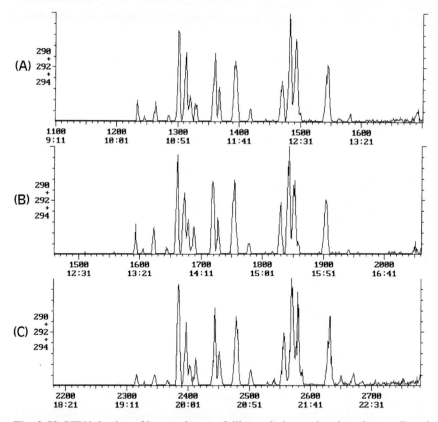

Fig. 2-52 PTV injection of large volumes of dilute solutions using the solvent split technique for a PCB sample (Aroclor 1230/1260). A section of the PCB chromatogram with the elution sequence of the Cl_7-PCBs is shown together with the mass chromatogram of m/z 290/292/294 (Finnigan Ion Trap System ITS40). The peak pattern is retained, but the retention times are shifted.
(A)   1 µl applied, dilution 20.0 ng/µl
(B) 10 µl applied, dilution   2.0 ng/µl
(C) 100 µl applied, dilution   0.2 ng/µl

The PTV solvent split mode also allows the derivatisation of substances in the injector. This procedure simplifies sample preparation considerably. The sample extract is treated with the derivatising agent (e.g. TMSH for methylation) and injected into the PTV. The excess solvent is blown out during the solvent split phase. The derivatisation reaction takes place during the heating phase and the derivatised substances pass on to the column.

PTV Cryo-enrichment

This injection technique is suitable for trace analysis and for concentrating volatile compounds, which are present in large volumes of gas, for example in gas sampling, thermodesorption or the headspace technique. Sample transfer into the injector is carried out while the PTV is in trapping mode. For this the insert is filled with a small quantity (ca. 1–2 mg, ca. 1–3 cm wide) of Tenax, Carbosieve or another thermally stable adsorbent. The PTV injector is cooled with liquid CO_2 or nitrogen until injection. During the injection the split valve is open as in the solvent split mode. Like the total sample transfer method, the oven temperature is kept correspondingly low in order to focus the components at the beginning of the column.

The gaseous sample is passed slowly through the injector and the organic substances retained on the adsorbent. Before injection the PTV can be heated to a low temperature with the split still open to dry the Tenax material if required. For transfer of the sample to the GC column, the split valve is closed and the PTV is heated to effect the total sample transfer of the concentrated components. After ca. 30–120 s the temperature program of the oven is started and the split valve is opened (Fig. 2-53). The PTV remains at the same temperature until the end of the analysis. This last step is particularly important in cryo-enrichment because the cooled adsorbent could become enriched with residual impurities from the carrier gas, which would lead to ghost peaks in the subsequent analysis.

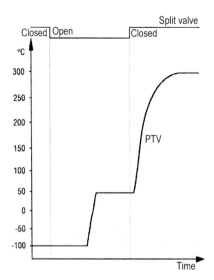

Fig. 2-53 PTV cryo-enrichment at –100 °C (with an optional fractionation step here at 50 °C) and injection with total sample transfer

2.2.1.3 Injection Volumes

For the discussion of the optimal injection volumes for hot and cold injection systems, the use of a 25 m long capillary column with an internal diameter of 0.32 mm is assumed, which operates with a linear carrier gas velocity of 30 cm/s. Under these conditions a carrier gas flow of ca. 2 ml/min through the column is expected. For the total sample transfer mode the carrier gas velocity for both types of injector can be determined by considering the insert volumes in each case. The expansion volumes of the solvents used must also be taken into account. These are shown in Table 2-15.

Table 2-15 Typical solvent expansion volumes

Injection volume	H$_2$O	CS$_2$	CH$_2$Cl$_2$	Hexane	iso-Octane
0.1 µl	142 µl	42 µl	40 µl	20 µl	16 µl
0.5 µl	710 µl	212 µl	200 µl	98 µl	78 µl
1.0 µl	1420 µl	423 µl	401 µl	195 µl	155 µl
2.0 µl	2840 µl	846 µl	802 µl	390 µl	310 µl
3.0 µl	4260 µl	1270 µl	1200 µl	585 µl	465 µl
4.0 µl	5680 µl	1690 µl	1600 µl	780 µl	620 µl
5.0 µl	7100 µl	2120 µl	2000 µl	975 µl	775 µl

The expansion volumes given here refer to an injection temperature of 250 °C and a column pressure of 0.7 bar (10 psi). The values for other temperatures and pressures can be calculated according to $V_{Exp.} = 1/P \cdot n \cdot R \cdot T$.

To determine the time required to flush the contents of the injector completely on to the column at least twice the insert volume is applied, corresponding to a yield of 90–95%. Table 2-16 gives details for hot and cold injectors.

These considerations make it clear that the choice of capillary column and the appropriate injection parameters and volumes only allow the maximum performance of both hot and cold injection systems to be exploited. In particular, cold injection systems function less well if they are improperly used as hot injectors with solvent volumes which are normally used with all commercial hot split or splitless injectors. The probability that the sample cloud will expand beyond the insert increases with the volatility of the solvent. This then leads to losses and tailing as a result of interactions with active surfaces. This would not occur if the PTV were used according to the instructions.

For hot injectors capillary columns with an internal diameter of 0.32 mm are more suitable, as here the transfer of the sample from the injector to the column takes place without diffusion and a more rapid injection is permitted. However, on coupling with MS detectors, the flow rate of ca. 2 ml/min can exceed the maximum compatible flow rate in some quadrupole systems. Columns with an internal diameter of 0.25 mm can be used as an alternative and then should be used in hot injectors with a narrow 2.0 mm internal diameter insert and an injection volume of 0.5 µl, to obtain optimal results. If inserts with a wide internal diameter (4 mm) are used, the solvent effect must be particularly exploited for focusing. It leads to accelerated emptying of the insert through solvent recondensation and a slip-stream in the direction of the injector. Furthermore, the

Table 2-16 Injection times for hot and cold injection (GC column 25 m × 0.32 mm)

Injector type	Hot injector	Cold injection system
Split valve	Closed	Closed
Internal diameter of insert	4 mm	2 mm
Interior volume	1 ml	300 μl
Flow rate	0.3 cm/s [1]	1 cm/s
Minimum time with split closed	60 s [2]	18 s
Maximum injection volume (250 °C)		
CH_2Cl_2	2 μl	(0.5 μl) [3]
Hexane	5 μl	(1 μl)

Remarks:
1. This low flow rate already causes diffusion of the analytes injected against the carrier gas stream. The peaks are broader and the signal/noise ratio is poorer.
2. On use of a 25 m × 0.25 mm internal diameter column the carrier gas rate in the insert of the hot injector is only ca. 0.1 cm/s (4 mm insert) and would require a period with the split closed of at least 3 min. On the other hand the cold injection system only requires the split to be closed for at least 54 s.
3. The given injection volumes are valid for the case where a sample is injected into a heated cold injection system which contradicts the use envisaged for the construction. The injection volumes for cold injection are much higher.

Table 2-17 Choice of a suitable injector system

Characteristics of the sample	(1) Hot split	(2) Hot splitless	(3) PTV split	(4) PTV splitless	(5) PTV solvent split	(6) On-column
Concentrated samples	+	–	+	–	–	–
Trace analysis	–	+	–	+	+	+
Extreme dilution	–	–	–	–	+	+
Narrow boiling range	+	+	+	+	+	+
Wide boiling range	–	≈	+	+	–	+
Volatile substances	+	+	+	+	–	–
Involatile substances	–	≈	+	+	+	+
With involatile matrix	+	+	+	+	+	–
Thermally labile substances	≈	–	≈	≈	≈	+
Can be automated	+	+	+	+	+	+

+ recommended; ≈ can be used; – not recommended

start temperature of the program must be well below the boiling point of the solvent and there should be an isotherm of 1–3 min at the beginning of the oven program.

The use of capillary columns of 25 m in length and 0.25 mm internal diameter can be recommended without limitations for cold injection systems. The coupling to a mass spectrometer is particularly favourable because of the generally low flow rates. Also, the injection of larger sample volumes is straightforward and does not impair the operation of the mass spectrometer. For these reasons the use of cold injection systems for GC/MS is particularly recommended. This applies all the more to the throughput of large numbers of samples, as here automation of the system is indispensable. Unlike on-column injectors, cold injection systems can be used with autosamplers without special modifications.

In spite of the many advantages of the cold injection system, in practice there are also limits to its use (Table 2-17). These include the analysis of particularly thermally labile substances. Because of the low injection temperature cold injection systems should be particularly suitable for labile substances. However, during the heating phase, the residence times of the substances in the insert are long enough to initiate thermal decomposition. In this case, only on-column injection can be used because it completely avoids external evaporation of the sample for transfer to the column (see Section 2.2.1.4). To test for thermal decomposition, Donike suggested the injection of a mixture of the same quantities of fatty acid TMS esters (C_{10} to C_{22} thermolabile) and n-alkanes (C_{12} to C_{32} thermally stable). If no thermal decomposition takes place, all the substances should appear with the same intensity.

2.2.1.4 On-column Injection

In the era of packed columns on-column injection (although not known as such at the time) was the state of the art. The difference between that and the present procedure lies in coping with the small diameters of capillary columns which have caused the term "filigree" to be applied to the on-column technique. Schomburg introduced the first on-column injector for capillary columns in 1977 under the designation of direct injection and described the process of sample injection very precisely. A year later a variant using syringe injection was developed by Grob. Today's commercial on-column injection involves injecting the sample directly on to the column (internal diameter 0.32 mm) in liquid form using a standard syringe with a 75 mm long steel canula of 0.23 mm external diameter. More favourable dimensioning is possible on use of a retention gap with an internal diameter of 0.53 mm, which can even be used with standard canulas. The use of retention gaps allows autosamplers to be employed for on-column sample injection.

The injector itself does not have a complicated construction and can be serviced easily and safely. The carrier gas feed is situated and the capillary fixed centrally in the lower section. The middle section carries a rotating valve as a seal and in the upper section there is the needle entry point which allows central introduction of the syringe needle. In all types of operation the whole injector block remains cold and warming by the oven is also prevented by a surrounding air stream.

Small sample volumes of less than 10 µl can be injected directly on to a capillary column. Larger volumes require a deactivated retention gap of appropriate dimensions (Fig. 2-54). The retention gap is connected to the column with connectors with no dead volume (e.g. press fit connectors). For injection, the column should be operated at a high flow rate (ca. 2–3 ml/min for He). During the injection the oven temperature must be below the effective boiling point of the solvent. This corresponds to a temperature of ca. 10–15 °C above the boiling point of the solvent under normal conditions because of the pressure ratios in the on-column injector. As a rule of thumb, the boiling point increases by 1 °C for every 0.1 bar of pre-pressure. The injection is carried out by pressing the syringe plunger down rapidly (ca. 15–20 µl/s) to avoid the li-

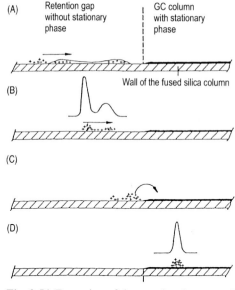

Fig. 2-54 Formation of the start band on use of a retention gap (after Grob).
(A) Starting situation after injection: the retention gap is wetted by the sample extract.
(B) The solvent evaporates and leaves an undefined substance distribution in the retention gap.
(C) The substance reaches the stationary phase of the GC column. Considerable retention corresponding to the capacity factor of the column begins.
(D) The analytes injected with the sample have been concentrated in a narrow band at the beginning of the column; this is the starting point for chromatographic separation.

quid being sucked up between the needle and the wall of the column through capillary effects, otherwise the subsequent removal of the syringe would result in loss of sample. For larger sample volumes, from ca. 20–30 μl, the injection rate must, however, be lowered (to ca. 5 μl/s) as the liquid plug causes increasing resistance and reversed movement of the liquid must be avoided (Fig. 2-55). The temperature program with the heating ramp can then begin when the solvent peak is clearly decreasing.

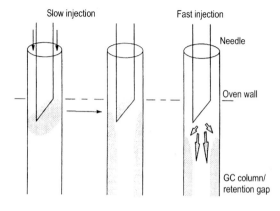

Fig. 2-55 Slow and fast on-column injections. For the slow injection the exterior of the needle is contaminated and part of the sample is lost.

> **On-column Injection**
> - Flow rate ca. 2-3 ml/min
> - Oven temperature at the effective boiling point of the solvent (normal boiling point plus 1 °C per 0.1 bar pre-pressure)
> - Rapid injection of small volumes
> - Oven temperature to be kept at the evaporation temperature
> - Only start the heating program after elution of the solvent peak

The on-column technique has the following advantages over the cold injection system:

- The danger of thermal decomposition during the injection is practically eliminated. A substance evaporates at the lowest possible temperature and is heated to its elution temperature at the maximum.
- Only on-column injection allows complete and reproducible sample injection on to the column without losses. Involatile substances, in particular, are transferred totally without discrimination.
- Defined volumes can be injected with high reproducibility. Standard deviations of ca. 1% can be achieved.
- The injection volumes can be varied within wide ranges without additional optimisation. At the beginning of the heating program, only the duration of elution of the solvent peak needs to be taken into account.

For use in GC/MS systems attention should be paid to the injection at the necessary high flow rates in connection with the maximum permitted carrier gas loading of the mass spectrometer. As the injection system is always cold, moisture can accumulate from samples or insufficiently purified carrier gas (Fig. 2-56). Because of this, the sensitivity of the mass spectrometer is permanently reduced. The effect is easy to detect because the mass spectrometer continually registers the water background and because the on-column injector can easily be opened at a particular time so that the carrier gas can pass out via the splitter.

If the water background decreases after the dead time, this effect must be taken into consideration during planning of the analysis, as the response behaviour of substances can change during the working day.

Other system-related limitations to the on-column technique for certain types of sample are:

- Concentrated samples may not be injected on-column. In these cases preliminary dilution is necessary or the use of the split injector is better (minimum on-column injection volume 0.2 µl).
- Samples containing a matrix rapidly lead to quality depletion. Here a retention gap is absolutely necessary. However, impurities in the retention gap quickly give rise to adsorption and peak broadening because of the small capacity, so that changing it regularly is necessary. For dirty samples hot or PTV sample injection is preferred.
- Samples containing volatile components are not easy to control, as focusing by the solvent effect is inadequate under certain circumstances. If changing the solvent does not help, changing to a hot or cold split or splitless system is recommended.

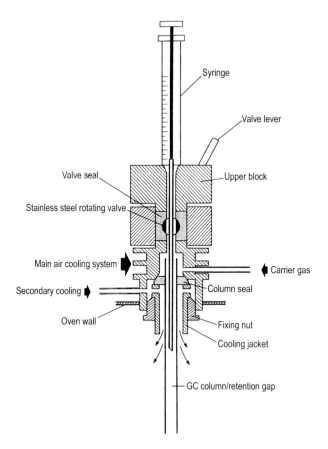

Fig. 2-56 On-column injector (Carlo Erba)

2.2.1.5 Cryofocusing

Cryofocusing should not be regarded as an independent injection system, but nevertheless should be treated individually in the list of injectors because the static headspace and purge and trap systems can be coupled directly to a cryofocusing unit as a GC injector. Furthermore, some thermodesorption systems already contain a cryofocusing unit. Many simpler instruments, however, do not, and require external focusing to ensure their proper function. Generally, for the direct analysis of air or gases from indoor rooms or at the workplace, or of emissions, a cryofocusing unit is required for concentration and injection on to the GC column.

Packed columns can effectively concentrate the substances contained in gaseous samples at the beginning of the column because of their high sample capacity. Capillary columns under normal working conditions cannot form sufficiently narrow initial bands for the substances to be detected from the gas volumes being handled in direct air analysis or on heating up traps (purge and trap, thermodesorption) because of their comparatively low sample capacity. Additional effective cooling is necessary. The entire oven space can be cooled, but this has a major disadvantage: the requirement in terms of time and cooling agents is immense.

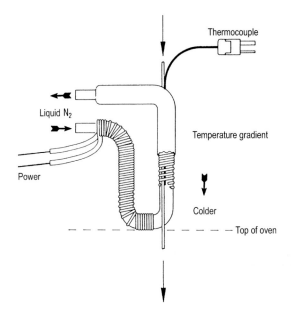

Fig. 2-57 Cut away diagram of a cryofocusing unit for use with fused silica capillary columns (Tekmar)

A special cryofocusing unit cools down the beginning of the GC column and allows on-column focusing of the analytes without requiring a retention gap. The column film present improves the efficiency at the same time by acting as an adsorbent. For this purpose the capillary column is inserted into a 1/16 inch stainless steel tube via an opening in the oven lid of the chromatograph and is connected via a connecting piece to the transfer line of the sampler (e.g. headspace, purge and trap, canister). This stainless steel tube is firmly welded to a cold finger and a thermoelement (Fig. 2-57). The tube is also surrounded by a differential heating coil which heats the inlet side more rapidly than the exit to the GC oven. In this way a uniform temperature gradient is guaranteed in all the phases of the operation. Because of the gradient, on cooling, e.g. with liquid nitrogen, all the analytes are focused into a narrow band at the beginning of the capillary column, as the substances migrate more slowly at the start of the band than at the end. After concentration and focusing, the chromatographic separation starts with the heating up of this region. On careful control of the gradient the heating rate does not affect the efficiency of the column (heating rates between 100 and 2000 °C/min).

Because of on-column focusing the cryotrap has the same sample capacity as the column has for these analytes. A breakthrough of the cryotrap as a result of too high an analyte concentration can be prevented by changing the column. Overloading the column would, in any case, result in poor separation. The diagram in Fig. 2-58 gives the temperatures at which a breakthrough of the analytes must be reckoned with in cryofocusing. Larger film thicknesses and internal diameters permit higher focusing temperatures which favour the mobilisation of the analytes from the column film.

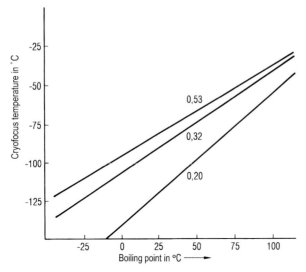

Fig. 2-58 Breakthrough temperatures in cryofocusing for various column internal diameters as a function of the boiling point of the analyte (Tekmar)
Film thicknesses: 0.25 µm at 0.20 mm internal diameter
1.0 µm at 0.32 mm internal diameter
3.0 µm at 0.53 mm internal diameter

2.2.2 Capillary Columns

There are no hard and fast rules for the choice of column for GC/MS coupling. The choice of the correct phase is made on the usual criteria: "like dissolves like". If substances exhibit no interaction with the stationary phase, there will be no retention and the substances leave the column in the dead time. The polarity of the stationary phase should correspond to the polarity of the substances being separated (Table 2-18). Less polar substances are better separated on nonpolar phases and vice versa.

Change to More Strongly Polar Stationary Phases

– Weaker retention of nonpolar compounds
– Stronger retention of polar compounds
– Shift of compounds with specific interactions

For coupling with mass spectrometry (GC/MS), the carrier gas flow and the specific noise of the column (column bleed) are included in the criteria governing the choice of column. When considering the optimal carrier gas flow, the maximum loading of the mass spectrometer must be taken into account. The limit for small benchtop mass spectrometers with quadrupole analysers is frequently ca. 1–2 ml/min. Larger instruments can generally tolerate higher loads because of their more powerful pumping systems. Ion trap mass spectrometers can be operated with a carrier gas flow of up to

3 ml/min and even higher with an external ion source. These conditions limit the column diameter which can be used. There can be no compromises concerning column bleed in GC/MS. Column bleed generally contributes to chemical noise where MS is used as the mass-dependent detector, and curtails the detection limits. The optimisation of a particular signal/noise ratio can also be effected in GC/MS by selecting particularly thermally stable stationary phases with a low tendency to bleed. For use in residue analysis, stationary phases for high-temperature applications have proved particularly useful (Figs. 2-59 and 2-60). Besides the phase itself, the film thickness also plays an important role. Thinner films and shorter columns have a lower tendency to cause column bleed.

2.2.2.1 Sample Capacity

The sample capacity is the maximum quantity of an analyte with which the phase can be loaded (Table 2-19). An overloaded column exhibits peak fronting, i.e. an asymmetrical peak which has a gentle gradient on one side and a sharp one on the other. This effect can increase until a triangular peak shape is obtained, a so-called "shark fin". Overloading occurs rapidly if a column of the wrong polarity is chosen. The capacity of a column depends on the internal diameter, the film thickness and the solubility of a substance in the phase.

Sample Capacity

– Increases with internal diameter
– Increases with film thickness
– Increases with solubility

2.2.2.2 Internal Diameter

The internal diameter of a column used in capillary gas chromatography varies from 0.1 mm (microbore capillary) via 0.18 mm and 0.25 mm (narrow bore) and the standard columns with 0.32 mm to 0.53 mm (megabore, halfmil). For direct coupling with mass spectrometers, in practice only the columns up to 0.32 mm internal diameter are used. Megabore columns are mainly used to replace packed columns in specially designed GC instruments.

The internal diameter affects the resolving power and the analysis time (Fig. 2-61). Basically, at constant film thicknesses lower internal diameters are preferred in order to achieve higher chromatographic resolution. As the flow per unit time decreases at a particular carrier gas velocity, the analysis time increases. In the case of complex mixtures, changing to a column with a smaller internal diameter gives better separation of critical pairs of compounds (Table 2-20). In practice it has been shown that even changing from 0.25 mm internal diameter to 0.20 mm allows an improvement in the separation of, for example, PCBs. For sample application to narrow bore columns certain conditions must be adhered to, depending on the type of injector (see Section 2.2.1).

(a)

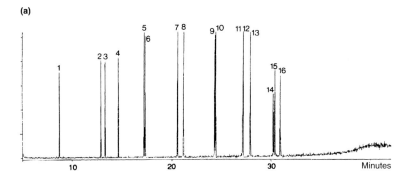

(b)

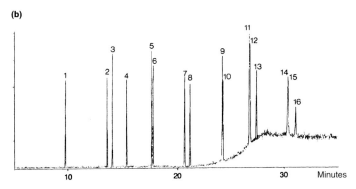

Polynuclear aromatic hydrocarbons

(a) Phase: HT5, 0.10 μm film thickness
  Column: 25 m × 0.22 mm ID
  Start temperature: 50 °C, 2 min
  Program rate: 10 °C/min
  Final temperature: 420 °C, 5 min
  Detector: HP 5971 MS
  Scan range: 35–550 amu
  Injection: Split 50:1

(b) Phase: BP1, 0.25 μm film thickness
  Column: 25 m × 0.22 mm ID
  Start temperature: 50 °C, 2 min
  Program rate: 10 °C/min
  Final temperature: 300 °C, 10 min
  Detector: HP 5971 MS
  Scan range: 35–550 amu
  Injection: Split 50:1

1. Naphthalene
2. Acenaphthylene
3. Acenaphthene
4. Fluorene
5. Phenanthrene
6. Anthracene
7. Pyrene
8. Fluoranthene
9. Chrysene
10. Benzo(a)anthracene
11. Benzo(b)fluoranthene
12. Benzo(k)fluoranthene
13. Benzo(a)pyrene
14. Indeno(1,2,3,-cd)pyrene
15. Dibenzo(ah)anthracene
16. Benzo(ghi)perylene
 (0.5 ng per component)

Fig. 2-59 Comparison of a conventional silicone phase (lower trace: SGE BP1, dimethylsiloxane) with a high temperature phase (upper trace: SGE HT5, siloxane-carborane) using a polyaromatic hydrocarbon standard (SGE). The long temperature program for the high temperature phase up to 420 °C allows the elution of components 14, 15 and 16 during the heating-up phase at an elution temperature of ca. 350 °C. Sharp narrow peaks with low column bleed give better detection conditions for high-boiling substances (GC/MS system HP-MSD 5971)

Smaller Internal Diameter (at Identical Film Thickness)

– Increases the resolution
– Increases the analysis times

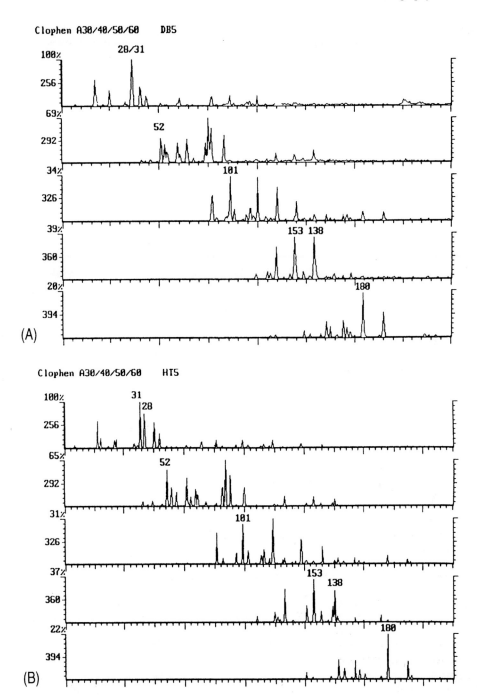

Fig. 2-60 An example of the different selectivities of stationary phases (separation of an Aroclor mixture, separation of PCBs 31, 28!)
(A) DB5, 25 m × 0.25 mm × 0.25 μm (dimethyl-diphenyl-polysiloxane phase)
(B) HT5, 25 m × 0.25 mm × 0.10 μm (siloxane-carborane phase)

Table 2-18 Composition of stationary phases for fused silica capillary columns with column designations for comparable phases from different manufacturers, arranged in order of increasing polarity, and columns with special phases (selection)

Stationary Phase	Polarity	Alltech www.alltechweb.com	Chrompack www.chrompack.com	Agilent (HP) www.chem.agilent.com
50% n-Octyl-50% Methyl-Polysiloxane	nonpolar	AT Petro	CP Sil 2 CB CP Squalane[8]	HP-PONA
100% Dimethylpolysiloxane	nonpolar	AT-1 EC-1 MC-1 MC-1HT SE-30	CP Sil 5 CB CP Sil 5 CB-MS[1]	HP-1 HP-1MS[1] HP Ultra-1
95% Dimethyl-5% Diphenyl-Polysiloxane	nonpolar	AT-5 EC-5 MC-5 MC-5HT SE-54	CP Sil 8 CB CP Sil 8 CB-MS[1]	HP-5 HP-5MS[1] HP-5TA HP Ultra-2
95% Dimethyl-5% Diphenyl-Polycarboranesiloxane	nonpolar	–	–	–
95%-Dimethyl-8% Diphenyl-Polycarboranesiloxane	slightly polar	–	–	–
6% Cyanopropylphenyl-94% Dimethyl-Polysiloxane	slightly polar	AT-1301 AT-624	CP 1301 CP Select 624 CB	HP-1301
80% Dimethyl-20% Diphenyl-Polysiloxane	slightly polar	AT-20 EC-20	CP Sil 13 CB[7,8]	–
65% Dimethyl-35% Diphenyl-Polysiloxane	moderately polar	AT-35	–	HP-35 HP-35MS[1]
50% Dimethyl-50% Diphenyl-Polysiloxane	moderately polar	AT-50	CP Sil 24 CB	HP-50+
35% Dimethyl-65% Diphenyl-Polysiloxane	moderately polar	–	CP TAB CB	–
Trifluoropropylmethyl-Polysiloxane	moderately polar	AT-210	–	HP-210
14% Cyanopropylphenyl-86% Dimethyl-Polysiloxane	moderately polar	AT-1701	CP Sil 19 CB	HP-1701
50% Cyanopropylphenyl-50% Dimethyl-Polysiloxane	moderately polar	AT-225	CP Sil 43 CB	HP-225 HP-23
10% Poly(ethylene glycol)-Dimethyl-polysiloxane	polar	–	–	–
50% Poly(ethylene glycol)-Dimethyl-polysiloxane	polar	–	–	–
85% Poly(ethylene glycol)-Dimethyl-polysiloxane	polar	–	–	–
Poly(ethylene glycol)	polar	AT-Wax EC-Wax MC-Wax MC-WaxHT Carbowax	CP Wax 52 CB CP Wax 57 CB	HP-20M HP-Wax HP-InnoWax

2.2 Gas Chromatography

J & W www.jandw.com	Ohio Valley www.win.net/ővsc.index.html	Quadrex www.quadrexcorp.com	Restek www.restekcorp.com	SGE www.sge.com	Supelco www.sigma-aldrich.com
DB-Petro	–	007-ODP	–	BPX-1 SimDist	SPB-Octyl
DB-1 DB-1HT[1] SE-30	OV-1	007-1	Rtx-1 Rtx-1MS[1] Rtx-2887	BP-1 BPX-1[1]	SPB-1 SPB-2100
DB-5 DB-5.625 DB-5HT[1] DB-5MS[1] SE-52[9] SE-54	OV-5	007-5 007-5MS[1]	Rtx-5 Rtx-5MS[1] XTI-5[1] Rtx-5Amine	BP-5 BPX-5[1,2]	SPB-5 PTE-5[1] PTA-5
	–	–	–	HT-5[1]	–
	–	–	–	HT-8[1]	–
DB-1301 DB-624	–	007-1301 007-624	Rtx-1301 Rtx-624	BP-624 BPX-624[1]	SPB-1301
–	–	007-20	Rtx-20	–	SPB-20 VOCOL[8]
DB-35 DB-35MS[1]	–	007-35 007-608[8]	Rtx-35 Rtx-35MS[1]	BPX-35[1,2]	SPB-35 SPB-608
DB-17 DB-17HT[1] DB-17MS[1] DB-608	OV-17	007-17	Rtx-50[4]	BPX-50[1] BPX-608[1]	SP-2250 SPB-50
–	–	400-65TG 007-65TG	Rtx-65 Rtx-65TG	–	–
DB-200 DB-210[6,8]	–	007-210	Rtx-200 Rtx-200MS[1]	–	–
DB-1701 DB-1701P	OV-1701	007-1701	Rtx-1701	BP-10 BPX-10[1] OV 1701 MTBE	SPB-1701
DB-225 DB-23[8]	OV-225	007-225	Rtx-225[5]	BP-225	
DX-1	–	–	–	–	–
DX-3	–	–	–	–	–
DX-4	–	–	–	–	–
DB-Wax DB-Wax etr Carbowax 20M	Carbowax 20M	007-CW BTR-CW	Stabilwax	BP-20 BP-Xylene	Supelcowax 10 Omegawax

Table 2-19 Sample capacities for common column diameters

Internal diameter	0.18 mm	0.25 mm	0.32 mm	0.53 mm
Film thickness	0.20 µm	0.25 µm	0.25 µm	1.00 µm
⇒ **Sample capacity**	<50 ng	50–100 ng	400–500 ng	1000–2000 ng
Theoretical plates per metre of column	5300	3300	2700	1600
Optimal flow rate at				
20 cm/s helium	0.3 ml/min	0.7 ml/min	1.2 ml/min	2.6 ml/min
40 cm/s hydrogen	0.6 ml/min	1.4 ml/min	2.4 ml/min	5.2 ml/min

Table 2-20 Effect of column diameter and linear carrier gas velocity on the flow rate

Internal diameter	Linear carrier gas velocity		Flow rate	
	He	H$_2$	He	H$_2$
0.18 mm	30–45 cm/s	45–60 cm/s	0.5–0.7 ml/min	0.7–0.9 ml/min
0.25 mm	**30–45 cm/s**	**45–60 cm/s**	**0.9–1.3 ml/min**	**1.3–1.8 ml/min**
0.32 mm	30–45 cm/s	45–60 cm/s	1.4–2.2 ml/min	2.2–2.8 ml/min
0.54 mm	30–45 cm/s	45–60 cm/s	4.0–6.0 ml/min	6.0–7.9 ml/min

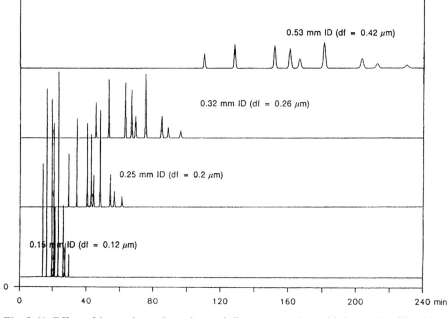

Fig. 2-61 Effect of increasing column internal diameter together with increasing film thickness on peak height and retention time. (The columns have the same phase ratio!) (Chrompack)

| J & W
www.jandw.com | Ohio Valley
www.win.net/övsc.index.html | Quadrex
www.quadrexcorp.com | Restek
www.restekcorp.com | SGE
www.sge.com | Supelco
www.sigma-aldrich.com |
|---|---|---|---|---|---|
| DB-Petro | – | 007-ODP | – | BPX-1 SimDist | SPB-Octyl |
| DB-1
DB-1HT[1]
SE-30 | OV-1 | 007-1 | Rtx-1
Rtx-1MS[1]
Rtx-2887 | BP-1
BPX-1[1] | SPB-1
SPB-2100 |
| DB-5
DB-5.625
DB-5HT[1]
DB-5MS[1]
SE-52[9]
SE-54 | OV-5 | 007-5
007-5MS[1] | Rtx-5
Rtx-5MS[1]
XTI-5[1]
Rtx-5Amine | BP-5
BPX-5[1,2] | SPB-5
PTE-5[1]
PTA-5 |
| – | – | – | – | HT-5[1] | – |
| – | – | – | – | HT-8[1] | – |
| DB-1301
DB-624 | – | 007-1301
007-624 | Rtx-1301
Rtx-624 | BP-624
BPX-624[1] | SPB-1301 |
| – | – | 007-20 | Rtx-20 | – | SPB-20
VOCOL[8] |
| DB-35
DB-35MS[1] | – | 007-35
007-608[8] | Rtx-35
Rtx-35MS[1] | BPX-35[1,2] | SPB-35
SPB-608 |
| DB-17
DB-17HT[1]
DB-17MS[1]
DB-608 | OV-17 | 007-17 | Rtx-50[4] | BPX-50[1]
BPX-608[1] | SP-2250
SPB-50 |
| – | – | 400-65TG
007-65TG | Rtx-65
Rtx-65TG | – | – |
| DB-200
DB-210[6,8] | – | 007-210 | Rtx-200
Rtx-200MS[1] | – | – |
| DB-1701
DB-1701P | OV-1701 | 007-1701 | Rtx-1701 | BP-10
BPX-10[1]
OV 1701 MTBE | SPB-1701 |
| DB-225
DB-23[8] | OV-225 | 007-225 | Rtx-225[5] | BP-225 | – |
DX-1	–	–	–	–	–
DX-3	–	–	–	–	–
DX-4	–	–	–	–	–
DB-Wax
DB-Wax etr
Carbowax 20M	Carbowax 20M	007-CW
BTR-CW	Stabilwax	BP-20
BP-Xylene	Supelcowax 10
Omegawax

Table 2-19 Sample capacities for common column diameters

Internal diameter	0.18 mm	0.25 mm	0.32 mm	0.53 mm
Film thickness	0.20 µm	0.25 µm	0.25 µm	1.00 µm
⇒ **Sample capacity**	<50 ng	50–100 ng	400–500 ng	1000–2000 ng
Theoretical plates per metre of column	5300	3300	2700	1600
Optimal flow rate at				
20 cm/s helium	0.3 ml/min	0.7 ml/min	1.2 ml/min	2.6 ml/min
40 cm/s hydrogen	0.6 ml/min	1.4 ml/min	2.4 ml/min	5.2 ml/min

Table 2-20 Effect of column diameter and linear carrier gas velocity on the flow rate

Internal diameter	Linear carrier gas velocity		Flow rate	
	He	H_2	He	H_2
0.18 mm	30–45 cm/s	45–60 cm/s	0.5–0.7 ml/min	0.7–0.9 ml/min
0.25 mm	**30–45 cm/s**	**45–60 cm/s**	**0.9–1.3 ml/min**	**1.3–1.8 ml/min**
0.32 mm	30–45 cm/s	45–60 cm/s	1.4–2.2 ml/min	2.2–2.8 ml/min
0.54 mm	30–45 cm/s	45–60 cm/s	4.0–6.0 ml/min	6.0–7.9 ml/min

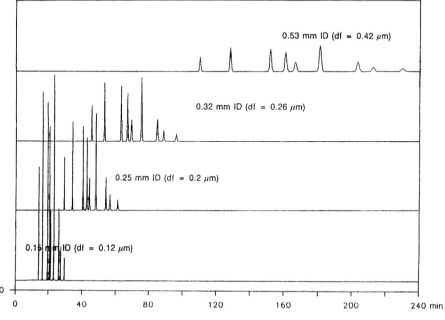

Fig. 2-61 Effect of increasing column internal diameter together with increasing film thickness on peak height and retention time. (The columns have the same phase ratio!) (Chrompack)

2.2.2.3 Film Thickness

The variation in the film thickness at a given internal diameter and column length gives the user the possibility of carrying out special separation tasks. As a rule, thick films are used for volatile compounds and thin films for high-boiling ones. Thick film columns with coatings of more than 1.0 µm can separate extremely low boiling compounds well, e.g. volatile halogenated hydrocarbons. Through the large increase in capacity with thicker films, it is even possible to dispense with additional oven cooling during injection in the analysis of volatile halogenated hydrocarbons using headspace or purge and trap. Start temperatures above room temperature are usual. However, thick film columns exhibit severe column bleed at elevated temperatures.

For the residue analysis of all other substances, thin film columns with coating thicknesses of ca. 0.1 µm have proved to be very effective in GC/MS. Thin film columns give narrow rapid peaks and can be used in higher temperature ranges without significant column bleed. The elution temperatures of the compounds decrease with thin films and, at the same program duration, the analysis can be extended to compounds with higher molecular weights. The duration of the analysis for a given compound becomes shorter, but the capacity of the column decreases (Fig. 2-62).

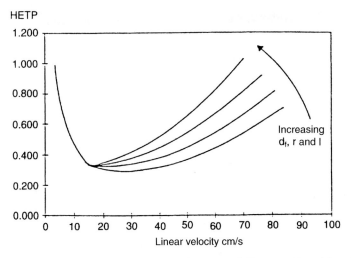

Fig. 2-62 Effect of film thickness (d_f), internal diameter (r) and length (l) on the van Deemter curves for helium as the carrier gas

---

Increasing Film Thicknesses

- Improve the resolution of volatile compounds
- Increase the analysis time
- Increase the elution temperatures

Table 2-21 Effect of column diameter and film thickness on the phase ratio

Internal diameter	\multicolumn{8}{c}{Film thickness}							
	0.10 μm	0.25 μm	0.50 μm	1.0 μm	1.50 μm	2.0 μm	3.0 μm	5.0 μm
0.18 mm	450	180	90	45	30	23	15	9
0.25 mm	**625**	**250**	**125**	**63**	**42**	**31**	**21**	**13**
0.32 mm	800	320	160	80	53	40	27	16
0.53 mm	1325	530	265	128	88	66	43	27

The Relationship between Film Thickness and Internal Diameter

The phase ratio of a capillary column is determined by the ratio of the volume of the gaseous mobile phase (internal volume) to the volume of the stationary phase (coating). From Table 2-21 the phase ratio can be read off for each combination of film thickness and internal diameter (assuming the same film and the same column length). High values mean good separation; the same values show combinations with the same separating capacity. For GC/MS optimal separations can be planned and also other conditions, such as carrier gas flow and column bleed, can be taken into consideration. To achieve better separation it is possible to change to a smaller film thickness at the same internal diameter or to keep the film thickness and choose a higher internal diameter.

2.2.2.4 Column Length

The column should be as short as possible. The most common lengths for standard columns are 30 or 60 m. Greater lengths are not necessary in residue analysis with GC/MS systems even for separating complex PCB or volatile halogenated hydrocarbon mixtures. Shorter columns would be desirable for simpler separations, but they are at the limit of the maximum flow for the mass spectrometer used (e.g. 15 m length, internal diameter 0.32 mm). Doubling the column length only results in an improvement in the separation by a factor of 1.4 ($\sqrt{2}$) while the analysis time is doubled (and the cost of the column also!). For isothermal chromatography, the retention time is directly proportional to the column length while with programmed operations the retention time is essentially determined by the elution temperature of a compound. On changing to a longer column, the temperature program should always be optimised again to achieve optimal retention times.

Doubling the Column Length

- Resolution only increases by a factor of 1.4
- Costs are doubled
- Retention times are doubled
- The temperature program must be optimised again

2.2.2.5 Adjusting the Carrier Gas Flow

The maximum separating capacity of a capillary column can only be exploited with an optimised carrier gas flow. With direct GC/MS coupling the carrier gas flow is only affected slightly by the vacuum on the detector side. In practice the adjustment is no different from that in classical GC systems. Only with instruments with electronic pressure control (EPC) is adjustment necessary for direct coupling with MS. The separating efficiency of a capillary column is given as theoretical plates in chromatographic terminology (see Section 2.2.3). A high analytical separating capacity is always accompanied by a large number of plates (number of separation steps) so the height equivalent to a theoretical plate (HETP) decreases. In van Deemter curves, the HETP is plotted against the carrier gas velocity (not flow!). The minimum of one of these curves gives the optimal adjustment for a particular carrier gas for isothermal operations.

For helium the optimal carrier gas velocity for standard columns is ca. 24 cm/s. As the viscosity of the carrier gas increases in the course of a temperature program, in practice at the lower start temperature a higher velocity is used (ca. 30 cm/s). Velocities which are too high lower the efficiency.

With hydrogen as the carrier gas a much higher velocity (>40 cm/s) can be used, which leads to significant shortening of the analysis time. Furthermore, with hydrogen the right hand branch of the van Deemter curve is very flat, so that a further increase in the gas velocity is possible without impairing the efficiency (Figs. 2-63 and 2-64). The separating efficiency is retained and time is gained. Because of the limited pumping capacity in commercial GC/MS systems (see Section 2.4.1), little use can be made of these advantages with hydrogen.

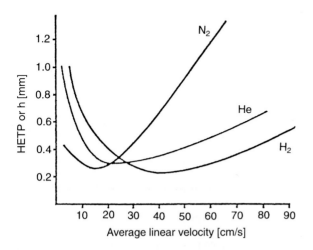

Fig. 2-63 van Deemter curves for nitrogen, helium and hydrogen as the carrier gas. The values refer to a standard column of 30 m length, 0.25 mm internal diameter and 0.25 μm film thickness (Restek)

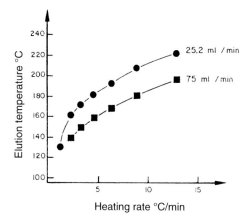

Fig. 2-64 The effect of heating rate and carrier gas flow on the elution temperature (retention temperature) (after Karasek)

2.2.2.6 Properties of Stationary Phases

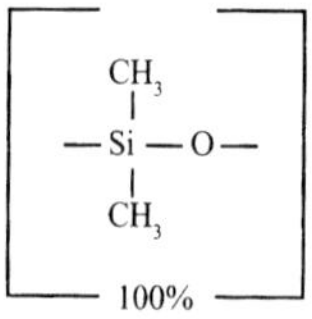

	Polarity:	least polar bonded phase
	Use:	boiling point separations for solvents, petroleum products, pharmaceuticals
	Properties:	minimum temperature $-60\,°C$ maximum temperature $340-430\,°C$ helix structure

Fig. 2-65 100% Dimethyl-polysiloxane e.g. RT_x-1 etc.

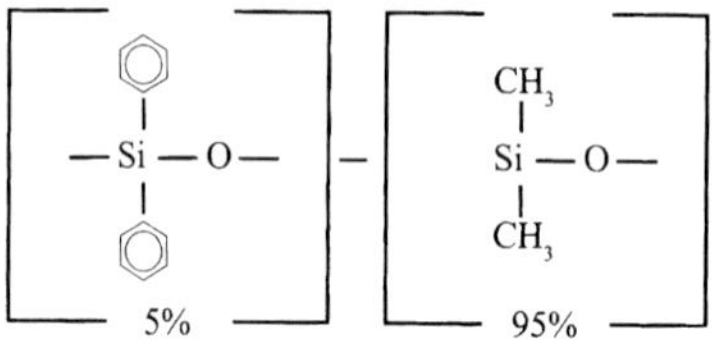

	Polarity:	nonpolar, bonded phase
	Use:	boiling point, point separations for aromatic compounds, environmental samples, flavours, aromatic hydrocarbons
	Properties:	minimum temperature $-60\,°C$ maximum temperature $340\,°C$

Fig. 2-66 5% Diphenyl-95% dimethyl-polysiloxane e.g. RT_x-5 etc.

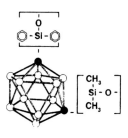

Polarity: nonpolar, similar to 5% phenylsiloxane

Use:
– ideal for GC/MS coupling because of very low bleeding
– all environmental samples
– all medium and high molecular weight substances
– polyaromatic hydrocarbons, PCBs, waxes, triglycerides

Properties: minimum temperature $-10\,°C$
maximum temperature $480\,°C$ (highest operating temperature of all stationary phases, aluminium coated, $370\,°C$ polyimide coated), high temperature phase

Fig. 2-67 Siloxane-carborane, comparable to 5% phenyl, e.g. HT 5

2.2 Gas Chromatography   101

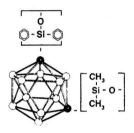

Polarity: weakly polar, similar to 8% phenylsiloxane
Use: – ideal for GC/MS coupling because of very low bleeding
             – all environmental samples, can be used universally
             – volatile halogenated hydrocarbons, solvents
             – polyaromatic hydrocarbons, PCBs, plant protection agents etc.
Properties: minimum temperature –20 °C
             maximum temperature 370 °C

Fig. 2-68 Siloxane-carborane, comparable to 8% phenyl, e.g. HT 8

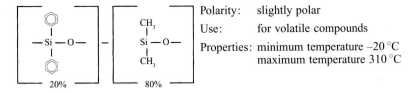

Polarity: slightly polar
Use: for volatile compounds
Properties: minimum temperature –20 °C
             maximum temperature 310 °C

Fig. 2-69 20% Diphenyl-80% dimethyl-polysiloxane, e.g. RT_x-20, etc.

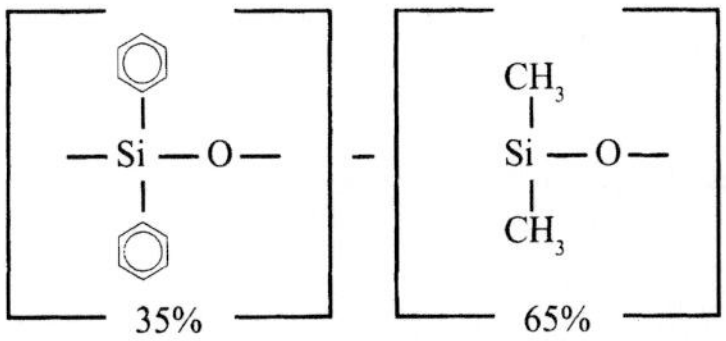

Polarity: intermediately polar
Use: pesticides, PCBs, amines
Properties: minimum temperature 20 °C
             maximum temperature 300 °C

Fig. 2-70 35% Diphenyl-65% dimethyl-polysiloxane, e.g. RT_x-35, etc.

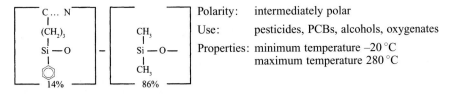

Polarity: intermediately polar
Use: pesticides, PCBs, alcohols, oxygenates
Properties: minimum temperature –20 °C
             maximum temperature 280 °C

Fig. 2-71 14% Cyanopropylphenyl-86% dimethyl-polysiloxane, e.g. RT_x-1701, etc.

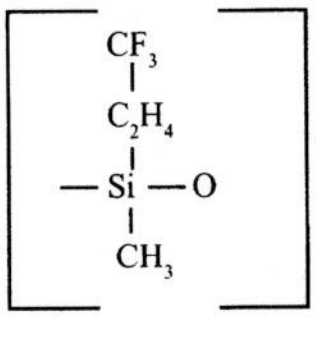

Polarity: selective for lone pair electrons
Use: environmental samples, solvents, freons
Properties: minimum temperature –20 °C
             maximum temperature 360 °C

Fig. 2-72 100% Trifluoropropylmethyl-polysiloxane, e.g. RT_x-200, etc.

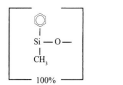

Polarity: intermediately polar
Use: triglycerides, phthalic acid esters etc.
Properties: minimum temperature 25 °C
              maximum temperature 340 °C

Fig. 2-73 50% Diphenyl-50% dimethyl-polysiloxane, e.g. RT_x-50, etc.

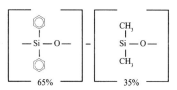

Polarity: medium polarity
Use: triglycerides, free fatty acids, terpenes
Properties: minimum temperature 50 °C
              maximum temperature 340 °C

Fig. 2-74 65% Diphenyl-35% dimethyl-polysiloxane, e.g. RT_x-65, etc.

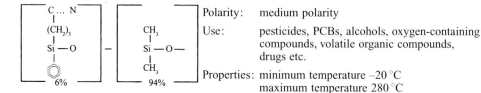

Polarity: medium polarity
Use: pesticides, PCBs, alcohols, oxygen-containing compounds, volatile organic compounds, drugs etc.
Properties: minimum temperature –20 °C
              maximum temperature 280 °C

Fig. 2-75 6% Cyanoprophylphenyl-94% dimethyl-polysiloxane, e.g. RT_x-1301, etc.

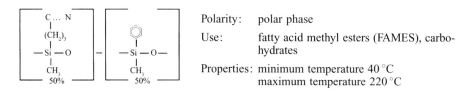

Polarity: polar phase
Use: fatty acid methyl esters (FAMES), carbohydrates
Properties: minimum temperature 40 °C
              maximum temperature 220 °C

Fig. 2-76 50% Cyanopropylmethyl-50% phenylmethyl-polysiloxane, e.g. RT_x-225, etc.

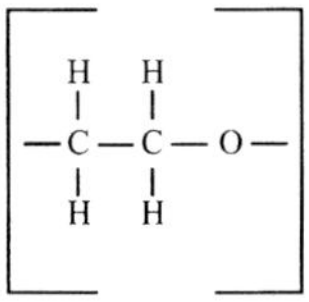

Polarity: polar phase
Use: fatty acid methyl esters (FAMES), terpenes, acids, amines, solvents
Properties: minimum temperature 40 °C
              maximum temperature 280 °C

Fig. 2-77 100% Carbowax polyethyleneglycol 20M, e.g. Stabilwax etc.

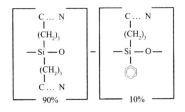

Polarity:	very polar
Use:	cis/trans isomers
Properties:	minimum temperature 0 °C
	maximum temperature 275 °C

Fig. 2-78 90% Biscyanopropyl-10% phenylcyanopropyl-polysiloxane, e. g. RT$_x$-2330, etc.

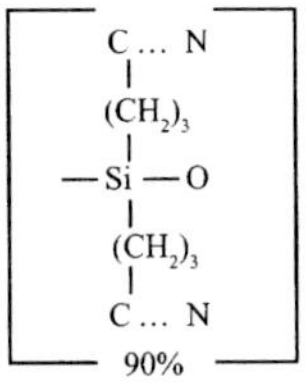

Polarity:	very polar
Use:	fatty acid methyl esters (FAMES)
Properties:	minimum temperature 20 °C
	maximum temperature 250 °C

Fig. 2-79 100% Biscyanopropyl-polysiloxane, e. g. RT$_x$-2340, etc.

2.2.3 Chromatography Parameters

All chromatography processes are based on the multiple repetition of a separation process, such as the continuous dynamic partition of the components between two phases.

In a model chromatography can be regarded as a continuous repetition of partition steps. The starting point is the partition of a substance between two phases in a separating funnel. Suppose a series of separating funnels is set up which all contain the same quantity of phase 1. As this phase remains in the separating funnels, it is known as the stationary phase.

The sample is placed in the first separating funnel dissolved in a second phase (the auxiliary phase). After establishing equilibrium through shaking, phase 2 is transferred to separating funnel 2. The auxiliary phase thereby becomes the mobile phase. Fresh mobile phase is placed in the first separating funnel etc. (Fig. 2-80).

The results of this type of partition with 100 vessels and two substances A and B are shown in Fig. 2-81. The prerequisite for this is the validity of the Nernst equation. For detection the concentrations of A and B in the vessels are determined.

With the model described, so many separating steps are carried out that the mobile phase leaves the system of 100 vessels and the individual components A and B are removed, one after the other, from the series of vessels. This process is known as elution.

2.2.3.1 The Chromatogram and its Meaning

The substances eluted are transported by the mobile phase to the detector and are registered as Gaussian curves (peaks). The peaks give qualitative and quantitative information on the mixture investigated.

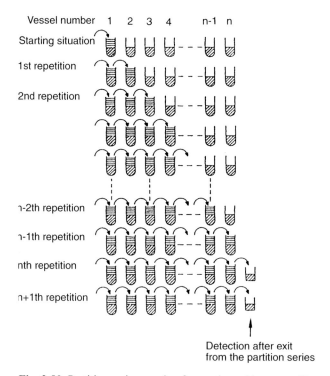

Fig. 2-80 Partition series: mode of operation with two auxiliary phases

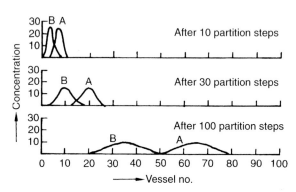

Fig. 2-81 Partition of substances A and B after 10, 30 and 100 separating steps.
After 10 steps A and B are hardly separated, after 30 steps quite well and after 100 steps practically completely. The two substances are partitioned among an ever increasing number of vessels and the concentrations decrease more and more ($\alpha_A = 2 : \alpha_B = 0.5$)

Qualitative: The retention time is the time elapsing between injection of the sample and the appearance of the maximum of the signal. The retention time of a component is always constant under the same chromatographic conditions. A peak can therefore be identified by a comparison of the retention time with a standard (pure substance).

Quantitative: The height and area of a peak is proportional to the quantity of substance injected. Unknown quantities of substance can be determined by a comparison of the peak areas (or heights) with known concentrations.

In the ideal case the peaks eluting are in the shape of a Gaussian distribution (bell-shaped curve, Fig. 2-82). A very simple explanation of this shape is the different paths taken by the molecules through the separating system (multipath effect), which is caused by diffusion processes (Eddy diffusion) (Fig. 2-83).

Under defined conditions the time required for elution of a substance A or B at the end of the separating system, the **retention time** t_R, is characteristic of the substance. It is measured from the start (sample injection) to the peak maximum (Fig. 2-84).

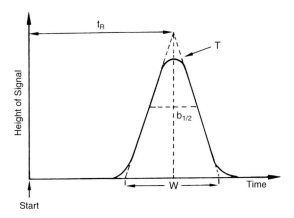

Fig. 2-82 Parameters determined for an elution peak
t_R Retention time
$b_{1/2}$ Half width
W Base width
T Tangent through the turning-point

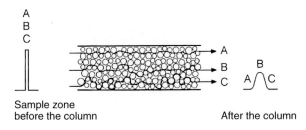

Fig. 2-83 Eddy diffusion in packed columns (multipath effect)

106   2 Basics

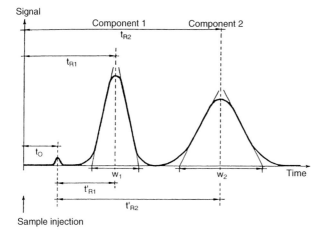

Fig. 2-84 The chromatogram and its parameters
W Peak width of a peak. $W = 4\sigma$ with $\sigma =$ the standard deviatiion of the Gaussian peak.
t_0 Dead time of the column; the time which the mobile phase requires to pass through the column. The linear velocity u of the solvent is calculated from

$$u = \frac{L}{t_0} \quad \text{with } L = \text{length of the column}$$

A substance which is not retarded, i.e. a substance which is not held by the stationary phase, appears at t_0 at the detector.
t_R Retention time; the time between the injection of a substance and the recording of a peak maximum.
t'_R Net retention time. From the diagram it can be seen that $t_R = t_0 + t'_R$.
t_0 is the same for all eluted substances and is therefore the residence time in the mobile phase. The substances separated differ in their residence times in the stationary phase t'_R. The longer a substance stays in the stationary phase, the later it is eluted.

At a constant flow rate t_R is directly proportional to the retention volume V_R.

$$V_R = t_R \cdot F \qquad \text{where } F = \text{flow rate in ml/min} \tag{10}$$

The retention volume shows how much mobile phase has passed through the separating system until half of the substance has eluted (peak maximum!).

2.2.3.2 Capacity Factor k'

The retention time t_R depends on the flow rate of the mobile phase and the length of the column. If the mobile phase moves slowly or the column is long, t_0 is large and so is t_R. Thus t_R is not suitable for the comparative characterisation of a substance, e.g. between two laboratories.

It is better to use the capacity factor, also known as the k' value, which relates the net retention time t'_R to the dead time:

$$k' = \frac{t'_R}{t_0} = \frac{t_R - t_0}{t_0} \tag{11}$$

Thus, the k' value is independent of the column length and the flow rate of the mobile phase and represents the molar ratio of a particular component in the stationary and mobile phases. Large k' values mean long analysis times.

The k' value is related to the partition coefficient K as follows:

$$k' = K \cdot \frac{V_1}{V_g} \qquad \text{where } V_1 = \text{volume of the stationary phase} \qquad (12)$$
$$V_g = \text{volume of the mobile phase}$$

The capacity factor is therefore directly proportional to the volume of the stationary phase (or for adsorbents, their specific surface area in m²/g).

α is a measure of the relative retention and is given by:

$$\alpha = \frac{k'_2}{k'_1} = \frac{K_2}{K_1} \qquad (k'_2 > k'_1) \qquad (13)$$

In the case where $\alpha = 1$, the two components 1 and 2 are not separated because they have the same k' values.

The relative retention α is thus a measure of the selectivity of a column and can be manipulated by choice of a suitable stationary phase. (In principle this is also true for the choice of the mobile phase, but in GC/MS helium or hydrogen are, in fact, always used.)

2.2.3.3 Chromatographic Resolution

A second model, the theory of plates, was developed by Martin and Synge in 1941. This is based on the functioning of a fractionating column, then as now a widely used separation technique. It is assumed that the equilibrium between two phases on each plate of the column has been fully established. Using the plate theory, mathematical relationships can be derived from the chromatogram, which are a practical measure of the sharpness of the separation and the resolving power.

The chromatography column is divided up into theoretical plates, i.e. into column sections in the flow direction, the separating capacity of each one corresponding to a theoretical plate. The length of each section of column is called the height equivalent to a theoretical plate (HETP). The HETP value is calculated from the length of the column L divided by the number of theoretical plates N:

$$\text{HETP} = \frac{L}{N} \qquad \text{in mm} \qquad (14)$$

The number of theoretical plates is calculated from the shape of the eluted peak. In the separating funnel model it is shown that with an increasing number of partition steps the substance partitions itself between a larger number of vessels. A separation system giving sharp separation concentrates the substance band into a few vessels or plates. The more plates there are in a separation system, the sharper the eluted peaks.

The number of theoretical plates N is calculated from the peak profile. The retention time t_R at the peak maximum and the width at the base of the peak measured as

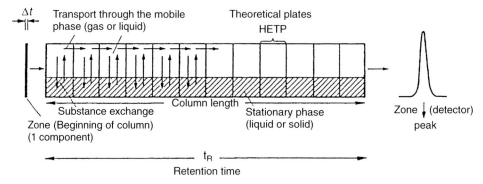

Fig. 2-85 Substance exchange and transport in a chromatography column are optimal when there are as many phase transfers as possible with the smallest possible expansion of the given zones (after Schomburg)

the distance between the cutting points of the tangents to the inflection points with the base line are determined from the chromatogram (Fig. 2-82).

$$N = 16 \cdot \left(\frac{t_g}{W}\right)^2 \qquad \text{where } t_R = \text{retention time} \qquad (15)$$
$$W = \text{peak width}$$

For asymmetric peaks the half width (the peak width at half height) is used:

$$N = 8 \ln 2 \cdot \left(\frac{t_g}{W_n}\right)^2 \qquad \text{where } t_R = \text{retention time} \qquad (16)$$
$$W_n = \text{peak width at half height}$$

Consequence: A column is more effective, the more theoretical plates it has (Fig. 2-85).

The width of a peak in the chromatogram determines the resolution of two components at a given distance between the peak maxima (Fig. 2-86). The resolution R is used to assess the quality of the separation:

$$R \approx \frac{\text{retention difference}}{\text{peak width}} \qquad (17)$$

The resolution R of two neighbouring peaks is defined as the quotient of the distance between the two peak maxima, i.e. the difference between the two retention times t_R and the arithmetic mean of the two peak widths:

$$R = 2 \cdot \frac{t_{R2} - t_{R1}}{W_1 + W_2} = 1.198 \cdot \frac{t_{R2} - t_{R1}}{W_{h1} + W_{h2}} \qquad \text{where } W_h = \text{peak width at half height}$$

Fig. 2-87 shows what one can expect optically from a value for R calculated in this way. At a resolution of 1 the peaks are not completely separated, but it can definitely be seen that there are two components. The tangents to the inflection points just touch each other and the peak areas only overlap by 2%.

2.2 Gas Chromatography   109

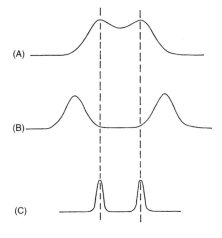

Fig. 2-86 Resolution
(A/C) Peaks with the same retention time
(A/B) Peaks with the same peak width
(B/C) Separation with the same resolution

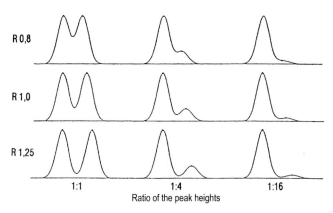

Fig. 2-87 Resolution of two neighbouring peaks (after Snyder, Kirkland)

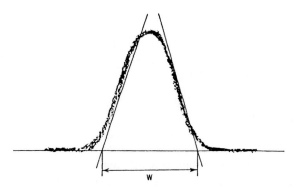

Fig. 2-88 Manual determination of the peak width using tangents to the inflection points

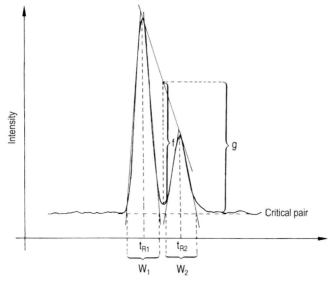

Fig. 2-89 Determination of the resolution and peak widths for a critical pair

For the precise determination of the peak width the tangents to the inflection points can be drawn in manually (Fig. 2-88). For a critical pair, e.g. stearic acid (C_{18-0}) and oleic acid (C_{18-1}) the construction of the tangents is shown in Fig 2-89.

2.2.3.4 Factors Affecting the Resolution

Rearranging the resolution equation and putting in the capacity factor $k' = (t_R - t_0)/t_0$, the selectivity factor $\alpha = k'_2/k'_1$ and the number of theoretical plates N gives an important basic equation for all chromatographic elution processes.

The resolution R is related to the selectivity α (relative retention), the number of theoretical plates N and the capacity factor k' by:

$$R = \tfrac{1}{4}(\alpha - 1) \cdot \underbrace{\frac{k'}{1+k'}}_{\text{II.}} \cdot \underbrace{\sqrt{N}}_{\text{III.}} \qquad (18)$$

$\underbrace{\phantom{R = \tfrac{1}{4}(\alpha - 1)}}_{\text{I.}}$

I. The Selectivity Term

R is directly proportional to $(\alpha - 1)$. An increase in the ratio of the partition coefficients leads to a sharp improvement in the resolution, which can be achieved, for example, by changing the polarity of the stationary phase for substances of different polarities.

As the selectivity generally decreases with increasing temperature, difficult separations must be carried out at as low a temperature as possible.

Table 2-22 Relationship between relative retention α and the chromatographic resolution R

Relative retention α	R = 1.0	R = 1.5
1.005	650 000 plates	1 450 000 plates
1.01	163 000	367 000
1.05	7 100	16 000
1.10	3 700	8 400
1.25	400	900
1.50	140	320
2.0	65	145

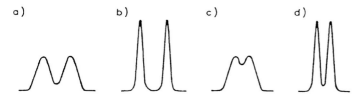

Fig. 2-90 Relative retention, number of plates and resolution

The change in the selectivity is the most effective of the possible measures for improving the resolution. As shown in Table 2-22, more plates are required to achieve the desired resolution when α is small.

Figure 2-90 shows the effect of relative retention and number of plates on the separation of two neighbouring peaks:

– At high relative retention the number of theoretical plates in the column does not need to be large to achieve satisfactory resolution (a). The column is poor but the system is selective.
– A high relative retention and large number of theoretical plates give a resolution which is higher than the optimum. The analysis is unnecessarily long (b).
– At the same (small) number of theoretical plates as in (a), but at a smaller relative retention, the resolution is strongly reduced (c).
– If the relative retention is small, a large number of theoretical plates are required to give a satisfactory resolution (d).

II. The Retardation Term

Here the resolution is directly proportional to the residence time of a component in the stationary phase based on the total retention time. If the components only stayed in the mobile phase ($k' = 0$!) there would be no separation.

For very volatile or low molecular weight nonpolar substances there are only weak interactions with the stationary phase. Thus at a low k' value the denominator $(1 + k')$ of the term is large compared with k' and R is therefore small.

2 Basics

This also applies to columns with a small quantity of stationary phase and column temperatures which are too high. To improve the resolution a larger content of stationary phase can be chosen (greater film thickness).

III. The Dispersion Term

The number of plates N characterises the performance of a column (Fig. 2-91). However the resolution R only increases with the square root of N. As N is directly proportional to the column length L, the performance is only proportional to the square root of the column length.

Doubling the column length therefore only increases the resolution by a factor of 1.4. Since the retention time t_R is proportional to the column length, for an improvement in the resolution by a factor of 1.4, the analysis time is doubled.

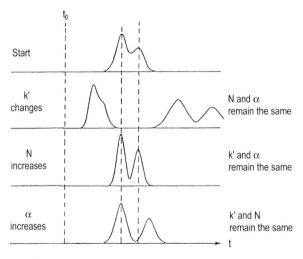

Fig. 2-91 Effect of capacity factor, number of plates and relative retention on the chromatogram (after L. R. Snyder and J. J. Kirkland)

2.2.3.5 Maximum Sample Capacity

The maximum sample capacity can be derived from the equations concerning the resolution. Under ideal conditions (see Fig. 2-89):

$$\frac{f}{g} = 100\% \quad \text{where } f = \text{the area under the line connecting the peak maxima} \quad (19)$$
$$g = \text{the height of the connecting line above the base line, measured in the valley between the peaks}$$

The maximum sample capacity of a column is reached if f/g falls below 90% for a critical pair. If too much sample material is applied to a column, the k' value and the

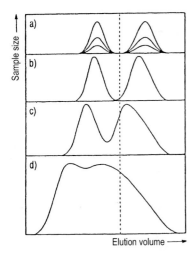

Fig. 2-92 Change in the chromatogram with increasing sample size (after L. R. Snyder and J. J. Kirkland)
(a)   Constant k' values
(b–d) Increasing changes to the retention behaviour through overloading

peak width are no longer independent of the size of the sample, which ultimately affects the identification and the quantitation of the results (Fig. 2-92).

2.2.3.6 Peak Symmetry

In exact quantitative work (integration of the peak areas) a maximum asymmetry must not be exceeded, otherwise there will be errors in determining the cut-off point of the peak with the base line.

For practical reasons the peak symmetry T is determined at a height of 10% of the total peak height (Fig. 2-93):

$$T = \frac{b_{0.1}}{a_{0.1}} \qquad (20)$$

where $a_{0.1}$ = distance from the peak front to the maximum measured at 0.1 h
$b_{0.1}$ = distance from the maximum to the end of the peak measured at 0.1 h

T should not be greater than 2.5. If the tailing exceeds the value of 3, there will be errors in the quantitative area measurement because the point where the peak reaches the base line is very difficult to find.

2.2.3.7 Optimisation of Flow

The flow of the mobile phase affects the rate of substance transport through the stationary phase. High flow rates allow rapid separation. However, the efficiency is reduced because of the slower exchange of substances between the stationary and mobile phases and leads to peak broadening. On the other hand, peak broadening caused by diffusion of the components within the mobile phase is only hindered by increasing the flow rate.

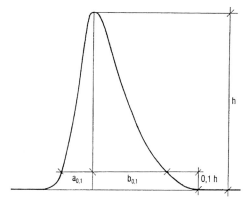

Fig. 2-93 Asymmetric peak

The aim of flow optimisation for given column properties at a given temperature and with a given carrier gas is to find the flow rate which gives either the maximum number of separation steps or at adequate efficiency the shortest possible analysis time.

The height equivalent to a theoretical plate (HETP) and the number of theoretical plates N depends on the flow rate of the mobile phase u according to van Deemter. The linear flow rate u of the mobile phase is calculated from the chromatogram:

$$u = \frac{L}{t_0} \qquad \text{with } L = \text{length of the column in cm} \qquad (21)$$
$$t_0 = \text{dead time in s}$$

The following affect the optimum flow rate:

1. The *Eddy diffusion* (Fig. 2-94) on the peak broadening. This effect is independent of flow and, naturally, for packed columns only, dependent on the nature of the packing material and the density of packing. For open capillary (tabular) columns, Eddy diffusion does not occur.
2. The *axial diffusion* on peak broadening. This diffusion occurs in and against the direction of flow and decreases with increasing flow rate.
3. *Incomplete partition equilibrium.* The transfer of analyte between the stationary and mobile phases only has a finite rate relative to that of the mobile phase, corresponding to the diffusion rates. The contribution to peak broadening increases with increasing flow rate of the mobile phase.

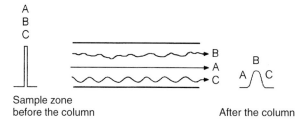

Fig. 2-94 Path differences for lamina flow in capillary columns (multipath effect, caused by turbulence at high flow rates)

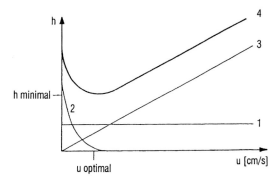

Fig. 2-95 The van Deemter curve
1 is the proportion of Eddy diffusion and flow distribution at band broadening
2 is the proportion of longitudinal diffusion
3 is the proportion of substance exchange phenomena
4 is the resulting curve H(u), called the van Deemter curve

For the maximum efficiency of the separation (Fig. 2-95) a flow rate u_{min} must be chosen as a compromise between these opposing effects. The position of the minimum is affected by:

- the quantity of the stationary phase (e.g. film thickness),
- the particle size of the packing material (for packed columns),
- the diameter of the column,
- the nature of the mobile phase (diffusion coefficient, viscosity).

Definition of chromatographic parameters		
Carrier gas velocity	$v = L/t_0$	The average linear carrier gas velocity has an optimal value for each column with the lowest possible height equivalent to a theoretical plate (see van Deemter); v is independent of temperature. L = length of column t_0 = dead time
Partition coefficient	$K = c_1/c_g$	Concentration of the substance in the stationary phase (liquid) divided by the concentration in the mobile phase (gas). K is constant for a particular substance in a given chromatographic system.
	$K = k' \cdot \beta$	K is also expressed as the product of the capacity ratio (k') and the phase ratio (β).

Capacity ratio (partition ratio)	$k' = K \cdot V_l/V_g$	Determines the retention time of a compound. V_l = volume of the stationary phase V_g = volume of the gaseous mobile phase
	$k' = t_R - t_0/t_0$	t_R = retention time of the substance t_0 = dead time
Phase ratio	$\beta = r/2d_f$	r = internal column radius d_f = film thickness
Number of theoretical plates	$N = 5.54 (t_R/W_h)^2$	The number of theoretical plates is a measure of the efficiency of a column. The value depends on the nature of the substance and is valid for isothermal work. N = number of theoretical plates t_R = retention time of the substance W_h = peak width at half height
Height equivalent to a theoretical plate (HETP)	$h = L/N$	Is a measure of the efficiency of a column independent of its length. L = length of the column N = number of theoretical plates
Resolution	$R = 2(t_j - t_i)/(W_j + W_i)$	Gives the resolving power of a column with regard to the separation of components i and j (isothermally). t_i = retention time of substance i t_j = retention time of substance j $W_{i,j}$ = peak width at half height of substances i, j
Separation factor	$\alpha = k'_j/k'_i$	Measure of the separation of the substances i, j
Trennzahl number	$TZ = \dfrac{t_{R(x+1)} - t_{R(x)}}{W_{h(x+1)} + W_{h(x)}} - 1$	
		The trennzahl number is, like the resolution, a means of assessing the efficiency of a column and is also used for temperature-programmed work. TZ gives the number of components which can be resolved between two homologous n-alkanes.
Effective plates	$N_{\text{eff.}} = 5.54((t_{R(i)} - t_0)/W_{h(i)})$	
		The effective number of theoretical plates takes the dead volume of the column into account.
Retention volume	$V_R = t_R \cdot F$	Gives the carrier gas volume required for elution of a given component. F = carrier gas flow

Kovats index	$KI = 100 \cdot c + 100 \, \dfrac{\log(t'_R)_x - \log(t'_R)_c}{\log(t'_R)_{c+1} - \log(t'_R)_c}$	
		The Kovats index is used for isothermal work. t'_R = corrected retention times for standards and substances $t'_R = t_R - t_0$
Modified Kovats index	$RI = 100 \cdot c + 100 \, \dfrac{(t'_R)_x - (t'_R)_c}{(t'_R)_{c+1} - (t'_R)_c}$	
		The modified Kovats index according to van den Dool and Kratz is used with temperature programming.

2.2.4 Classical Detectors for GC/MS Systems

Classical detectors are important for the consideration of GC/MS coupling if an additional specific means of detection is to be introduced parallel to mass spectrometry. The parallel coupling of a flame ionisation detector (FID) does not lead to results which are complementary to those of mass spectrometry, as both detection processes give practically identical chromatograms when the response factors are comparable. Parallel detection with a thermal conductivity detector is not used in practice as the mass spectrometric analysis of gases is generally carried out with special instruments (mass range < 100 u).

Additional information can, however, be obtained with element-specific detectors. The detection limits which can be achieved with an electron capture detector (ECD) or a nitrogen/phosphorus detector (NPD) are usually better than those attainable using a mass spectrometer. On dividing up the carrier gas flow, the ratio of the two parts must be considered when planning such a process. Normally a larger proportion is passed into the mass spectrometer so that in residue analysis low concentrations of substances do not fall below the detection limit. The use of such flow dividers for quantitative determinations must be checked in an individual case, as a constant division cannot be expected for all boiling point ranges.

2.2.4.1 FID

With FID the substances to be detected are burned in a hydrogen flame and are thus partially ionised (Table 2-23). As the jet is at a negative potential, positive ions are neutralised. The corresponding electrons are captured at the ring-shaped collector electrode to give a signal current (Fig. 2-96). The electrode is at a potential which is ca. 200 V more positive than the jet.

Table 2-23 Reactions in the FID (Perkin Elmer)

Pyrolysis:	CH_3^o, CH_2^o, CH^o, C^o
Excited radicals:	O_2^*, OH^*
Ionisation:	$CH_2^o + OH^* \rightarrow CH_3O^+ + e^-$
	$CH^o + OH^* \rightarrow CH_2O^+ + e^-$
	$CH^o + O_2^* \rightarrow CHO_2^+ + e^-$
	$C^o + OH^* \rightarrow CHO^+ + e^-$

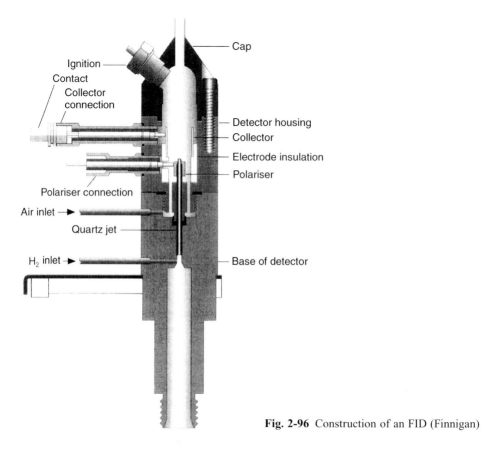

Fig. 2-96 Construction of an FID (Finnigan)

Provided that only hydrogen burns in the flame, only radical reactions occur. No ions are formed. If organic substances with C-H and C-C bonds get into the flame, they are first pyrolysed. The carbon-containing radicals are oxidised by oxygen and the OH radicals formed in the flame. The excitation energy leads to ionisation of the oxidation products. Only substances with at least one C-H or C-C bond are detected, but not permanent gases, carbon tetrachloride or water.

If a reactor (hydrogenator, methaniser) is connected before the FID, the latter can be converted into an extremely sensitive detector for permanent gases such as CO and

CO_2. The oxygen specific detector (O-FID) uses two reactors. In the first hydrocarbons are decomposed into carbon, hydrogen and carbon monoxide at above 1300 °C. CO is then converted into methane in the hydrogenation reactor and detected with the FID. With the O-FID, for example, oxygen-containing components in fuels can be detected.

FID Flame Ionisation Detector

Universal detector

Advantages:	High dynamics High sensitivity Robust
Use:	Hydrocarbons, e.g. fuels, odorous substances, BTX, polyaromatic hydrocarbons etc Comparison of diesel with petrol Important all round detector
Limits:	As it is universal, its performance is poor for trace analysis in complex matrices Low response for highly chlorinated or brominated substances.

2.2.4.2 NPD

An NPD is a modified FID which contains a source of alkali situated between the jet and the collector electrode on a Pt wire, for the specific detection of nitrogen or phosphorus (Fig. 2-97).

The alkali beads are heated to red heat both electrically and in the flame and are excited to alkali emission. They are always at a negative potential compared with the collector electrode. For the detection of phosphorus the jet is earthed. The electrons emitted by the hydrocarbon parts of the molecule cannot exceed the negative potential of the beads and do not reach the collector electrode, but are earthed. The electrons from the specific alkali reaction reach the collector electrode unhindered (Table 2-24).

Phosphorus-containing substances are first converted in the flame into phosphorus oxides with an uneven number of electrons. Anions formed in the alkali reaction by the addition of an electron are now oxidised by OH radicals. The electrons added are now released and produce a signal current.

Like phosphorus, nitrogen has an uneven number of electrons. Under the reducing conditions of the flame cyanide and cyanate radicals are formed, which can undergo the alkali reaction (Table 2-25). For this the input of hydrogen and air are reduced. Instead of the flame the hydrogen burns in the form of a cold plasma around the electrically heated alkali beads.

In order to form the required cyanide and cyanate radicals the C-N structure must already be present in the molecule. Nitro compounds are detected, but not nitrate esters, ammonia or nitrogen oxides. By taking part in the alkali reaction the cyanide radical receives an electron. Cyanide ions are formed, which react with other radicals to give neutral species. The electron released provides the detector signal.

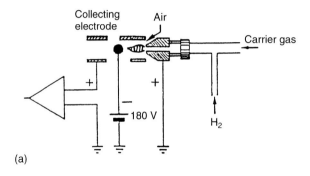

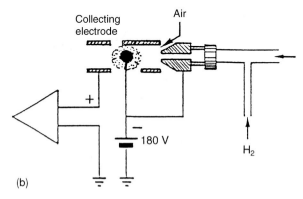

Fig. 2-97 (a) Diagram of an NPD (P operation) (Perkin Elmer); (b) diagram of an NPD (N operation) (Perkin Elmer)

Table 2-24 Reactions with P compounds in the NPD (Perkin Elmer)

$\bar{O} = \dot{P}$	+	A*	→	$[\bar{O} = \bar{P}]^-$	+	A⁺
$\bar{O} = \dot{P} = \bar{O}$	+	A*	→	$[\bar{O} = P = \bar{O}]^-$	+	A⁺
$[\bar{O} = \bar{P}]^-$	+	OH°	→	HPO₂	+	e⁻
$[\bar{O} = P = \bar{O}]^-$	+	OH°	→	HPO₃	+	e⁻
HPO₃	+	H₂O	→	H₃PO₄		

A = alkali

Table 2-25 Reactions with N compounds in the NPD (Perkin Elmer)

Pyrolysis of CNC compounds			→	C ≡ N|°		
CN°	+	A*	→	CN⁻	+	A*
CN⁻	+	H°	→	HCN	+	e⁻
CN⁻	+	OH°	→	HCNO	+	e⁻

A = alkali

NPD **N**itrogen/**P**hosphorus **D**etector

Specific detector

Advantages:	High selectivity and sensitivity Ideal for trace analyses
Use:	Only for N- and P-containing compounds Plant protection agents Chemical warfare gases, explosives Pharmaceuticals
Limits:	Additional detector for ECD or MS Quantitative measurements with an internal standard are recommended To some extent time-consuming optimisation of the Rb beads

2.2.4.3 ECD

The ECD (electron capture detector) consists of an ionisation chamber, which contains a nickel plate, on the surface of which a thin layer of the radioactive isotope ^{63}Ni has been applied (ca. 10–15 mC, Fig. 2-98). The carrier gas (N_2 or Ar/10% methane) is first ionised by the β radiation. The free electrons migrate towards the collector electrode and provide the background current of the detector. Substances with electronegative groups reduce the background current by capturing electrons and forming negative molecular ions. The main reactions in the ECD are dissociative electron capture (Table 2-26) and electron capture (Table 2-27, see also Section 2.3.2.2). Negative molecular ions can recombine with positive carrier gas ions.

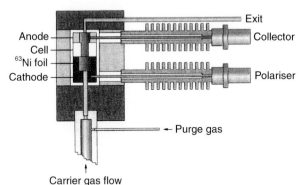

Fig. 2-98 Construction of an ECD (Finnigan)

Table 2-26 Substance reactions in the ECD (Perkin Elmer)

1. Dissociative electron capture
    AB + e$^-$ → A^0 + B$^-$

2. Addition of an electron
    AB + e$^-$ → (AB$^-$)*

Table 2-27 Basic reactions in the ECD (Perkin Elmer)

CG	$\xrightarrow{\beta}$	CG^+ + e^-
Electron capture		
M + e^-	$\rightarrow$	M^-
Recombination		
M^- + CG^+	$\rightarrow$	M^o + CG

CG = Carrier gas

Table 2-28 Conversion rates in the ECD for molecules with different degrees of chlorination

Molecule	Conversion rate [$cm^3 \cdot mol^{-1} \cdot s^{-1}$]	Main product
CH_2Cl_2	1×10^{-11}	Cl^-
$CHCl_3$	4×10^{-9}	Cl^-
CCl_4	4×10^{-7}	Cl^-
CH_3CCl_3	1×10^{-8}	Cl^-
$CH_2ClCHCl_2$	1×10^{-10}	Cl^-
CF_4	7×10^{-13}	–
CF_3Cl	4×10^{-10}	Cl^-
CF_3Br	1×10^{-8}	Br^-
CF_2Br_2	2×10^{-7}	Br^-
C_6F_6	9×10^{-8}	M^-
$C_6F_5CF_3$	2×10^{-7}	M^-
$C_6F_{11}CF_3$	2×10^{-7}	M^-
SF_6	4×10^{-7}	M^-
Azulene	3×10^{-8}	M^-
Nitrobenzene	1×10^{-9}	M^-
1,4-Naphthoquinone	7×10^{-9}	M^-

Electron capture is more effective, the slower the electrons move. For this reason a more sensitive ECD is now operated using pulsed DC voltage. By changing the pulse frequency the background current generated by the electrons is kept constant. The pulse frequency thus becomes the actual detector signal.

ECD	Electron Capture Detector

Specific detector

The ECD reacts with all electronegative elements and functional groups, such as -F, -Cl, -Br, -OCH$_3$, -NO$_2$ with a high response. All hydrocarbons (generally the matrix) remain transparent, although present.

Advantages:	Selectivity for Cl, Br, methoxy, nitro groups
	Transparency of all hydrocarbons (= matrix)
	High sensitivity
Disadvantages:	Radioactive radiator, therefore handling authorisation necessary
	Sensitive to misuse
	Mobile only under limited conditions
Use:	Typical detector for environmental analysis
	Ideal for trace analysis
	Plant protection agents
	PCBs, dioxins,
	Volatile halogenated hydrocarbons, freons
Limits:	Substances with low halogen contents (Cl$_1$, Cl$_2$) only have a low response (Table 2-28)
	Volatile halogenated hydrocarbons with low chlorine levels are better detected with FID or MS
	Limited dynamics
	Multipoint calibration necessary

2.2.4.4 PID

PID	Photoionisation Detector

Universal and selective detector

The energy-rich radiation of a UV lamp (Fig. 2-99) ionises the substances to be analysed more or less selectively, depending on the energy content (Table 2-29); measurement of the overall ion flow.

Advantages:	The selectivity can be chosen
	High sensitivity
	No gas supply is required
	Robust, no maintenance required
	Ideal for mobile and field analyses
Use:	Field analysis of volatile halogenated hydrocarbons, BTEX, polyaromatic hydrocarbons etc.
Limits:	Only for substances with low ionisation potentials (≤ 10 eV)
	Consult manufacturers' data (Fig. 2-100)

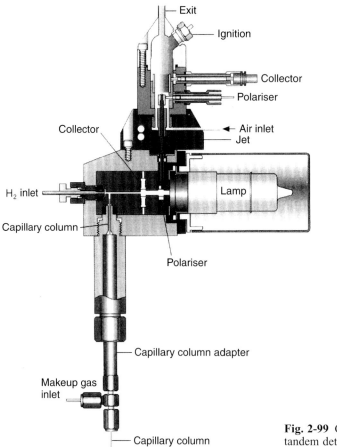

Fig. 2-99 Construction of a PID/FID tandem detector (Finnigan)

Table 2-29 Ionisation potentials of selected analytes (HNU systems)

Substance	Ionisation potential [eV]	Substance	Ionisation potential [eV]
Helium	24.59	Isobutyraldehyde	9.74
Argon	15.76	Propene	9.73
Nitrogen	15.58	Acetone	9.69
Hydrogen	15.43	Benzene	9.25
Methane	12.98	Methyl isothiocyanate	9.25
Ethane	11.65	N, N-Dimethylformamide	9.12
Ethylene	11.41	2- Iodobutane	9.09
2-Chlorobutane	10.65	Toluene	8.82
Acetylene	10.52	n-Butylamine	8.72
n-Hexane	10.18	o-Xylene	8.56
2-Bromobutane	9.98	Phenol	8.50
n-Butyl acetate	9.97		

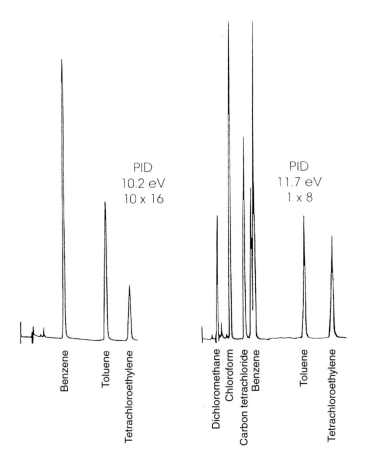

Fig. 2-100 Selectivity of the PID at 10.2 eV and 11.7 eV for a mixture of aromatic and chlorinated hydrocarbons (HNU)

2.2.4.5 ELCD (Electrolytical Conductivity Detector)

In ELCD the eluate from a GC passes into a Ni reactor in which all substances are completely oxidised or reduced at an elevated temperature (ca. 1000 °C).

The products dissociate in circulating water which flows through a cell where the conductivity is measured between two Pt spirals (Fig 2-101). The change in conductivity is the measurement signal. Ionic compounds can be measured without using the reactor.

The Hall detector has a comparable function to the ELCD and is a typical detector for packed or halfmil columns (0.53 mm internal diameter) because of the large volume of its measuring cell. Because of the latter significant peak tailing occurs with capillary columns. Special constructions for use with normal bore columns (<0.32 mm internal diameter) are available with special instructions.

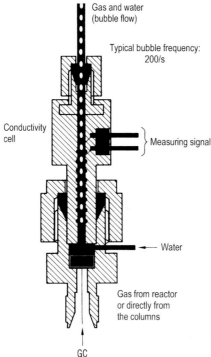

Fig. 2-101 Construction of an ELCD (patent Fraunhofer Gesellschaft, Munich)

ELCD Electrolytical **C**onductivity **D**etector	
Selective detector	
Advantages:	Can be used for capillary chromatography
	Selectivity can be chosen for halogens, amines, nitrogen, sulfur
	High sensitivity at high selectivity
	Can be used for halogen detection without a source of radioactivity, therefore no authorisation necessary
	Simple calibration, as the response is directly proportional to the number of heteroatoms in the analyte, e.g. the proportion of Cl in the molecule
Use:	Environmental analysis, e.g. volatile halogenated hydrocarbons, PCBs
	Selective detection of amines, e.g. in packagings or foodstuffs
	Determination of sulfur-containing components
Limits:	The sensitivity in the halogen mode just reaches that of ECD, so that its use is particularly favourable in association with concentration procedures, such as purge and trap or thermodesorption.

2.2.4.6 FPD (Flamephotometric Detector)

FPDs are used as one- or two-flame detectors. In a hydrogen-rich flame P- and S-containing radicals are in an excited transition state. On passing to the ground state a characteristic band spectrum is emitted (S: 394 nm, P: 526 nm). The flame emissions initiated by the eluting analytes (chemiluminescence) are determined using an optical filter and magnified by a photomultiplier (Fig. 2-102).

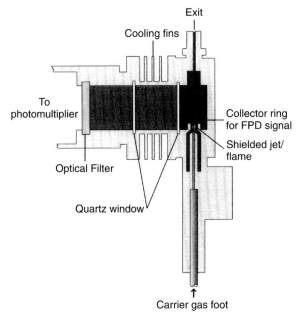

Fig. 2-102 Construction of an FPD (Finnigan)

FPD Flamephotometric Detector	
Selective detector	
Advantages:	In phosphorus mode, comparable selectivity to FID with high dynamics High selectivity in sulfur mode
Use:	Mostly selective for sulfur and phosphorus compounds (e.g. plant protection agents) Detection of sulfur compounds in a complex matrix
Limits:	Adjustment of the combustion gases important for reproducibility and selectivity In sulfur mode quenching effect possible because of too high a hydrocarbon matrix (double flame necessary) Sensitivity in sulfur mode not always sufficient for trace analysis

2.2.4.7 Connection of Classical Detectors Parallel to the Mass Spectrometer

In principle there are two possibilities for operating another detector parallel to the mass spectrometer. The sample can be already divided in the injector and passed through two identical columns. The second more difficult but controllable solution is the split of the eluate at the end of the column.

The division of the sample on to two identical columns can be realised easily and carried out very reliably. Since the capacities of the two columns are additive, the quantity of sample injected can be adjusted in order to make use of the operating range of the mass spectrometer. Two standard columns can be installed for most injectors without further adaptation being necessary. In the simplest case the connection can be made using ferrules with two holes in them. It must be ensured that there is a good seal. A better connection involves an adaptor piece with a separate screw-in joint for each column. With this construction reliable positioning of the columns in the injector is also possible. Suitable adaptors can be obtained for all common injectors.

In GC/MS the division of the flow at the end of the column is considerably more difficult because the direct coupling of the branch to the mass spectrometer causes reduced pressure in the split. For this reason the use of a simple Y piece is seldom possible with standard columns. The consequence would be reversed flow through the detector connected in parallel. The effect is equivalent to a leakage of air into the mass spectrometer. A split at the end of the column must therefore be carried out with high flow rates or using a makeup gas. Precise split of the eluent is possible, for example, using the glass cap cross divider (Fig. 2-103). Here the column, the transfer capillaries to the mass spectrometer and the parallel detector and a makeup gas inlet all meet. By choosing the internal diameter and the position of the end of the column in the glass cap (also known as the glass dome) the ratio of the split can easily be adjusted. The advantage of this solution lies in the free choice of column so that small internal diameters with comparatively low flow rates can also be used.

The calculation of gas flow rates through outlet splitters with fixed restrictors is desirable to have a means of estimating the rate of gas flow through a length of fixed restrictor tubing of specified dimensions. Conversely, for other applications it is necessary to estimate the length and ID of restriction tubing required to yield a desired flow rate at some specified head pressure. Within limits, the following calculation can be used for this purpose.

$$V = \frac{\pi \cdot P \cdot r^4 \cdot (7.5)}{l \cdot \eta}$$

with: V = volumetric gas flow rate in cm^3/min
P = pressure differential across the tube in dynes/cm
  = (PSI) · (68 947.6)
r = tube radius in cm
l = tube length in cm
η = gas viscosity in poise (dyne-seconds/cm^2)

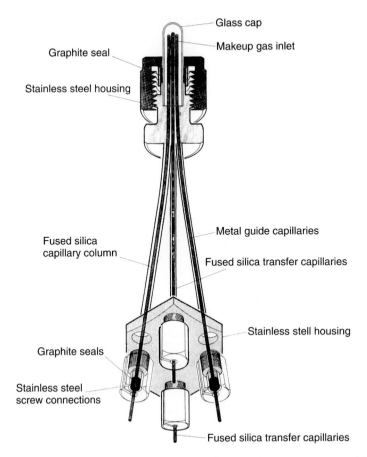

Fig. 2-103 Flow divider after Kaiser, Rieder (Seekamp company, Werkhoff Splitter, Glass Cap Cross)

One of the obvious limitations to this calculation is the critical nature of the radius measurement. Since this is a fourth power term, small errors in the ID measurement can result in relatively large errors in the flow rate. This becomes more critical with smaller diameter tubes (i.e. less than 100 micron). A 100x microscope (if available) is a convenient tool for determining the exact ID of a particular length of restriction tubing.

The gas viscosity values can be determined from the graph of gas viscosity against temperature for hydrogen, helium and nitrogen (Fig. 2-103A).

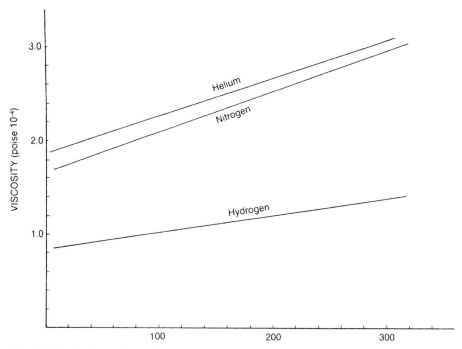

Fig. 2-103A Graphs to determine viscosity for hydrogen, helium and nitrogen

2.3 Mass Spectrometry

> "Looking back on my work in MS over the last 40 years, I believe that my major contribution has been to help convince myself, as well as other mass spectrometrists and chemists in general, that the things that happen to a molecule in the mass spectrometer are in fact chemistry, not voodoo; and that mass spectrometrists are, in fact, chemists and not shamans".
>
> Seymour Meyerson
> Research Dept., Amoco Corp.

Mass spectrometers are instruments for producing and analysing mixtures of ions for components of differing mass and for exact mass determination. The substances to be analysed are fed to an ion source and ionised. In GC/MS systems there is continuous transport of substances by the carrier gas into the ion source. Mass spectrometers basically differ in the construction of the analyser as a beam or ion storage instrument. The performance of a mass spectrometric analyser is determined by the resolving power for differentiation between masses with small differences, the mass range and the transmission required to achieve high detection sensitivity.

2.3.1 Resolution in Mass Spectrometry

In mass spectrometry the resolving power is the capacity of an MS analyser to separate mass signals (ions) which are close to one another according to their m/z (mass/charge) ratios.

To determine the resolution the nature of the separating system must be considered. High resolution indicates a resolving power which differentiates between C, H, N, and O multiplets (resolution > 10 000) and low resolution indicates the separation according to nominal masses (whole mass numbers, unit mass resolution).

2.3.1.1 High Resolution

The paths of ions with different m/z values follow a different course as a result of the magnetic and electric field in a magnetic sector instrument (Fig. 2-104). Splitting systems can mask ion beams. Spectra can be recorded by continually changing the parameters of the instrument, e.g. the acceleration voltage. The width of the ion beam is determined by the source slit at the ion source. The beams must not overlap, or only to a very small extent so that ions of different masses can be registered consecutively (Fig. 2-105).

The resolution A of neighbouring signals (Fig. 2-106) for magnetic sector instruments is calculated according to

$$A = \frac{m}{\Delta m} \quad \text{with} \quad \begin{array}{l} m = \text{mass and} \\ \Delta m = \text{distance between neighbouring masses} \end{array} \tag{22}$$

According to this formula the resolution is dimensionless.

By determination of the exact mass the empirical formula of a molecular ion (and of fragment ions) can be determined if the precision of the measurement is high enough (Fig 2-107). The more the mass increases, the more interference can arise.

Example of the Calculation of the Necessary Minimum Resolution

What mass spectroscopic resolution is required to obtain the signals of carbon monoxide (CO), nitrogen (N_2) and ethylene (C_2H_4) which are passed through suitable tubing into the ion source of a mass spectrometer?

Substance	Nominal mass	Exact mass
CO	m/z 28	m/z 27.994910
C_2H_4	m/z 28	m/z 28.006148
N_2	m/z 28	m/z 28.031296

For MS separation of CO and N_2, which appear at the same time in accordance with the formula given above, a resolving power of at least 2500 is necessary.

All MS systems with low resolution need preliminary GC separation of the components (CO, C_2H_4 and N_2 would then arrive one after the other in the ion source of an MS). This is the case for all ion trap and quadrupole instruments.

132   2 Basics

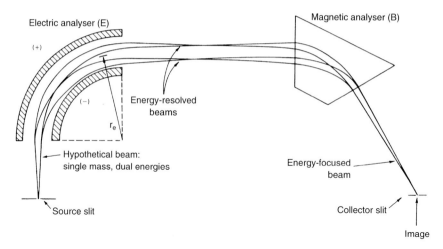

Fig. 2-104 Principle of the double focusing magnetic sector mass spectrometer (Nier Johnson geometry: EB)

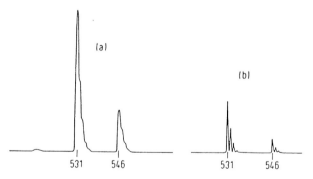

Fig. 2-105 Section of a poorly resolved spectrum (a) and of a better, more highly resolved one (b) which, however, results in a lowering of intensity of the signals (after Budzikiewicz 1980)

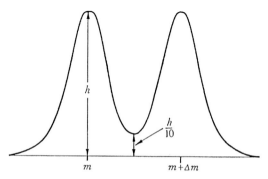

Fig. 2-106 General resolution conditions for two mass signals – 10% trough definition (after Budzikiewicz 1980)

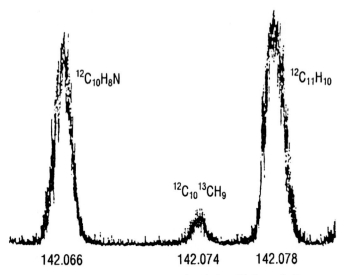

Fig. 2-107 Diagram showing mass signals in a high resolution spectrum at $A = 61\,000$ (Finnigan MAT 90)

A selection of the large number of realistic chemical formulae for the mass 310 ($C_{22}H_{46}$, M 310.3599) is shown in Fig. 2-108. To differentiate between the individual signals a precision in the mass determination of 2 ppm would be necessary.

It is characteristic for a mass spectrum produced by a magnetic sector instrument that the resolution A is constant over the whole mass range. According to the formula for the resolution, the distance between the mass signals, Δm, for water ($m = 17/18$) is much greater than for signals in the upper mass range. The maximum resolution that can be achieved characterises the slit system and the quality of the ion optics of the magnetic sector instrument.

2.3.1.2 Unit Mass Resolution

A mass spectrum obtained with an ion trap or quadrupole mass spectrometer (Figs. 2-109 and 2-112) shows another characteristic: the distance between two mass signals and their signal widths are constant over the whole mass range! How far the instrument can scan to higher masses is therefore unimportant. The quadrupole/ion trap peaks of water (m/z = 17/18) are the same width and have the same separation as the masses in the upper mass range (Fig. 2-110).

Since the distance between two signals, Δm, is constant over the whole mass range for these instruments, the formula $A = m/\Delta m$ would have the result that the resolution A would be directly proportional to the highest possible mass m (Fig. 2-111). At a peak separation of one mass unit the resolving power in the lower mass range would be small using the formula for the resolution (e.g. for water $A \approx 18$) and in the upper mass range higher (e.g. $A \approx 614$ for FC43).

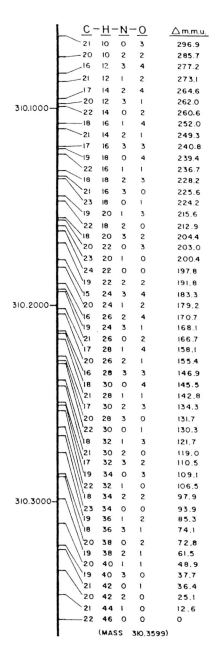

Fig. 2-108 Exact masses of chemically realistic empirical formulae consisting of C, H, N (≤ 3), O (≤ 5) given in deviations (Δ mu) from the molecular mass of $C_{22}H_{46}$ m/z 310 (after McLafferty 1993)

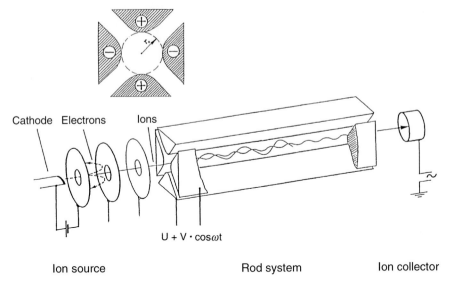

Fig. 2-109 Diagram of a quadrupole mass spectrometer

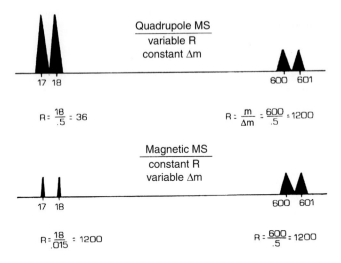

Fig. 2-110 Comparison of the principles of the mass spectra obtained from quadrupole/ion trap and magnetic sector analysers

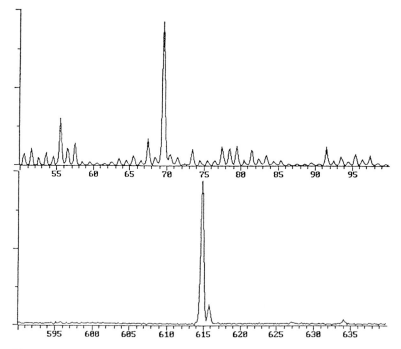

Fig. 2-111 Diagram of the mass signals obtained using an ion trap analyser in the lower and upper mass ranges (calibration standard FC43)

The following conclusions may be drawn from these facts concerning the assessment of quadrupole and ion trap instruments:

1. The formula $A = m/\Delta m$ does not give any meaningful figures for quadrupole and ion trap instruments and therefore cannot be used in every case.
2. The visible optical resolution (on the screen) is the same in the upper and lower mass ranges. It can easily be seen that the signals corresponding to whole numbers are well separated. This corresponds in practice to the maximum possible resolution for quadrupole and ion trap instruments.
3. This resolving power, which is constant over the whole mass range, is set up by the manufacturer in the electronics of the instrument and is the same for all types and manufacturers. The peak width is chosen in such a way that the distance between two neighbouring nominal mass signals corresponds to one mass unit (1 u = 1000 mu). High resolution, as in the magnetic sector instrument, is not possible for quadrupole and ion trap instruments within the framework of the scan technique used. Because all quadrupole and ion trap instruments have the same resolving power, they are said to have unit mass resolution.
4. The mass range of quadrupole and ion trap instruments varies but still has no effect on the resolving power.

Frequently the terms mass range and unit mass resolution are mixed up when giving a quality criterion for a mass range above 1000 u for a quadrupole instrument (it is not obvious that a mass range up to 4000 u, for example, is always accompanied by unit mass resolution). The effective attainable resolution for a real measurable signal of a reference compound is accurate and meaningful.

The different types of analyser for quadrupole rod and quadrupole ion trap instruments function on the same mathematical basis (Paul/Steinwedel, 1953, and patent specification 1956) and therefore show the same resolution properties (Fig. 2-112).

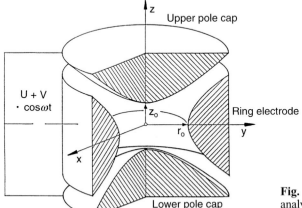

Fig. 2-112 Diagram of an ion trap analyser (Finnigan)

The recording of line spectra on the screen must be considered completely separately from the resolution of the analyser. By definition a mass peak with unit mass resolution has a base width of one mass unit or 1000 mu. On the other hand the position of the top of the mass peak (centroid) can be calculated exactly. Data sometimes given in terms of 1/10 of a mass unit gives the false impression of a resolution higher than unit mass resolution. Components appearing at the same time with signals of the same nominal mass, which can naturally occur in GC/MS (as a result of co-eluates, the matrix, column bleed etc) cannot be separated at unit mass resolving power of the quadrupole and ion trap analysers (see Section 2.3.1.2). The position of the centroid can therefore not be used for any sensible evaluation. *In no case is this the basis for the calculation of a possible empirical formula!* Depending on the manufacturer, the labelling of the spectra can be found with pure nominal masses to several decimal places and can usually be altered by the user.

It has been observed with many benchtop instruments that the unavoidable soiling of ion volumes and lenses, the changes in the carrier gas flow as a result of temperature programs, and temperature drift of the ion source during the operation cause the position of the peak maximum to shift by several tenths of a mass during a short time. Adequate mass stability is therefore only given for the requirement of unit mass resolution. Warning: what appears on the screen has nothing to do with the actual resolution of the MS separating system! Each MS data system uses whole number mass spectra (nominal masses) for internal processing by background subtraction and library searching. The spectra of all MS libraries also only contain whole mass numbers!

Very recent technologies introduced high resolution capabilities (R 10.000–30.000 at 1.000 u) with ion trap analysers and hyperbolic rod quadrupole instruments.

2.3.1.3 High and Low Resolution in the Case of Dioxin Analysis

Can dioxin analysis be carried out with quadrupole and ion trap instruments?

Yes! But with what sort of quality? *The main problem is the resolving power.* The sensitivity of quadrupole and ion trap instruments is adequate for many applications. However, the selectivity is not.

It is possible to make mistakes on using selected ion monitoring techniques (SIM). These types of mistakes are smaller for full scan data collection with ion traps and on using MS/MS techniques, as usually a complete spectrum and a product ion spectrum are available for further confirmation. Errors in peak integration and poor detection limits as a result of matrix overlap, which cannot be totally excluded, generally limit the use of low resolution mass spectrometers. For screening tests positive results should be followed by high resolution GC/MS for confirmation.

The high resolution and mass precision of the magnetic sector instrument allows the exact mass to be recorded (e.g. 2,3,7,8-TCDD at m/z 321.8937 instead of a nominal mass of 322) and thus masks the known interference effects (Table 2-30). As a result extremely low detection limits of <10 fg are achieved, which gives the necessary assurance for making decisions with serious implications (Fig. 2-113).

High resolution is therefore also required to verify positive screening results in dioxin analysis (see also Section 4.27). A comparison of the two spectrometric methods of the lower and higher resolution SIM techniques is shown in Fig. 2-114 for the mass traces in the detection of TCDF traces.

Table 2-30 Possible interference with the masses m/z 319.8965 and 321.8936 relevant for 2,3,7,8-TCDD and the minimum analyser resolution required for separation

Compound	Formula	m/z*	Resolution needed for separation
Tetrachlorobenzyltoluene	$C_{12}H_8Cl_4$	319.9508	5 900
Nonachlorobiphenyl	$C_{12}HCl_9$	321.8491	7 300
Pentachlorobiphenylene	$C_{12}H_3Cl_5$	321.8677	12 500
Heptachlorobiphenyl	$C_{12}H_3Cl_7$	321.8678	13 000
Hydroxytetrachlorodibenzofuran	$C_{12}H_4O_2Cl_4$	321.8936	Cannot be resolved
DDE	$C_{14}H_9Cl_5$	321.9292	9 100
DDT	$C_{14}H_9Cl_5$	321.9292	9 100
Tetrachloromethoxybiphenyl	$C_{13}H_8OCl_4$	321.9299	8 900

* of the interfering ion

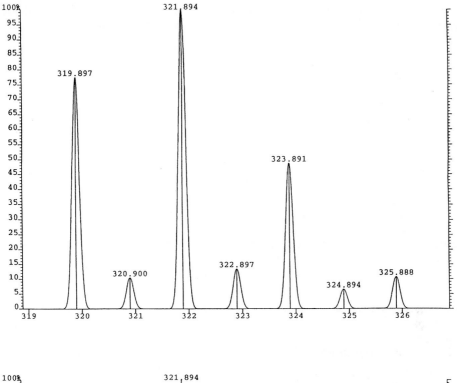

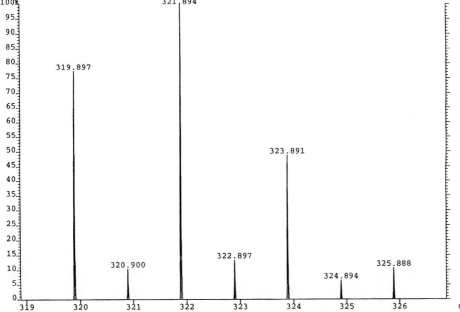

Fig. 2-113 The isotope pattern for 2,3,7,8-TCDD using a double focusing magnetic sector instrument with peak widths for resolutions of 1000 (above) and 10000 (below) (after Fürst)

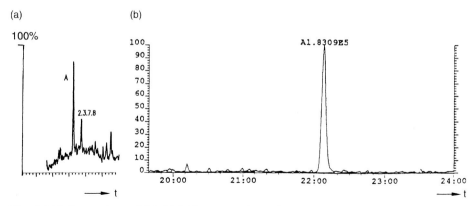

Fig. 2-114 Comparison of 2,3,7,8-TCDF traces in analyses using a low resolution quadrupole GC/MS system (a = m/z 319.9) and a high resolution GC/MS system (b = m/z 319.8965). Both chromatograms were run on an identical human milk sample. The component marked with A in the quadrupole chromatogram is an interfering component (after Fürst)

2.3.2 Ionisation Procedures

2.3.2.1 Electron Impact Ionisation

Electron impact ionisation (EI) is the standard process in all GC/MS instruments. An ionisation energy of 70 eV is currently used in all commercial instruments. Only a few benchtop instruments still allow the user to choose the ionisation energy for specific purposes. In particular, for magnetic sector instruments the ionisation energy can be lowered to ca. 15 eV. This allows EI spectra with a high proportion of molecular information to be obtained (Fig. 2-115) (*low voltage ionisation*). The technique has decreased in importance since the introduction of chemical ionisation.

The process of electron impact ionisation can be explained by a wave or a particle model. The current theory is based on the interaction between the energy-rich electron beam with the outer electrons of a molecule. Energy absorption initially leads to the formation of a molecular ion M^+ by loss of an electron. The excess energy causes excitation in the rotational and vibrational energy levels of this radical cation. The subsequent processes of fragmentation depend on the amount of excess energy and the capacity of the molecule for internal stabilisation. The concept of localised charge according to Budzikiewicz empirically describes the fragmentations. The concept was developed from the observation that bonds near heteroatoms (N, O, S, Fig. 2-116) or π electron systems are cleaved preferentially in molecular ions. This is attributed to the fact that a positive or negative charge is stabilised by an electronegative structure element in the molecule or one favouring mesomerism. Bond breaking can be predicted by subsequent electron migrations or rearrangement. These types of process include α-cleavage, allyl cleavage, benzyl cleavage and the McLafferty rearrangement (see also Section 3.2.5).

The energy necessary for ionisation of organic molecules is lower than the effective applied energy of 70 eV and is usually less than 15 eV (Table 2-31). The EI operation

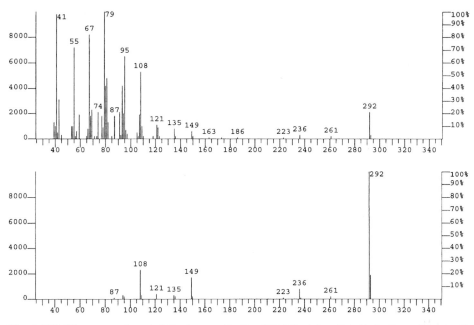

Fig. 2-115 EI spectra of methyl linolenate, $C_{17}H_{29}COOCH_3$ (after Spiteller). Recording conditions: direct inlet, above 70 eV, below 17 eV

Table 2-31 First ionisation potentials [eV] of selected substances

Helium	24.6	Pentane	10.34
Nitrogen	15.3	Nitrobenzene	10.18
Carbon dioxide	13.8	Benzene	9.56
Oxygen	12.5	Toluene	9.18
Propane	11.07	Chlorobenzene	9.07
1-Chloropropane	10.82	Propylamine	8.78
Butane	10.63	Aniline	8.32

of all MS instruments at the high ionisation of 70 eV was established with regard to sensitivity (ion yield) and the comparability of the mass spectra obtained.

Assuming a constant substance stream into an ion source, Fig. 2-117 shows the change in intensity of the signal with increasing ionisation energy. The steep rise of the signal intensity only begins when the ionisation potential (IP) is reached. Low measurable intensities just below it are produced as a result of the inhomogeneous composition of the electron beam. Generally the increase in signal intensity continues with increasing ionisation energy until a plateau is reached. A further increase in the ionisation energy is now indicated by a slight decrease in the signal intensity. An electron with an energy of 50 eV has a velocity of 4.2×10^6 m/s and crosses a molecular diameter of a few Ångstroms in ca. 10^{-16} s!

Further increases in the signal intensity are therefore not achieved via the ionisation energy with beam instruments, but by using measures to increase the density and the

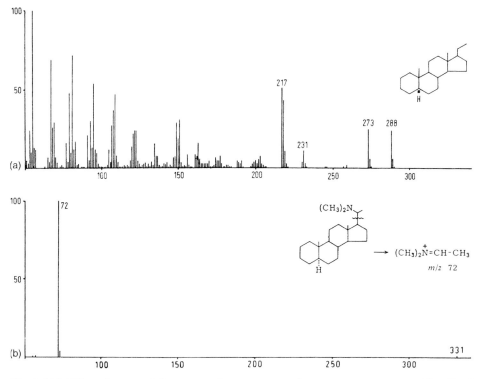

Fig. 2-116 Effect of structural features on the appearance of mass spectra – concept of localised charge (after Budzikiewicz).
(A) Mass spectrum of 5α-pregnane
(B) Mass spectrum of 20-dimethylamino-5α-pregnane. The α-cleavage of the amino group dominates in the spectrum; information on the structure of the sterane unit is completely absent!

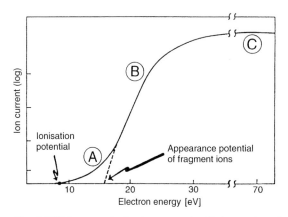

Fig. 2-117 Increase in the ion current with increasing electron energy (after Frigerio).
(A) Threshold region after reaching the appearance potential; molecular ions are mainly produced here
(B) Build-up region with increasing production of fragment ions
(C) Routine operation, stable formation of fragment ions

dispersion of the electron beam. The application of pairs of magnets to the ion source can be used, for example.

The standard ionisation energy of 70 eV which has been established for many years is also aimed at making mass spectra comparable. At an ionisation energy of 70 eV energy, which is in excess of that required for ionisation to M^+, remains as excess energy in the molecule, assuming maximum energy transfer. As a result fragmentation reactions occur and lead to an immediate decrease in the concentration of M^+ ions in the ion source. At the same time stable fragment ions are increasingly formed.

The fragmentation and rearrangement processes are now extensively known. They serve as fragmentation rules for the manual interpretation of mass spectra and thus for the identification of unknown substances. Each mass spectrum is the quantitative analysis by the analyser system of the processes occurring during ionisation. It is recorded as a line diagram.

Introduction

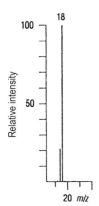

Learning how to identify a simple molecule from its electron-ionization (EI) mass spectrum is much easier than from other types of spectra. The mass spectrum shows the mass of the molecule and the masses of pieces from it. Thus the chemist does not have to learn anything new – the approach is similar to an arithmetic brain-teaser. Try one and see.

In the bar-graph form of a spectrum the abscissa indicates the mass (actually m/z, the ratio of mass to the number of charges on the ions employed) and the ordinate indicates the relative intensity. If you need a hint, remember the atomic weigths of hydrogen and oxygen are 1 and 16 respectively."

Prof. McLafferty, Interpretation of Mass Spectra (1993)

Fig. 2-118 McLafferty's unknown spectrum

What does a mass spectrum mean? In the graphical representation the mass to charge ratio m/z is plotted along the horizontal line. As ions with unit charge are generally involved in GC/MS with a few exceptions (e.g. polyaromatic hydrocarbons), this axis is generally taken as the mass scale and gives the molar mass of an ion. The intensity scale shows the frequency of occurrence of an ion under the chosen ionisation conditions. The scale is usually given both in percentages relative to the base peak (100 % intensity) and in measured intensity values (counts). As neutral particles are lost in the fragmentation or rearrangement of a molecular ion M^+ and cannot be detected by the analyser, the mass of these neutral particles is deduced from the difference between the fragment ions and the molecular ion (or precursor fragments) (Fig. 2-119).

In EI at 70 eV the extent of the fragmentation reactions observed for most organic compounds is independent of the construction of the ion source. For building up libraries of spectra the comparability of the mass spectra produced is thus ensured. All commercially available libraries of mass spectra are run under these standard conditions and allow the fragmentation pattern of an unknown substance to be compared with the spectra available in the library (see Section 3.2.4).

The Time Aspect in the Formation and Determination of Ions in Mass Spectrometry

- Formation of the molecular ion M$^+$ (EI) — 10^{-12} s
- Fragmentation reactions finished — 10^{-9} s
- Rearrangement reactions finished — 10^{-6}–10^{-7} s
- Lifetime of metastable ions — 10^{-3}–10^{-6} s

Flight times of ions:
- magnetic sector analyser — 10^{-5} s
- quadrupole analyser — 10^{-4} s
- ion trap analyser (storage times) — 10^{-2}–10^{-6} s

Types of Ions in Mass Spectrometry

- Molecular ion:
  The unfragmented positive or negatively charged ion with a mass equal to the molecular mass and of a radical nature because of the unpaired electron.

- Quasimolecular ion:
  Ions associated with the molecular mass which are formed through chemical ionisation e.g. as $(M + H)^+$, $(M - H)^+$ or $(M - H)^-$ and are not radicals.

- Adduct ions:
  Ions which are formed through addition of charged species e.g. $(M + NH_4)^+$ through chemical ionisation with ammonia as the CI gas.

- Fragment ions:
  Ions formed by cleavage of one or more bonds.

- Rearrangement products:
  Ions which are formed following bond cleavage and migration of an atom (see McLafferty rearrangement).

- Metastable ions:
  Ions (m_1) which lose neutral species (m_2) during the time of flight through the magnetic sector analyser and are detected with mass $m^* = (m_2)^2/m_1$.

- Base ion:
  This ion gives the highest signal (100%) (base peak) in a mass spectrum.

2.3.2.2 Chemical Ionisation

In electron impact ionisation (EI) molecular ions M$^+$ are first produced through bombardment of the molecule M by high energy electrons (70 eV). The high excess energy in M$^+$ (the ionisation potential of organic molecules is below 15 eV) leads to unimolecular fragmentation into fragment ions (F1$^+$, F2$^+$, ...) and uncharged species.

The EI mass spectrum shows the fragmentation pattern. The nature (m/z value) and frequency (intensity %) of the fragments can be read directly from the line spectrum. The loss of neutral particles is shown by the difference between the molecular ion and the fragments formed from it.

Which line in the EI spectrum is the molecular ion? Only a few molecules give dominant M$^+$ ions, e.g. aromatics and their derivatives, such as PCBs and dioxins.

2.3 Mass Spectrometry 145

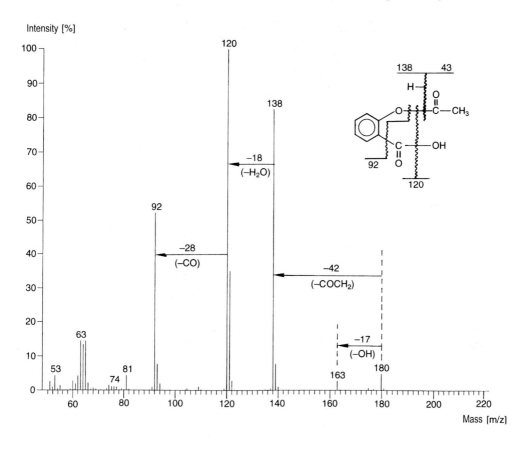

Fig. 2-119 A typical line spectrum in organic mass spectrometry with a molecular ion, fragment ions and the loss of neutral particles (acetylsalicylic acid, $C_9H_8O_4$, M 180)

The molecular ion is frequently only present with a low intensity and with the small quantities of sample applied, as is the case with GC/MS, can only be identified with difficulty among the noise (matrix), or it fragments completely and cannot be seen in the spectrum.

Figure 2-120, which shows the EI/CI spectra of the phosphoric acid ester Tolclofos-methyl, is an example of this. The base peak in the EI spectrum shows a Cl atom. Loss of a methyl group $(M - 15)^+$ gives m/z 250. Is m/z 265 the nominal molecular mass? The CI spectrum shows m/z 301 for a protonated ion so the nominal molecular mass could be 300 u. The isotope pattern of two Cl atoms is also visible.

How can both EI and CI spectra be completed? Obviously Tolclofos fragments completely in EI by loss of a Cl atom to m/z 265 as $(M - 35)^+$. With CI this fragmentation does not occur. The attachment of a proton retains the complete molecule with formation of the quasimolecular ion $(M + H)^+$.

The importance of EI spectra for identification and structure confirmation is due to the fragmentation pattern. All searches through libraries of spectra are based on EI spectra. With the introduction of Magnum CI (advanced chemical ionisation) for ion trap systems, a commercial CI library of spectra with more than 300 pesticides was first introduced by Finnigan.

The term chemical ionisation, unlike EI, covers all soft ionisation techniques which involve an exothermic chemical reaction in the gas phase mediated by a reagent gas

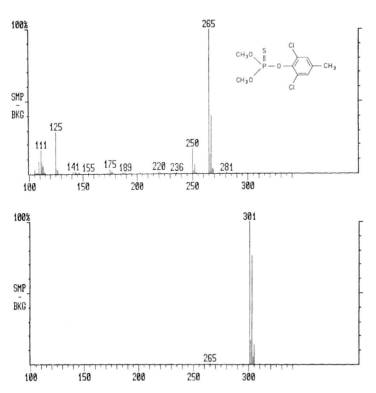

Fig. 2-120 EI and CI (NH$_3$) spectra of Tolclofos-methyl

and its reagent ions. Stable positive or negative ions are formed as products. Unlike the molecular ions of EI ionisation, the quasimolecular ions of CI are not radicals.

The principle of chemical ionisation was first described by Munson and Field in 1966. CI has now developed into a widely used technique for structural determination using sector instruments. Instead of an open, easily evacuated ion source, a closed ion volume is necessary for carrying out chemical ionisation. In the high vacuum environment of the ion source, a reagent gas pressure of ca. 1 Torr must be maintained to achieve the desired CI reactions. Depending on the construction of the instrument, either changing the ion volume, an expensive change of the whole ion source or only a software switch is necessary (Fig. 2-121). Through the straightforward technical realisation in the case of combination ion sources and through the broadening of the use of ion trap mass spectrometers, CI has now become established in residue and environmental analysis, even for routine methods.

Chemical ionisation uses considerably less energy for ionising the molecule M. CI spectra therefore have fewer or no fragments and thus generally give important information on the molecule itself.

The use of chemical ionisation is helpful in structure determination, confirmation or determination of molecular weights, and also in the determination of significant substructures. Additional selectivity can be introduced into mass spectrometric detection by using the CI reaction of certain reagent gases, e. g. the detection of active substances with a transparent hydrocarbon matrix. Analyses can be quantified selec-

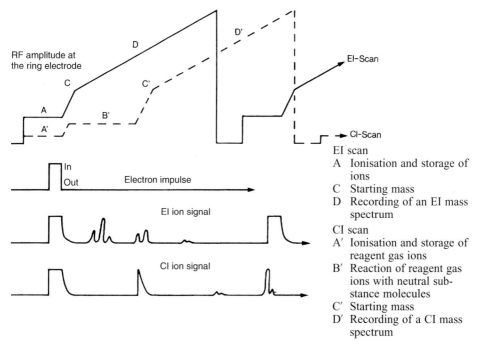

EI scan
A   Ionisation and storage of ions
C   Starting mass
D   Recording of an EI mass spectrum

CI scan
A′ Ionisation and storage of reagent gas ions
B′ Reaction of reagent gas ions with neutral substance molecules
C′ Starting mass
D′ Recording of a CI mass spectrum

Fig. 2-121 Switching between the EI and CI scan functions in the case of an ion trap analyser with internal ionisation (Finnigan)

tively, with high sensitivity and unaffected by the low molecular weight matrix, by the choice of a quantitation mass in the upper molecular weight range. The spectrum of analytical possibilities with CI is not limited to the basic reactions described briefly here. Furthermore, it opens up the whole field of chemical reactions in the gas phase.

The Principle of Chemical Ionisation

In CI two reaction steps are always necessary. In the *primary reaction* a stable cluster of reagent ions is produced from the reagent gas through electron bombardment. The composition of the reagent gas cluster is typical for the gas used. The cluster formed usually shows up on the screen for adjustment.

In the *secondary reaction* the molecule M in the GC eluate reacts with the ions in the reagent gas cluster. The ionic reaction products are detected and displayed as the CI spectrum. It is the secondary reaction which determines the appearance of the spectrum. Only exothermic reactions give CI spectra. In the case of protonation this means that the proton affinity PA of M must be higher than that of the reagent gas PA(R) (Fig. 2-122, Table 2-32). Through the choice of the reagent gas R, the quantity of energy transferred to the molecule M and thus the degree of possible fragmentation and the question of selectivity can be controlled. If PA(R) is higher than PA(M), no protonation occurs. When a nonspecific hydrocarbon matrix is present, this leads to transparency of the background, while active substances, such as plant protection agents, appear with high signal/noise ratios.

For chemical ionisation many types of reaction can be used analytically. In gas phase reactions, not only positive, but also negative ions can be formed. GC/MS systems currently commercially available usually only detect positive ions (positive chemical ionisation, PCI). To detect negative ions (negative chemical ionisation, NCI), special equipment to reverse the polarity of the analyser potential and a multiplier with a conversion dynode are required. Specially equipped instruments also allow the simultaneous detection of positive and negative ions produced by CI. The reversal of polarity during scanning (pulsed positive ion negative ion chemical ionisation, PPI-NICI) produces two complementary data files from one analysis.

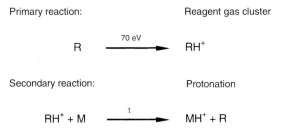

Fig. 2-122 Primary and secondary reactions in protonation

2.3 Mass Spectrometry

Table 2-32 Proton affinities of some simple compounds

Aliphatic amines

NH_3	857	n-Pr_2NH	951
$MeNH_2$	895	i-Pr_2NH	957
$EtNH_2$	907	n-Bu_2NH	955
n-$PrNH_2$	913	i-Bu_2NH	956
i-$PrNH_2$	917	s-Bu_2NH	965
n-$BuNH_2$	915	Me_3N	938
i-$BuNH_2$	918	Et_3N	966
s-$BuNH_2$	922	n-Pr_3N	976
t-$BuNH_2$	925	n-Bu_3N	981
n-Amyl-NH_2	918	Me_2EtN	947
Neopentyl-NH_2	920	$MeEt_2N$	957
t-Amyl-NH_2	929	Et_2-n-PrN	970
n-Hexyl-NH_2	920	Pyrrolidine	938
Cyclohexyl-NH_2	925	Piperidine	942
Me_2NH	922	N-methylpyrrolidine	952
$MeEtNH$	930	N-methylpiperidine	956
Et_2NH	941	$Me_3Si(CH_2)_3NMe_2$	966

Oxides and sulfides

H_2O	723	n-Bu_2O	852
MeOH	773	i-PrO-t-Bu	873
EtOH	795	n-$Pentyl_2O$	858
n-PrOH	800	Tetrahydrofuran	834
t-BuOH	815	Tetrahydropyran	839
Me_2O	807	H_2S	738
MeOEt	844	MeSH	788
Et_2O	838	Me_2S	839
i-PrOEt	850	MeSEt	851
n-Pr_2O	848	Et_2S	859
i-Pr_2O	861	i-Pr_2S	875
t-BuOMe	852	H_2Se	742

Disubstituted alkanes

$NH_2CH_2CH_2NH_2$	947	$Me_2N(CH_2)_6NMe_2$	1 041
$NH_2CH_2CH_2CH_2NH_2$	973	$NH_2CH_2CH_2OMe$	933
$NH_2CH_2CH_2CH_2CH_2NH_2$	995	$NH_2CH_2CH_2CH_2OH$	952
$NH_2(CH_2)_5NH_2$	986	$NH_2(CH_2)_6OH$	966
$NH_2(CH_2)_6NH_2$	989	$NH_2CH_2CH_2CH_2F$	928
$Me_2NCH_2CH_2NMe_2$	996	$NH_2CH_2CH_2CH_2Cl$	928
$Me_2NCH_2CH_2CH_2NH_2$	1 002	$OHCH_2CH_2CH_2CH_2OH$	886
$Me_2NCH_2CH_2CH_2NMe_2$	1 012	2,4-Pentanedione	886
$Me_2N(CH_2)_4NMe_2$	1 028		

Substituted alkylamines and alcohols

$CH_2FCH_2NH_2$	890	CF_3NMe_2	815
$CHF_2CH_2NH_2$	871	CHF_2CH_2OH	755
$CF_3CH_2NH_2$	850	CF_3CH_2OH	731
$(CF_3)_3CNH_2$	800	CCl_3CH_2OH	760
$CF_3CH_2NHCH_3$	880	Piperazine	936
$CF_3CH_2CH_2NH_2$	885	1,4-Dioxan	811
$CF_3CH_2CH_2CH_2NH_2$	900	Morpholine	915
$CH_2FCH_2CH_2NH_2$	914		

Table 2-32 (continued)

Unsaturated amines and anilines

$CH_2=CHCH_2NH_2$	905	m-$CH_3OC_6H_4NH_2$	906
Cyclo-$C_3H_5NH_2$	899	p-$CH_3OC_6H_4NH_2$	899
$CH_2=C(CH_3)CH_2NH_2$	912	p-$ClC_6H_4NH_2$	876
$HC\equiv CCH_2NH_2$	884	m-$ClC_6H_4NH_2$	872
$(CH_2=CCH_3CH_2)_3N$	964	m-$FC_6H_4NH_2$	870
$(CH_2=CHCH_2)_3N$	958	C_6H_5NHMe	912
$HC\equiv CCH_2)_3N$	916	$C_6H_5NMe_2$	935
$C_6H_5NH_2$	884	$C_6H_5CH_2NH_2$	918
m-$CH_3C_6H_4NH_2$	896	$C_6H_5CH_2NMe_2$	953
p-$CH_3C_6H_4NH_2$	896		

Other N, O, P, S compounds

Aziridine	902	$CH_2=CHCN$	802
Me_2NH	922	ClCN	759
N-Methylaziridine	926	BrCN	770
Me_3N	938	CCl_3CN	760
MeNHEt	930	CH_2ClCN	773
2-Methylaziridine	916	CH_2ClCH_2CN	795
Pyridine	921	$NCCH_2CN$	757
Piperidine	942	i-PrCN	819
MeCH=NEt	931	n-BuCN	818
n-PrCH=NEt	942	cyclo-C_3H_6CN	824
$Me_2C=NEt$	959	Ethylene oxide	793
HCN	748	Oxetane	823
CH_3NH_2	895	CH_2O	741
MeCN	798	MeCHO	790
EtCN	806	$Me_2C=O$	824
n-PrCN	810	Thiirane	818

Carbonyl compounds, iminoethers and hydrazines

EtCHO	800	CF_3CO_2-n-Bu	782
n-PrCHO	809	$HCO_2CH_2CF_3$	767
n-BuCHO	808	$NCCO_2Et$	767
i-PrCHO	808	$CF_3CO_2CH_2CH_2F$	764
Cyclopentanone	835	$(MeO)_2C=O$	837
HCO_2Me	796	HCO_2H	764
HCO_2Et	812	$MeCO_2H$	797
HCO_2-n-Pr	816	$EtCO_2H$	808
HCO_2-n-Bu	818	FCH_2CO_2H	781
$MeCO_2Me$	828	$ClCH_2CO_2H$	779
$MeCO_2Et$	841	CF_3CO_2H	736
$MeCO_2$-n-Pr	844	n-PrNHCHO	878
CF_3CO_2Me	765	Me_2NCHO	888
CF_3CO_2Et	777	$MeNHNH_2$	895
CF_3CO_2-n-Pr	781		

Table 2-32 (continued)

Substituted pyridines

Pyridine	921	4-CF$_3$-pyridine	890
4-Me-pyridine	935	4-CN-pyridine	880
4-Et-pyridine	939	4-CHO-pyridine	900
4-t-Bu-pyridine	945	4-COCH$_3$-pyridine	909
2,4-diMe-pyridine	948	4-Cl-pyridine	910
2,4-di-t-Bu-pyridine	967	4-MeO-pyridine	947
4-Vinylpyridine	933	4-NH$_2$-pyridine	961

Bases weaker than water

HF	468	CO$_2$	530
H$_2$	422	CH$_4$	536
O$_2$	423	N$_2$O	567
Kr	424	CO	581
N$_2$	475	C$_2$H$_6$	551
Xe	477		

All data in [kJ/mol] calculated for proton affinities PA(M) at 25 °C corresponding to the reaction M + H$^+$ → MH$^+$ (after Aue, Bowers 1979)

1. Positive Chemical Ionisation

Essentially four types of reaction contribute to the formation of positive ions. As in all CI reactions, reaction partners meet in the gas phase and form a transfer complex M · R$^+$. In the following types of reaction the transfer complex is either retained or reacts further.

Protonation

Protonation is the most frequently used reaction in positive chemical ionisation. Protonation leads to the formation of the quasimolecular ion (M + H)$^+$, which can then undergo fragmentation:

M + RH$^+$ → MH$^+$ + R

Normally methane, water, methanol, isobutane or ammonia are used as protonating reagent gases (Table 2-33). Methanol occupies a middle position with regard to fragmentation and selectivity. Methane is less selective and is designated a hard CI gas. Isobutane and ammonia are typical soft CI gases.

The CI spectra formed through protonation show the quasimolecular ions (M + H)$^+$. Fragmentations start with this ion. For example, loss of water shows up as M-17 in the spectrum, formed through (M + H)$^+$–H$_2$O!

The existence of the quasimolecular ion is often indicated by low signals from addition products of the reagent gas. In the case of methane, besides (M + H)$^+$, (M + 29)$^+$ and (M + 41)$^+$ appear (see methane), and for ammonia, besides (M + H)$^+$, (M + 18)$^+$ with varying intensity (see ammonia).

Table 2-33 Reagent gases for proton transfer

Gas	Reagent ion	PA [kJ/mol]
H_2	H_3^+	422
CH_4	CH_5^+	527
H_2O	H_3O^+	706
CH_3OH	$CH_3OH_2^+$	761
$i\text{-}C_4H_{10}$	$t\text{-}C_4H_9^+$	807
NH_3	NH_4^+	840

Hydride Abstraction

In this reaction a hydride ion (H^-) is transferred from the substance molecule to the reagent ion:

$$M + R^+ \rightarrow RH + (M-H)^+$$

This process is observed, for example, in the use of methane when the C_2H_5 ion (m/z 29) contained in the methane cluster abstracts hydride ions from alkyl chains.

With methane as the reagent gas, both protonation and hydride abstraction can occur, depending on the reaction partner M. The quasimolecular ion obtained is either $(M-H)^+$ or $(M+H)^+$. In charge exchange reactions M^+ is also formed.

Charge Exchange

The charge exchange reaction gives a radical molecular ion with an odd number of electrons as in electron impact ionisation. Accordingly, the quality of the fragmentation is comparable to that of an EI spectrum. The extent of fragmentation is determined by the ionisation potential IP of the reagent gas.

$$M + R^+ \rightarrow R + M^+$$

The ionisation potentials of most organic compounds are below 15 eV. Through the choice of reagent gas it can be controlled whether only the molecular ion appears in the spectrum or whether, and how extensively fragmentation occurs. In the extreme case spectra similar to those with EI are obtained. Common reagent gases for ionisation by charge exchange are benzene, nitrogen, carbon monoxide, nitric oxide or argon (Table 2-34).

Table 2-34 Reagent gases for charge exchange reactions

Gas	Reagent ion	IP [eV]
C_6H_6	$C_6H_6^+$	9.3
Xe	Xe^+	12.1
CO_2	CO_2^+	13.8
CO	CO^+	14.0
N_2	N_2^+	15.3
Ar	Ar^+	15.8
He	He^+	24.6

In the use of methane, charge exchange reactions as well as protonations can be observed, in particular for molecules with low proton affinities.

Adduct Formation

If the transition complex described above does not dissociate, the adduct is visible in the spectrum:

$$M + R^+ \rightarrow (M + R)^+$$

This effect is seldom made use of in GC/MS analysis, but must be taken into account on evaluating CI spectra. The enhanced formation of adducts is always observed with intentional protonation reactions where differences in the proton affinity of the participating species are small. High reagent gas pressure in the ion source favours the effect by stabilising collisions.

Frequently an $(M + R)^+$ ion is not immediately recognised, but can give information which is as valuable as that from the quasimolecular ion formed by protonation. Cluster ions of this type nevertheless sometimes make interpretation of spectra more difficult, particularly when the transition complex does not lose immediately recognisable neutral species.

2. Negative Chemical Ionisation

Negative ions are also formed during ionisation in mass spectrometry even under EI conditions, but their yield is so extremely low that it is of no use analytically. The intentional production of negative ions can take place by addition of thermal electrons (analogous to an ECD), by charge exchange or by extraction of acidic hydrogen atoms.

Charge Transfer

The ionisation of the sample molecule is achieved by the transfer of an electron between the reagent ion and the molecule M

$$M + R^- \rightarrow M^- + R$$

The reaction can only take place if the electron affinity EA of the analyte M is greater than that of the electron donor R.

$$EA(M) > EA(R)$$

In practical analysis charge transfer to form negative ions is less important (unlike the formation of positive ions by charge exchange, see above).

Proton Abstraction

Proton transfer in negative chemical ionisation can be understood as proton abstraction from the sample molecule. In this way all substances with acidic hydrogens, e.g. alcohols, phenols or ketones, can undergo soft ionisation.

$$M + R^- \rightarrow (M-H)^- + RH$$

Proton abstraction only occurs if the proton affinity of the reagent gas ion is higher than that of the conjugate base of the analyte molecule. A strong base, e.g. OH^-, is used as the reagent for ionisation. Substances which are more basic than the reagent are not ionised.

Reagent gases and organic compounds were arranged in order of gas phase acidity by Bartmess and McIver in 1979. The order corresponds to the reaction enthalpy of the dissociation of their functional groups into a proton and the corresponding base. Table 2-35 can be used for controlling the selectivity via proton abstraction by choosing suitable reagent gases.

Extensive fragmentation reactions can be excluded by proton abstraction. The energy released in the exothermic reaction is essentially localised in the new compound RH. The anion formed does not contain any excess energy for extensive fragmentation.

Reagent Ion Capture

The capture of negative reagent gas ions was described in the early 1970s by Manfred von Ardenne and coworkers for the analysis of long-chain aliphatic hydrocarbons with hydroxyl ions.

$$M + R^- \rightarrow M \cdot R^-$$

Besides associative addition, with weak bases adduct formation can lead to a new covalent bond. The ions formed are more stable than the comparable association products.

Substitution reactions, analogous to an S_N2-substitution in solution, occur more frequently in the gas phase because of poor solvation and low activation energy. Many aromatics give a peak at $(M + 15)^-$, which can be attributed to substitution of H by an O^- radical. Substitution reactions are also known for fluorides and chlorides. Fluoride is the stronger nucleophile and displaces chloride from alkyl halides.

Electron Capture

Electron capture with formation of negative ions is the NCI ionisation process most frequently used in analysis. There is a direct analogy with the behaviour of substances in ECD and the areas of use may also be compared. The recently named ECD-MS indicates the parallel mechanisms and applications. With negative chemical ionisation the lowest detection limits in organic mass spectrometry have been reached (Fig. 2-123). The detection of 100 ag octafluoronaphthalene corresponds to the detection of ca. 200 000 molecules!

Table 2-35 Scale of gas phase acidities (after Bartmess, McIver 1979)

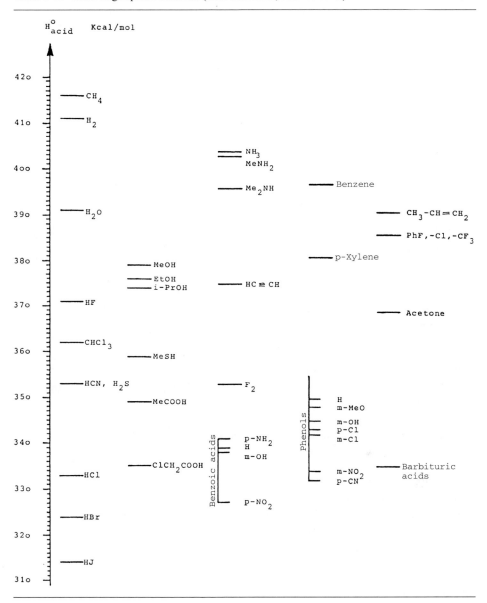

At the same energy electrons in the CI plasma have a much higher velocity (mobility) than those of the heavier positive reagent ions.

$$E = m/2 \cdot v^2 \qquad m(e^-) \ = 9.12 \cdot 10^{-28} \text{ g}$$
$$m(CH_5^+) = 2.83 \cdot 10^{-22} \text{ g}$$

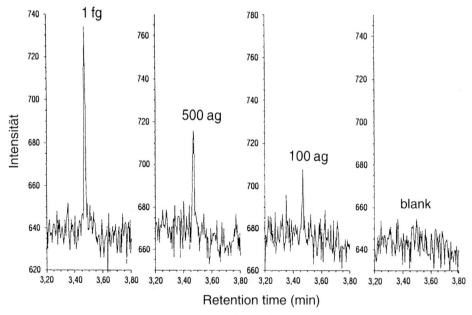

Fig. 2-123 GC/MS detection of traces of 10^{-15} to 10^{-16} g octafluoronaphthalene by NCI detection of the molecular ion m/z 272 (after McLafferty, and Michnowicz 1992)

Electron capture as an ionisation method is 100–1000 times more sensitive than ion/molecule reactions limited by diffusion. For substances with high electron affinities, higher sensitivities can be achieved than with positive chemical ionisation. NCI permits the detection of trace components in complex biological matrices (Fig. 2-124). Substances which have a high NCI response typically have a high proportion of halogen or nitro groups. In practice it has been shown that from ca. 5–6 halogen atoms in the molecule detection with NCI gives a higher specific response than that using EI (Fig. 2-125). For this reason, in analysis of polychlorinated dioxins and furans, although chlorinated, EI is the predominant method, in particular in the detection of 2,3,7,8-TCDD in trace analysis.

Another feature of NCI measurements is the fact that, like ECD, the response depends not only on the number of halogen atoms, but also on their position in the molecule. Precise quantitative determinations are therefore only possible with defined reference systems via the determination of specific response factors.

The key to sensitive detection of negative ions lies in the production of a sufficiently high population of thermal electrons. The extent of formation of M^- at sufficient electron density depends on the electron affinity of the sample molecule, the energy spectrum of the electron population and the frequency with which molecular anions collide with neutral particles and become stabilised (collision stabilisation). Even with an ion trap analyser, using external ionisation additions to give M^- can be utilised analytically. The storage of electrons in the ion trap itself is not possible because of their low mass (internal ionisation).

2.3 Mass Spectrometry

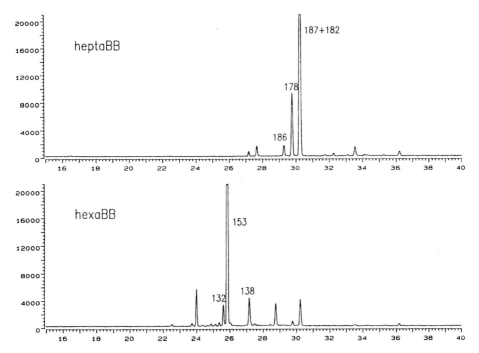

Fig. 2-124 Detection of heptabromobiphenylene (above) and hexabromobiphenylene (below) in human milk by NCI (after Fürst 1994)

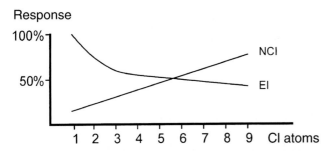

Fig. 2-125 Response dependence in the EI and NCI modes of the increasing number of Cl atoms in the molecule. The decrease in the response in the EI mode with increasing Cl content is caused by splitting of the overall signal by individual isotopic masses. In the NCI mode the response increases with increasing Cl content through the increase in the electronegativity for electrons capture (analogous to an ECD)

For residue analysis, derivatisation with perfluorinated reagents (e.g. heptafluorobutyric anhydride, perfluorobenzoyl chloride) in association with NCI is becoming important. Besides being easier to chromatograph, the compounds concerned have higher electron affinities allowing sensitive detection.

3. Reagent Gas Systems

Methane

Methane is one of the longest known and best studied reagent gases. As a hard reagent gas it has now been replaced by softer ones in many areas of analysis.

The reagent gas cluster of methane (Fig. 2-126) is formed by a multistep reaction, which gives two dominant reagent gas ions with m/z 17 and 29, and in lower intensity an ion with m/z 41.

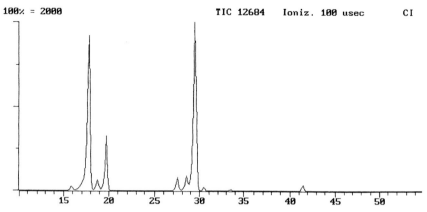

Fig. 2-126 Reagent gas cluster with methane (Finnigan)

CH_4 at 70 eV $\rightarrow$ CH_4^+, CH_3^+, CH_2, CH^+	and others		
$CH_4^+ + CH_4$ $\rightarrow$ $CH_5^+ + CH_3$	m/z 17	50%
$CH_3^+ + CH_4$ $\rightarrow$ $C_2H_5^+ + H_2$	m/z 29	48%
$CH_2^+ + CH_4$ $\rightarrow$ $C_2H_3^+ + H_2 + H$		
$C_2H_3^+ + CH_4$ $\rightarrow$ $C_3H_5^+ + H_2$	m/z 41	2%

Good CI conditions are achieved if a ratio of m/z 17 to m/z 16 of 10:1 is set up. Experience shows that the correct methane pressure is that at which the ions m/z 17 and 29 dominate in the reagent gas cluster and have approximately the same height with good resolution. The ion m/z 41 should also be recognisable with lower intensity.

Methane is mainly used as the reagent gas in protonation reactions, charge exchange processes (PCI), and in pure form or as a mixture with N_2O in the formation of negative ions (NCI). In protonation methane is a hard reagent gas. For substances with lower proton affinity methane frequently provides the final possibility of obtaining CI spectra. The adduct ions $(M + C_2H_5)^+ = (M + 29)^+$ and $(M + C_3H_5)^+ = (M + 41)^+$ formed by the methane cluster help to confirm the molecular mass interpretation.

Methanol

Because of its low vapour pressure, methanol is ideal for CI in ion trap instruments with internal ionisation. Neither pressure regulators nor cylinders or a long tubing system are required. The connection of a glass flask or a closed tube containing methanol directly on to the CI inlet is sufficient. In addition every laboratory has methanol available.

$(CH_3OH \cdot H)^+$ is formed as the reagent ion, which is adjusted to high intensity with good resolution (Fig. 2-127). The appearance of a peak at m/z 47 shows the dimer formed by loss of water (dimethyl ether), which is only produced at sufficiently high methanol concentrations. It does not function as a protonating reagent ion, but its appearance shows that the pressure adjustment is correct.

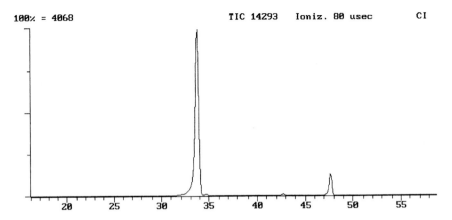

Fig. 2-127 Methanol as reagent gas (Finnigan)

Methanol is used exclusively for protonation. Because of its medium proton affinity, methanol allows a broad spectrum of classes of compounds to be determined. It is therefore suitable for a preliminary CI measurement of compounds not previously investigated. The medium proton affinity does not give any pronounced selectivity. However, substances with predominantly alkyl character remain transparent. Fragments have low intensities.

Water

For most mass spectroscopists water is a problematic substance. However, as a reagent gas, water has extraordinary properties. Because of the high conversion rate into H_3O^+ ions, and its low proton affinity, water achieves a high response for many compounds when used as the reagent gas. The spectra obtained usually have few fragments and concentrate the ion beam on a dominant ion.

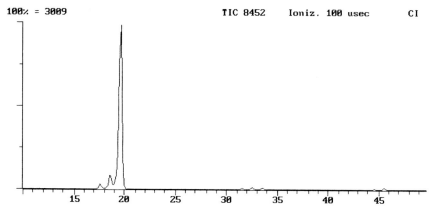

Fig. 2-128 Water as reagent gas (Finnigan)

When water is used as the reagent gas (Fig. 2-128), the intensity of the H_3O^+ ion should be as high as possible. With ion trap instruments with internal ionisation, no additional equipment is required. However a short tube length should be used for good adjustment. For instruments with high pressure CI ion sources, the use of a heated reservoir and completely heated inlet tubing are imperative.

The use of water is universal. In the determination of polyaromatic hydrocarbons considerable increases in response compared with EI detectors are found. Analytical procedures have even been published for nitroaromatics. Water can also be used successfully as a reagent gas for screening small molecules, e.g. volatile halogenated hydrocarbons (industrial solvents), as it does not interfere with the low scan range for these substances.

Isobutane

Like that of methane, chemical ionisation with isobutane has been known and well documented for years. The t-butyl cation (m/z 57) is formed in the reagent gas cluster and is responsible for the soft character of the reagent gas (Fig 2-129).

Isobutane is used for protonation reactions of multifunctional and polar compounds. Its selectivity is high and there is very little fragmentation. In practice, significant coating of the ion source through soot formation has been reported, which can even lead to dousing of the filament.

This effect depends on the adjustment and on the instrument. In such cases ammonia can be used instead.

Ammonia

To supply the CI system with ammonia a steel cylinder with a special reducing valve is necessary. Because of the aggressive properties of the gas, the entire tubing system

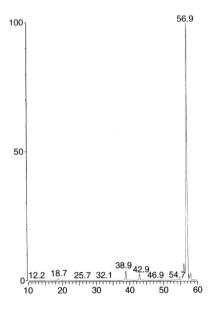

Fig. 2-129 iso-Butane as reagent gas (Finnigan GCQ)

must be made of stainless steel. In ion trap instruments only the ammonium ion NH_4^+ with mass m/z 18 is formed in the reagent gas cluster. At higher pressures adducts of the ammonium ion with ammonia $(NH_3)_n \cdot NH_4^+$ can be formed in the ion source.

Warning: very often the ammonium ion is confused with water. Freshly installed reagent gas tubing generally has an intense water background with high intensities at m/z 18 and 19 as H_2O^+/H_3O^+. Clean tubing and correctly adjusted NH_3 CI gas shows no intensity at mass m/z 19 (Fig 2-130)!

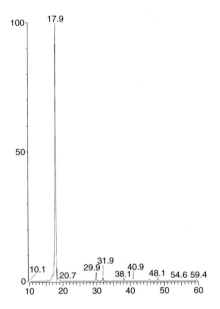

Fig. 2-130 Ammonia as reagent gas (Finnigan GCQ)

Ammonia is a very soft reagent gas for protonation. The selectivity is correspondingly high, which is made use of in the residue analysis of many active substances. Fragmentation reactions only occur to a small extent with ammonia CI.

Adduct formation with NH_4^+ occurs with substances where the proton affinity differs little from that of NH_3 and can be used to confirm the molecular mass interpretation. In these cases the formation and addition of higher $(NH_3)_n \cdot NH_4^+$ clusters is observed with instruments with threshold pressures of ca. 1 Torr. Interpretation and quantitation can thus be impaired with such compounds.

4. Aspects of Switching between EI and CI

Quadrupole and Magnetic Sector Instruments

To initiate CI reactions and to guarantee a sufficient conversion rate, an ion source pressure of ca. 1 Torr in an environment of 10^{-5}–10^{-7} Torr is necessary for beam instruments. For this, the EI ion source is replaced by a special CI source, which must have a gastight connection to the GC column, the electron beam and the ion exit in order to maintain the pressure in these areas.

Combination sources with mechanical devices for sealing the EI to the CI source have so far only proved successful with magnetic sector instruments. With the very small quadrupole sources, there is a significant danger of small leaks. As a consequence the response is below optimised sources and EI/CI mixed spectra may be produced.

Increased effort is required for conversion, pumping out and calibrating the CI source in beam instruments. Because of the high pressure, the reagent gas also leads to rapid contamination of the ion source and thus to additional cleaning measures in order to restore the original sensitivity of the EI system. Readily exchangeable ion volumes have been shown to be ideal for CI applications. This permits a high CI quality to be attained and, after a rapid exchange, unaffected EI conditions to be restored.

Ion Trap Instruments

Ion trap mass spectrometers can be used immediately for CI without conversion. Because of their mode of operation as storage mass spectrometers, only an extremely low reagent gas pressure is necessary for instruments with internal ionisation. The pressure is adjusted by means of a special needle valve which is operated at low leakage rates and maintains a partial pressure of only about 10^{-5} Torr in the analyser. The overall pressure of the ion trap analyser of about 10^{-4}–10^{-3} Torr remains unaffected by it. CI conditions thus set up give rise to the term low pressure CI. Compared to the conventional ion source used in high pressure CI, in protonation reactions, for example, a clear dependence of the CI reaction on the proton affinities of the reaction partners is observed. Collision stabilisation of the products formed does not occur with low pressure CI. This explains why CI-typical adduct ions are not formed here, which would confirm the identification of the (quasi)molecular ion (e.g. with methane besides $(M + H)^+$, also $M + 29$ and $M + 41$ are expected). The determination of ECD-active substances by electron capture is not possible with low pressure CI.

Switching between EI and CI modes in an ion trap analyser with internal ionisation takes place with a keyboard command or through the scheduled data acquisition sequence in automatic operations. All mechanical devices necessary in beam instruments are dispensed with completely. The ion trap analyser is switched to a CI scan function internally without effecting mechanical changes to the analyser itself.

The CI reaction is initiated when the reagent ions are made ready by changing the operating parameters and a short reaction phase has taken place in the ion trap analyser. The scan function used in the CI mode with ion trap instruments (see Fig. 2-121) clearly shows two plateaus which directly correspond to the primary and secondary reactions. After the end of the secondary reaction the product ions, which have been produced and stored, are determined by the mass scan and the CI spectrum registered. In spite of the presence of the reagent gas, typical EI spectra can therefore be registered in the EI mode. The desired chemical ionisation is made possible by simply switching to the CI operating parameters.

On using autosamplers it is therefore possible to switch alternately between EI and CI data acquisition and thus use both ionisation processes routinely in automatic operations. The danger of additional contamination by CI gas does not occur with ion trap instruments because of the extremely small reagent gas input and allows this mode of operation to be run without impairing the quality.

Ion trap instruments with external ionisation have an ion source with a conventional construction. Changing between EI and CI ionisation takes place by changing the ion volume. Chemical ionisation in the classical manner of high pressure CI is thus carried out and thus the formation of negative ions by electron capture (NCI) in association with an ion trap analyser is made possible (Fig. 2-131).

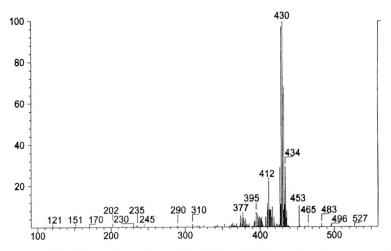

Fig. 2-131 NCI spectrum of octachlorobiphenyl (Finnigan GCQ). The typical fragmentation of electron impact ionisation is absent; the total ion current is concentrated by the addition of electrons in the molecular ion region

2.3.3 Measuring Techniques in GC/MS

In data acquisition by the mass spectrometer there is a difference between detection of the complete spectrum (full scan) and the recording of individual masses (SIM, selected ion monitoring; MID multiple ion detection). Particularly with continually operating spectrometers (ion beam instruments: magnetic sector MS, quadrupole MS) there are large differences between these two recording techniques with respect to sensitivity and information content. For spectrometers with storage facilities (ion storage: ion trap MS, ICR-MS) these differences are less strongly pronounced. Besides one-stage types of analyser (GC/MS), multistage mass spectrometers (GC/MS/MS) are playing an increasingly important role in residue analysis and structure determination. With the MS/MS technique (multidimensional mass spectrometry), which is now available in both beam instruments and ion storage mass spectrometers, much more analytical information can be obtained from a sample.

2.3.3.1 Detection of the Complete Spectrum (Full Scan)

The continuous recording of mass spectra (full scan) and the simultaneous determination of the retention time allow the identification of analytes by comparison with libraries of mass spectra. With beam instruments it should be noted that the sensitivity required for recording the spectrum depends on the efficiency of the ion source, the transmission through the analyser and, most particularly, on the dwell time of the ions. The dwell time per mass is given by the width of the mass scan (e.g. 50–550 u) and the scan rate of the chromatogram (e.g. 500 ms). From this a scan rate of 1000 masses/s is calculated. Each mass from the selected mass range is measured only once during a scan over a short period (here 1 ms/u, Fig. 2-132). All other ions formed from the substance in parallel in the ion source are masked during the mass scan (quadrupole as mass filter). Typical sensitivities for most compounds with benchtop quadrupole systems lie in the region of 1 ng of substance or less. Prolonging the scan time can increase the sensitivity of these systems for full scan operation (Fig. 2-132). However, there is an upper limit because of the rate of the chromatography. In practice

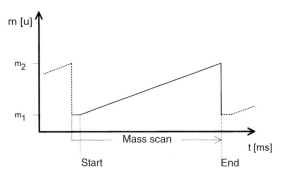

Fig. 2-132 Scan function of the quadrupole analyser: each mass between the start of the scan (m_1) and the end (m_2) is only registered once during the scan

for coupling with capillary gas chromatography, scan rates of 0.5–1 s are used. For quantitative determinations it should be ensured that the scan rate of the chromatographic peak is adequate in order to determine the area and height correctly (see also Section 3.3). The SIM/MID mode is usually chosen to increase the sensitivity and scan rates of quadrupole systems for this reason (see also Section 2.3.3.2).

In ion storage mass spectrometers all the ions produced on ionisation of a substance are detected in parallel. The mode of function is opposite to the filter character of beam instruments (Table 2-36). All the ions formed are first stored in the analyser. At the end of the storage phase the ions, sorted according to mass during the scan, are directed to the multiplier. This process can take place very rapidly (Fig. 2-133). The scan rates of ion trap mass spectrometers are higher than 11 000 masses/s. Typical sensitivities for the production of full spectra in ion trap mass spectrometers are 10 pg or even less.

Table 2-36 Duty cycle for ion trap and beam instruments

Scan range	Dwell time per mass		Quotient
	Ion trap	Beam instrument	
[u/s]	[s/u]	[s/u]	
1	0.83	1	0.8
3	0.82	0.33	2.5
10	0.79	0.1	8
30	0.71	0.033	21
100	0.52	0.01	52
300	0.3	0.0033	91

The longer dwell time per mass because of the system leads to the highest sensitivity in the recording of complete mass spectra with ion trap instruments. Compared with beam instruments an increase in the duty cycle is achieved, depending on the mass range, of up to a factor of 100 and higher.

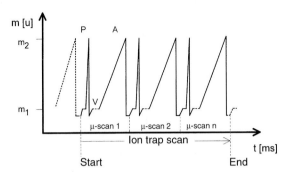

Fig. 2-133 Scan function of the ion trap analyser with internal ionisation: within an ion trap scan, several μ-scans (three μ-scans shown here) are carried out and their spectra added.
P = pre-scan
A = analytical scan
V = variable ionisation time (AGC, automatic gain control)
(With ion trap instruments with external ionisation V stands for the length of the storage phase, ion injection time)

> **Dwell times per ion in full scan data recording with ion trap and quadrupole MS**
>
> Mass range 50–550 u (500 masses wide), scan rate 500 ms
>
> | **Ion trap MS** | **Quadrupole MS** |
> |---|---|
> | Ion storage during ionisation. Storage time can vary, typically up to 25 ms at simultaneous storage of *all* ions formed. | $t = 500\ u/500\ ms$
 $= 1\ ms/u$ |
> | Detection of *all* stored ions | In a scan each type of ion is measured for only 1 ms; only a minute quantity of the ions formed are detected |

2.3.3.2 Recording Individual Masses (SIM/MID)

In the use of conventional mass spectrometers (beam instruments), the detection limit in the full scan mode is frequently insufficient for residue analysis because the analyser only has a very short dwell time per ion available during the scan. Additional sensitivity is achieved by dividing the same dwell time between a few selected ions by means of individual mass recording (SIM, MID) (Table 2-37, Fig. 2-134).

At the same time a higher scan rate can be chosen so that chromatographic peaks can be plotted more precisely. The SIM technique is used exclusively for quantifying data on known compounds, e.g. also in the area of sample screening.

Table 2-37 Dwell times per ion and relative sensitivity in SIM analysis (for beam instruments) at constant scan rates

Number of SIM ions	Total scan time[1] [ms]	Total voltage setting time[2] [ms]	Effective dwell time per ion[3] [ms]	Relative sensitivity[4] [%]
1	500	5	495	100
2	500	10	245	49
3	500	15	162	33
5	500	25	95	19
10	500	50	45	9
20	500	100	20	4
30	500	150	12	2
40	500	200	8	2
50	500	250	5	1
For comparison full scan:				
500	500	5	1	0.2

[1] The total scan time is determined by the necessary scan rate of the chromatogram and is held constant.
[2] Total voltage setting times are necessary in order to adjust the mass filter for the subsequent SIM masses. The actual times necessary can vary slightly depending on the type of instrument.
[3] Duty cycle/ion
[4] The relative sensitivity is directly proportional to the effective dwell time per ion.

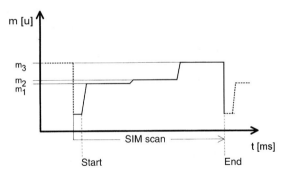

Fig. 2-134 SIM scan with a quadrupole analyser: the total scan time is divided here into the three individual masses m_1, m_2, and m_3 with correspondingly long dwell times

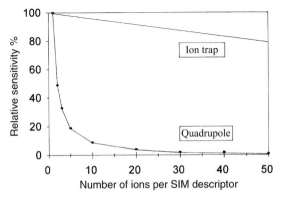

Fig. 2-135 Comparison of the SIS (ion trap analyser) and the SIM (quadrupole analyser) techniques based on the effective dwell time per ion (relative sensitivity)

The mode of operation of a GC/MS system as a mass-selective detector requires the selection of certain ions (fragments, molecular ions), so that the desired analytes can be detected selectively. Other compounds contained in the sample besides those chosen for analysis remain undetected. Thus the matrix present in large quantities in residue analysis is masked, as are analytes whose appearance is not expected or planned. In the choice of masses required for detection it is assumed that for three selective signals in the fragmentation pattern per substance a secure basis for a yes/no decision can be found in spite of variations in the retention times (SIM, selected ion monitoring, MID, multiple ion detection). Identification of substances by comparison with spectral libraries is no longer possible. The relative intensities of the selected ions serve as quality criteria (qualifiers) (1 ion – no criterion, 2 ions – 1 criterion, 3 ions – 3 criteria!). This process for detecting compounds can be affected by errors through shifts in retention times caused by the matrix. In residue analysis it is known that with SIM analysis false positive findings occur in ca. 10% of the samples. Recently positive SIM data have been safeguarded in the same way as positive results from classical GC detectors by running a complete mass spectrum of the analytes suspected. In many laboratories full scan safeguarding is already routine.

Gain in sensitivity using SIM/MID

A typical SIM data acquisition of 5 selected masses at a scan rate of 0.5 s is given as a typical example:

	Ion trap MS	**Quadrupole MS**
Dwell time per ion:	Identical ionisation procedure to that with full scan, however selective and parallel or sequential storage of the selected SIM ions.	At a scan time of 500 ms the effective dwell time per SIM mass is divided up as t = 500 ms/5 masses = 100 ms/mass.
Function:	The ion trap is filled exclusively with the ions with the selected masses. If the capacity of the ion trap is not used up completely, the storage phase ends after a given time (ms).	To measure an SIM mass the quadrupole spends 100 times longer on one mass compared with full scan, and thus permits a dwell time which is 100 times longer for the selected ions to be achieved.
Sensitivity:	The gain in sensitivity is most marked with matrix-containing samples, as the length of the storage phase still mainly depends on the appearance of the selected SIM ions in the sample and is not shortened by a high concentration of matrix ions.	Theoretically the sensitivity increases by a factor of 100. In practice for real samples a factor of 30–50 compared with full scan is achieved.
Consequences for trace analysis:	Ion trap systems already give very high sensitivity in the full scan mode. Samples with high concentrations of matrix and detection limits below the pg level require the SIM technique (MS/MS is recommended).	Quadrupole systems require the SIM mode to achieve adequate sensitivity.
Confirmation:	For 3 SIM masses by 3 intensity criteria (qualifiers), with MS/MS by means of the product ion spectra.	For 3 SIM masses by 3 intensity criteria (qualifiers), check of positive results after further concentration by the full scan technique or external confirmation with an ion trap instrument (Table 2-37).

SIM Set-up

1. Choice of column and program optimisation for optimal GC separation, paying particular attention to analytes with similar fragmentation patterns.
2. Full scan analysis of an average substance concentration to determine the selective ions (SIM masses, 2–3 ions/component); special matrix conditions are to be taken into account.
3. Determination of the retention times of the individual components.
4. Establishment of the data acquisition interval (time window) for the individual SIM descriptors.
5. Test analysis of a low standard (or better, a matrix spike) and possible optimisation (SIM masses, separation conditions).

Planning an analysis in the SIM/MID mode first requires a standard run in the full scan mode to determine both the retention times and the mass signals necessary for the SIM selection (Tables 2-38 and 2-39). As the gain in sensitivity achieved in individual mass recording with beam instruments is only possible on detection of a few ions, for the analysis of several compounds the group of masses detected must be adjusted. The more components there are to be detected, the more frequently and precisely must the descriptors be adjusted. Multicomponent analyses, such as the MAGIC-60 analysis with purge and trap (volatile halogenated hydrocarbons and BTEX, see Section 4.5), cannot be dealt with in the SIM mode.

The use of SIM analysis with ion trap mass spectrometers has also been developed. Through special control of the analyser (waveform ion isolation) during the ionisation phase only the preselected ions of analytical interest are stored (SIS, selective ion storage). This technique allows the detection of selected ions in ion storage mass spectrometers in spite of the presence of complex matrices or the co-elution of another component in high concentration. As the storage capacity of the ion trap analyser is only used for a few ions instead of for a full spectrum, extremely low detection limits are possible (<1 pg/component) and the usable dynamic range of the analyser is extended considerably. Unlike conventional SIM operations with beam instruments, the detection sensitivity only alters slightly with the number of selected ions using the ion trap SIM technique (Fig. 2-135). For the SIM technique the sensitivity depends almost exclusively on the ionisation time. The SIM technique with ion trap instruments is regarded as a necessary prerequisite for carrying out MS/MS detection.

As an SIM/MID analysis is limited to the detection of certain ions in small mass ranges and the comparability of a complete spectrum for a library search is not required, on tuning the analyser a mass-based optimisation to a transmission which is as high as possible is undertaken. Standard tuning aims to produce a balanced spectrum which corresponds to the data in a reference list for the reference substance FC43 (perfluorotributylamine), in order to guarantee good comparability of the spectra run with those in the library. The position of the standard does not need to be adhered to in SIM analysis as no comparisons are made between spectra, but the relative intensities, e.g. of isotope patterns, must be evaluated. In the optimisation of the ion source special attention should therefore be paid to the masses (or the mass range) involved in SIM data acquisition. The source and lens potentials should then be selected manually so that a nearby FC43 fragment or an ion produced by column bleed (GC temperature ca. 200 °C) can be detected with the highest intensity but good resolution. In this way a significant additional increase in sensitivity can be achieved with quadrupole analysers for the SIM mode.

Figure 2-138 shows the chromatogram of a PCB standard as the result of a typical SIM routine analysis. In this case two masses are chosen as SIM masses for each PCB isomer. The switching points of the individual descriptors are recognised as steps in the base line. The different base line heights arise as a result of the different contributions of the chemical noise to these signals. To control the evaluation, the substance signals can be represented as peaks in the expected retention time windows (Fig. 2-139). A deviation from the calibrated retention time (Fig. 2-139, right segment with the masses m/z 499.8 and 497.8) leads to a shift of the peak from the middle to the edge of the window and should be a reason for further checking. If qualifiers are

Table 2-38 m/z values for selected polycondensed aromatics and their alkyl derivatives (in the elution sequence for methylsilicone phases)

Benzene-d_6	92, 94	Pyrene	202
Benzene	77, 78	Methylfluoranthene	215, 216
Toluene-d_8	98, 100	Benzofluorene	215, 216
Toluene	91, 92	Phenylanthracene	252, 253, 254
Ethylbenzene	91, 106	Benzanthracene	228
Dimethylbenzene	91, 106	Chrysene-d_{12}	240
Methylethylbenzene	105, 120	Chrysene	228
Trimethylbenzene	105, 120	Methylchrysene	242
Diethylbenzene	105, 119, 134	Dimethylbenz[a]anthracene	239, 241, 256
Naphthalene-d_8	136	Benzo[b]fluoranthene	252
Naphthalene	128	Benzo[j]fluoranthene	252
Methylnaphthalene	141, 142	Benzo[k]fluoranthene	252
Azulene	128	Benzo[e]pyrene	252
Acenaphthene	154	Benzo[a]pyrene	252
Biphenyl	154	Perylene-d_{12}	264
Dimethylnaphthalene	141, 155, 156	Perylene	252
Acenaphthene-d_{10}	162, 164	Methylcholanthrene	268
Acenaphthene	152	Diphenylanthracene	330
Dibenzofuran	139, 168	Indeno[1,2,3-cd]pyrene	276
Dibenzodioxin	184	Dibenzanthracene	278
Fluorene	165, 166	Benzo[b]chrysene	278
Dihydroanthracene	178, 179, 180	Benzo[g,h,i]perylene	276
Phenanthrene-d_{10}	188	Anthanthrene	276
Phenanthrene	178	Dibenzo[a,l]pyrene	302
Anthracene	178	Coronene	300
Methylphenanthrene	191, 192	Dibenzo[a,i]pyrene	302
Methylanthracene	191, 192	Dibenzo[a,h]pyrene	302
Phenylnaphthalene	204	Rubicene	326
Dimethylphenanthrene	191, 206	Hexaphene	328
Fluoranthene	202	Benzo[a]coronene	350

Table 2-39 Main fragments and relative intensities for pesticides and some of their derivatives (DFG 1992)

Compound	Molar mass	Main fragment m/z (intensities)					
		1	2	3	4	5	6
Acephate	183	43 (100)	44 (88)	136 (80)	94 (58)	47 (56)	95 (32)
Alaclor	269	45 (100)	188 (23)	160 (18)	77 (7)	146 (6)	224 (6)
Aldicarb	190	41 (100)	86 (89)	58 (85)	85 (61)	87 (50)	44 (50)
Aldrin	362	66 (100)	91 (50)	79 (47)	263 (42)	65 (35)	101 (34)
Allethrin	302	123 (100)	79 (40)	43 (32)	81 (31)	91 (29)	136 (27)
Atrazine	215	43 (100)	58 (84)	44 (75)	200 (69)	68 (43)	215 (40)
Azinphos-methyl	317	77 (100)	160 (77)	132 (67)	44 (30)	105 (29)	104 (27)
Barban	257	51 (100)	153 (76)	87 (66)	222 (44)	52 (43)	63 (43)
Benzazolin methyl ester	257	170 (100)	134 (75)	198 (74)	257 (73)	172 (40)	200 (31)
Bendiocarb	223	151 (100)	126 (58)	166 (48)	51 (19)	58 (18)	43 (17)
Bromacil	260	205 (100)	207 (75)	42 (25)	70 (16)	206 (16)	162 (12)
Bromacil N-methyl derivative	274	219 (100)	221 (68)	41 (45)	188 (41)	190 (40)	56 (37)
Bromophos	364	331 (100)	125 (91)	329 (80)	79 (57)	109 (53)	93 (45)
Bromophos-ethyl	392	97 (100)	65 (35)	303 (32)	125 (28)	359 (27)	109 (27)

Table 2-39 (continued)

Compound	Molar mass	Main fragment m/z (intensities)					
		1	2	3	4	5	6
Bromoxynil methyl ether	289	289 (100)	88 (77)	276 (67)	289 (55)	293 (53)	248 (50)
Captafol	347	79 (100)	80 (42)	77 (28)	78 (19)	151 (17)	51 (13)
Captan	299	79 (100)	80 (61)	77 (56)	44 (44)	78 (37)	149 (34)
Carbaryl	201	144 (100)	115 (82)	116 (48)	57 (31)	58 (20)	63 (20)
Carbendazim	191	159 (100)	191 (57)	103 (38)	104 (37)	52 (32)	51 (29)
Carbetamid	236	119 (100)	72 (54)	91 (44)	45 (38)	64 (37)	74 (29)
Carbofuran	221	164 (100)	149 (70)	41 (27)	58 (25)	131 (25)	122 (25)
Chlorbromuron	292	61 (100)	46 (24)	62 (11)	63 (10)	60 (9)	124 (8)
Chlorbufam	223	53 (100)	127 (20)	51 (13)	164 (13)	223 (13)	70 (10)
cis-Chlordane	406	373 (100)	375 (84)	377 (46)	371 (39)	44 (36)	109 (36)
trans-Chlordane	406	373 (100)	375 (93)	377 (53)	371 (47)	272 (36)	237 (30)
Chlorfenprop-methyl	232	125 (100)	165 (64)	75 (46)	196 (43)	51 (43)	101 (37)
Chlorfenvinphos	358	81 (100)	267 (73)	109 (55)	269 (47)	323 (26)	91 (23)
Chloridazon	221	77 (100)	221 (60)	88 (37)	220 (35)	51 (26)	105 (24)
Chloroneb	206	191 (100)	193 (61)	206 (60)	53 (57)	208 (39)	141 (35)
Chlorotoluron	212	72 (100)	44 (29)	167 (28)	132 (25)	45 (20)	77 (11)
3-Chloro-4-methylaniline (GC degradation product of Chlorotoluron)	141	141 (100)	140 (37)	106 (68)	142 (36)	143 (28)	77 (25)
Chloroxuron	290	72 (100)	245 (37)	44 (31)	75 (21)	45 (19)	63 (16)
Chloropropham	213	43 (100)	127 (49)	41 (35)	45 (20)	44 (18)	129 (16)
Chlorpyrifos	349	97 (100)	195 (59)	199 (53)	65 (27)	47 (23)	314 (21)
Chlorthal-dimethyl	330	301 (100)	299 (81)	303 (47)	332 (29)	142 (26)	221 (24)
Chlorthiamid	205	170 (100)	60 (61)	171 (50)	172 (49)	205 (35)	173 (29)
Cinerin I	316	123 (100)	43 (35)	93 (33)	121 (27)	81 (27)	150 (27)
Cinerin II	360	107 (100)	93 (57)	121 (53)	91 (50)	149 (35)	105 (33)
Cyanazine	240	44 (100)	43 (60)	68 (60)	212 (48)	41 (47)	42 (34)
Cypermethrin	415	163 (100)	181 (79)	165 (68)	91 (41)	77 (33)	51 (29)
2,4-DB methyl ester	262	101 (100)	59 (95)	41 (39)	162 (36)	69 (28)	63 (25)
Dalapon	142	43 (100)	61 (81)	62 (67)	97 (59)	45 (59)	44 (47)
Dazomet	162	162 (100)	42 (87)	89 (79)	44 (73)	76 (59)	43 (53)
Demetron-S-methyl	230	88 (100)	60 (50)	109 (24)	142 (17)	79 (14)	47 (11)
Desmetryn	213	213 (100)	57 (67)	58 (66)	198 (58)	82 (44)	171 (39)
Dialifos	393	208 (100)	210 (31)	76 (20)	173 (17)	209 (12)	357 (10)
Di-allate	269	43 (100)	86 (62)	41 (38)	44 (25)	42 (24)	70 (19)
Diazinon	304	137 (100)	179 (74)	152 (65)	93 (47)	153 (42)	199 (39)
Dicamba methyl ester	234	203 (100)	205 (60)	234 (27)	188 (26)	97 (21)	201 (20)
Dichlobenil	171	171 (100)	173 (62)	100 (31)	136 (24)	75 (24)	50 (19)
Dichlofenthion	314	97 (100)	279 (92)	223 (90)	109 (67)	162 (53)	251 (46)
Dichlofluanid	332	123 (100)	92 (33)	224 (29)	167 (27)	63 (23)	77 (22)
2,4-D isooctyl ester	332	43 (100)	57 (98)	41 (76)	55 (54)	71 (41)	69 (27)
2,4-D methyl ester	234	199 (100)	45 (97)	175 (94)	145 (70)	111 (69)	109 (68)
Dichlorprop isooctyl ester	346	43 (100)	57 (83)	41 (61)	71 (48)	55 (47)	162 (41)
Dichlorprop methyl ester	248	162 (100)	164 (80)	59 (62)	189 (56)	63 (39)	191 (35)
Dichlorvos	220	109 (100)	185 (18)	79 (17)	187 (6)	145 (6)	47 (5)
Dicofol	368	139 (100)	111 (39)	141 (33)	75 (18)	83 (17)	251 (16)
o,p'-DDT	352	235 (100)	237 (59)	165 (33)	236 (16)	199 (12)	75 (12)
p,p'-DDT	352	235 (100)	237 (58)	165 (37)	236 (16)	75 (12)	239 (11)
Dieldrin	378	79 (100)	82 (32)	81 (30)	263 (17)	77 (17)	108 (14)
Dimethirimol methyl ether	223	180 (100)	223 (23)	181 (10)	224 (3)	42 (2)	109 (2)
Dimethoate	229	87 (100)	93 (76)	125 (56)	58 (40)	47 (39)	63 (33)

Table 2-39 (continued)

Compound	Molar mass	Main fragment m/z (intensities)					
		1	2	3	4	5	6
DNOC methyl ether	212	182 (100)	165 (74)	89 (69)	90 (57)	212 (48)	51 (47)
Dinoterb methyl ether	254	239 (100)	209 (41)	43 (36)	91 (35)	77 (33)	254 (33)
Dioxacarb	223	121 (100)	122 (62)	166 (46)	165 (42)	73 (35)	45 (31)
Diphenamid	239	72 (100)	167 (86)	165 (42)	239 (21)	152 (17)	168 (14)
Disulfoton	274	88 (100)	89 (43)	61 (40)	60 (39)	97 (36)	65 (23)
Diuron	232	72 (100)	44 (34)	73 (25)	42 (20)	232 (19)	187 (13)
Dodine	227	43 (100)	73 (80)	59 (52)	55 (47)	72 (46)	100 (46)
Endosulfan	404	195 (100)	36 (95)	237 (91)	41 (89)	24 (79)	75 (78)
Endrin	378	67 (100)	81 (67)	263 (59)	36 (58)	79 (47)	82 (41)
Ethiofencarb	225	107 (100)	69 (48)	77 (29)	41 (26)	81 (21)	45 (17)
Ethirimol	209	166 (100)	209 (17)	167 (14)	96 (12)	194 (4)	55 (2)
Ethirimol methyl ether	223	180 (100)	223 (23)	85 (14)	181 (12)	55 (10)	96 (9)
Etrimfos	292	125 (100)	292 (91)	181 (90)	47 (84)	153 (84)	56 (73)
Fenarimol	330	139 (100)	107 (95)	111 (40)	219 (39)	141 (33)	251 (31)
Fenitrothion	277	125 (100)	109 (92)	79 (62)	47 (57)	63 (44)	93 (40)
Fenoprop isooctyl ester	380	57 (100)	43 (94)	41 (85)	196 (63)	71 (60)	198 (59)
Fenoprop methyl ester	282	196 (100)	198 (89)	59 (82)	55 (36)	87 (34)	223 (31)
Fenuron	164	72 (100)	164 (27)	119 (24)	91 (22)	42 (14)	44 (11)
Flamprop-isopropyl	363	105 (100)	77 (44)	276 (21)	106 (18)	278 (7)	51 (5)
Flamprop-methyl	335	105 (100)	77 (46)	276 (20)	106 (14)	230 (12)	44 (11)
Formothion	257	93 (100)	125 (89)	126 (68)	42 (49)	47 (48)	87 (40)
Heptachlor	370	100 (100)	272 (81)	274 (42)	237 (33)	102 (33)	
Iodofenphos	412	125 (100)	377 (78)	47 (64)	79 (59)	93 (54)	109 (49)
Ioxynil isooctyl ether	483	127 (100)	57 (96)	41 (34)	43 (33)	55 (26)	37 (16)
Ioxynil methyl ether	385	385 (100)	243 (56)	370 (41)	127 (13)	386 (10)	88 (9)
Isoproturon	206	146 (100)	72 (54)	135 (28)	128 (29)	45 (28)	161 (25)
Jasmolin I	330	123 (100)	43 (52)	55 (34)	93 (25)	91 (24)	81 (23)
Jasmolin II	374	107 (100)	91 (69)	135 (69)	93 (67)	55 (66)	121 (58)
Lenacil	234	153 (100)	154 (20)	110 (15)	109 (15)	152 (13)	136 (10)
Lenacil N-methyl derivative	248	167 (100)	166 (45)	168 (12)	165 (12)	124 (9)	123 (6)
Lindane	288	181 (100)	183 (97)	109 (89)	219 (86)	111 (75)	217 (68)
Linuron	248	61 (100)	187 (43)	189 (29)	124 (28)	46 (28)	44 (23)
MCPB isooctyl ester	340	87 (100)	57 (81)	43 (62)	71 (45)	41 (42)	69 (29)
MCPB methyl ester	242	101 (100)	59 (70)	77 (40)	107 (25)	41 (22)	142 (20)
Malathion	330	125 (100)	93 (96)	127 (75)	173 (55)	158 (37)	99 (35)
Mecoprop isooctyl ester	326	43 (100)	57 (94)	169 (77)	41 (70)	142 (69)	55 (52)
Mecoprop methyl ester	228	169 (100)	143 (79)	59 (58)	141 (57)	228 (54)	107 (50)
Metamitron	202	104 (100)	202 (66)	42 (42)	174 (35)	77 (24)	103 (19)
Methabenzthiazuron	221	164 (100)	136 (73)	135 (69)	163 (42)	69 (30)	58 (25)
Methazole	260	44 (100)	161 (44)	124 (36)	187 (31)	159 (24)	163 (23)
Methidathion	302	85 (100)	145 (90)	93 (32)	125 (22)	47 (21)	58 (20)
Methiocarb	225	168 (100)	153 (84)	45 (40)	109 (37)	91 (31)	58 (21)
Methomyl	162	44 (100)	58 (81)	105 (69)	45 (59)	42 (55)	47 (52)
Metobromuron	258	61 (100)	46 (43)	60 (15)	91 (13)	258 (13)	170 (12)
Metoxuron	228	72 (100)	44 (27)	183 (23)	228 (22)	45 (21)	73 (15)
Metribuzin	214	198 (100)	41 (78)	57 (54)	43 (39)	47 (38)	74 (36)
Mevinphos	224	127 (100)	192 (30)	109 (27)	67 (20)	43 (8)	193 (7)
Monocrotophos	223	127 (100)	67 (25)	97 (23)	109 (14)	58 (14)	192 (13)
Monolinuron	214	61 (1003	126 (63)	153 (42)	214 (34)	46 (29)	125 (25)
Napropamide	271	72 (100)	100 (81)	128 (62)	44 (55)	115 (41)	127 (36)
Nicotine	162	84 (100)	133 (21)	42 (18)	162 (17)	161 (15)	105 (9)

Table 2-39 (continued)

Compound	Molar mass	Main fragment m/z (intensities)					
		1	2	3	4	5	6
Nitrofen	283	283 (100)	285 (67)	202 (55)	50 (55)	139 (37)	63 (37)
Nuarimol	314	107 (100)	235 (91)	203 (85)	139 (60)	123 (46)	95 (35)
Omethoat	213	110 (100)	156 (83)	79 (39)	109 (32)	58 (30)	47 (21)
Oxadiazon	344	43 (100)	175 (92)	57 (84)	177 (60)	42 (35)	258 (22)
Parathion	291	97 (100)	109 (90)	291 (57)	139 (47)	125 (41)	137 (39)
Parathion-methyl	263	109 (100)	125 (80)	263 (56)	79 (26)	63 (18)	93 (18)
Pendimethalin	281	252 (100)	43 (53)	57 (43)	41 (41)	281 (37)	253 (34)
Permethrin	390	183 (100)	163 (100)	165 (25)	44 (15)	184 (15)	91 (13)
Phenmedipham	300	133 (100)	104 (52)	132 (34)	91 (34)	165 (31)	44 (27)
Phosalone	367	182 (100)	121 (48)	97 (36)	184 (32)	154 (24)	111 (24)
Pirimicarb	238	72 (100)	166 (85)	42 (63)	44 (44)	43 (24)	238 (23)
Pirimiphos-ethyl	333	168 (100)	318 (94)	152 (88)	304 (79)	180 (73)	42 (71)
Pirimiphos-methyl	305	290 (100)	276 (93)	125 (69)	305 (53)	233 (44)	42 (41)
Propachlor	211	120 (100)	77 (66)	93 (36)	43 (35)	51 (30)	41 (27)
Propanil	217	161 (100)	163 (70)	57 (64)	217 (16)	165 (11)	219 (9)
Propham	179	43 (100)	93 (88)	41 (42)	120 (24)	65 (24)	137 (23)
Propoxur	209	110 (100)	152 (47)	43 (28)	58 (27)	41 (21)	111 (20)
Pyrethrin I	328	123 (100)	43 (62)	91 (58)	81 (47)	105 (45)	55 (43)
Pyrethrin II	372	91 (100)	133 (70)	161 (55)	117 (48)	107 (47)	160 (43)
Quintozene	293	142 (100)	237 (96)	44 (75)	214 (67)	107 (62)	212 (61)
Resmethrin	338	123 (100)	171 (67)	128 (52)	143 (49)	81 (38)	91 (28)
Simazine	201	201 (100)	44 (96)	186 (72)	68 (63)	173 (57)	96 (40)
Tecnazene	259	203 (100)	201 (69)	108 (69)	215 (60)	44 (57)	213 (51)
Terbacil	216	160 (100)	161 (99)	117 (69)	42 (45)	41 (41)	162 (37)
Terbacil N-methyl derivative	230	56 (100)	174 (79)	175 (31)	57 (24)	176 (23)	41 (20)
Tetrachlorvinphos	364	109 (100)	329 (48)	331 (42)	79 (20)	333 (14)	93 (9)
Tetrasul	322	252 (100)	254 (67)	324 (51)	108 (49)	75 (40)	322 (40)
Thiabendazole	201	201 (100)	174 (72)	63 (12)	202 (11)	64 (11)	65 (9)
Thiofanox	218	57 (100)	42 (75)	68 (39)	61 (38)	55 (34)	47 (33)
Thiometon	246	88 (100)	60 (63)	125 (56)	61 (52)	47 (49)	93 (47)
Thiophanat-methyl	342	44 (100)	73 (97)	159 (89)	191 (80)	86 (72)	150 (71)
Thiram	240	88 (100)	42 (25)	44 (20)	208 (18)	73 (15)	45 (10)
Tri-allate	303	43 (100)	86 (73)	41 (43)	42 (31)	70 (23)	44 (21)
Trichlorfon	256	109 (100)	79 (34)	47 (26)	44 (20)	185 (17)	80 (8)
Tridemorph	297	128 (100)	43 (26)	42 (18)	44 (13)	129 (11)	55 (5)
Trietazine	229	200 (100)	43 (81)	186 (52)	229 (52)	214 (50)	42 (48)
Trifluralin	335	43 (100)	264 (33)	306 (32)	57 (7)	42 (6)	290 (5)
Vamidothion	287	87 (100)	58 (47)	44 (40)	61 (29)	59 (26)	60 (25)
Vinclozolin	285	54 (100)	53 (93)	43 (82)	124 (65)	212 (63)	187 (61)

The data refer to EI ionisation at 70 eV. The relative intensities can depend in individual cases on the type of mass spectrometer or mass-selective detector used. For confirmation mass spectra should be consulted which were run under identical instrumental conditions.

present (e.g. isotope patterns), these should be checked using the relative intensities as a line spectrum, if possible (Fig. 2-140), or as superimposed mass traces (Fig. 2-141). In the region of the detection limit the noise width should be taken into account in the test of agreement.

174   2 Basics

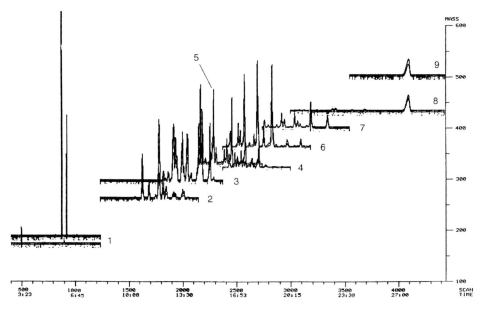

Fig. 2-136 MID chart of a PCB/Ugilec analysis
In each case two degrees of chlorination of the PCBs and Ugilec T were detected in parallel, each with three masses. The overlapping MID descriptors were switched in such a way that each degree of chlorination was detected in two consecutive time windows (see text)

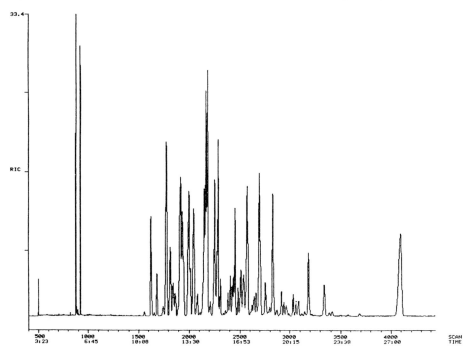

Fig. 2-137 Chromatogram (RIC) compiled after termination of the PCB/Ugilec T analysis in the MID mode from Fig. 2-136

Example of selected ion monitoring
for PCB analysis taking into account the PCB replacement product Ugilec T

The analysis strategy shown here has the aim of determining a PCB pattern as completely as possible in the relevant degrees of chlorination and to test in parallel for the possible presence of Ugilec T (tetrachlorobenzyltoluenes/trichlorobenzene). Three individual masses per time window for every two degrees of chlorination are planned for the selected SIM descriptors (scan width ± 0.25 u based on the centroid determined in the full scan mode) to determine the overlapping retention ranges of the individual degrees of chlorination (Fig. 2-136). A staggered mass determination is thus obtained (Fig. 2-137).

No.	Substance	SIM masses [m/z]	Staggered time window [min:s]
1	Trichlorobenzenes	180/182/184	Start – 8:44
2	Cl_3-PCBs	256/258/260	8:44–14:30
3	Cl_4-PCBs	290/292/294	8:44–15:52
4	Cl_5-PCBs	324/326/328	14:30–18:34
5	Ugilec	318/320/322	15:52–20:15
6	Cl_6-PCBs	358/360/362	15:52–21:36
7	Cl_7-PCBs	392/394/396	18:34–23:38
8	Cl_8-PCBs	428/430/432	20:15–30:00
9	Cl_{10}-PCB	496/498/500	23:58–30:00

(The times refer to a capillary column J&W DB5 30 m × 0.25 mm × 0.25 μm. Program: 60 °C – 2 min, 20 °/min to 180 °C, 5 °/min to 290 °C – 5 min. Splitless injection at 280 °C, column pressure 1 bar He. GC/MS system Finnigan INCOS 50. Sample: Aroclor 1240/1260 + Ugilec T, total concentration of the mixture ca. 10 ng/μl.)

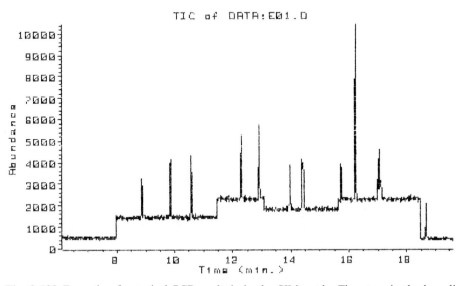

Fig. 2-138 Example of a typical PCB analysis in the SIM mode. The steps in the base line show the switching of the SIM descriptors

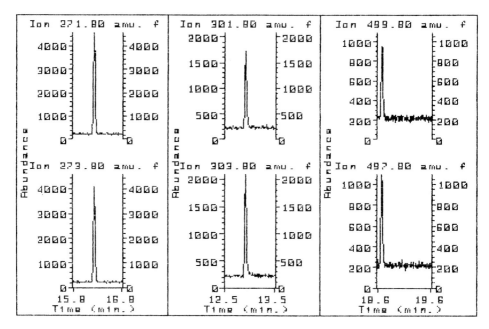

Fig. 2-139 Evaluation of the PCB analysis from Fig. 2-138 by showing the peaks at specific mass traces (see text)

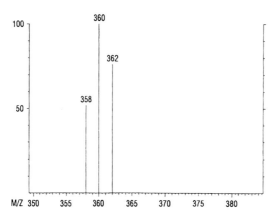

Fig. 2-140 Evaluation of isotope patterns from an MID analysis of hexachlorobiphenyl by comparison of relative intensities (shown as a bar graph spectrum)

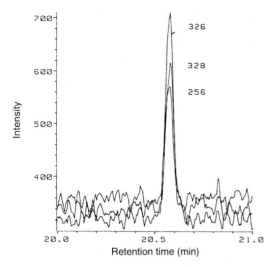

Fig. 2-141 Test of PCB isotope patterns (PCB 101, pentachlorobiphenyl) in the range of the detection limit (10 pg, S/N ca. 4 : 1) after SIM analysis (shown as mass traces)

2.3.3.3 MS/MS – Tandem Mass Spectrometry

> "Can atomic particles be stored in a cage without material walls?
>
> This question is already quite old. The physicist Lichtenberg from Göttingen wrote in his notebook at the end of the 18th century: "I think it is a sad situation that in the whole area of chemistry we cannot freely suspend the individual components of matter." This situation lasted until 1953. At that time we succeeded, in Bonn, in freely suspending electrically charged atoms, i.e. ions, and electrons using high frequency electric fields, so-called multipole fields. We called such an arrangement an ion cage."
>
> Prof. Wolfgang Paul
> in a lecture at the Cologne Lindenthal Institute in 1991
>
> (from: Wolfgang Paul, A Cage for Atomic Particles – a basis for precision measurements in navigation, geophysics and chemistry, Frankfurter Allgemeine Zeitung, Wednesday 15th December 1993 (291) N4)

As part of the further development of instrumental techniques in mass spectrometry, MS/MS analysis has become the method of choice for trace analysis in complex matrices. Most of the current applications involve the determination of substances in the ppb and ppt ranges in samples of urine, blood and animal or plant tissues and in many environmental analyses. In addition the determination of molecular structures is an important area of application for multidimensional mass spectrometry.

As an analytical background to the use of the GC/MS/MS technique in residue analysis it should be noted that the signal/noise ratio increases with the number of analytical steps (Fig. 2-142). Clean-up steps lower the potential signal intensity. The se-

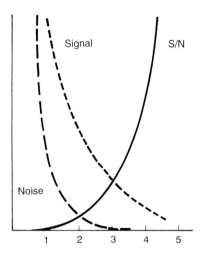

Fig. 2-142 Relationship between signal, noise and the number of analytical steps

quence of wet chemical or instrumental sample preparation steps can easily lead to the situation whereby, as a consequence of processing losses, a substance can no longer be detected. From this consideration, the first separation step (MS^1) can be regarded as a mass-specific clean-up in the analysis of extracts with large quantities of matrix by GC/MS/MS. After the subsequent induced fragmentation of selected ions, an analyte is determined using the characteristic mass spectrum of the product ions or it is quantified using substance-specific signals.

When is MS/MS used?

– The sample matrix contributes to chemical noise.
– Co-elution with impurities occurs.
– The structures of the compounds are unknown.
– Quantitation with the highest possible sensitivity in difficult matrices is necessary.
– The SIM analysis requires confirmation.

MS/MS scan techniques

Scan mode		Result	Application
MS_1	MS_2		
SIM	Scan	Product ion spectrum	Identification and confirmation of compounds, structure determination
SIM	SIM	Individual intensities of product ions	Highly specific and highly sensitive quantitation with complex matrices (selected reaction monitoring, SRM)
Scan	SIM	Precursor masses of certain fragments (precursor ion scan)	Specific analysis of compounds (classes of substance) with common structural features, screening
Scan	Scan-NL	Precursor ions, which undergo loss of neutral particles with NL u (neutral loss scan)	Specific analysis of compounds (classes of substance) with common functional groups/structural features, e.g. loss of COCl from PCDD/PCDF

Soft ionisation techniques of electron impact ionisation are the preferred ionisation processes for MS/MS analysis. Although fragmentation in the EI mode of GC/MS is desirable for substance identification, frequently only low selectivity and sensitivity are achieved in complex matrices. Soft ionisation techniques, such as chemical ionisation, concentrate the ion flow to a few intense ions which can form a good starting point for MS/MS analysis. For this reason the HPLC coupling techniques atmospheric pressure CI (APCI) and electrospray (ESI) are also in the forefront of the development and extension of MS/MS analysis. The use of the GC/MS/MS technique, which can now be carried out in association with positive and negative chemical ionisation (PCI and NCI) using ion trap benchtop instruments, will in the future give this methodology a new status in residue analysis.

The information content of the GC/MS/MS technique was already evaluated in 1983 by Richard A. Yost (University of Florida, codeveloper of the MS/MS technique). Based on the theoretical task of detecting one of the five million substances catalogued at that time by the Chemical Abstracts Service, a minimum information content of 23 bits ($\log_2(5 \cdot 10^6)$) was required for the result of the chosen analysis procedure and the MS procedures available were evaluated accordingly. The calculation showed that capillary GC/MS/MS can give 1000 times more information than the traditional GC/MS procedure (Table 2-40)!

Table 2-40 Information content of mass spectroscopic techniques (after Yost 1983, Kaiser 1978)

Technique	P	Factor
MS[a]	$1.2 \cdot 10^4$	0.002
Packed GC/MS[b]	$7.8 \cdot 10^5$	0.12
Capillary GC/MS[c]	$\mathbf{6.6 \cdot 10^6}$	1
MS/MS	$1.2 \cdot 10^7$	2
Packed GC/MS/MS	$7.8 \cdot 10^8$	118
Capillary GC/MS/MS	$6.6 \cdot 10^9$	1000

[a] MS: 1000 u, unit mass resolution, maximum intensity 2^{12}
[b] Packed GC: $2 \cdot 10^3$ theoretical plates, 30 min separating time
[c] Capillary GC: $1 \cdot 10^5$ theoretical plates, 60 min separating time

For the instrumental technique required for tandem mass spectrometry, as in the consideration of the resolving power, the main differences lie between the performances of the magnetic sector and quadrupole ion trap instruments (see Section 2.3.1). For coupling with GC and HPLC, triple-quadrupole instruments have been used since the 1980s. Ion trap MS/MS instruments have been used in research since the middle of the 1980s and are now being used in routine residue analysis. Magnetic sector instruments are mainly used in research and development for tandem mass spectrometry. Much higher energies (keV range) are used to induce fragmentation. For the selection of precursor ions and/or the detection of the product ion spectrum, high resolution can be used. Magnetic sector instruments are preferably used for high molecular weight compounds (e.g. biopolymers) in association with suitable ionisation processes, and coupling with GC is only of marginal interest.

> **MS/MS Tandem Mass Spectrometry**
>
> – Ionisation of the sample (EI, CI and other methods)
> – Selection of a precursor ion
> – Collision-induced dissociation (CID) to product ions
> – Mass analysis of product ions (product ion scan)
> – Detection as a complete product ion spectrum or in the form of individual masses (selected reaction monitoring, SRM)

Ion trap mass spectrometry offers a new extension to the instrumentation used in tandem mass spectrometry. The methods for carrying out MS/MS analyses differ significantly from those involving triple-quadrupole instruments and reflect the mode of operation as storage mass spectrometers rather than beam instruments.

Tandem mass spectrometry consists of several consecutive processes. The GC peak with all the components (analytes, co-eluents, matrix, column bleed etc) reaching the ion source is first ionised. From the resulting mixture of ions the precursor ion with a particular m/z value is selected in the first step (MS^1). This ion can in principle be formed from different molecules (structures) and, even with different empirical formulae, they can be of the same nominal m/z value. The MS^1 step is identical to SIM analysis in traditional GC/MS, which, for the reasons mentioned above, cannot rule out false positive signals. Fragmentation of the precursor ions to product ions occurs through energy-rich collisions of the selected ions with neutral gas molecules, (CID, collision induced dissociation, or CAD, collision activated decomposition). In this process the kinetic energy of the precursor ions (quadrupole: ca. 10–100 eV, magnetic sector: >1 keV) is converted into internal energy by the collisions (in an ion trap typically <6 eV), which leads to a substance-specific fragmentation by cleavage or rearrangement of bonds and the loss of neutral particles. A mixture of product ions with low m/z values is formed. Following this fragmentation step, a second mass spectrometric separation (MS^2) is necessary for the mass analysis of the product ions. The spectrum of the product ions is then detected.

With regard to existing analysis procedures, for quantitative residue analysis a certain flexibility in the choice of precursor ions is necessary. For the analysis with internal standards (e.g. deuterated standards), which co-elute with the analytes, the alternate MS/MS analysis of two different ions is necessary. Without a mass spectrometer which allows the precursor mass to be changed from scan to scan, the particularly favourable use of isotopically labelled standards in MS/MS analysis has to be dispensed with and a new methodology using new standard substances must be developed.

Besides recording a spectrum for the product ions (precursor-product ion scan), or of an individual mass, two other MS/MS scan techniques give valuable analysis data. By linking the scans in MS_2 and MS_1, very specific and targeted analysis routes are possible. In the precursor ion scan, the first mass analyser (MS_1) is scanned over a preselected mass range. All the ions in this mass range reach the collision chamber and form product ions (CID). The second mass analyser (MS_2) is held constant for a specific fragment. Only emerging ions which form the selected fragment are recorded. This recording technique makes the identification of substances of related structure, which lead to common fragments in the mass spectrometer, easier (e.g. biomarkers in crude oil characterisation, drug metabolites).

In neutral loss scan all precursor ions, which lose a particular neutral particle, are detected. Both mass analysers scan, but with a constant preadjusted mass difference, which corresponds to the mass of the neutral particle lost. This analysis technique is particularly meaningful if molecules contain the same functional groups (e.g. metabolites as glucuronides or sulfates). In this way it is possible to identify the starting ions which are characterised by the loss of a common structural element. Both MS/MS scan techniques can be used for substance-class-specific detection in triple-quadrupole systems. Ion trap systems allow the mapping of these processes by linking the scans between separate stages of MS in time.

Mode of Operation of Tandem Mass Spectrometers

In tandem mass spectrometry using quadrupole or magnetic sector instruments, the various steps take place in different locations in the beam path of the instrument. The term "tandem-in-space" (R. Yost) has been coined to show how they differ from ion storage mass spectrometers (Fig. 2-143). In the ion trap analyser, a typical storage mass spectrometer, these processes take place in the same location, but consecutively. Richard Yost has described these as "tandem-in-time".

In a triple-quadrupole mass spectrometer (or another beam instrument, Fig. 2-144), an ion beam is passed continuously through the analyser from the ion source. The selection of precursor ions and collision-induced dissociation take place in special compartments of the analyser (Q_1, Q_2) independent of time. The only time-dependent process is the mass scan in the second mass analyser (Q_3, MS_2) for the recording of the product ion spectrum (Fig 2-145). An enclosed quadrupole or octapole rod system (Q_2) is used as the collision chamber, which is only operated at an alternating voltage (radio frequency, RF) to focus the ion beam. (The mass filters MS_1 (Q_1) and MS_2 (Q_3) are operated at a constant RF/DC ratio!). Helium, nitrogen, argon or xenon at pressures of ca.

Tandem-in-space

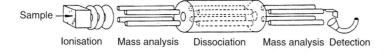

Ionisation   Mass analysis   Dissociation   Mass analysis   Detection

Tandem-in-time

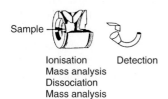

Ionisation   Detection
Mass analysis
Dissociation
Mass analysis

Fig. 2-143 GC/MS/MS techniques:
Tandem-in-space: triple-quadrupole technique
Tandem-in-time: ion trap technique

Fig. 2-144 Curved CID section in a triple-quadrupole system for reducing nonspecific noise by masking rapid neutral particles and photons from the radiation path (Finnigan, TSQ 7000)

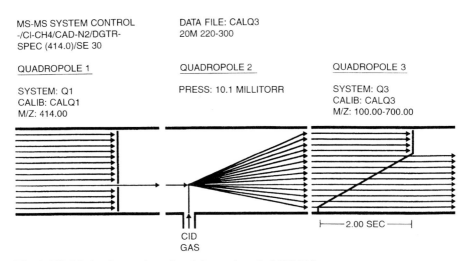

Fig. 2-145 Mode of operation of a triple-quadrupole MS/MS instrument
Q_1 = Mass selection of the precursor ion
Q_2 = Collision induced dissociation
Q_3 = Product ion scan

10^{-3} Torr are used as the collision gas. Heavier collision gases increase the yield of product ions. The collision energies are of the order of up to 100 eV and are calculated as an offset voltage of the collision chamber (Q_2) to the ion source.

In instruments with ion trap analysers, the selection of the precursor ions, the collision-induced dissociation and the analysis of the product ions occur in the same place, but with time control via a sequence of frequency and voltage values at the end caps and ring electrode of the analyser (scan function see Fig. 2-146). The systems with internal or external ionisation differ in complexity and ease of calibration. In the case of internal ionisation the sample spectrum is first produced in the ion trap analyser and stored. The precursor ion m_p is then selected by ejecting ions above and below m_p from the trap by applying a multifrequency signal at the end caps (waveform). Ion trap systems with external ionisation can employ waveforms during injection of the ions into the analyser from the external source for isolation of the precursor ion in MS/MS mode or for isolation of the desired mass scan range in other scan modes. In this way a longer storage phase and a more rapid scan rate can be used for GC/MS in the MS/MS and SIM modes.

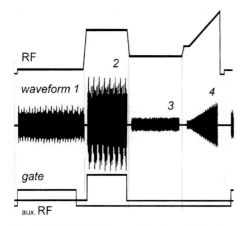

Fig. 2-146 MS/MS scan function of the ion trap analyser with external ionisation (Finnigan GCQ)
1. A gate switches the ion beam for transfer to the analyser, selection of the precursor ion m_p by a special frequency spectrum (ionisation waveform), variable ion injection time up to maximum use of the storage capacity.
2. One or more m/z values are then isolated using a synthesised frequency spectrum (isolation waveform). This phase corresponds to the SIM mode.
3. Collision induced dissociation (CID) by selective excitation to the secular frequency (activation waveform) of the precursor ion, storage of the product ions formed at a low RF value without exciting them further.
4. Product ion scan at a scan rate of ca. 11 000 u/s (resonance ejection waveform), detection of the ions by the multiplier.

Collision-induced dissociation is initiated by an additional AC voltage at the end cap electrodes of the ion trap analyser. If the frequency of the AC voltage corresponds to the secular frequency of the selected ions, there is an uptake of kinetic energy by resonance. The collisions of the precursor ions m_p excited in this way with the helium molecules present are sufficiently energy-rich to effect fragmentation. The collision energy is determined by the level of the applied AC voltage. In the collisions the kinetic energy is converted into internal energy and used up in bond cleavage. The product ions formed are stored in this phase at low RF values of the ring electrode and are detected at the end of the CID phase by a normal mass scan. The time required for these processes is in the lower ms range and, at a high scan rate for the GC, allows the separate monitoring of the deuterated internal standard at the same time. With the realisation of tandem-in-time MS/MS an aspect of particular importance for the efficiency of the process lies in the reliable choice of the frequencies for precursor ion selection and excitation. Instruments with external ion sources permit a calibration of the frequency scale which is generally carried out during the automatic tuning. In this way the MS/MS operation is analogous to that of the SIM mode in that only the masses of the desired precursor ions m_p need to be known. In the case of instruments with internal ionisation, where the GC carrier can cause pressure fluctuations within the analyser, the empirical determination of excitation frequencies and the broad band excitation of a mass window is necessary.

The efficiency of the CID process is of critical importance for the use of GC/MS/MS in residue analysis. In beam instruments optimisation of the collision energy (via the potential of Q_2) and the collision gas pressure is necessary. The optimisation is limited by scattering effects at high chamber pressures and subsequent fragmentation of the product ions. With the ion trap analyser the energy absorbed by the ions depends on the duration of the resonance conditions for the absorption of kinetic energy and on the voltage level. The pressure of the helium buffer gas can play a role in the collision energy, but is not typically adjusted. Typical values for the induction phase are 3–15 ms and 500–1000 mV. With the ion trap technique there is higher efficiency in the fragmentation and transmission to product ions (Johnson, Yost 1990). In addition the ion trap technique has clear advantages because of the sensitivity resulting from the storage technique, which allows recording of complete product ion spectra even below the pg range. The triple-quadrupole technique gives SIM data acquisition with the detection of the strongest fragments instead of the full mass spectrum in this low concentration range, like traditional GC/MS.

Today comparable results are available for both the triple-quadrupole and ion trap analysers. It is typical for the product ion spectra obtained with ion trap mass spectrometers that the intensities of the precursor ions are significantly reduced due to effi-

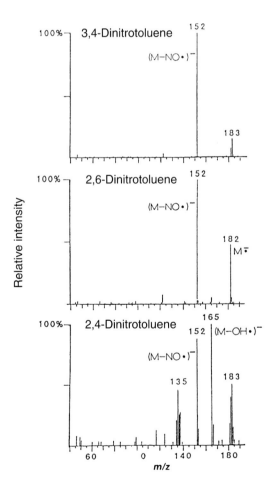

Fig. 2-147 NCI-MS/MS: Differentiation between dinitrotoluene isomers by comparison of the product ion spectra (ion trap analyser, reagent gas water, detection of negative ions; after Brodbelt, Cooks 1988)

cient CID. The product ion spectra show only a few well defined but intense product ions derived directly from the precursor. Further fragmentation of product ions usually does not occur as the m/z of the precursor ion is excited exclusively. Triple-quadrupole instruments show less efficiency in the CID process resulting in a higher precursor ion signal. Primary product ions fragment consecutively in the collision cell and produce additional signals in the product ion spectrum. The efficiency of ion trap instruments is also reported to be higher than that of magnetic sector instruments. The mass range and the scan techniques of precursor scan and neutral loss scan make triple-quadrupole instruments suitable for applications outside the range of pure GC/MS use.

When evaluating MS/MS spectra it should be noted that no isotope intensities appear in the product ion spectrum (independent of the type of analyser used). During selection of the precursor ion for the CID process, naturally occurring isotope mixtures are separated and isolated. The formation of product ions is usually achieved by the loss of common neutral species. The interpretation of these spectra is generally straightforward and less complex compared with EI spectra. When comparing product ion spectra of different instruments the acquisition parameter used must be taken into account. In particular with beam instruments the recording parameters can be reflected in the relative intensities of the spectra. On the other hand, rearrangements and isomerisations are possible, which can lead to the same product ion spectra in spite of different starting structures.

2.3.4 Mass Calibration

To operate a GC/MS system calibration of the mass scale is necessary. The calibration converts the voltage or time values into m/z values by controlling the analyser during a mass scan. The mass spectrum of a chemical compound is run where both the fragments (m/z values) and their intensities are known and are stored in the data system in the form of a reference table.

With modern GC/MS systems performing an up-to-date mass calibration is generally the final process in a tuning or autotuning program. This is preceded by a series of necessary adjustments and optimisations of the ion source, focusing, resolution and detection, which affect the position of the mass calibration. Tuning the lens potential particularly affects the transmission in individual mass areas. In particular, with beam instruments focusing must be adapted to the intensities of the reference substances in order to obtain the intensity pattern of the reference spectrum (Section 2.3.3.2). On carrying out the subsequent mass calibration, the spectrum of the reference substance is measured on-line or is stored in the form of time/intensity data. The m/z values contained in the reference table are localised by the calibration program in the spectrum of the reference compound measured. The relevant centroid of the reference peak is calculated and correlated with the operation of the analyser. Using the stored reference table a precise calibration function for the whole mass range of the instrument can be calculated. The actual state of the mass spectrometer at the end of the tuning procedure is thus taken into account. The data are plotted graphically and are available to the user for assessment and documentation. For quadrupole and ion trap instruments the calibration graph is linear (Fig. 2-148), whereas with magnetic sector instruments the graph is exponential (Fig. 2-149).

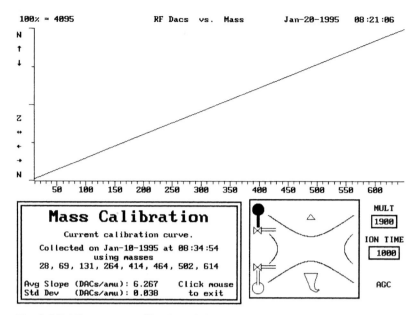

Fig. 2-148 Linear mass calibration of the quadrupol ion trap analyser (voltage of ring electrode against m/z value)

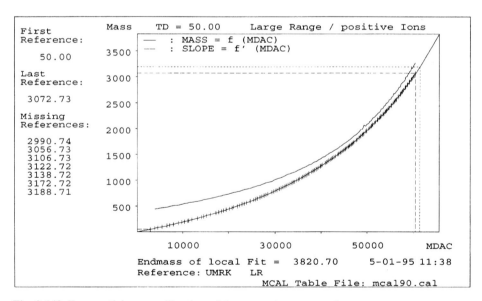

Fig. 2-149 Exponential mass calibration of the magnetic sector analyser

Table 2-41 Calibration substances and their areas of use

Name	Formula	M	m/z max.	Instrument used Magnetic sector	Quadrupole/ion trap
FC43 [1]	$C_{12}F_{27}N$	671	614	Up to 620	Up to more than 1000
PFK [2]	–	–	1017	Up to 1000	Up to more than 1000
Perfluorinated triazines					
(C_7) [3]	$C_{24}F_{45}N_3$	1185	1185	800–1200	–
(C_9) [4]	$C_{30}F_{57}N_3$	1485	1485	800–1500	–
Fomblin [5]	$(OCF(CF_3)CF_2)_x–(OCF_2)_y$			Up to 2500	–
CsI [6]	–	–	15 981	High mass range	–

Remarks: [1] Perfluorotributylamine, [2] Perfluorokerosene, [3] Perfluorotriheptylazine, [4] Perfluorotrinonyltriazine, [5] Poly(perfluoropropylene oxide) (also used as diffusion pump oil), [6] Caesium iodide

Perfluorinated compounds are usually used as calibration standards (Table 2-41). Because of their high molecular weights the volatility of these compounds is sufficient to allow controllable leakage into the ion source. In addition, fluorine has a negative mass defect ($^{19}F = 18.9984022$), so that the fragments of these standards are below the masses with whole numbers and can easily be separated from a possible background of hydrocarbons with positive mass defects. The requirements of the reference substance are determined by the type of analyser. Quadrupole and ion trap instruments are calibrated with FC43 (PFTBA, perfluorotributylamine, Table 2-42, Fig. 2-150) independent of their available mass range. The precision of the mass scan and the linearity of the calibration allow the line to be extrapolated beyond the highest fragment which can be determined, m/z 614. For magnetic sector instruments the use of PFK (perfluorokerosene, Table 2-43, Fig. 2-151) has proved successful besides FC43. It is particularly suitable for magnetic field calibration as it gives signals at regular intervals up to over m/z 1000. Higher mass ranges can be calibrated using perfluorinated

Table 2-42 Reference table for FC43 (Finnigan TSQ 7000, EI intensities >1%)

Exact mass	Intensity [%]	Formula	Exact mass	Intensity [%]	Formula
68.9952	100	CF_3	225.9903	2	C_5F_8N
92.9952	2	C_3F_3	263.9871	27	$C_5F_{10}N$
99.9936	19	C_2F_4	313.9839	3	$C_6F_{12}N$
113.9967	11	C_2F_4N	351.9807	4	$C_6F_{14}N$
118.9920	16	C_2F_5	363.9807	1	$C_7F_{14}N$
130.9920	72	C_3F_5	375.9807	1	$C_8F_{14}N$
149.9904	4	C_3F_6	401.9775	2	$C_7F_{16}N$
168.9888	7	C_3F_7	413.9775	9	$C_8F_{16}N$
175.9935	3	C_4F_6N	425.9775	1	$C_9F_{16}N$
180.9888	3	C_4F_7	463.9743	3	$C_9F_{18}N$
213.9903	2	C_4F_8N	501.9711	8	$C_9F_{20}N$
218.9856	78	C_4F_9	613.9647	2	$C_{12}F_{24}N$

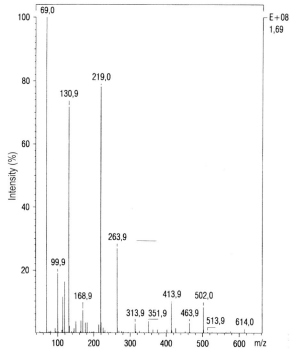

Fig. 2-150 FC43 spectrum (Finnigan TSQ 700, EI ionisation, Q3)

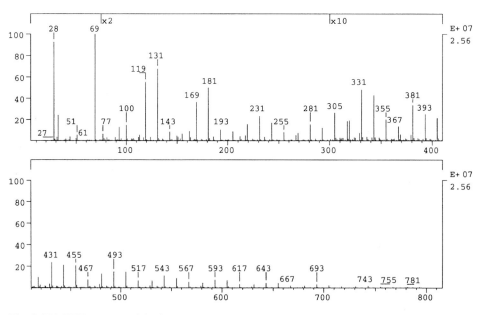

Fig. 2-151 PFK spectrum (Finnigan MAT 95, EI ionisation, B scan)

Table 2-43 Reference table for PFK (Perfluorokerosene), EI ionisation (Finnigan MAT 95)

No.	Mass	Intensity	No.	Mass	Intensity
1	4.00960	100	44	504.96966	520
2	14.01565	10	45	516.96966	300
3	18.01056	1 000	46	530.96646	400
4	28.00615	1 500	47	542.96646	440
5	30.99840	184	48	554.96646	450
6	31.98983	297	49	566.96646	410
7	39.96238	18	50	580.96327	270
8	51.00463	1 493	51	592.96327	365
9	68.99521	30 000	52	604.96327	430
10	80.99521	75	53	616.96327	320
11	92.99521	605	54	630.96007	200
12	99.99361	1 772	55	642.96007	200
13	118.99201	8 570	56	654.96007	390
14	130.99202	8 330	57	666.96007	250
15	142.99202	510	58	680.95688	120
16	154.99202	330	59	692.95688	170
17	168.98882	7 110	60	704.95688	340
18	180.98882	6 666	61	716.95688	200
19	192.98882	1 050	62	730.95369	100
20	204.98882	630	63	742.95369	225
21	218.98563	4 000	64	754.95369	220
22	230.98563	3 650	65	766.95369	80
23	242.98563	1 715	66	780.95049	130
24	254.98563	650	67	792.95049	150
25	268.98243	1 610	68	804.95049	100
26	280.98243	2 680	69	816.95049	40
27	292.98243	1 333	70	830.94730	70
28	304.98243	500	71	842.94730	40
29	318.97924	850	72	854.94730	10
30	330.97924	1 777	73	866.94730	5
31	342.97924	1 000	74	880.94411	10
32	354.97924	440	75	892.94410	10
33	366.97924	270	76	904.94411	5
34	380.97605	1 450	77	916.94410	5
35	392.97604	735	78	930.94091	5
36	404.97604	500	79	942.94091	5
37	416.97605	255	80	954.94091	5
38	430.97285	1 000	81	966.94091	5
39	442.97285	666	82	980.93772	5
40	454.97285	600	83	992.93772	5
41	466.97285	250	84	1 004.93772	5
42	480.96966	640	85	1 016.93772	5
43	492.96966	540			

alkyltriazines or caesium iodide. Reference tables frequently take account of the masses from the lower mass range, such as He, N_2, O_2, Ar or CO_2, which always form part of the background spectrum.

To record high resolution data for manual work peak matching is employed. At a given magnetic field strength one or two reference peaks and an ion of the substance being analysed are alternately shown on a screen. By changing the acceleration voltage (and the electric field coupled to it) the peaks are superimposed. From the known mass and voltage difference the exact mass of the substance peak is determined. This process is an area of the solid probe technique, as here the substance signal can be held constant over a long period. For GC/MS systems an internal scanwise calibration (scan-to-scan) by control of the data system is employed. At a given resolution (e.g. 10 000), a known reference is used which has been fed in at the same time. The analyser is positioned on the exact mass of the substance ion to be analysed relative to the measured centroid of the known reference. At the beginning of the next scan the exact position of the centroid of the reference mass is determined again and is used as a new basis for the next scan.

The usability of the calibration depends on the type of instrument and can last for a period of a few hours to several days or weeks. All tuning parameters, in particular the adjustment of the ion source, affect the calibration as described above. Special attention should be paid to the temperature of the source. Ion sources which do not contain any internal heating are heated by the cathode in the course of the operation. This results in a significant drift in the calibration over one day. Regular mass calibration is therefore recommended. Heated ion sources exhibit high stability which lasts for weeks and months. The carrier gas flow of the GC has a pronounced effect on the position of the calibration. Ion sources with small volumes and also ion trap instruments with internal ionisation show a significant drift of several tenths of a mass unit if the carrier gas flow rate is significantly changed by a temperature program. Calibration at an average elution temperature, the use of an open split, or equipping the gas chromatograph with electronic pressure programming (EPC, electronic pressure control) for analysis at constant flow are imperative in this case. Severe contamination of the ion source or the ion optics also affects the calibration. However reduced transmission of such an instrument should force cleaning to be carried out in good time.

For analyses with differing scan rates using magnetic instruments calibrations are carried out at the different rates which are required for the subsequent measurements. For scan rates which differ significantly, a mass drift can otherwise occur between calibration and measurement. Calibrations of quadrupole and ion trap instruments are practically independent of the scan rate.

Depending on the type of instrument, a new mass calibration is required if the ionisation process is changed. While for ion trap instruments switching from EI to CI ionisation is possible without alterations to the analyser, with other types switching of the ion source or changing the ion volume are required. For an optimised CI reaction a lower source temperature is frequently used compared with the EI mode. After these changes have been made, a new mass calibration is necessary. This involves running a CI spectrum of the reference substance and consulting the CI reference table (Table 2-44). The perfluorinated reference substances can be used for both positive and negative chemical ionisation (PCI and NCI) (Figs. 2-152 and 2-153). The intensities given in the CI reference tables can also be used here to optimise the adjustment of the reagent gas.

For reasons of quality control mass calibration should be carried out regularly and should be documented with a print-out (see Figs. 2-148 and 2-149).

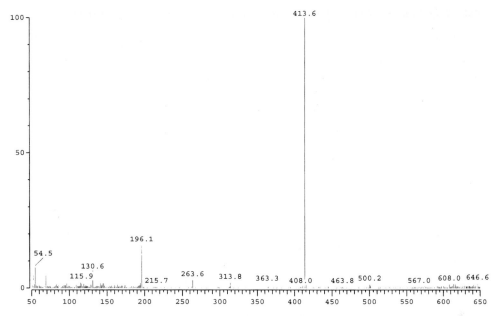

Fig. 2-152 FC43 spectrum in PCI mode (CI gas methane, Finnigan GCQ)

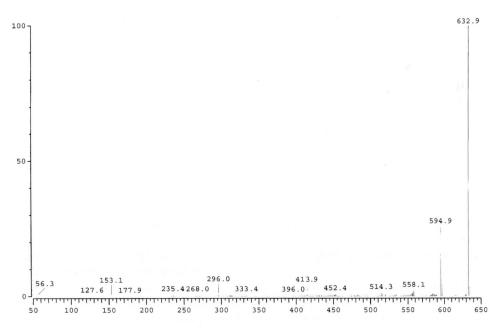

Fig. 2-153 FC43 spectrum in NCI mode (CI gas methane, Finnigan GCQ)

Table 2-44 Reference table for PFK (perfluorokerosene) in NCI mode (CI gas ammonia)

Exact mass	Intensity [%]	Formula	Exact mass	Intensity [%]	Formula
168.9888	2	C_3F_7	530.9664	11	$C_{11}F_{21}$
211.9872	5	C_5F_8	535.9680	8	$C_{13}F_{20}$
218.9856	8	C_4F_9	542.9664	13	$C_{12}F_{21}$
230.9856	26	C_5F_9	554.9664	21	$C_{13}F_{21}$
249.9840	3	C_5F_{10}	573.9648	17	$C_{13}F_{22}$
261.9840	40	C_6F_{10}	585.9648	20	$C_{14}F_{22}$
280.9824	81	C_6F_{11}	604.9633	31	C_4F_{23}
292.9824	12	C_7F_{11}	611.9617	13	$C_{13}F_{24}$
311.9808	51	C_7F_{12}	623.9617	42	$C_{14}F_{24}$
330.9792	100	CF_{13}	635.9617	32	$C_{15}F_{24}$
342.9792	32	C_8F_{13}	654.9601	21	$C_{15}F_{25}$
361.9776	35	C_8F_{14}	661.9585	13	$C_{14}F_{26}$
380.9760	63	C_8F_{15}	673.9585	37	$C_{15}F_{26}$
392.9760	48	C_9F_{15}	685.9585	21	$C_{16}F_{26}$
411.9744	16	C_9F_{16}	699.9553	7	$C_{14}F_{28}$
423.9744	15	$C_{10}F_{16}$	704.9569	6	$C_{16}F_{27}$
430.9728	39	C_9F_{17}	711.9553	8	$C_{15}F_{28}$
442.9728	36	$C_{10}F_{17}$	723.9553	12	$C_{16}F_{28}$
454.9728	16	$C_{11}F_{17}$	735.9553	6	$C_{17}F_{28}$
473.9712	11	$C_{11}F_{18}$	750.9601	4	$C_{23}F_{25}$
480.9696	19	$C_{10}F_{19}$	761.9521	2	$C_{16}F_{30}$
492.9696	23	$C_{11}F_{19}$	773.9521	3	$C_{17}F_{30}$
504.9696	18	$C_{12}F_{19}$	787.9489	3	$C_{15}F_{32}$
516.9696	7	$C_{13}F_{19}$	799.9489	1	C_6F_{32}
523.9680	7	$C_{12}F_{20}$			

2.4 Special Aspects of GC/MS Coupling

2.4.1 Vacuum Systems

Many benchtop quadrupole GC/MS systems are designed for flow rates of about 1 ml/min and are therefore suited for use with normal bore capillary columns (internal diameter 0.25 mm). Larger column diameters, however, can usually only be used with limitations. Even higher performance vacuum systems cannot improve on this value, as the design of the ion sources may be optimised for a particular carrier gas flow. Ion trap GC/MS systems are designed for a carrier gas flow of up to 3 ml/min. Normal and wide bore capillaries can be used with these instruments (see also Section 2.2.2).

The carrier gas for GC/MS is either helium or hydrogen. It is well known that the use of hydrogen significantly improves the performance of the GC, lowers the elution temperatures of compounds, and permits shorter analysis times because of higher flow rates. The more favourable van Deemter curve for hydrogen (see Fig. 2-62) accounts for these improvements in analytical performance. As far as mass spectrometry is concerned, when hydrogen is used the mass spectrometer requires a higher vacuum capacity and thus a more powerful pumping system. The type of analyser and the pumping system determine the advantages and disadvantages.

For the turbo molecular pumps mostly used at present the given rating is defined for pumping out a nitrogen atmosphere. The important performance data of turbo molecular pumps are the suction capacity and the compression. The compression ratio describes the ratio of the outlet pressure (forepump) of a particular gas to the inlet pressure (MS). The turbo molecular pump gives a completely background-free high vacuum and exhibits excellent start-up properties, which is important for benchtop instruments or those for mobile use. The use of helium lowers the performance of the pump. The compression ratio (e.g. for the Balzer TPH 062) decreases from 10^8 for nitrogen to $7 \cdot 10^3$ for helium, then to $6 \cdot 10^2$ for hydrogen. The reason for the much lower performance with hydrogen is its low molecular weight and high diffusion rate.

A turbo molecular pump essentially consists of the rotor and a stator (Fig. 2-154). Rotating and stationary discs are arranged alternately. All the discs have diagonal channels, whereby the channels on the rotor disc are arranged so that they mirror the positions of the channels of the stator discs. Each channel of the disc forms an elementary molecular pump. All the channels on the disc are arranged in parallel. A rotor disc together with a stator disc forms a single pump stage which produces a certain compression. The pumping process is such that a gas molecule which meets the rotor acquires a velocity component in the direction of rotation of the rotor in addition to its existing velocity. The final velocity and the direction in which the molecule continues to move are determined from the vector sum of the two velocities (Fig. 2-155). The thermal motion of a molecule, which is initially undirected, is converted into di-

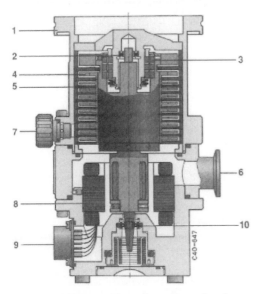

1 High vacuum connection (MS)
2 Emergency bearing
3 Permanent magnet bearing
4 Rotor
5 Stator
6 Forevacuum connection
7 Flooding connection
8 Motor
9 Electrical connection (control instrument)
10 High precision ball bearings with ceramic spheres

Fig. 2-154 Construction of a turbo molecular pump with ceramic ball bearings and permanent magnet bearings for use in mass spectrometry (after Balzers)

194   2 Basics

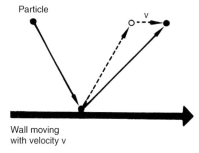

Wall moving
with velocity v

Fig. 2-155 Principle of the molecular pump (after Balzers)

rected motion when the molecule enters the pump. An individual pumping step only produces a compression of about 30. Several consecutive pumping steps, which reinforce each other's action, lead to very high compression rates.

The reduction in the performance of the pump on using hydrogen leads to a measurable increase in pressure in the analyser. Through collisions of substance ions with gas particles on their path through the analyser of the mass spectrometer, the transmission and thus the sensitivity of the instrument is reduced in the case of beam instruments (Fig. 2-156). The mean free path L of an ion is calculated according to:

$$L = p^{-1} \cdot 5 \cdot 10^{-3} \text{ [cm]} \tag{23}$$

The effect can be compensated for by using higher performance pumps or additional pumps (differentially pumped systems for source and analyser). Ion trap instruments do not exhibit this behaviour because of their storage systems.

When hydrogen is used as the carrier gas in gas chromatography the use of oil diffusion pumps is an advantage. The pump capacity is largely independent of the molecular weight and is therefore very suitable for hydrogen and helium. Oil diffusion pumps also have long service lives and are economical as no movable parts are required. The pump

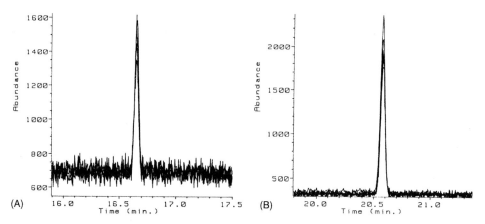

Fig. 2-156 Effect of the carrier gas on the signal/noise ratio in the quadrupole GC/MS (after Schulz). PCB 101, 50 pg, SIM m/z 256, 326, 328
(A) Carrier gas hydrogen, S/N 4:1
(B) Carrier gas helium, S/N 19:1

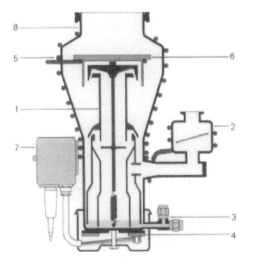

1 Jet system made of pressed aluminium components
2 Forevacuum baffle
3 Fuel top-up/measurement
4 Heating
5 Water cooling
6 Optically sealed baffle
7 Thermal protection switch/connection
8 Stainless steel pump casing

Fig. 2-157 Construction of an oil diffusion pump (after Balzers)

operates using a propellant which is evaporated on a heating plate (Fig. 2-157). The propellant vapour is forced downwards via a baffle and back into the propellant bath. Gas molecules diffuse into the propellant stream and are conveyed deeper into the pump. They are finally sucked out by a forepump. Perfluoropolyethers (e. g. Fomblin) or poly (phenyl ethers) (e. g. Santovac) are now used exclusively as propellants in mass spectrometry. However, the favourable operation of the pump results in disadvantages for the operation of the mass spectrometer. Because a heating plate is used, the diffusion pump starts sluggishly and can only be aerated again after cooling. While older models require water cooling, modern diffusion pumps for GC/MS instruments are air-cooled by a ventilator so that heating and cool-down times of 15–30 min are achieved. Propellant vapour can easily lead to a permanent background in the mass spectrometer, which makes the use of a cooled baffle necessary, depending on the construction. The detection of negative ions in particular can be affected by the use of fluorinated polymers.

Furthermore, it should be noted that, as a reactive gas, hydrogen can hydrogenate the substances being analysed. This effect has already been known from gas chromatography for many years. Reactions in hot injectors can lead to the appearance of by-products. In GC/MS ion sources reactions leading to hydrogenation products are also known. These cases are easy to identify by changing to helium as the carrier gas.

Both turbo molecular pumps and oil diffusion pumps require a forepump, as the compression is not sufficient to work against atmospheric pressure. Rotating vane pumps are generally used as forepumps (Fig. 2-158). Mineral oils are used as the operating liquid. Oil vapours from the rotating vane pump passing into the vacuum tubing to the turbo pump are visible in the mass spectrometer as a hydrocarbon background. In particular, on frequent aeration of the system, special devices for separation or removal are necessary. As an alternative for the production of a hydrocarbon-free forevacuum, spiro molecular pumps can be used. These pumps can be run without additional oil. Their function involves centrifugal acceleration of the gas molecules through several pumping steps like the turbo molecular pump. For the start-up

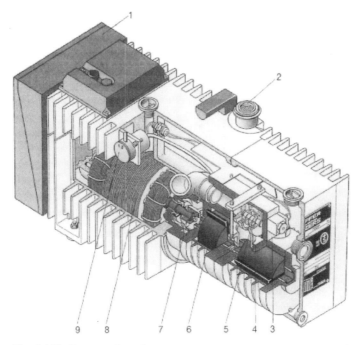

Fig. 2-158 Cross-section of a two-stage rotary vane vacuum pump (after Balzers)
1 Start-up control
2 Exhaust
3 Aeration valve
4 Pump stage 1
5 High vacuum safety valve
6 Pump stage 2
7 Motor coupling
8 Motor
9 Gas ballast valve

an integrated membrane pump is used. In particular for mobile use of GC/MS systems and in other cases where systems are frequently disconnected from the mains electricity, spiro molecular pumps have proved useful.

2.4.2 GC/MS Interface Solutions

2.4.2.1 Open Split Coupling

Open split coupling has been known to many spectroscopists from the earliest days of the GC/MS technique. The challenge of open split coupling lay in the balancing of incompatible flow rates, particularly in the cutting out of the solvent peak or other unwanted main components (Fig. 2-159). The cathode was thus protected from damage and the ion source from contamination and the associated reduction in sensitivity thus prevented. These initial problems have now receded completely into the background thanks to fused silica capillaries and easily maintained ion source constructions.

2.4.2 GC/MS Interface Solutions

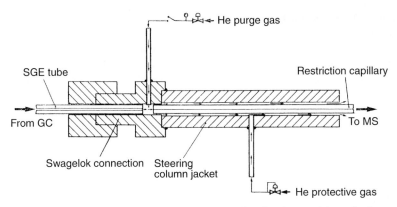

Fig. 2-159 Lengthways section of an open coupling for the expulsion of solvents (after Abraham)

Many advantages of open split coupling are nevertheless frequently exploited in various applications.

The rapid change of GC columns and the desire of the analyst to have a further detector besides the mass spectrometer after the split are now priorities. Because of the different pressure ratios a makeup gas now usually has to be added (Fig. 2-160). If the carrier gas stream through the column is not sufficient, a considerable air leak can appear if the detector is only coupled by means of a T-piece. In this case the mass spectrometer sucks air through the detector into the ion source via the T-piece. By adding a makeup gas at the split point, these situations can be reliably prevented. The split situated after the column can thus be adapted rapidly to different column types and split ratios.

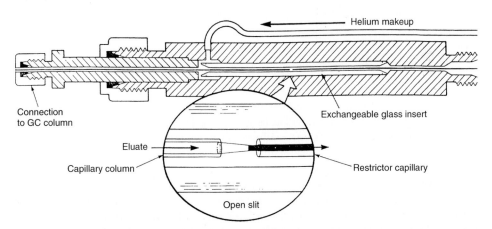

Fig. 2-160 Lengthways section of an open coupling for complete transfer of substance with controllable He makeup (ITD 800, Finnigan)

Advantages of Open Split Coupling

- The retention times of classical detectors, such as FID or ECD, are retained in the GC/MS and allow the direct comparison of chromatograms.
- At the end of the column a split can allow the use of an additional element-specific detector, such as ECD or NPD, and thus give information additional to that given by the mass spectrometer.
- The capillary column can be readily exchanged, usually without venting the mass spectrometer, allowing rapid resumption of work.
- The choice of GC conditions (column length, diameter, flow rate) can be optimised to give the best possible separation independent of the mass spectrometer.
- The use of wide bore, megabore and even packed columns (gas analysis) is possible. The excess eluate not sucked into the mass spectrometer is passed out via the split.
- A constant flow of material, which is independent of the oven temperature of the gas chromatograph, reaches the mass spectrometer. This allows precise optimisation of the ion source.

Disadvantages of Open Split Coupling

- The split point is at atmospheric pressure. To prevent the penetration of air the split point must be flushed with carrier gas. The additional screw joints can give rise to damaging leaks.
- If, because of the balance of column flow and suction capacity of the mass spectrometer, there is a positive split ratio, the sensitivity of the system decreases.
- On improper handling, e.g. on penetration of particles from the sealing ferrule, or a poor cut area at the end of the column, the quality of the GC is impaired (tailing).

2.4.2.2 Direct Coupling

If the end of the column is inserted right into the ion source of the mass spectrometer, the eluate reaches the mass spectrometer directly and unhindered. The entire eluate thus reaches the ion source undivided (Fig. 2-161). For this reason direct coupling is now regarded as ideal for residue analysis. The pumping capacity of modern mass spectrometers is adjusted to be compatible with the usual flow rates of capillary columns.

The end part of the capillary column which is inserted into the ion source requires particular attention to be paid to the construction of the GC/MS interface. The effect of the vacuum on the open end of the column causes molecular flow (as opposed to viscous flow in the column itself) with an increased number of wall collisions. According to the systematic investigations of Henneberg and Schomburg, there is possible adsorption on the column walls at the entry of the capillary into the ion source. It should therefore be ensured that there is uniform heating at this location in particular. Heating the interface too strongly can cause thermal decomposition and a reduction in the quality of the spectra.

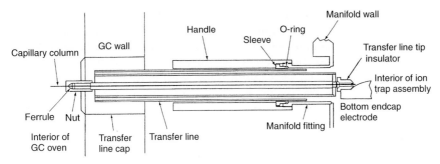

Fig. 2-161 Cross-section of a direct coupling (Finnigan)

Advantages of Direct Coupling

– Uncomplicated construction and handling.
– Single route for the substance from the GC injector to the ion source via a capillary column.
– Reliable vacuum seal at the point where the column is screwed in.

Disadvantages of Direct Coupling

– To change the column the mass spectrometer must be cooled and vented.
– The carrier gas flow into the ion source is not constant as it depends on the GC temperature program chosen and the column dimensions. Working under constant flow conditions (EPC) is recommended.
– The vacuum in the MS system affects the GC column and shortens the retention times in comparison with FID or ECD, for example. A direct comparison with these chromatograms is not straightforward.
– Constantly high interface temperatures limit the lifetime and inertness of sensitive types of column film, such as Carbowax 1701. It is essential to be aware of the maximum continuous temperature of a particular GC column.
– The choice of column and adjustment of optimal flow rate is limited by the maximum carrier gas flow for the mass spectrometer.

2.4.2.3 Separator Techniques

For the coupling of capillary gas chromatography with mass spectrometry, special separators are no longer necessary. Only GC systems, which still involve packed columns with flow rates of ca. 10 ml/min or the use of halfmil capillary columns (internal diameter 0.53 mm), cannot be coupled to mass spectrometers without a separator. Separators are inserted between the column and the mass spectrometer to selectively lower the carrier gas load. This procedure, however, involves considerable loss of analyte.

The one-step jet separator (Biemann-Watson separator) frequently used with packed columns, operates according to the principle of diffusion of the smaller carrier gas molecules away from the transmission axis. These are sucked off by an additional rotating vane pump. A reduction in the carrier gas flow of up to ca. 1:100 can be used. Heavier molecules reach the ion source of the mass spectrometer via a transfer capillary. The use of separators involves the loss of ca. 40–80% of the sample material and since capillary column techniques have become widely used for GC/MS residue analysis, they should only be regarded as having historical interest.

References for Chapter 2

Section 2.1.1

Arthur, C. L., Potter, D. W., Buchholz, K. D., Motlagh, S., Pawliszyn, J., Solid-Phase Microextraction for the Direkt Analysis of Water: Theory and Practice, LCGC 10 (1992), 656–661.

Berlardi, R., Pawliszyn, J., The Application of Chemically Modified Fused Silica Fibers in the Extraction of Organics from Water Matrix Samples and Their Rapid Transfer to Capillary Columns, Water Pollution Research J. of Canada, 1989, 24, 179.

Boyd-Boland, A., Magdic, S., Pawleszyn, J., Simultaneous Determination of 60 Pecticides in Water Using Solid Phase Microextraction and Gas Chromatography-Mass Spectrometry, Analyst 121, 929–937 (1996).

Boyd-Boland, A., Pawleszyn, J., Solid Phase Microextraction of Nitrogen Containing Herbicides, J. Chromatogra., 1995, 704, 163–172.

Brenk, F. R., Bestimmung von polychlorierten Biphenylen in Ölproben, LaborPraxis 3 (1986), 222–224.

Brenk, F. R., GC-Auswertung und Quantifizierung PCB-haltiger Öle. LaborPraxis 4 (1986), 332–340.

Bundt, J., Herbel, W., Steinhart, H., Franke, S., Francke, W., Structure Type Separation of Diesel Fuels by Solid Phase Extraction and Identification of the Two- and Three-Ring Aromatics by Capillary GC-Mass Spectrometry, J. High Res. Chrom. 14 (1991), 91–98.

Ciupe, R., Spangenberg, J., Meyer, T., Wild, G., Festphasenextraktionstrennsystem zur gaschromatographischen Bestimmung von PAK's in Mineralöl, GIT Fachz. Lab. 8 (1994), 825–829.

Deutsche Gesellschaft für Mineralölwissenschaft und Kohlechemie e.V., DGMK-Projekt 387 – Bestimmung von PCB's in Ölproben, DGMK-Berichte, Hamburg 1985.

Dünges, W., Muno, H., Unckell, F., Rationelle prä-chromatographische Mikromethoden für die ppb- und ppt-Analytik, DVGW-Schriftenreihe Wasser Nr. 108, Eschborn 1990.

Eisert, R., Pawliszyn, J., New Trends in Solid Phase Microextraction, Critical Reviews in Analytical Chemistry 27, 103–135 (1997).

Festphasen-Mikroextraktion – Extraktion organischer Komponenten aus Wasser ohne Lösungsmittel, SUPELCO International 1992 (Firmenschrift), 3–5.

Gorecki, T., Martos, P., Pawliszyn, J., Strategies for the Analysis of Polar Solvents in Liquid Matrices, Anal. Chem. 70, 19–27 (1998).

Gorecki, T., Pawleszyn, J., Sample introduction approaches for solid phase microextraction/rapid GC, Anal. Chem., 1995, 34, 3265–3274.

Gorecki, T., Pawliszyn, J., The Effect of Sample Volume on Quantitative Analysis by SPME. Part I: Theoretical Considerations, Analyst 122, 1079–1086 (1997).

Hein, H., Kunze, W., Umweltanalytik mit Spektrometrie und Chromatographie, VCH: Weinheim 1994.

Kicinski, H. G., Adamek, S., Kettrup, A., Trace Enrichment and HPLC Analysis of Polycyclic Aromatic Hydrocarbons in Environmental Samples Using Solid Phase Extraction in Connection with UV/VIS Diode-Array and Fluorescence Detection. Chromatographia 28 (1989), 203–208.

Kicinski, H. G., Kettrup, A., Festphasenextraktion und HPLC-Bestimmung von polycyclischen Aromaten aus Trinkwasser, Vom Wasser 71 (1988), 245–254.

Lord, H. L., Pawliszyn, J., Recent Advances in Solid Phase Microextraction, LCGC International 1998 (12) 776–785.

Luo, Y., Pan, L., Pawliszyn, J., Determination of Five Benzodiazepines in Aqueous Solution and Biological Fluids using SPME with Carbowax/DVB Fibre Coating, J. Microcolumn. Sep. 10, 193–201 (1998).

MacGillivray, B., Pawleszyn, J., Fowlei, P., Sagara, C., Headspace Solid-Phase Microextraction Versus Purge and Trap for the Determination of Substituted Benzene Compounds in Water, J. Chromatogr. Sci., 1994, 32, 317–322.

Magdic, S., Boyd-Boland, A., Jinno, K., Pawleszyn, J., Analysis of Organophosphorus Insecticides from Environmental Samples by Solid Phase Micro-extraction, J. Chromatogr. A 736, 219–228 (1996).

Martos, P., Saraullo, A., Pawliszyn, J., Estimation of Air/Coating Distribution Coefficients for Solid Phase Microextraction Using Retention Indexes from Linear Temperature-Programmed Capillary Gas Chromatography. Application to the Sampling and Analysis of Total Petroleum Hydrocarbons in Air, Anal. Chem. 69, 402–408 (1997).

Nielsson, T., Pelusio, F., et al., A Critical Examination of Solid Phase Micro-Extraction for Water Analysis, In: 16th Int. Symp. Cap. Chrom. Riva del Garda, P. Sandra (Ed.), Huethig 1994, 1148–1158.

Nolan, L., Ziegler, H., Shirey, R., Hastenteufel, S., Festphasenextraktion von Pestiziden aus Trinkwasserproben, LaborPraxis 11 (1991), 958–960.

Pan, L., Chong, J. M., Pawliszyn, J., Determination of Amines in Air and Water using Derivatization Combined with SPME, J. Chromatogr. 773, 249–260 (1997).

Poerschmann, Z., Zhang, Z., Kopinke, F.-D., Pawliszyn, J., Solid Phase Microextraction for Determining the Distribution of Chemicals in Aqueous Matrices, Anal. Chem. 69, 597–600 (1997).

Potter, D., Pawliszyn, J., Detection of Substituted Benzenes in Water at the pg/ml Level Using Solid Phase Microextraction, J. Chromatogr., 1992, 625, 247–255.

Potter, D., Pawliszyn, J., J. Chromatogr. 625 (1992), 247–255.

Reupert, R., Brausen, G., Trennung von PAK's durch HPLC und Nachweis durch Fluoreszenzdetektion, GIT Fachz. Lab. 11 (1991), 1219–1221.

Reupert, R., Plöger, E., Bestimmung stickstoffhaltiger Pflanzenbehandlungsmittel in Trink-, Grund- und Oberflächenwasser: Analytik und Ergebnisse, Vom Wasser 72 (1989), 211–233.

Saraullo, A., Martos, P., Pawliszyn, J., Water Analysis by SPME Based on Physical Chemical Properties of the Coating, Anal. Chem. 69, 1992–1998 (1997).

Shirey, R. E., Eine neue Polyacrylatfaser zur Festphasenmikroextraktion von polaren halbflüchtigen Verbindungen aus Wasser, SUPELCO Reporter 13 (1994), Firmenschrift, 8–9.

Shirey, R. E., Mani, Venkatachalem, New Carbon-coated Solid Phase Microextraction (SPME) Fibers for Improved Analyte Recovery, SUPELCO talk 497015, given at the 1997 Pittcon conference in Atlanta, Georgia, USA.

Steffen, A., Pawleszyn, J., The Analysis of Flavour Volatiles using Headspace Solid Phase Micro-extraction, J. Agric. Food Chem. 44, 2187-2193 (1996).

SUPELCO, Festphasen-Mikroextraktion organischer Komponenten aus wäßrigen Proben ohne Lösungsmittel, Die SUPELCO-Reporter 12 (1993), Firmenschrift, 3–5.

Van der Kooi, M. M. E., Noij, Th. H. M., Evaluation of Solid Phase Micro-Extraction for the Analysis of Various Priority Pollutants in Water, In: 16th Int. Symp. Cap. Chrom. Riva del Garda, P. Sandra (Ed.). Huethig 1994. 1087–1098.

Van Horne, K. C., Handbuch der Festphasenextraktion, ICT: Frankfurt, Basel, Wien 1993.

Yang, M., Orton, M., Pawliszyn, J., Quantitative Determination of Caffeine in Beverages Using a Combined SPME-GC/MS Method, J. Chem. Ed. 74, 1130–1132 (1997).

Zhang, Z., Pawliszyn, J., Headspace Solid Phase Microextraction, Anal. Chem. 65, 1843–1852, 1993.

Zhang, Z., Yang, M. J., Pawleszyn, J., Solid Phase Microextraction – A New Solvent-Free Alternative for Sample Preparation, Anal. Chem. 66, 844A–853A, 1994.

Section 2.1.2

Abraham, B., Rückstandsanalyse von anabolen Wirkstoffen in Fleisch mit Gaschromatographie-Massenspektrometrie, Dissertation, Technische Universität Berlin, 1980.
Bartle, K., Pullen, F., A happy coupling? – Aspects of the ongoing debate about the place of supercritical fluid chromatography coupled to MS, Analysis Europa 4 (1995) 44–45.
Bittdorf, H., Kernforschungszentrum Karlsruhe, Institut für Radiochemie Abt. Wassertechnologie, Überprüfung der Einsetzbarkeit der Supercritical Fluid Extraction (SFE) bei der Probenvorbereitung von Nitroaromaten, (pers. Mitteilung).
Blanch, G. P., Ibanez, E., Herraiz, M., Reglero, G., Use of a Programmed Temperature Vaporizer for Off-Line SFE/GC Analysis in Food Composition Studies, Anal. Chem. 66 (1994) 888–892.
Bowadt, S., Johansson, B., et al., Independent Comparison of Soxhlet and Supercritical Fluid Extraction for the Determination of PCBs in an Industrial Soil, Anal. Chem. 67 (1995) 2424–2430.
Burford, M. D., Hawthorne. S. B., Miller, D. J., Extraction Rates of Spiked versus Native PAH's from Heterogeneous Environmental Samples Using Supercritical Fluid Extraction and Sonication in Methylene Chloride, Anal. Chem. 65 (1993), 1497–1505.
Cammann, K., Kleiböhmer, W., Meyer, A., SFE of PAH's from Soil with a High Carbon Content and Analyte Collection via Combined Liquid/Solid Trapping, In: P. Sandra (Ed.), 15th Int. Symp. Cap. Chrom., Riva del Garda May 1993, Huethig Verlag 1993, 105–110.
Croft, M. Y., Murby, E. J., Wells, R. J., Simultaneous Extraction and Methylation of Chlorophenoxyacetic Acids from Aqueous Solution Using Supercritical Carbon Dioxide as a Phase Transfer Solvent, Anal. Chem. 66 (1994), 4459–4465.
Hartonen, K., Riekkola, M.-L., Determination of Beta-Blockers from Urine by SFE and GC/MS, In: 16th Int. Symp. Cap. Chrom. Riva del Garda, P. Sandra (Ed.), Huethig 1994, 1729–1739.
Häufel, J., Creutznacher, H., Weisweiler, W., Analyse von PAK in Bodenproben – Extraktion mit organischen Lösungsmitteln gegen überkritische Fluide, GIT Fachz. Lab. 7 (1994). 764–768.
Hawthorne, S. B., Analytical-Scale Supercritical Fluid Extraction, Anal. Chem. 62 (1990), 633A.
Hawthorne, S. B., Methodology for Off-Line Supercritical Fluid Extraction, In: S. A. Westwood (Ed.), Supercritical Fluid Extraction and its Use in Chromatographic Sample Preparation, Blackie Academic & Professional: London 1993, 39–64.
Klein, E., Hahn, J., Gottesmann, P., Extraktion mit überkritischem CO_2 (SFE) – Applikationen aus der Lebensmittelüberwachung, Lebensmittelchemie 47 (1993), 84.
Langenfeld, J. J., Hawthorne, S. B., Miller, D. J., Pawliszyn, J., Role of Modifiers for Analytical-Scale Supercritical Fluid Extraction of Environmental Samples, Anal. Chem. 66 (1994), 909–916.
Lehotay, S. J., Eller, K. I., Development of a Method of Analysis for 46 Pesticides in Fruits and Vegetables by Supercritical Fluid Extraction and Gas Chromatography/Ion Trap Mass Spectrometry, J. AOAC Int. 78 (1995) 821–830.
Levy, J. M., Interfacing SFE On-Line to GC, SUPREX Technology Focus, SFE 91–6. Suprex Corporation: Pittsburg 1991.
Levy, J. M., Houck, R. K., Developments in Off-Line Collection for Supercritical Fluid Extraction, Am. Laboratory 4 (1993).
Lopez-Avila, V., Young, R., er al., Mini-round-robin study of a supercritical fluid extraction method for polynuclear aromatic hydrocarbons in soil with dichloromethane as a static modifier, J. Chromatogr. A 672 (1994) 167–175.
Luque de Castro, M. D., Valcarel, M., Tena, M. T., Analytical Supercritical Fluid Extraction, Springer: Berlin 1994.
Meyer, A., Kleiböhmer. W., Rapid Determination of PCP in Leather Using SFE with in Situ Derivatization, In: 16th Int. Symp. Cap. Chrom. Riva del Garda, P. Sandra (Ed.), Huethig 1994, 1752–1753.

Onuska, F. I., Terry, K. A., Supercritical Fluid Extraction of Polychlorinated Dibenzo-p-dioxins from Municipal Fly Ash, In: P. Sandra (Ed.). 13th Int. Symp. Cap. Chrom., Riva del Garda, May 1991, Huethig Verlag 1991.

Onuska, F. I., Terry, K. A., Supercritical Fluid Extraction of 2,3,7,8-Tetrachlorodibenzo-p-dioxin from Sediment Samples, J. High Res. Chrom. (1989), 537.

Oostdyk, T. S., Grob, R. L., Snyder, J. L., McNally, M. E., Study of Sonication and Supercritical Fluid Extraction of Primary Aromatic Amines, Anal. Chem. 65 (1993), 596–600.

Paschke, Th., SFE zur einfachen Fett- und Ölbestimmung, LaborPraxis 7 (1995) 24–26.

Paschke, Th., Untersuchungen zur Umweltrelevanz nitrierter polyzyklischer aromatischer Verbindungen in der Luft. Dissertation, Universität Gesamthochschule Siegen, Fachbereich 8, Siegen 1994.

Paschke, Th., Hawthorne, S. B., Miller, D. J., Supercritical Fluid Extraction of Nitrated Polycyclic Aromatic Hydrocarbons and Polycyclic Aromatic Hydrocarbons from Diesel Exhaust Particulate Matter. J. Chromatogr. 609 (1992), 333–340.

Pawliszyn, J., Kinetic Model of SFE Based on Packed Tube Extractor, Proc. 3rd Int. Symp. Supercritical Fluids, Strasbourg (France) 17.–19. Oct. 1994.

Raynie, D. E., Warning Concerning the Use of Nitrous Oxide in Supercritical Fluid Extractions, Anal. Chem. 65 (1993), 3127–3128.

Sachs, H., Uhl, M., Opiat-Nachweis in Haar-Extrakten mit Hilfe der GC/MS/MS und Supercritical Fluid Extraction (SFE), Toxichem + Krimichem 59 (1992), 114–120.

Smith, R. M., Nomenclature for Supercritical Fluid Chromatography and Extraction (IUPAC Recommendations 1993), Pure & Appl. Chem. 65 (1993), 2397–2403.

Snyder, J. L., Grob, R. L., McNally, M. E., Oostdyk, T. S., Comparison of Supercritical Extraction with Classical Sonication and Soxhlet Extractions for Selected Pesticides, Anal. Chem. 64 (1992), 1940–1946.

Taylor, L. T., Strategies for Analytical SFE, Anal. Chem. (1995) 364A-370A.

Towara, J., Extraktion von polychlorierten Dibenzo-p-Dioxinen und Dibenzofuranen (PCDD/F) aus Klärschlammproben mit Supercritical Fluid Extraction (SFE), Lehrstuhl für Ökologische Chemie und Geochemie der Universität Bayreuth, Studienarbeit 1993.

US EPA Methode Nr. 3560, Revision 1, December 1992.

Wenclawiak, B., (Ed.): Analysis with Supercritical Fluids: Extraction and Chromatography, Springer: Berlin 1992.

Section 2.1.3

Ashley, D. L., Bonin, M. A., et al., Determining Volatile Organic Compounds in Human Blood from a Large Sample Population by Using Purge and Trap Gas Chromatography/Mass Spectrometry, Anal. Chem. 64 (1992), 1021–1029.

Belouschek, P., Brand, H., Lönz, P., Bestimmung von chlorierten Kohlenwasserstoffen mit kombinierter Headspace- und GC/MS-Technik, Vom Wasser 79 (1992), 1–8.

Deutsche Einheitsverfahren zur Wasser-, Abwasser- und Schlammuntersuchung, Gemeinsam erfaßbare Stoffgruppen (Gruppe F), DIN 38407 (Teil 5), Beuth Verlag: Berlin 1990.

Eichelberger, J. W., Bellar, T. A., Behymer, T. D., Budde, W. L., Analysis of Organic Wastes from Solvent Recycling Operations, Finnigan MAT Application Report No. 212.

Ettre, L. S., Kolb, B., Headspace Gas Chromatography: The Influence of Sample Volume on Analytical Results, Chromatographia 32(1/2), July 1991, 5–12.

Hachenberg, H., Die Headspace Gaschromatographie als Analysen- und Meßmethode, DANI-Analysentechnik: Mainz 1988.

Hachenberg, H., Schmidt, A. P., Gaschromatographic Headspace Analysis, J. Wiley and Sons, Chichester 1977.

Johnson, E., Madden, A., Efficient Water Removal for GC/MS Analysis of Volatile Organic Compounds with Tekmar's Moisture Control Module, Finnigan MAT Technical Report No. 616, 1990.

Kolb, B., Angewandte Gaschromatographie, Bodenseewerk Perkin Elmer 1981 Heft 38.

Kolb, B., Applications of Headspace Gas Chromatography, Perkin Elmer GC Applications Laboratory 1991.

Kolb, B., Applied Headspace Gas Chromatography, J. Wiley and Sons, Chichester 1980.
Kolb, B., Die Bestimmung des Wassergehaltes in Lebensmitteln und Pharmaka mittels der gaschromatographischen Headspace-Technik, Lebensmittel- & Biotechnol. 1 (1993), 17–20.
Kolb, B., HSGC mit Kapillar-Trennsäulen, LaborPraxis 1986.
Kolb, B., Bichler, Chr., Auer, M., Simultaneous Determination of Volatile Aromatic and Halogenated Hydrocarbons in Water and Soil Samples by Dual-Channel ECD/PID Equilibrium Headspace Analysis, In: P. Sandra: 15th Int. Symp. Cap. Chrom., Riva del Garda, May 24–27 1993, Huethig Verlag 1993, 358–364.
Krebs, G., Schneider, E., Schumann, A., Head Space GC – Analytik flüchtiger aromatischer und halogenierter Kohlenwasserstoffe aus Bodenluft, GIT Fachz. Lab. 1 (1991) 19–22.
Lin, D. P., Falkenberg, C., Payne, D. A., et al., Kinetics of Purging for the Priority Volatile Organic Compounds in Water, Anal. Chem. 1993 (65), 999–1002.
Madden, A. T., Lehan, H. J., The Effects of Condensate Traps on Polar Compounds in Purge & Trap Analysis, Pittsburgh Conference 1991.
Maggio, A., Milana, M. R., et. al., Multiple Headspace Extraction Capillary Gas Chromatography (MHE-CGC) for the Quantitative Determination of Volatiles in Contaminated Soils, In: P. Sandra: 13th Int. Symp. Cap. Chrom., Riva del Garda, May 1991, Huethig Verlag 1991, 394–405.
Matz, G., Kesners, P., Spray and Trap Method for Water Analysis by Thermal Desorption Gas Chromatography/Mass Spectrometry in Field Applications, Anal. Chem, 65 (1993), 2366–2371.
Pigozzo, F., Munari, F., Trestianu ,S., Sample Transfer from Head Space into Capillary Columns, In: P. Sandra: 13th Int. Symp. Cap. Chrom., Riva del Garda, May 1991, Huethig Verlag 1991, 409–416.
TekData, Fundamentals of Purge and Trap, Tekmar Technical Documentation B 121 988, ohne Jahresangabe.
Westendorf, R. G., Automatic Sampler Concepts for Purge and Trap Gas Chromatography, American Laboratory 2 (1989).
Westendorf, R. G., Design and Performance of a Microprozessor-based Purge and Trap Concentrator, American Laboratory 10 (1987).
Westendorf, R. G., Performance of a Third Generation Cryofocussing Trap for Purge and Trap Gas Chromatography, Pittsburgh Conference 1989.
Westendorf, R. G., A Quantitation Method for Dynamic Headspace Analysis Using Multiple Runs, J. Chrom.Science 11 (1985).
Willemsen, H. G., Gerke, Th., Krabbe, M. L., Die Analytik von LHKW und BTX im Rahmen eines DK-ARW(AWBR)-IKSR-Projektes, 17. Aachener Werkstattgespräch am 28. und 29. Sept. 1993, Zentrum für Aus- und Weiterbildung in der Wasser- und Abfallwirtschaft Nordrhein-Westfalen GmbH (ZAWA): Essen 1993.
Wylie, P. L., Comparing Headspace with Purge and Trap for Analysis of Volatile Priority Pollutants, Research and Technology 8 (1988), 65–72.
Wylie, P. L., Comparison of Headspace with Purge and Trap Techniques for the Analysis of Volatile Priority Pollutants, In: P. Sandra (Ed.), 8th Int. Symp. Cap. Chrom., Riva del Garda, May 19th–21st 1987, Huethig 1987, 482–499.

Section 2.1.4

Betz, W. R., Hazard, S. A., Yearick, E. M., Characterization and Utilization of Carbon-Based Adsorbents for Adsorption and Thermal Desorption of Volatile, Semivolatile and Non-Volatile Organic Contaminants in Air, Water and Soil Sample Matrices, Int. Labmate XV (1989), 1.
Brown, R. H., Purnell, C. J., Collection and Analysis of Trace Vapour Pollutants in Ainbient Atmospheres, J. Chromatogr. 178 (1979), 79–90.
Carbotrap – an Excellent Adsorbent for Sampling Many Airborne Contaminants, GC Bulletin 846C, Supelco Firmenschrift 1986.
Efficently Monitor Toxic Compounds by Thermal Desorption, GC Bulletin 849C, Supelco Firmenschrift 1988.

Figg, K., Rubel, W., Wieck, A., Adsorptionsmittel zur Anreicherung von organischen Luftinhaltsstoffen, Fres. Z. Anal. Chem. 327 (1987), 261–278.

Föhl, A., Basnier, P., Untersuchung der Löschverfahren und Löschmittel zur Bekämpfung von Bränden gefährlicher Güter – GC/MS Rauchgasanalyse –, Forschungsbericht Nr. 81, Forschungsstelle für Brandschutztechnik an der Universität Karlsruhe (TH) 1992.

Knobloch, Th., Efer, J., Engewald, W., Adsorptive Anreicherung an Kohlenstoffadsorbentien und Thermodesorption in: SUPELCO Deutschland GmbH (Hrsg), Themen der Umweltanalytik, VCH: Weinheim 1993, 103–110.

Knobloch, Th., Engewald, W., Sampling and Gas Chromatographic Analysis of Volatile Organic Compounds (VOCs) in Hot and Extremely Humid Emissions, In: 16th Int. Symp. Cap. Chrom. Riva del Garda. P. Sandra (Ed.), Huethig 1994, 472–484.

MDHS 3 – Generation of Test Atmospheres of Organic Vapours by Syringe Injection Technique, September 1984 (HSE-Methode).

Mülle, A., Tschickard, M., Herstellung von Kalibriergasen im Emissions- und Immissionsbereich, Symposium Probenahme und Analytik flüchtiger organischer Gefahrstoffe, München/Neuberberg, 2./3. Dez.1993, Abstracts.

New Adsorbent Trap for Monitoring Volatile Organic Compounds in Wastewater, Environmental Notes, Supelco Firmenschrift: Bellafonte (USA) 1992.

Niebel, J., Analyse von Bodenluft durch Konzentration auf TENAX und thermischer Desorption mit dem mobilen GC/MS-System SpektraTrak 620, Applikationsschrift. Axel Semrau GmbH: Sprockhövel 1993.

Niebel, J., Analyse von Bodenproben auf BTX durch Thermodesorption mit dem mobilen GC/MSSystem SpektraTrak 620, Applikationsschrift, Axel Semrau GmbH: Sprockhövel 1993.

Niebel, J., Untersuchung von lösungsmittelbelasteten Bodenproben mit dem mobilen GC/MSD VIKING SpektraTrak 620. Applikationsschrift, Axel Semrau GmbH: Sprockhövel 1994.

Packing Traps for Tekmar Automatic Desorbers, Tekmar Firmenschrift, TekData B012589.

Perkin Elmer, Zusammenfassung des User Meetings Thermodesorption 18./19.9.91 in Mainz, Tagungsunterlagen.

Thermal Desorption, Technique and Applications, Perkin Elmer Applikation Report No. L-1360, 1990.

Tschickard, M., Analyse von polychlorierten Biphenylen in Innenraumluft mit Thermodesorption und Ion Trap Detektor, Perkin Elmer User Meeting, Landesamt für Umweltschutz und Gewerbeaufsicht, Mainz 1991, Tagungsunterlagen.

Tschickard, M., Analytik von leicht- und schwerflüchtigen Luftschadstoffen mit GC-Thermodesorption, Symposium Probenahme und Analytik flüchtiger organischer Gefahrstoffe, München/Neuherberg, 2./3. Dez. 1993. Abstracts.

Tschickard, M., Bericht über eine Prüfgasapparatur zur Herstellung von Kalibriergasen nach dem Verfahren der kontinuierlichen Injektion, Gesch.Zchn: 35–820 Tsch, Mainz: Landesamt für Umweltschutz und Gewerbeaufsicht 25.5.1993.

Tschickard, M., Routineeinsatz des Thermodesorbers ATD50 in der Gefahrstoffanalytik, Methodenbericht des Landesamtes für Umweltschutz und Gewerbeaufsicht Rheinland-Pfalz, Mainz 1991, In: Perkin Elmer, Zusammenfassung des User Meetings Thermodesorption 18./19.9.91 in Mainz, Tagungsunterlagen.

Tschickard, M., Hübschmann, H.-J., Bestimmung von MAK-Werten und TRK-Werten mit dem Ion Trap-Detektor, Finnigan MAT Application Report No. 67, Bremen 1988.

VDI-Richtlinie 3490, Blatt 8, Prüfgase – Herstellung durch kontinuierliche Injektion, Januar 1981.

Vorholz, P., Hübschmann, H.-J., Einsatz des mobilen VIKING SpektraTrak 600 GC/MS-Systems zur Vor-Ort-Analyse von Geruchsemissionen auf einer Kompostierungsanlage. Applikationsschrift, Axel Semrau GmbH: Sprockhövel 1994.

Section 2.1.5

Brodda, B.-G., Dix, S., Fachinger, J., Investigation of the Pyrolytic Degradation of Ion Exchange Resins by Means of Foil Pulse Pyrolysis Coupled with Gas Chromatography/Mass Spectrometry, Sep. Science Technol. 28 (1993). 653–673.

Ericsson, I., Determination of the Temperature-Time Profile of Filament Pyrolyzers, J. Anal. Appl. Pyrolysis 2 (1980), 187–194.

Ericsson, I., Influence of Pyrolysis Parameters on Results in Pyrolysis-Gas Chromatography, J. Anal. Appl. Pyrolysis 8 (1985), 73–86.

Ericsson, I., Sequential Pyrolysis Gas Chromatographic Study of the Decomposition Kinetics of cis-1,4-Polybutadiene. J. Chrom. Science 16 (1978), 340–344.

Ericsson, I., Trace Determination of High Molecular Weight Polyvinylpyrrolidone by Pyrolysis-Gas-Chromatography, J. Anal. Appl. Pyrolysis 17 (1990), 251–260.

Ericsson, I., Lattimer, R. P., Pyrolysis Nomenclature, J. Anal. Appl. Pyrolysis 14 (1989), 219–221.

Fischer, W. G., Kusch, P., An Automated Curie-Point Pyrolysis-High Resolution Gas Chromatography System, LCGC 6 (1993), 760–763.

Galletti, G. C., Reeves, J. B., Pyrolysis-Gas Chromatography/Mass Spectrometry of Lignocellulosis in Forages and By-Products, J. Anal. Appl. Pyrolysis 19 (1991), 203–212.

Hancox, R. N., Lamb, G. D., Lehrle, R. S., Sample Size Dependence in Pyrolysis: An Embarrassment or an Utility?, J. Anal. Appl. Pyrolysis 19 (1991), 333–347.

Hardell, H. L., Characterization of Spots and Specks in Paper Using PY-GC-MS Including SPM, Poster at the 10th Int. Conf. Fund. Aspects Proc. Appl. Pyrolysis, Hamburg 28.9.–2.10.1992.

Irwin, W. J., Analytical Pyrolysis, Marcel Dekker: New York 1982.

Klusmeier, W., Vögler, P., Ohrbach, K. H., Weber, H., Kettrup, A., Thermal Decomposition of Pentachlorobenzene, Hexachlorobenzene and Octachlorostyrene in Air, J. Anal. Appl. Pyrolysis 14 (1988), 25–36.

Matney, M. L., Limero, Th. F., James, J. T., Pyrolysis-Gas Chromatography/Mass Spectrometry Analyses of Biological Particulates Collected during Recent Space Shuttle Missions, Anal. Chem. 66 (1994), 2820–2828.

Richards, J. M., McClennen, W. H., Bunger, J. A., Meuzelaar, H. L. C., Pyrolysis Short-Column GC/MS Using the Ion Trap Detektor (ITD) and Ion Trap Mass Spectrometer (ITMS). Finnigan MAT Application Report No. 214, 1988.

Schulten, H.-R., Fischer, W., Wallstab, HRC & CC 10 (1987), 467.

Simon, W., Giacobbo, H., Chem. Ing. Techn. 37 (1965), 709.

Snelling, R. D., King, D. B., Worden, R., An Automated Pyrolysis System for the Analysis of Polymers. Poster ,10P, Proc. Pittsburgh Conference, Chicago 1994.

Uden, P. C., Nomenclature and Terminology for Analytical Pyrolysis (IUPAC Recommendations 1993), Pure & Appl. Chem. 65 (1983). 2405–2409.

Zaikin, V. G., Mardanov, R. G., et al., J. Anal. Appl. Pyrolysis 17 (1990) 291.

Section 2.2.1

Bergna, M., Banfi, S., Cobelli, L., The Use of a Temperature Vaporizer as Preconcentrator Device in the Introduction of Large Amount of Sample, In: 12th Int. Symp. Cap. Chrom. Riva del Garda, P. Sandra, G. Redant (Eds.), Huethig 1989. 300–309.

Blanch, G. P., Ibanez, E., Herraiz, M., Reglero, G., Use of a Programmed Temperature Vaporizer for Off-Line SFE/GC Analysis in Food Composition Studies. Anal. Chem. 66 (1994), 888–892.

David, F., Sandra. P., Large Volume Sampling by PTV Injection. Application in Dioxine Analysis, In: 14th Int. Symp. Cal). Chrom. Riva del Garda, P. Sandra (Ed.), Huethig 1991, 380–382.

David, W., Gaschromatographie: Alte Ideen – erfolgreich neu kombiniert! LABO 7–8 (1994), 62–68.

Donike, M., Die temperaturprogrammierte Analyse von Fettsäuremethylsilylestern: Ein kritischer Qualitätstest für gas-chromatographische Trennsäulen, Chromatographia 6 (1973) 190.

Efer, J., Müller, S., Engewald, W., Möglichkeiten der PTV-Technik für die gaschromatographische Trinkwasseranalytik, GIT Fachz. Lab. 7 (1995) 639–646.

Färber, H., Peldszus, S., Schöler, H. F., Gaschromatographische Bestimmung von aziden Pestiziden in Wasser nach Methylierung mit Trimethylsulfoniumhydroxid, Vom Wasser 76 (1991), 13–20.
Färber, H., Schöler, H. F., Gaschromatographische Bestimmung von Harnstoffherbiziden in Wasser nach Methylierung mit Trimethylaniliniumhydroxid oder Trimethylsulfoniumhydroxid, Vom Wasser 77 (1991), 249–262.
Färber, H., Schöler, H. F., Gaschromatographische Bestimmung von OH- und NH-aziden Pestiziden nach Methylierung mit Trimethylsulfoniumhydroxid im „Programmed Temperature Vaporizer" (PTV), Lebensmittelchemie. 46 (1992), 93–100.
Grob, K., Classical Split and Splitless Injection in Capillary Gas Chromatography with some Remarks on PTV Injection, Huethig: Heidelberg 1986.
Grob, K., Einspritztechniken in der Kapillar-Gaschromatographie, Huethig: Heidelberg 1995.
Grob, K., Guidelines on How to Carry Out On-Column Injections, HRC & CC, 6 (1983), 581–582.
Grob, K., Injection Techniques in Capillary GC, Anal. Chem. 66 (1994), 1009A-1019A.
Grob, K., Grob, K. jun., HRC & CC, 1 (1978), 1.
Grob, K., Grob, K. jun., J. Chromatogr., 151 (1978). 31 1.
Hinshaw, J. W., Splitless Injection: Corrections and Further Information, LCGC Int., 5 (1992), 20–22.
Karasek, F. W., Clement, R. E., Basic Gas Chromatography-Mass Spectrometry, Elsevier: Amsterdam 1988.
Klemp, M. A., Akard, M. L., Sacks, R. D., Cryofocussing Inlet with Reversed Flow Sample Collection for Gas Chromatography, Anal. Chem. 65 (1993) 2516 2521.
Matter, L., Lebensmittel- und Umweltanalytik mit der Kapillar-GC, VCH: Weinheim 1994.
Matter, L., Poeck, M., Kalte Injektionsmethoden, GIT Fachz.Labor, 1 1 (1987), 1031–1039.
Matter, L., Poeck, M., Probeaufgabetechniken in der Kapillar-GC, GIT Suppl.Chrom., 3 (1987), 81–86.
Mol, H. G. J., Janssen, H. G. M., Crainers, C. A., Use of a Temperature Programmed Injector with a Packed Liner for Direct Water Analysis and On-Line Reversed Phase LC-GC, In: 15th Int. Symp. Cap. Chrom. Riva del Garda, P. Sandra (Ed.), Huethig 1993, 798–807.
Mol, H. G. J., Janssen, H. G. M., Cramers, C. A., Brinkman, K. A. Th., Large Volume Sample Introduction Using Temperature Programmable Injectors-Implication of Line Diameter, In: 16th Int. Symp. Cap. Chrom. Riva del Garda, P. Sandra (Ed.), Huethig 1994, 1124–1136.
Müller, H.-M., Stan, H.-J., Pesticide Residue Analysis in Food with CGC. Study of Long-Time Stability by the Use of Different Injection Techniques, In: 12th Int. Symp. Cap. Chrom. Riva del Garda, P. Sandra, G. Redant (Eds.), Huethig 1989, 582–587.
Müller, H.-M., Stan, H.-J., Thermal Degradation Observed with Varying Injection Techniques: Quantitative Estimation by the Use of Thermolabile Carbamate Pesticides, In: 8th Int. Symp. Cap. Chrom. Riva del Garda, P. Sandra (Ed.), Huethig 1987, 588–596.
Müller, S., Efer, J., Wennrich. L., Engewald, W., Levsen, K., Gaschromatographische Spurenanalytik von Methamidophos und Buminafos im Trinkwasser – Einflußgrößen bei der PTV-Dosierung großer Probenvolumina, Vom Wasser 81 (1993), 135–150.
Pretorius, V., Bertsch, W., HRC & CC 6 (1983), 64.
Saravalle, C. A., Munari, F., Trestianu, S., Multipurpose Cold Injector for High Resolution Gas Chromatography. HRC & CC 10 (1987) 288–296.
Schomburg, G., In: Capillary Chromatography, 4th Int. Symp. Hindelang, R. Kaiser (Ed.). Inst. F. Chromatographie, 371 und A921 (1981).
Schomburg, G., Praktikumsversuch GDCh-Fortbildungskurs 305/89 Nr. 1, Erhöhung der Nachweisgrenze von Spurenkomponenten mit Hilfe des PTV-Injektors durch Anreicherung mittels Mehrfachaufgabe, Mühlheim 1989.
Schomburg, G., Gaschromatographie, VCH: Weinheim 1987.
Schomburg, G., Praktikumsversuch GDCh-Fortbildungskurs 305/89 Nr. 5, Gaschromatographie in Kapillarsäulen, Mühlheim 1989.
Schomburg, G., Probenaufgabe in der Kapillargaschromatographie, LaBo, 7 (1983), 2–6.
Schomburg, G., Praktikumsversuch GDCh-Fortbildungskurs 305/89 Nr. 7, Schnelle automatische Split-Injektion mit der Spritze von Proben mit größerem Flüchtigkeitsbereich, Mühlheim 1989. Schomburg, G., Behlau, H., Dielmann, R., Weeke, F., Husmann, H., J. Chromatogr., 142 (1977), 87.

Stan, H.-J., Müller, H.-M., Evaluation of Automated and Manual Hot-Splitless, Cold-Splitless (PTV) and On-Column Injektion Technique Using Capillar Gas Chromatography for the Analysis of Organophosphorus Pesticides, In: 8th Int. Symp. Cap. Chrom. Riva del Garda, P. Sandra (Ed.), Huethig 1987, 406–415.

Staniewski, J., Rijks, J. A., Potentials and Limitations of the Liner Design for Cold Temperature Programmed Large Volume Injection in Capillary GC and for LC-GC Interfacing, In: 14th Int. Symp. Cap. Chrom. Riva del Garda, P. Sandra (Ed.), Huethig 1991, 1334–1347.

Sulzbach, H., Magni, P., Quantitative Ultraspurengaschromatographie, Huethig: Heidelberg 1995.

Tipler, A., Johnson, G. L., Optimization of Conditions for High Temperature Capillary Gas Chromatography Using a Split-Mode Programmable Temperature Vaporizing Injektion System, In: 12th Int. Symp. Cap. Chrom. Riva del Garda, P. Sandra, G. Redant (Eds.), Huethig 1989, 986–1000.

Poy, F., Chromatographia, 16 (1982), 345.

Poy, F., 4th Int. Symp. Cap. Chrom., Hindelang 1981, Vorführung.

Poy, F., Cobelli, L., In: Sample Introduction in Capillar Gas Chromatography Vol 1, P. Sandra (Ed.), Huethig: Heidelberg 1985, 77–97.

Poy, F., Visani, S., Terrosi, F., HRC & CC 4 (1982), 355.

Poy, F., Visani, S., Terrosi, F., J. Chromatogr. 217 (1981), 81.

Vogt, W., Jacob, K., Obwexer, H. W., J. Chromatogr. 174 (1979), 437.

Vogt, W., Jacob, K., Ohnesorge, A. B., Obwexer, H. W., J. Chromatogr. 186 (1979), 197.

Section 2.2.2

Jennings, W., Gas Chromatography with Glass Capillary Columns, Academic Press: New York 1978.

Knapp, D. R., Handbook of Analytical Derivatization Reactions, John Wiley & Sons: New York 1979.

Matter, L., Anwendung der chromatographischen Regeln in der Kapillar-GC, In: Matter L., Lebensmittel- und Umweltanalytik mit der Kapillar-GC, VCH: Weinheim 1994, 1–27.

Pierce, A. E., Silylation of Organic Compounds, Pierce Chemical Company: Rockford, III, 1982.

Restek Corporation, A Capillary Chromatography Seminar, Restek: Bellafonte PA (USA), 1993 (Firmenschrift).

Schomburg, G., Gaschromatographie, VCH: Weinheim 1987.

van Ysacker, P. G., Janssen, H. G., Snijders, H. M. J., van Cruchten, H. J. M., Leclercq, P. A., Cramers, C. A., High-Speed-Narrow-Bore-Capillary Gas Chromatography with Ion-Trap Mass Spectrometric Detection. 15th Int. Symp. Cap. Chromatogr. Riva del Garda 1993, In: P. Sandra (Ed.), Huethig 1993.

Section 2.2.3

Bock, R., Methoden der analytischen Chemie 1, Verlag Chemie, Weinheim 1974.

Desty, D. H., LC-GC Intl. *4(5)*, 32 (1991).

Martin, A. J. P., Synge R. L. M., J. Biol. Chem. *35,* 1358 (1941).

Schomburg, G., Gaschromatographie, Verlag Chemie, Weinheim 1987.

Section 2.2.4

Bretschneider, W., Werkhoff, P., Progress in All-Glass Stream Splitting Systems in Capillary Gas Chromatography Part I, HRC & CC 11 (1988), 543–546.

Bretschneider, W., Werkhoff, P., Progress in All-Glass Stream Splitting Systems in Capillary Gas Chromatography Part 11, HRC & CC 11 (1988), 589–592.

Ewender, J., Piringer O., Gaschromatographische Analyse flüchtiger aliphatischer Amine unter Verwendung eines Amin-spezifischen Elektrolytleitfähigkeitsdetektors, Dt.Lebensm.Rundschau 87 (1991), 5–7.
Hill, H. H., McMinn, D. G., Detectors for Capillary Gas Chromatography, John Wiley & Sons Inc., New York 1992.
Kolb, B., Otte, E., Gaschromatographische Detektoren. Manuskript. Technische Schule Bodenseewerk Perkin Elmer GmbH.
Piringer, O., Wolff, E., New Electrolytic Conductivity Detector for Capillary Gas Chromatography – Analysis of Chlorinated Hydrocarbons. J Chromatogr. 284 (1984), 373–380.
Schneider, W., Frohne, J. Ch., Brudderreck, H., Selektive gaschromatographische Messung sauerstoffhaltiger Verbindungen mittels Flammenionisationsdetektor, J. Chromatogr. 245 (1982), 71.
Wentworth, W. E., Chen, E. C. M. in: Electron Capture Theory and Practice in Chromatography, Zlatkis, A., Poole, C. F. (Eds.). Elsevier, New York 1981, 27.

Section 2.3.1

Balzers Fachbericht. Das Funktionsprinzip des Quadrupol-Massenspektrometers, Firmenschrift DN 9272.
Brunnée, C., The Ideal Mass Analyzer: Fact or Fiction?, Int. J. Mass Spectrom. Ion Proc. 76 (1987) 121–237.
Brunnée, C., Voshage, H., Massenspektroskopie, Thiemig: München 1964.
Feser, K., Kögler, W., The Quadrupole Mass Filter for GC/MS Applications, J. Chromatogr. Science 17 (1979), 57–63.
Miller, P. E., Denton, M. B., The Quadrupol Mass Filter: Basic Operating Concepts, J.Chem.Education 63 (1986) 617–622.
Paul, W., Reinhard, H. P., von Zahn, U., Das elektrische Massenfilter als Massenspektrometer und Isotopentrenner. Z. Physik 152 (1958), 143–182.
Paul, W., Steinwedel, H., Apparat zur Trennung von geladenen Teilchen mit unterschiedlicher spezifischer Ladung, Deutsches Patent 944900, 1956 (U. S. Patent 2939952 v. 7.6.1960).
Paul, W., Steinwedel, H., Ein neues Massenspektrometer ohne Magnetfeld. Z. Naturforschg. 8a (1953), 448–450.
Todd, J. F. J., Instrumentation in Mass Spectrometry, In: Advances in Mass Spectrometry, J. F. J. Todd (Ed.), Wiley: New York 1986, 35–70.
Van Ysacker, P. G., Janssen, H. M., Leclerq, P. A., Wollnik, H., Cramers, C. A., Comparison of Different Mass Spectrometers in Combination with High-Speed Narrow-Bore Capillary Gas Chromatography, In: 16th Int. Symp. Cap. Chrom. Riva del Garda, P. Sandra (Ed.), Huethig 1994, 785–796.
Wang, J., The Determination of Elemental Composition by Low Resolution MS/MS, Proceedings of the 43rd ASMS Conference. Atlanta 21–26 May 1995, 722.

Section 2.3.2

Ardenne, M., Steinfelder, K., Tümmler, R., Elektronenanlagerungsmassenspektrometrie organischer Substanzen, Springer Verlag: Berlin 1971.
Aue, D. H., Bowers, M. T., Stability of Positive Ions from Equilibrium Gas-Phase Basicity Measurements, in: M. T. Bowers (Ed.): Gas Phase Ion Chemistry Vol. 2, Academic Press: New York 1979.
Bartmess, J., McIver, R., The Gas Phase Acidity Scale, in: M. T. Bowers (Ed.): Gas Phase Ion Chemistry Vol. 2, Academic Press: New York 1979.
Beck, H., Eckart, K., Mathar, W., Wittkowski, R., Bestimmung von polychlorierten Dibenzofuranen (PCDF) und Dibenzodioxinen (PCDD) in Lebensmitteln im ppq-Bereich, Lebensmittelchem. Gerichtl. Chem. 42 (1988), 101–105.

Bowadt, Frandsen, E., et al., Combined Positive and Negative Ion Chemical Ionisation for the Analysis of PCB'S, In: P.Sandra (Ed.), 15th Int. Symp. Cap. Chrom., Riva del Garda, May 1993, Huethig 1993.

Budzikiewicz, H., Massenspektrometrie, 3. erw. Aufl., VCH: Weinheim 1993.

Budzikiewicz, H., Massenspektrometrie negativer Ionen, Angew.Chem. 93 (1981), 635–649.

Buser, H.-R., Müller, M., Isomer- and Enantiomer-Selective Analyses of Toxaphene Components Using Chiral High-Resolution Gas Chromatography and Detection by Mass Spectrometry/Mass Spectrometry, Environ. Sci. Technol. 28 (1994), 119–128.

Class, T. J., Determination of Pyrethroids and their Degradation Products in Indoor Air and on Surfaces by HRGC-ECD and HRGC-MS (NCI), In: P. Sandra (Ed.), 12th Int. Symp. Cap. Chrom., Riva del Garda, May 1991, Huethig 1991.

Crow, F. W., Bjorseth, A., et al., Anal. Chem. 53 (1981), 619.

DePuy, C. H., Grabowski, J. J., Bierbaum, V. M., Chemical Reactions of Anions in the Gas Phase, Science 218 (1982), 955–960.

Dorey, R. C., Williams, K., Rhodes, C. L., Fossler, C. L., Heinze, T. M., Freeman, J. P., High Kinetic Energy Chemical Ionization in the Quadrupole Ion Trap: Methylamine CIMS of Amines, Presented at the 42nd ASMS Conference on Mass Spectrometry and Allied Topics, Chicago. June 1–6, 1994.

Dougherty, R. C., Detection and Identification of Toxic Substances in the Environment, In: C. Merritt, C. McEwen (Eds.): Mass Spectrometry Part B, Marcel Dekker: New York 1980, 327.

Dougherty, R. C., Negative Chemical Ionization Mass Spectrometry, Anal.Chem. 53 (1981) 625A–636A.

Frigerio, A., Essential Aspects of Mass Spectrometry, Halsted Press: New York 1974.

Hainzl, D., Burhenne, J., Parlar, H., Isolierung von Einzelsubstanzen für die Toxaphenanalytik, GIT Fachz. Lab. 4 (1994) 285–294.

Harrison, A. G., Chemical Ionization Mass Spectrometry, CRC Press: Boca Raton 1983.

Horning, E. C., Caroll, D. I., Dzidic, I., Stillwell, R. N., Negative Ion Atmospheric Pressure Ionization Mass Spectrometry and the Electron Capture Detector. In: A. Zlatkis, C. F. Poole (Eds.): Electron Capture, J. Chromatogr. Library Vol. 20, Elsevier: Amsterdam 1981.

Howe, I., Williams, D. H., Bowen, R., Mass Spectrometry, 2nd Ed., McGraw Hill: New York 1981.

Hübschmann, H.-J., Einsatz der chemischen Ionisierung zur Analyse von Pflanzenschutzmitteln, Finnigan MAT Application Report No. 75, 1990.

Hunt, D. F., Stafford, G. C., Crow, F., Russel, J., Pulsed Positive Negative Ion Chemical Ionization Mass Spectrometry, Anal.Chem. 48 (1976), 2098–2105.

Keller, P. R., Harvey, G. J., Foltz, D. J., GC/MS Analysis of Fragrances Using Chemical Ionization on the Ion Trap Detector: An Easy-to-Use Method for Molecular Weight Information and Low Level Detection, Finnigan MAT Application Report No. 220, 1989.

McLafferty, F. W., Michnowicz, J. A., Chem. Tech. 22 (1992), 182.

McLafferty, F. W., Turecek, F., Interpretation of Mass Spectra. 4th Ed., University Science Books: Mill Valley 1993.

Munson, M. S. B., Field, F. H., J. Am. Chem. Soc. 88 (1966), 4337.

Schröder, E., Massenspektroskopie, Springer: Berlin 1991.

Smit, A. L. C., Field, F. H., Gaseous Anion Chemistry. Formation and Reaction of OH^-; Reactions of Anions with N_2O; OH^- Negative Chemical Ionization, J Am. Chem. Soc. 99 (1977), 6471–6483.

SPECTRA, Analytical Applications of Ion Trap Mass Spectrometry, Vol 11 (2), 1988.

Spiteller, M., Spiteller, G., Massenspektrensammlung von Lösungsmitteln, Verunreinigungen. Säulenbelegmaterialien und einfachen aliphatischen Verbindungen, Springer: Wien 1973.

Stan, H.-J., Kellner, G., Analysis of Organophosphoric Pesticide Residues in Food by GC/MS Using Positive and Negative Chemical Ionisation, In: W. Baltes, P. B. Czedik-Eysenberg, W. Pfannhauser (Eds.), Recent Developments in Food Analysis, Verlag Chemie: Weinheim 1981.

Stout, S. J., Steller, W. A., Application of Gas Chromatography Negative Ion Chemical Ionization Mass Spectrometry in Confirmatory Procedures for Pesticide Residues, Biomed. Mass Spectrom. 11 (1984), 207–210.

Section 2.3.3

Brodbelt, J. S., Cooks, R. G., Ion Trap Tandem Mass Spectrometry, Spectra 11/2 (1988) 30–40.
Brunée, C., New Instrumentation in Mass Spectrometry, Spectra 9, 2/3 (1983) 10–36.
Brunée, C., The Ideal Mass Analyzer: Fact or Fiction?, Int. J. Mass Spec. Ion Proc. 76 (1987) 121–237.
Busch, K. L., Glish, G. L., McLuckey, S. A., Mass Spectrometry/Mass Spectrometry: Techniques and Applications of Tandem Mass Spectrometry, VCH Publishers: New York 1988.
Dawson, P. H., In: Dawson, P. H., Ed.: Quadrupole Mass Spectrometry and its Applications, Elsevier: Amsterdam 1976, 19–70.
DFG Deutsche Forschungsgemeinschaft, Manual of Pesticide Residue Analysis Vol. II, Thier, H.P., Kirchhoff, J., Eds., VCH: Weinheim 1992, 25–28.
Johnson, J. V., Yost, R. A., Anal. Chem. 57 (1985) 758A.
Johnson, J. V., Yost, R. A., Kelley, P. E., Bradford, D. C., Tandem-in-Space and Tandem-in-Time Mass Spectrometry: Triple Quadrupoles and Quadrupole Ion Traps, Anal. Chem. 62 (1990) 2162–2172.
Julian, R. K., Nappi, M., Weil, C., Cooks, R. G., Multiparticle Simulation of Ion Motion in the Ion Trap Mass Spectrometer: Resonant and Direct Current Pulse Excitation, J. Am. Soc. Mass spectrom. 6 (1995) 57–70.
Kaiser, H., Foundations for the Critical Discussion of Analytical Methods, Spectrochim. Acta Part B, 33b (1978) 551.
March, R. E., Hughes, R. J., Quadrupole Storage Mass Spectrometry, John Wiley: New York 1989.
McLafferty, F. E., Ed.: Tandem Mass Spectrometry, John Wiley: New York 1983.
Noble, D., MS/MS, Flexes Ist Muscles, Anal. Chem. 67 (1995) 265A–269A.
Plomley, J. B., Koester, C. J., March, R. E., Determination of N-Nitrosodimethylamine in Complex Environmental Matrices by Quadrupole Ion Storage Tandem Mass Spectrometry Enhanced by Unidirectional Ion Ejection, Anal. Chem. 66 (1994) 4437–4443.
Slayback, J. R. B., Taylor, P. A., Analysis of 2.3,7,8-TCDD and 2,3,7,8-TCDF in Environmental Matrices Using GC/MS/MS Techniques. Spectra 9/4 (1983) 18–24.
Soni, M. H., Cooks, R. G., Selective Injection and Isolation of Ions in Quadrupole Ion Trap Mass Spectrometry Using Notched Waveforms Created Using the Inverse Fourier Transform. Anal. Chem. 66 (1994) 2488–2496.
Strife, R. J., Tandem Mass Spectrometry of Prostaglandins: A Comparison of an Ion Trap and a Reversed Geometry Sector Instrument, Rap. Comm. Mass Spec. 2 (1988) 105–109.
Wagner-Redeker, W., Schubert, R., Hübschmann, H.-J., Analytik von Pestiziden und polychlorierten Biphenylen mit dem Finnigan 5100 GC/MS-System, Finnigan MAT Application Report No. 58, 1985.
Yasek, E., Advances in Ion Trap Mass Spectrometry, Finnigan MAT Technical Report No. 614, Finnigan Corporation: San Jose USA 1989.
Yost, R. A., MS/MS: Tandem Mass Spectrometry, Spectra 9/4 (1983) 3–6.

Section 2.4.2

Abraham, B., Rückstandsanalyse von anabolen Wirkstoffen in Fleisch mit Gaschromatographie-Massenspektrometrie, Dissertation, Technische Universität Berlin, 1980.
Cramers, C. A., Scherpenzeel, G. J., Leclercq, P. A., J. Chromatogr. 203 (1981), 207.
Gohlke, R. S., Time-of-Flight Mass Spectrometry and Gas-Liquid Partition Chromatography, Anal. Chem. 31 (1959), 535–541.
Henneberg, D., Henrichs, U., Husmann, H., Schomburg, G., J. Chromatogr. 187 (1978), 139.
Schulz, J., Nachweis und Quantifizieren von PCB mit dem Massenselektiven Detektor, Labor-Praxis 6 (1987), 648–667.
Tiebach, R., Blaas, W., Direct Coupling of a Gas Chromatograph to an Ion Trap Detectoir, J. Chromatogr. 454 (1988), 372–381.

3 Evaluation of GC/MS Analyses

3.1 Display of Chromatograms

Chromatograms obtained by GC/MS are plots of the intensity of the signal against the retention time, as with classical GC detectors. Nevertheless, there are considerable differences between the two types of chromatogram arising from the fact that data from GC/MS analyses are in three dimensions. Figure 3-1 shows a section of the chromatogram of the total ion current in the analysis of volatile halogenated hydrocarbons. The retention time axis also shows the number of continually registered mass spectra (scan no.). The mass axis is drawn above the time axis at an angle. The elution of each individual substance can be detected by evaluating the mass spectra using a 'maximising masses peak finder' program and can be shown by a marker. Each substance-specific ion shows a local maximum at these positions, which can be determined by the peak finder. The mass spectra of all the analytes detected are shown in a three dimensional

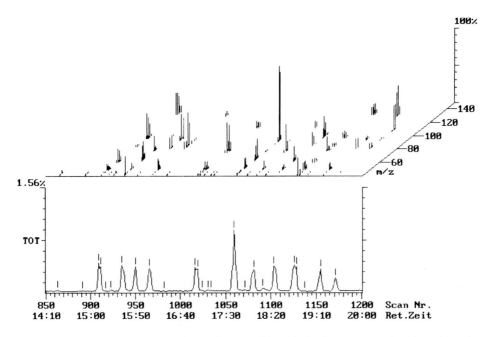

Fig. 3-1 Three-dimensional data field of a GC/MS analysis showing retention time, intensity and mass axis

representation for the purposes of screening. For further evaluation the spectra can be examined individually.

3.1.1 Total Ion Current Chromatograms

The intensity axis in GC/MS analysis is shown as a total ion current (TIC) or as a calculated ion chromatogram (reconstructed ion chromatogram, RIC). The intensity scale may be given in absolute values, but a percentage scale is more frequently used. Both terms describe the mode of representation characteristic of the recording technique. At constant scan rates the mass spectrometer plots spectra over the pre-selected mass range and thus gives a three-dimensional data field arising from the retention time, mass scale and intensity. A signal size equivalent to FID detection is not directly available. (Magnetic sector mass spectrometers were equipped with a total ion current detector directly at the ion source until the end of the 1970s!). A total signal intensity comparable to the FID signal at a particular point in the scan can, however, be calculated from the sum of the intensities of all the ions at this point. All the ion intensities of a mass spectrum are added together by the data system and stored as a total intensity value (total ion current) together with the spectrum. The total ion current chromatogram thus constructed is therefore dependent on the scan range used for data acquisition. When making comparisons it is essential to take the data acquisition conditions into consideration.

SIM/MID analyses give a chromatogram in the same way but no mass spectrum is retrievable. The total ion current in this case is composed of the intensities of the selected ions. Analyses where switching from individual masses to fixed retention times is planned often show clear jumps in the base line (see Fig. 2-138).

The appearance of a GC/MS chromatogram (TIC/RIC) showing the peak intensities is therefore strongly dependent on the mass range shown. The repeated GC/MS analysis of one particular sample employing mass scans of different widths leads to peaks of different heights above the base line of the total ion current. The starting mass of the scan has a significant effect here. The result is a more or less strong recording of an unspecific background which manifests itself in a higher or lower base line of the TIC chromatogram. Peaks of the same concentration are therefore shown with different signal/noise ratios in the total ion current at different scan ranges. In spite of differing representation of the substance peaks, the detection limit of the GC/MS system naturally does not change. Particularly in residue analysis the concentration of the analytes is usually of the same order of magnitude or even below that of the chemical noise (matrix) in spite of good sample processing so that the total ion current cannot represent the elution of these analytes. Only the use of selective information from the mass chromatogram (see Section 3.1.2) brings the substance peak sought on to the screen for further evaluation.

In the case of data acquisition using selected individual masses (SIM/MID), only the changes in intensity of the masses selected before the analysis are shown. Already during data acquisition only those signals (ion intensities) are recorded from the total ion current which correspond to the prescriptions of the user. The greater part of the total ion current is therefore not detected using the SIM/MID technique (see Section 2.3.3). Only substances which give signals in the region of the selected masses as a result of fragment or molecular ions are shown as peaks. A mass spectrum for the purpose of checking identity is therefore not available. For confirmation this should be measured separately in a subsequent analysis. The retention time and the relative intensities of two or three specific lines are used as qualifying features. In residue analysis unambiguous detection is never possible using this method. Positive results of an SIM/MID analysis basically require additional confirmation by a mass spectrum.

3.1.2 Mass Chromatograms

A meaningful assessment of signal/noise ratios of certain substance peaks can only be carried out using mass chromatograms of substance-specific ions (fragment/molecular ions). The three-dimensional data field of GC/MS analyses in the full scan mode does not only allow the determination of the total ion intensity at a point in the scan. To show individual analytes selectively the intensities of selected ions (masses) from the total ion current are shown and plotted as an intensity/time trace (chromatogram).

The evaluation of these mass chromatograms allows the exact determination of the detection limit above the signal/noise ratio of the substance-specific ion produced by a compound. With the SIM/MID mode this ion would be detected exclusively, but a complete mass spectrum for confirmation would not be available. In the case of complex chromatograms of real samples mass chromatograms offer the key to the isolation of co-eluting components so that they can be integrated perfectly and quantified.

In the analysis of lemons for residues from plant protection agents a co-elution situation was discovered by data acquisition in the full scan mode of the ion trap detector and was evaluated using a mass chromatogram.

The routine testing with an ion trap GC/MS system gives a trace which differs from that using an element-specific NPD detector (Fig. 3-2). A large number of different peaks appear in the retention region which indicates the presence of Quinalphos as the active substance in the NPD evaluation (Fig. 3-3). The Quinalphos peak has a shoulder on the left side and is closely followed by another less intense component. In the mass chromatogram of the characteristic individual masses (fragment ions) it can be deduced from the total ion current that another eluting active substance is present (Fig. 3-4). Un-

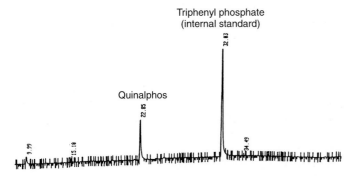

Fig. 3-2 Analysis of a lemon extract using the NPD as detector. The chromatogram shows the elution of a plant protection agent component as well as the internal standard

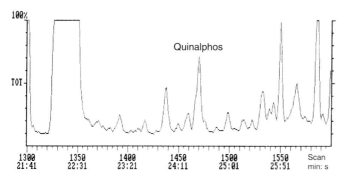

Fig. 3-3 Confirmation of the identity of a lemon extract by GC/MS. The total ion current clearly shows the questionable peak with shoulder mass spectra. A co-eluting second active substance, Chlorfenvinphos, gives rise to the shoulder.

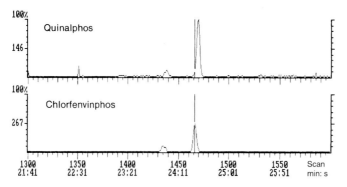

Fig. 3-4 Mass chromatograms for the specific masses show the co-elution of Quinalphos (m/z 146) and Chlorfenvinphos (m/z 267.). The retention time range is identical with that in Fig. 3-3. The selective plot of the mass signals can be identified as part of the total ion current.

like NPD detection, with GC/MS analysis it becomes clear after evaluating the mass chromatogram and mass spectra that the co-eluting substance is Chlorfenvinphos.

In routine analysis this evaluation is carried out by the computer. If the information on the retention time of an analyte, the mass spectrum, the selective quantifying mass and a valid calibration are supplied, a chromatogram can be evaluated in a very short time for a large number of components (Fig. 3-5, see Section 3.3).

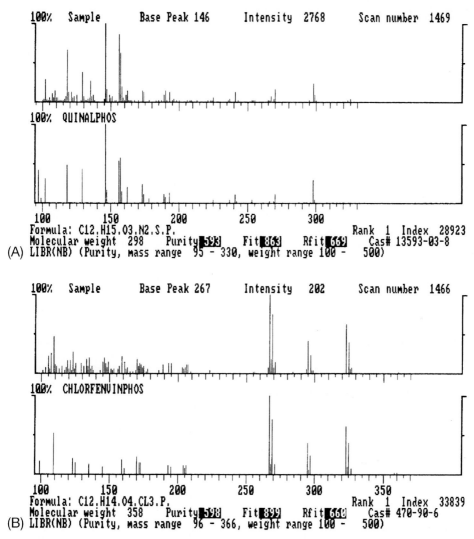

Fig. 3-5 The phosphoric acid esters are successfully identified by a library comparison after extraction of the spectra by a background subtraction.
(A) Quinalphos is confirmed by comparison with the NBS library (FIT value 863)
(B) Chlorfenvinphos is confirmed by a comparison with the NBS library (FIT value 899)

3.2 Substance Identification

3.2.1 Extraction of Mass Spectra

One of the great strengths of mass spectrometry is the immediate provision of direct information about an eluting component. The careful extraction of the substance-specific signals from the chromatogram is critical for reliable identity determination. For identification or confirmation of individual GC peaks recording mass spectra which are as complete as possible is an important basic prerequisite.

By plotting mass chromatograms co-elution situations can be discovered, as shown in Section 3.1.2. The mass chromatograms of selected ions give important information via their maximising behaviour. Only when maxima are shown at exactly the same time can it be assumed that the fragments observed originate from a single substance, i.e. from the same chemical structure. The only exception is the ideal simultaneous co-elution of compounds. If peak maxima with different retention times are shown by various ions, it must be assumed that there are co-eluting components (Fig. 3-4).

By subtraction of the background or the co-elution spectra before or behind a questionable GC peak the mass spectrum of the substance sought is extracted from the chromatogram as free as possible from other signals. All substances co-eluting with an unknown substance including the matrix components and column bleed are described in this context as chemical background. The differentiation between the substance signals and the background and its elimination from the substance spectrum is of particular im-

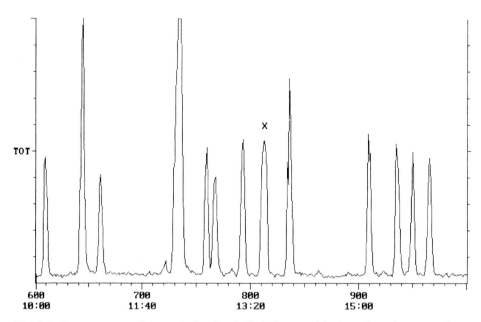

Fig. 3-6 Chromatogram of an analysis of volatile halogenated hydrocarbons by purge & trap GC/MS. The component marked with X has a larger half width than the neighbouring peaks

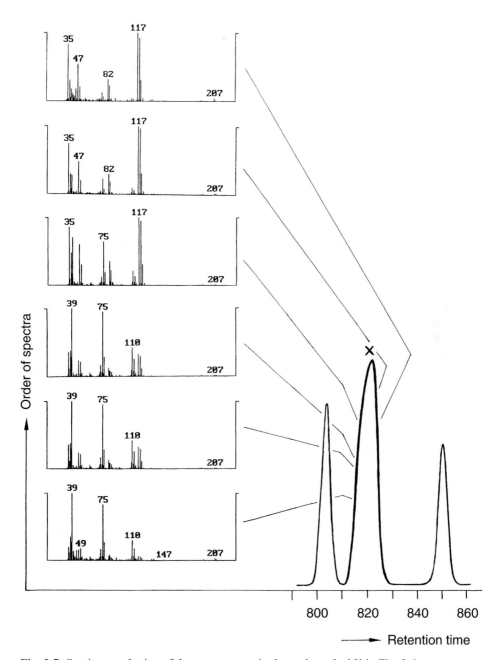

Fig. 3-7 Continuous plotting of the mass spectra in the peak marked X in Fig. 3-6

portance for successful spectroscopic comparison in a library search. In the example of the GC/MS analysis of lemons for plant protection agents described above this procedure is used to determine the identity of the active substances.

The possibilities for subtraction of mass spectra are shown in the following real example of the analysis of volatile halogenated hydrocarbons by purge and trap-GC/MS. Figure 3-6 shows part of a total ion current chromatogram. The peak marked with X shows a larger half-width than that of the neighbouring components. On closer inspection of the individual spectra in the peak it can be seen that in the rising slope of the peak ions with m/z 39, 75, 110 and 112 dominate. As the elution of the peak continues other ion signals appear. The ions with m/z 35, 37, 82, 84, 117, 119 and 121 appear in increased strength, while the previously dominant signals decrease. Figure 3-7 shows this situation using the continuing presentation of individual mass spectra in a characterised GC peak. (The signal m/z 207 arises from the stationary silicone phase).

From the individual mass spectra (Fig. 3-7) it can be recognised that some signals obviously belong together. Figure 3-8 shows the mass chromatograms of the ions m/z 110/112 and m/z 117/119 (as a sum in each case) above the total ion current from the detected mass range of m/z 33-260. The mass chromatograms show an intense GC peak at the questionable retention time in each case. The peak maxima are not superimposed and are slightly shifted towards each other. This is an important indication of the co-elution of two components (Fig. 3-9).

If other ions are included in this first mass analysis, it can be concluded that the fragments belong together from their common maximising behaviour.

After the individual mass signals have been assigned to the two components, the extraction of the spectrum of each compound can be performed. Figure 3-10 shows the di-

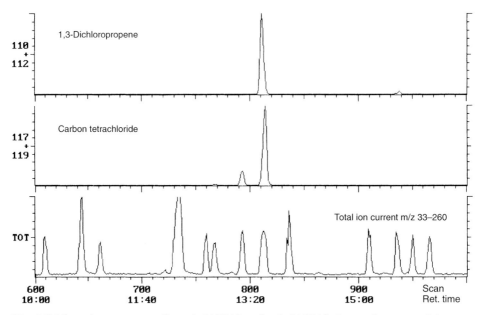

Fig. 3-8 Mass chromatograms for m/z 110/112 and m/z 117/119 shown above a total ion current chromatogram

3.2 Substance Identification   221

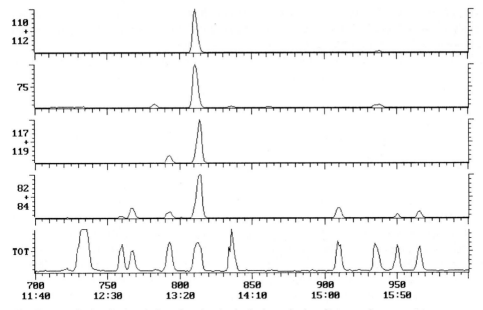

Fig. 3-9 Analysis of a co-elution situation by inclusion of other fragment ions

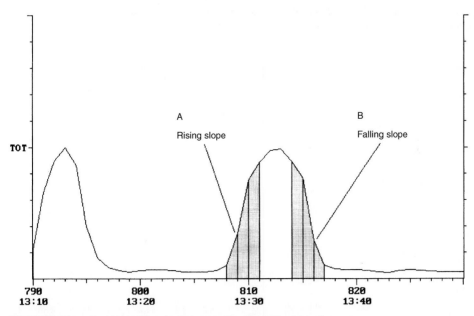

Fig. 3-10 Plot of a peak with selected areas for spectral subtraction

vision of the peak into the front peak slope A and the back peak slope B. With the background subtraction function contained in all data systems the spectra in the areas A and B are added and subtracted from one another.

The subtraction of the areas A and B gives the clean spectra of the co-eluting analytes. In Fig. 3-11 the subtraction A – B shows the spectrum of the first component, which is shown to be 1,3-dichloropropene (Fig. 3-12) from a library comparison. The reverse procedure, i.e. the subtraction B – A, gives the identity of the second component (Figs. 3-13 and 3-14).

A further frequent use of spectrum subtraction allows the removal of background signals caused by the matrix or column bleed. Figure 3-15 shows the elution of a minor component from the analysis of volatile halogenated hydrocarbons, which elutes in the region where column bleed begins. For background subtraction the spectra in the peak and from the region of increasing column bleed are subtracted from one another.

The result of the background subtraction is shown in Fig. 3-16. While the spectrum clearly shows column bleed from the substance peak with m/z 73, 207 and a weak CO_2 background at m/z 44, the resulting substance spectrum is free from the signals of the interfering chemical background after the subtraction. This clean spectrum can then be used for a library search in which it can be confirmed as 1,2-dibromo-3-chloropropane.

In the subtraction of mass spectra it should generally be noted that in certain cases substance signals can also be reduced. In these cases it is necessary to choose another background area. If changes in the substance spectrum cannot be prevented in this way,

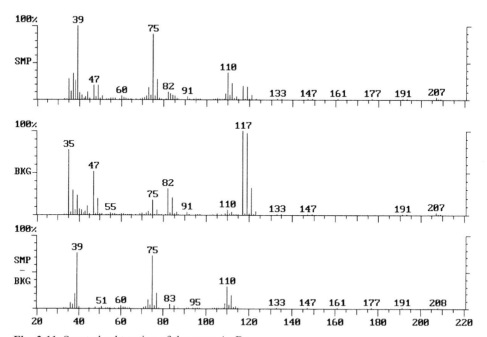

Fig. 3-11 Spectral subtraction of the areas A–B:
SMP   = Spectra of the rising peak slope (A), sample
BKG   = Spectra of the falling peak slope (B), background
SMP-BKG = Resulting spectrum of the component eluting first

3.2 Substance Identification   223

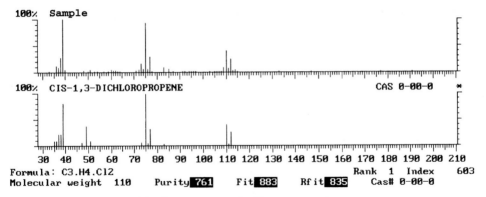

Fig. 3-12 Identification of the first component by library comparison

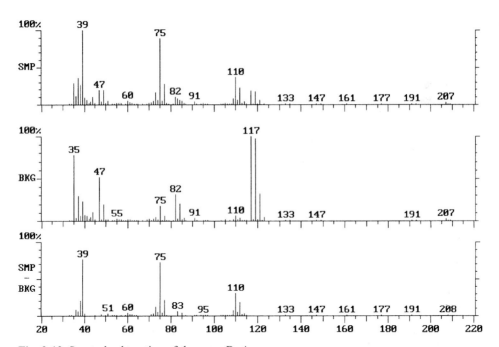

Fig. 3-13 Spectral subtraction of the areas B–A:
SMP   = Spectra of the falling peak slope (B), sample
BKG   = Spectra of the rising peak slope (A), background
SMP-BKG = Resulting spectrum of the component eluting second

the library search should be carried out with a minimal proportion of chemical noise. In the library search programs of individual manufacturers there is also the possibility of editing the spectrum before the start of the search. In critical cases this option should also be followed to remove known interfering signals resulting from the chemical noise from the substance spectrum.

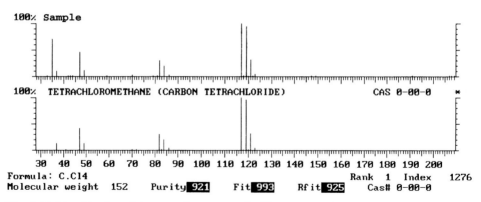

Fig. 3-14 Identification of the second component by library comparison

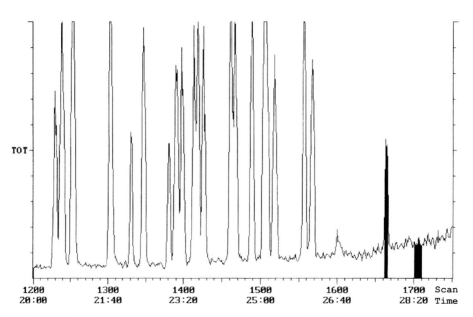

Fig. 3-15 The total ion current chromatogram of an analysis of volatile halogenated hydrocarbons shows the elution of a minor component at the beginning of column bleed (the areas of background subtraction are shown in black)

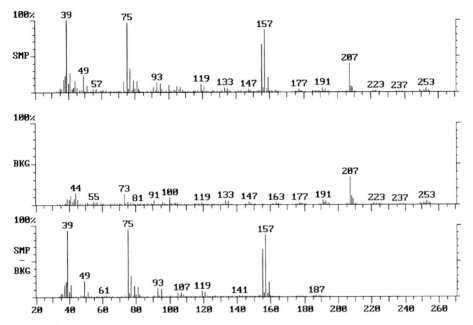

Fig. 3-16 Result of background subtraction from Fig. 3-15:
SMP   = Spectra from the substance peak, sample
BKG   = Spectra from the background (column bleed)
SMP-BKG = Resulting substance spectrum

3.2.2 The Retention Index

If the chromatographic conditions are kept constant, the retention times of the compounds remain the same. All identification concepts using classical detectors function on this basis. The retention times of compounds, however, can change through ageing of the column and more particularly through differing matrix effects.

The measurement of the retention times relative to a co-injected standard can help to overcome these difficulties. Fixed retention indices (RI) are assigned to these standards. An analyte is included in a retention index system with the RI values of the standards eluting before and after it. It is assumed small variances in the retention times affect both the analyte and the standards so that the RI values calculated remain constant.

The first retention index system to become widely used was developed by Kovats. In this system a series of n-alkanes is used as the standard. Each n-alkane is assigned the value of the number of carbon atoms multiplied by 100 as the retention index (pentane 500, hexane 600, heptane 700 etc.). For isothermal operations the RI values for other substances are calculated as follows:

$$\text{Kovats index} \quad RI = 100 \cdot c + 100 \; \frac{\log(t'_R)_x - \log(t'_R)_c}{\log(t'_R)_{c+1} - \log(t'_R)_c} \tag{24a}$$

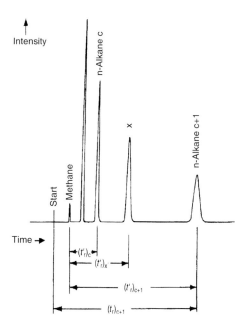

Fig. 3-17 Determination of the Kovats index for a substance X by interpolation between two n-alkanes (after Schomburg)

The t'_R values give the retention times of the standards and the substance corrected for the dead time t_0 ($t'_R = t_R - t_0$). As the dead time is constant in the cases considered, uncorrected retention times are mostly used. The determination of the Kovats indices (Fig. 3-17) can be carried out very precisely and on comparison between various different laboratories is reproducible within ± 10 units. In libraries of mass spectra the retention indices are also given (see the terpene library by Adams, the pesticide library by Ockels, the toxicology library by Pfleger/Maurer/Weber).

On working with linear temperature programs a simplification is used which was introduced by Van den Dool and Kratz, whereby direct retention times are used instead of the logarithmic terms used by Kovats:

$$\text{Modified Kovats index} \quad RI = 100 \cdot c + 100 \, \frac{(t'_R)_x - (t'_R)_c}{(t'_R)_{c+1} - (t'_R)_c} \tag{24b}$$

The weakness of retention index systems lies in the fact that not all analytes are affected by variances in the measuring system to the same extent. For these special purposes homologous series of substances which are as closely related as possible have been developed. For use in residue analysis in environmental chemistry and particularly for the analysis of plant protection agents and chemical weapons, the homologous M-series (Fig. 3-18) of n-alkylbis(trifluoromethyl)phosphine sulfides has been synthesised.

The molecule in the M-series contains active groups which also respond to the selective detectors ECD, NPD, FPD and PID and naturally also give good responses in FID and MS (detection limits: ECD ca. 1 pg, FID ca. 300 pg, Fig. 3-19). In the mass spectrometer all components of the M-series show intense characteristic ions at M-69 and M-101 and a typical fragment at m/z 147 (Fig. 3-20). The M-series can be used with positive and negative chemical ionisation.

3.2 Substance Identification 227

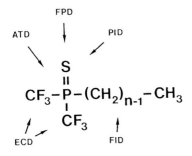

Fig. 3-18 n-Alkylbis(trifluoromethyl)phosphine sulfides (M series with n = 6, 8, 10 ... 20)

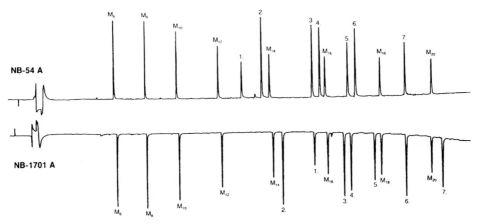

Fig. 3-19 Chromatograms of the M series and of pesticides (phosphoric acid esters) on columns of different polarities (after HNU/Nordion).
Carrier gas He, detector NPD, program: 50 °C (2 min), 150 °C (20°/min), 270 °C (6°/min)
Components: M series M_6, M_8, ... M_{20}, 1 Dimethoate, 2 Diazinon, 3 Fenthion, 4 Trichloronate, 5 Bromophos-ethyl, 6 Ditalimphos, 7 Carbophenothion

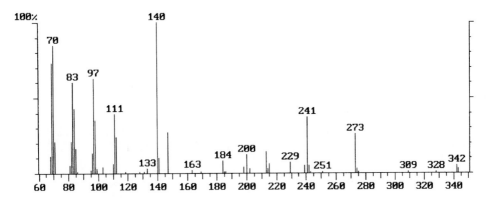

Fig. 3-20 M Series: EI Mass Spectrum of the component M 10 (HNU)

Comparison of calculated retention indices with empirically determined values

Substance	Calculated RI	Determined RI
Atrazine	1500	1716
Parathion-ethyl	2000	1970
Triadimefon	2000	1979

Examples of retention index calculation

Parathion-ethyl: $C_{10}H_{14}NO_5PS$

Number	Element	Contribution	Total
10	C	100	1000
14	H	–	–
1	N	100	100
5	O	100	500
1	P	200	200
1	S	200	200
		Sum	2000

Atrazine: $C_8H_{14}ClN_5$

Number	Element	Contribution	Total
8	C	100	800
14	H	–	–
1	Cl	200	200
5	N	100	500
		Sum	1500

Triadimefon: $C_{14}H_{14}ClN_3O_2$

Number	Element	Contribution	Total
14	C	100	1400
14	H	–	–
1	Cl	200	200
3	N	100	300
2	O	100	200
1	tert-C	–100	–100
		Sum	2000

3.2 Substance Identification

The use of retention indices in spite of, or perhaps because of, the wide use of GC/MS systems is now becoming more important again as a result of the outstanding stability of fused silica capillaries and the good reproducibility of gas chromatographs now available. The broadening of chromatography data systems with optional evaluation routines is just beginning. These are especially dedicated to the processing of retention indices, e. g. for two-column systems.

If the retention index of a compound is not known, it can also be estimated from empirical considerations of the elements and partial structures present in the molecule (Tables 3-1 and 3-2). A first approximation can already be made using the empirical formula of an analyte. This is particularly valuable for assessing suggestions from the GC/MS library search because, besides good correspondence to a spectrum, plausibility with regard to the retention behaviour can be tested (see also Section 4-14). According to Weber the values determined give a correct estimation to within 10%.

Polar groups with hydrogen bonding increase the boiling point of a compound and are thus responsible for stronger retention. For the second and every additional polar group the retention index increases by 150 units. Branches in the molecule increase the volatility. For each quaternary carbon atom present in a t-butyl group the retention index is reduced by 100 units. Values can be estimated with higher precision from retention indices for known structures by calculating structure elements according to Tables 3-1 and 3-2.

Table 3-1 Contributions to the determination of the retention index from the empirical formula (after Weber)

Element	Index contribution
H, F	0
C, N, O	100
Si in $Si(CH_3)_3$	0
P, S, Cl	200
Br	300
J	400

Table 3-2 Retention behaviour of structural isomers

Alkyl branches:	tertiary < secondary < n-alkyl
Disubstituted aromatics:	ortho < meta, para

3.2.3 Libraries of Mass Spectra

In electron impact ionisation (70 eV) a large number of fragmentation reactions take place with organic compounds. These are independent of the manufacturer's design of the ion source. The focusing of the ion source has a greater effect on the characteristics of a mass spectrum, which leads to a particularly wide adjustment range, especially in the case of quadrupole analysers. The relative intensities of the upper and lower mass ranges can easily be reversed. In the early days of use of quadrupole instruments this possibility was highly criticised by those using the established magnetic instruments. The problem is resolved in so far as both the manual and the automatic tuning of the instruments are aimed at giving the intensities of a reference compound (in contrast SIM tuning aims to give high sensitivity within a specific mass range, see Section 2.3.3.2). Perfluorotributylamine (FC43) is now used as the reference substance in all GC/MS systems. Other influences on the mass spectrum in GC/MS systems are caused by changing the substance concentration during the mass scan (beam instruments). On running spectra over a large mass range (e.g. in the case of methyl stearate with a scan of 50 to 350 u in 1 s, which corresponds to a standard adjustment in routine analysis) sharp GC peaks lead to a reversal of intensities (skewing) between the front and back slopes of the peak. The reversal of intensities is thus the opposite of the true situation. This effect can only be counteracted by the use of fast scan rates, which, however, result in loss of sensitivity. In practice standardised spectra must be used for these systems in order to calculate the compensation (for background subtraction see Section 3.2.1). Ion trap mass spectrometers do not show this reversal of intensities because there is parallel storage of all the ions formed. A mass spectrum should therefore not be regarded as naturally constant, but the result of an extremely complex process.

In practice the variations observed affect the relative intensities of particular groups of ions in the mass spectrum. The fragmentation processes are not affected (the same fragments are found with all GC/MS instruments) nor are the isotope ratios which result from natural distribution. Only adherence to a parameter window which is as narrow as possible (so-called standard conditions) during data acquisition creates the desired independence from the external influences described.

The comparability of the mass spectra produced is thus ensured for building up libraries of mass spectra. All commercially available libraries were run under the standard conditions mentioned and allow the comparison of the fragmentation pattern of an unknown substance with those available from the library. For the large universal libraries it should be assumed that most of the spectra were not run with GC/MS systems, and that today many reference spectra are still run using a solid sample inlet (connecting rod) or similar inlet techniques. For example, the reference spectrum of Aroclor 1260 (a mixture of PCBs with a 60% degree of chlorination) can only be explained in this way. Information on the inlet system used is rarely found in the libraries.

EI spectra are naturally particularly informative because of their fragmentation patterns. All search processes through libraries of spectra are still based on EI spectra. With the introduction of the highly reproducible advanced chemical ionisation into ion trap mass spectrometers, the first commercial CI library with over 300 pesticides was produced (Finnigan, 1992). The introduction of substructure libraries (MS/MS product ion spectra) is currently being discussed. The commercially available libraries are divided into general extensive collections and special task-related collections with a narrow range of applications.

3.2.3.1 Universal Libraries of Mass Spectra

The most extensive commercial library is undoubtedly the combined Wiley/NIST Registry of Mass Spectral Data, the origins of which date back to 1963. The sixth edition contains more than 260 000 spectra (Table 3-3). Because this library is mainly used for the PBM search procedure (see Section 3.2.4.2), in some of the entries there are several spectra available for each substance, involving different scanning conditions and spectrum qualities (Fig. 3-21). Besides the mass spectra, the database contains structural formulae, names of commercial products, trivial names and details of the origin of the spectrum. It is the only library to give information on the inlet procedure, the instrument and literature references in a reference table for each spectrum. Further work on the database has been undertaken by Prof. Fred W. McLafferty, Cornell University, Ithaca, New York. The collection of spectra is also available in book form as Version 5.0 and consists of seven volumes with the spectra of ca. 123 000 compounds.

The very broad based NIST/EPA/NIH Mass Spectral Database is equally important. It started as a project of the EPA (US Environmental Protection Agency) and the NIH (National Institute of Health) in the early 1970s (Table 3-4). It was first published in

Table 3-3 Technical data of the Wiley library, 6. Ed., 1994 (Palisade)

Total number of spectra	262 000
of which NIST spectra	29 000
CAS registry numbers	214 000
CAS structures	120 000
Chemical names	845 000
PBM search times	1 s

Table 3-4 NIST/EPA/NIH Mass Spectral Library – NIST 98 Version

Compounds	107 886
Chemical structures	107 829
Spectra	129 136
Replicate spectra	21 250
Compounds with replicate spectra	13 205
Average peaks/spectrum	93
Median/peaks/spectrum	78

Major practical collections included:
Chemical Concepts (including Dr. Henneberg's industrial chemicals collection)
Georgia and Virginia Crime Laboratories
TNO Flavors and Fragrances
AAFS Toxicology Section Drug Library
Association of Official Racing Chemists
St. Louis University Urinary Acids
VERIFIN & CBDCOM Chemical Weapons

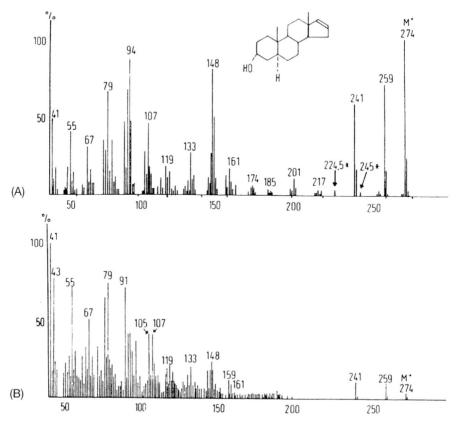

Fig. 3-21 Mass spectrum of 3-hydroxy-5α-androst-16-ene (after Budzikiewicz 1998)
(A) solid probe inlet
(B) GC/MS analysis

1978 by S. R. Heller and G. W. A Milne with ca. 25 000 spectra. The current version has more than 129 000 spectra. The quality of the spectra in the NIST library is constantly checked. Nine factors are taken into account for determining the quality index. In new versions spectra with a higher quality index replace earlier entries following manual checking by experienced spectroscopists.

Other collections of spectra suitable for general consultation are the library of the Max Planck Institute, Mühlheim with 14 000 spectra of terpenes, methylene and cyclopropane compounds, boron-containing and other derivatised sugars, paraffins and olefins and organometallic compounds and the ETH library, Zurich compiled by J. Seibl, which contains a representative cross section of organic chemical compounds in the form of ca. 5000 spectra.

3.2.3.2 Application-orientated Libraries of Mass Spectra

A series of special libraries attempt to cover clearly defined areas of use. The advantage of these intentionally limited libraries is the concentration on substances which only lead to sensible suggestions within a special area of use. The spectra in these libraries are frequently produced using the same instrumentation for a particular application so that good comparability is guaranteed. Updating these collections is easier and can be completed more rapidly.

The Pfleger/Maurer/Weber toxicology library with more than 4300 entries in the second edition (drugs and poisons: 1500, pesticides and environmental contaminants: 800, metabolites: 2000) is well known in clinical, forensic and pharmacological-toxicological laboratories. This library achieves its importance through the accompanying four volumes which contain important additional information on retention indices, formulae, metabolites, indications, sample processing, derivatisation and gas chromatographic analysis. This collection is available in electronic format for the data systems of various GC/MS manufacturers.

The toxicology library TX supplied by ThermoQuest as part of the basic equipment for the ion trap mass spectrometer is also dedicated to the areas of toxicology and pharmaceuticals (Fig. 3-22). This library contains ca. 2200 entries of pharmaceutically active substances, drugs and their metabolites. Dr J. Zamecnik, Bureau of Drug Research, Ottawa, Canada, drew up and updates the library.

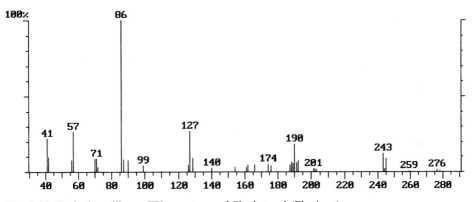

Fig. 3-22 Toxicology library TX: spectrum of Clenbuterol (Finnigan)

A collection of terpene spectra worthy of note was published by R. P. Adams. It contains ca. 1500 compounds (situation in 1996) with details of the retention indices on the silicone phases DB5 and OV1 (Fig. 3-23). The book gives additional information on methods and structural formulae and contains lists compiled using different sorting criteria. This library is available in the electronic format required for ThermoQuest mass spectrometers.

Volatile compounds in foodstuffs form the subject of a library compiled by the Dutch TNO Institute in Zeist. 1630 mass spectra of fragrances, preservatives, nitrosamines, solvents, pesticides and veterinary pharmaceuticals are included. The TMS derivatives of involatile compounds are also included (Chemical Concepts, Weinheim, Germany).

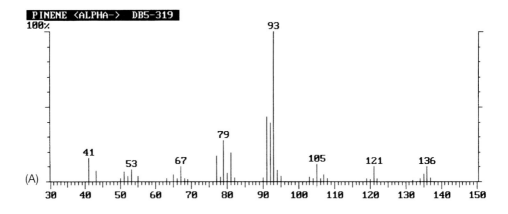
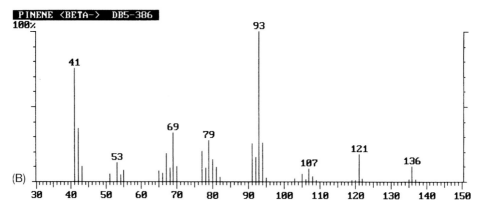

Fig. 3-23 Terpene library after Adams: spectra of (A) α-pinene and (B) β-pinene with retention indices (print-out from Finnigan data system)

Other special libraries have collections of spectra of plant protection agents. The most extensive current collection is the pesticide library with ca. 750 entries by W. Ockels. This contains the reference spectra of the Riedel-de-Haën-Pestanal collection, a collection of pesticide standards of worldwide importance. The mass spectroscopic data are available in electronic format compatible with the GC/MS data systems of various producers. The accompanying handbook contains the structural and empirical formula, the CAS no. and GC retention index for methyl-silicone capillary columns for each substance (Fig. 3-24, Axel Semrau, Sprockhövel).

Stan and Lipinsky have compiled the MS pesticide library for the mass selective detector in PBM format. This library contains 340 pesticide spectra of active substances and metabolites which were registered and are regulated by threshold values. Besides the description of the processing procedure, the measuring conditions are given and for each substance brief accompanying information together with the structure is given on a data sheet (Agilent Technologies, Palo Alto, CA).

Other commercial libraries available in electronic format for special areas of application are the geo- and petrochemical collection of spectra of the Technical University in

METHOPROTRYN

Code-No.:   35856
CAS-No.:   [000841-06-5]
Formula: C11 H21 N5 O S
Mol.-Weight: 271
Density:
Mp.:   69-71
Ret.-Index: 2173

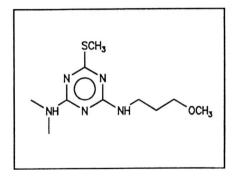

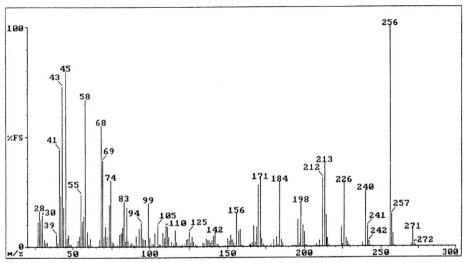

Fig. 3-24 Extract from the pesticide library (SpectralService)

Delft with 1100 spectra of saturated, unsaturated, and aromatic hydrocarbons, aromatic and nonaromatic sulfur and oxygen compounds and terpenes, steroids and their degradation products and a steroid library with 500 entries. Here the spectra were run by G. Spiteller, University of Bayreuth and compiled by K. Varmuza, Technical University, Vienna (Chemical Concepts, Weinheim).

The use of the spectral libraries available has up till now been considerably limited by the incompatibility of the data formats of the computer systems used in GC/MS. Only very recently have conversion programs with suitable flexibility become available (e.g. MassTransit, Palisade Corp., Newfield, NY) which can read and write the library formats of leading instrument manufacturers and thus permit the transfer into an available data system. In particular the use and further development of collections of spectra belonging to a particular laboratory will profit from these transfer possibilities.

3.2.4 Library Search Procedures

In general it is expected that the identity of an unknown compound will be found in a library search procedure. However it is better to consider the results of a search procedure from the aspect of similarity between the reference and the unknown spectrum. Other information for confirmation of identity, such as retention time, processing procedure, and other spectroscopic data should always be consulted. A short review in the journal Analytical Chemistry (W. Warr, 1993) began with the sentence 'Library searching has limitations and can be dangerous in novice hands'. Examples of critical cases are different compounds which have the same spectra (isomers), the same compounds with different spectra (measuring conditions, reactivity, decomposition) or the fact that a substance being searched for is not in the library but similar spectra are suggested. In particular the limited scope of the libraries must be taken into account. According to a short press publication by the Chemical Abstract Service in 1994 the total of CAS registry numbers had passed the 12 million mark. Every year ca. 600 000 new compounds are added!

As a result of the different software equipment used in current benchtop GC/MS systems two search procedures have become widely established: INCOS and PBM. The SISCOM procedure (Search for Identical and Similar Compounds) developed by Henneberg/Weimann is now also available on stand-alone NT work-stations. It stands out on account of its excellent performance for data-system-supported interpretation of mass spectra. The procedures for determination of similarity between spectra are based on very different considerations. The INCOS and PBM procedures aim to give suggestions of possible substances to explain an unknown spectrum. Both algorithms dominate in the qualitative evaluation using magnetic sector, quadrupole and ion trap GC/MS systems. Other search procedures, such as the Biemann search, have been replaced by newer developments and broadening of the algorithms by the manufacturers of spectrometers.

The newest development in the area of computer-supported library searches, the further development of the INCOS procedure, has been presented by Steven Stein (NIST) through targeted optimisation of the weighting and combination with probability values. An improvement in the hit rate was demonstrated in a comparison of test procedures with more than 12 000 spectra.

3.2.4.1 The INCOS/NIST Search Procedure

At the beginning of the 1970s the INCOS company (Integrated Control Systems) presented a search procedure which operated both on the principles of pattern recognition and with the components of classical interpretation techniques and which could reliably process data from different types of mass spectrometer. The early years of GC/MS were characterised by the rapid development of quadrupole instruments which were ideal for coupling with gas chromatographs, because of their scan rates, which were high compared with the magnetic sector instruments of that time. The spectral libraries then available had been drawn up from spectra run on magnetic sector instruments. The spectral databases of today are also built up on the same basis.

3.2 Substance Identification 237

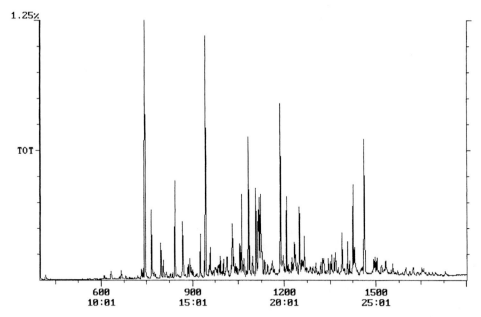

Fig. 3-25 (A) Complex chromatogram from the analysis of the soil at the location of a coking plant

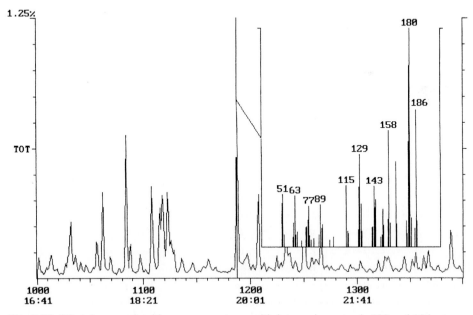

Fig. 3-25 (B) A large peak with a mass spectrum with intense ions at m/z 180 and 186 appears at a retention time of ca. 20 min

238   3 Evaluation of GC/MS Analyses

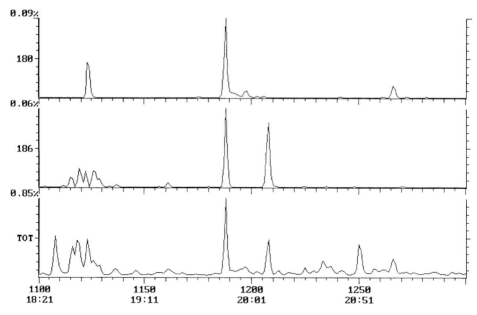

Fig. 3-25 (C) The mass chromatograms m/z 180 and 186 show peak maxima at the same retention time. Both peaks show the same intensity pattern

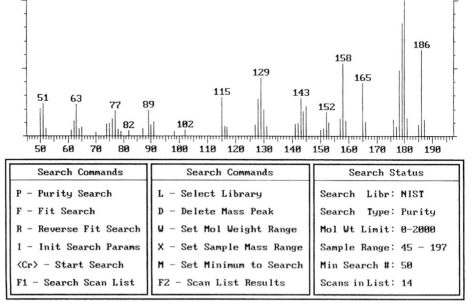

Fig. 3-25 (D) The spectrum of the peak is compared with the NIST library: INCOS sorting according to purity, all molecular weights permitted

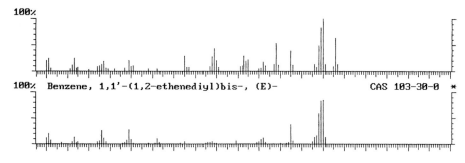

MATCHES	COUNT	TOTAL		RANK	INDEX	PURITY	FIT	RFIT
8	0	0		1	15297	562	821	599
7	8	8		2	15292	558	900	586
6	37	45		3	15293	556	806	590
5	162	207**		4	40669	545	875	575
4	476	683		5	15296	539	806	582
3	1446	2129		6	15300	535	865	555
2	3874	6003		7	15291	534	844	555
1	11352	17355		8	15295	532	855	578
0	36638	53993		9	15298	526	847	566
				10	16674	524	790	552

Fig. 3-25 (E) The first suggestion of the INCOS search with tables showing the pre-search (left) and the spectrum comparison (main search, right). The pre-search table shows that 207 candidates were taken over into the main search. The main search table shows the order of rank of the first 10 hits sorted according to purity. All suggestions have high FIT values and low RFIT values

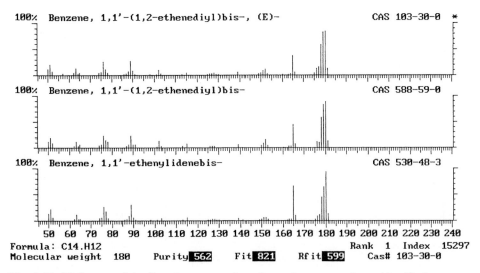

Fig. 3-25 (F) Spectra of the first three suggestions. Isomeric compounds are identified

240   3 Evaluation of GC/MS Analyses

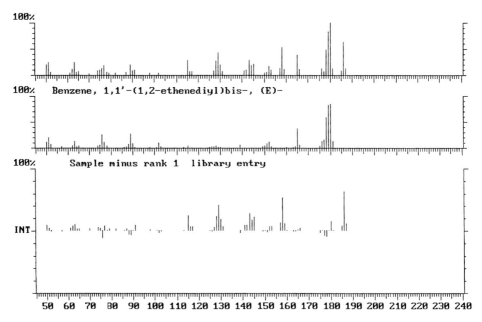

Fig. 3-25 (G) Difference spectrum of the first suggestion compared with the unknown spectrum. The positive part of the difference can be used for a new search

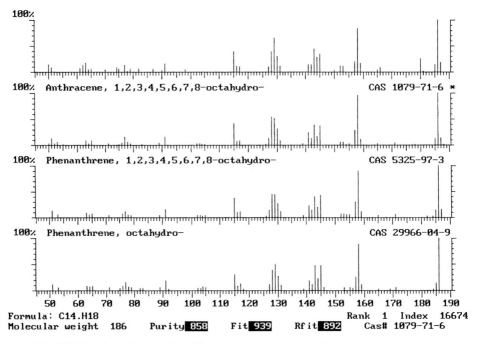

Fig. 3-25 (H) Search result from the difference spectrum with a high FIT and PURity value

From the beginning the INCOS procedure was able to take into account the relatively low intensities of the higher masses in spectra run on quadrupole systems, besides the typical magnetic sector spectra. The INCOS search has remained virtually unchanged since the 1970s. The search is known for its high hit probability, even with mass spectra with a high proportion of matrix noise obtained in residue analysis, and its complete independence from the type of instrument.

After a significance weighting (square root of the product of the mass and the intensity) and data reduction by a noise filter and a redundancy filter, the extensive reference database is searched for suitable candidates for a pattern comparison in a rapid pre-search. The presence of up to eight of the most significant masses counts as an important starting criterion. The intensity ratios are not yet considered. It is required that only those reference spectra which contain at least eight of the most significant masses of the unknown spectrum are considered. Depending on the requirements of the user, spectra with less than eight matching masses are also further processed (Fig. 3-25 E: pre-search report, the number of candidates is marked with **). At the start of the search the parameter 'minimum number to search' must be adapted by the user. Reference spectra which only contain a small number or no matching masses or whose molecular weight does not match an optional suggestion, are excluded from the list of possible candidates and are not further processed.

The main search is the critical step in the INCOS algorithm, in which the candidates found in the pre-search are compared with the unknown spectrum and arranged in a prioritised list of suggestions. Of critical importance for the tolerance of the INCOS procedure for different types of mass spectrometer and marginal conditions of data acquisition (and thus for the high hit rate) is a process known as local normalisation.

Local normalisation introduces an important component into the search procedure which is comparable to the visual comparison of two patterns (Fig. 3-26). Individual clusters of ions and isotope patterns are compared with one another in a local mass window. The central mass of such a window from the reference spectrum is compared with the intensity of this mass in the unknown spectrum in order to assess the matching of the line pattern in windows a few masses to the left and right. In this way the nearby

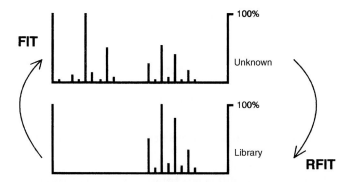

Fig. 3-26 Diagram showing local normalisation.
FIT value high: all masses in the library spectrum are present in the unknown spectrum and the isotope pattern also fits
RFIT value low: only a few masses from the unknown spectrum are present in the library spectrum

A significance weighting of the mass signals (here, a sum of the mass m and the intensity I) is also carried out first with the PBM search procedure. In addition the frequency of individual mass signals based on their appearance in the whole spectral library is also taken into account (the reference is the latest version of the Wiley library). The pre-search of the PBM procedure is also orientated towards the method of significance weighting. In the database used for PBM the reference spectra are sorted according to the values of maximum significance. For a given maximum significance of the weighted masses of an unknown spectrum a set of reference spectra can thus be selected. The choice of several sets of mass spectra is specified as the search depth, which, starting from $(m + I)_{max}$, can reach search depths of 3 to $(m + I)_{max} - 3$. The number of possible candidates for the main search is thus broadened.

The main search in the PBM procedure can be carried out in two ways. In the pure search mode only the fragments of the unknown spectrum are searched for in the reference spectra and compared (forward search). This procedure requires mass spectra which are free from overlap and matrix signals, which is the case with simple separations at a medium concentration range.

The mixture search mode first tests whether the mass signals of the reference spectrum are present in the unknown spectrum (reverse search). Local normalisation analogous to the INCOS procedure is also carried out from version 3.0 (1993) onwards. With every spectrum selected from the pre-search, a subtraction from the unknown spectrum is carried out in the course of the procedure. The result of the subtraction is, in turn, compared with the candidates remaining from the pre-search and matching criteria are met. In this way account is taken of the possibility that, even with high resolution gas chromatography, mixed spectra are detected because of the matrix or co-elution. A successful search requires that the second (or third) component of a mixture is contained in the pre-search result (hit list). Combinations of spectra with markedly different highest significances do not appear together in the pre-search hit list because the pre-search is limited by the depth of the search. Using this procedure only the probability values of the hit list are improved. This constitutes a major difference between this and the SISCOM search (see Section 3.2.4.3).

The sorting of results from the PBM search procedure takes place on the basis of probabilities which are determined in the course of spectrum comparison (Table 3-6). In the forefront is the aim of giving a statement about the identity of a suggestion. The assignment to class I gives information as to which degree a spectrum is the same or stereoisomeric with the suggestion. Here maximum values can reach 80%, but not higher. An extension of the definition to the ring positions of positional isomers, homologous compounds, the change in position of individual C-atoms or of double bonds gives a probability defined as class IV. A higher value implies that a compound with

Table 3-6 Results of the benchtop/PBM spectral comparison

Class I	The probability that the suggestion is **identical** or a *stereoisomer*
Class IV	Extension of the probability from class I to compounds having *structural differences compared with the reference*, which only have a small effect on the mass spectrum (homologues, positional isomers).
% Contamination	Gives the proportion of ions which are *not present in the reference*

structural features is present which has little or no effect on the appearance of the mass spectrum. In such cases the mass spectroscopic procedure shows itself to be insensitive to the differences in structural details between individual molecules. The user is thus given a value for assessment which indicates that the mass spectrum being considered matches the spectrum well on account of its few specific fragments, but probably cannot be identified conclusively. The literature should be consulted for details of the calculation of the probability values in PBM.

The clear grading of the probabilities is typical for the PBM procedure (Fig. 3-27). Sensible suggestions with high values of 70 to over 90% (class IV) are given. Then in the hit list there is the rapid lowering of the values to below 20%, which avoids the occurrence of false positive suggestions. It is obvious that, as a result of the procedure in individual cases, recognisably correct suggestions are given with low probability values (false negative suggestions). Therefore poorly placed suggestions must also be included in the discussion of the search results (Fig. 3-28). In such cases the framework conditions, such as measuring range, quality of the library spectrum or possibly a large quantity of noise in the extracted spectrum should be investigated.

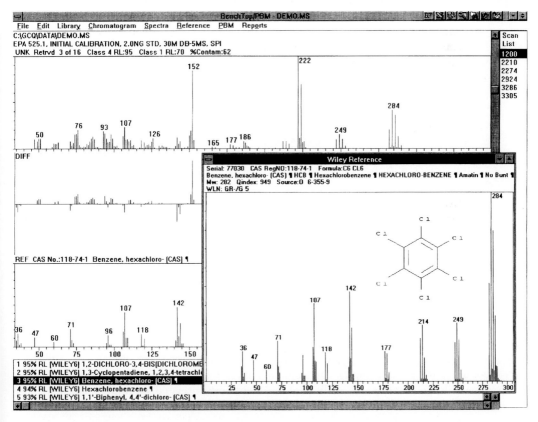

Fig. 3-27 Screen display of suggestions arising from a PBM library search (above: unknown spectrum; middle: difference spectrum; below and in the reference window: library suggestion; underneath: list of suggestions with probabilities)

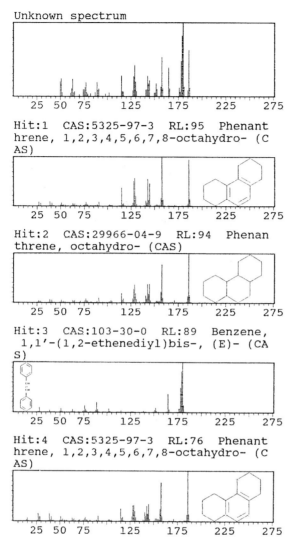

Fig. 3-28 Suggestions from a PBM search with probabilities (RL). The unknown spectrum is the same as that shown in Fig. 3-25 B. The mixture search mode gives suggestions for both co-eluting compounds

In practice the hit quota of the PBM search depends on the spectrum quality, which is also affected by the data acquisition parameters of the instrument, the choice of the spectrum in the peak (rising/falling slope, maximum), scan time and on the type of instrument. In this context it is useful that in PBM libraries several spectra per substance are available from different sources.

PBM Library Search

Principle   Statistical mathematical tests after Prof. McLafferty

Course

Modi:	pure search or mixture search
Both modi:	significance weighting $(m + I)$
Both modi:	pre-search by search depths 1 to 3 $(m + I)_{max}$ to $(m + I)_{max} - 3$

Pure Search Mode	**Mixture Search Mode**
Forward search	Reverse search
Fragments of the unknown spectrum must be present in the reference	Fragments of the reference spectrum must be present in the unknown spectrum
	Spectrum subtraction with each reference from the pre-search
	Comparison of results by forward search in the remaining references
Sorting	Sorting

Advantages

+ the mixture search is more meaningful
+ information on multiple occurrence of similar spectra
+ suggestions with regard to structural formulae and trivial names
+ available as complete additional software for many MS and GC/MS systems

Limitations

− pure search is less useful for complex GC/MS analyses
− very dependent on the quality of spectra and thus on recording parameters, choice of spectrum in the GC peak and on the type of instrument
− therefore more spectra per substance in PBM libraries

Note

There are different PBM programs among the data systems of MS producers, some are more powerful than others; the mixture search mode is often omitted for reasons of speed.

3.2.4.3 The SISCOM Procedure

The SISCOM procedure developed by Henneberg and Weimann (Max Planck Institute for Coal Research, Mülheim a. d. Ruhr, Germany) is used for structural determination in industrial MS laboratories. The procedure is commercially available in the form of MassLib in association with various libraries. Its primary goal is to give the most plausible suggestion for the structure of an unknown component and offers information to support the interpretation of the spectra from a large basis of knowledge, i.e. the collection of spectra which have already been explained: the Search for Identical and Similar Compounds.

One of the great strengths of the procedure is the deconvolution of the mass spectrum of an analyte from chromatographic data using an automatic background correction.

248   3 Evaluation of GC/MS Analyses

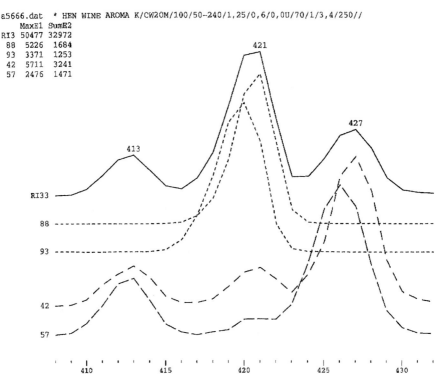

Fig. 3-29 Automatic peak purity control showing the relevant mass chromatograms on co-elution using the SISCOM procedure (MassLib, Chemical Concepts)

Within this process the variation of mass intensities during a GC peak is used to determine a reliable substance spectrum (Fig. 3-29). In addition to the identity search, the search procedure offers a similarity search. Through this, account is taken of the fact that mass spectra can differ somewhat from a reference spectrum depending on the conditions of measurement, the compound can be contaminated by other compounds or by the matrix or the quality of the reference spectrum can be in doubt. A similarity search also allows the limitation caused by the limited scope of spectral libraries to be counteracted. All the above-named search systems are limited by this handicap. If the aim of the search is only identification, the results of the similarity search can be rearranged in the direction of identity using a special algorithm. Complete spectra are used here and weightings with regard to mass and intensity are undertaken. A mixture correction procedure is also used to determine the nature of co-eluates. The retention index can also be included in the evaluation of GC/MS analyses.

An essential part of the recognition of similarities between an unknown spectrum and a reference spectrum is the simultaneous use of different methods of comparison, which evaluate different aspects of the similarities between two spectra. In the SISCOM procedure the results of these methods are combined finally in such a way that the contribution of the part of the comparison which does not agree can be compensated for by the part which does agree. The differences between an unknown spectrum and the reference spectrum can thus be tolerated and do not hinder the selection of a relevant reference. The results of the different methods of comparison are represented by factors. The order of the hit list is based on a weighted function of these comparison factors. The calculation of a comparison value as a combination of factors with different relative importance is a multidimensional consideration. The ordering of references in a one-dimensional list of suggestions can therefore only be a compromise. The recognition of relevant reference spectra, which are arranged as subsequent hits in the similarity evaluation, is possible for an experienced user when employing the whole range of comparison factors.

The libraries used for the SISCOM search are divided into 14 sections. The assignment of a reference spectrum to one of these sections is based on the strongest peak in the ion series spectrum. Ion series spectra consist of 14 peaks. Each of these peaks (or ion series) represents the sum of all the intensities of the original spectrum and with a mass of m/z modulo 14 = n. Thus, for example, the ion series 1 corresponds to the sum of the intensities of the masses 15, 29, 43, ... The strongest ion series in the spectrum of the unknown substance decides which section of the library is to be used for further searching. As the ion series spectra of related compounds (similar structures) are almost identical (even when the intensity ratios in the spectra differ), the ion series spectra are particularly suitable as a filter for a similarity search.

To compare the unknown spectrum (U) with the reference (R) from the library, SISCOM uses coded spectra, which are calculated from the original spectra by reduction to characteristics. To be determined as characteristic in the SISCOM sense, a mass peak must be larger than the arithmetic mean of its neighbouring homologues (in ± 14 u intervals). The characteristics used in a spectroscopic comparison are divided into three groups. One group contains the characteristics present in U and R; and the other two groups contain characteristics present only in U and only in R respectively. Three comparison factors arise from these three groups:

NC: The number of common characteristics contained in both the unknown spectrum and the reference spectrum. High NC values indicate similar structures.

NR: The number of characteristics from the reference spectrum which are not present in the unknown spectrum. NR is a measure in relation to NC of the extent to which the reference spectrum is represented in the unknown spectrum or is part of it.

NU: The number of characteristics in the unknown spectrum which are not present in the reference spectrum. NU gives the part of the unknown spectrum which is not explained by the reference spectrum. Contamination or a mixture may be present.

These three factors NC, NR, and NU do not contain any intensity parameters. 150 candidates result from them (pre-search).

Three other factors are derived from these first three which use intensities. All six factors are used together to order the 150 candidates in the pre-search for the hit list.

IR: The relative intensity in % of the characteristics of NR. IR completes or differentiates the importance of NR. A few characteristics not contained can have high peak intensities or many characteristics which are absent only consist of peaks of low intensity.

IU: The relative intensity as a % of the characteristics of NC. IU is equivalent to IR and is a measure of the purity of the spectrum, in the case where the suggestion is contained as part of a mixture.

PC: The measure of the correlation of the peak patterns of the relative intensities in the characteristics of NC. As PC is essentially calculated from the occurrence of intense peaks, high values are of only limited importance in spectra with few dominant masses. The PC only makes a limited contribution to the similarity index SI, i.e. deviations, for example as a result of measuring conditions or different types of instrument, are tolerated better.

The evaluation of the hit list requires experience and sound knowledge of mass spectrometry. In the case of identical reference spectra NR and IR are small but PC is high. For a pure spectrum IU is small. If a mixture is present IU is large. Components of a mixture can also appear in the suggestion list using SISCOM, provided that NR and IR are very small and PC correspondingly high. Substances with similar structures are assigned comparable factors, as are components of mixtures. Particularly high NC values indicate high similarity even when PC is negligibly small. Isomeric compounds are an example of this. In the evaluation it should also be noted that a search result depends on the start of the mass range measured in the unknown mass spectrum and on the reference spectrum as well as on the presence of reduced reference spectra. For optimal search results the unknown and the reference spectra should both contain the lower mass region from m/z 25 and be present in nonreduced form. Many important characteristics from this region (e.g. m/z 27, 30, 31, 35 ...) can contribute to similarity searches or identification of spectra.

SISCOM provides a powerful automatic mixture correction. The subtraction procedure is based on the assumption that specific ions exist for a substance which is removed. These ions disappear completely if the exact percentage proportion is removed. If larger proportions are removed negative peaks are formed. The mixture correction uses an iterative procedure based on these which is orientated towards the sum of the negative intensities.

The SISCOM search with mixture correction subtracts the spectrum of the best hits from the unknown spectrum. According to experience this is almost always one of the components present (if a corresponding spectrum is present in the library) or an isomer with almost the same spectrum. With the spectrum thus corrected a completely new search is carried out! In this way a second component is then found if it was not present in the first hit list. The best hit in this second search is also subtracted from the original unknown spectrum and a third complete search is carried out so that in the case of a mixture a purified spectrum is now also present for the first component, which leads to higher identity values. This threefold search gives high identity values from the purified spectra. For identity values of less than 80% identification is very improbable. This makes it easier for only the relevant results to be produced in automatic evaluation processes. A further advantage of this method is that the result of the correction procedure is independent of whether the unknown spectrum is pure or the spectrum of a mixture. In the case of mixtures the MassLib program package (Chemical Concepts, Weinheim, Germany) has a module available for determining the ratio of components in a GC peak using regression analysis (Beynon 1982).

3.2.5 Interpretation of Mass Spectra

There are no hard and fast rules about fully interpreting a mass spectrum. Unlike spectra obtained using other spectroscopic procedures, such as UV, IR, NMR, or fluorescence, mass spectra do not show uptake or emission of energy by the compound (i.e. the intact molecule), but reflect the qualitative and quantitative analysis of the processes accompanying ionisation (fragment formation, rearrangements, chemical reaction) (see Section 2.3.2). The time factor and the energy required for ionisation (electron beam, temperature, pressure) also play a role. With all other spectroscopic procedures the features of certain functional groups or other structural elements always appear in the same way. In mass spectrometry, however, the appearance of certain details of a structure always depends on the total structure of the compound and may not occur at all or only under certain specific conditions. The failure of expected signals to appear generally does not prove in mass spectrometry that certain structural elements are not present; only positive signals count. It is also true that a mass spectrum cannot be associated with a particular chemical structure without additional information.

Nevertheless procedures are recommended for deciphering information hidden in a mass spectrum. (Certain users accuse experienced mass spectrometrists of having a criminological feel for the subject – and they are justified!). In this spectroscopic discipline the experience of a frequently investigated class of substances is rapidly built up. New groups of substances usually require new methods of resolution. It is therefore of extremely great importance that other parameters relating to the substance besides mass spectrometry (spectrum and high resolution data), such as UV, IR, NMR spectra, solubility, elution temperature, acidic or basic clean-up, synthesis reaction equations or those of conversion processes, should be incorporated into the interpretation of the spectrum.

The procedure shown in the following scheme has proved to be effective:

1. Spectrum display
   Does the mass spectrum originate from a single substance or are there signals which do not appear to belong to it? Can the subtraction of the background give a clearer representation or is the spectrum falsified by the subtraction? What information do the mass chromatograms give about the most significant ions?

2. Library search
   As all GC/MS systems are now connected to very powerful computer systems, each interpretation process for EI spectra should begin with a search through available spectral libraries (see Section 3.2.4). Spectral libraries are an inestimable source of knowledge, which can give information as to whether the substance belongs to a particular class or on the appearance of clear structural features, even when identification seems improbable. Careful use of the database spares time and gives important suggestions. The different search procedures can all help, although to differing degrees.

3. Molecular ion
   Which signal could be that of the molecular ion? Is an $M^{+/-}$, $(M+H)^+$ or an $(M-H)^{+/-}$ present? Which signals are considered to be noise or chemical background? Mass chromatograms can also help here. If the molecular weight is known, e.g. from CI results, the library search should be carried out again limited to the molecular mass.

4. Isotope pattern
   Is there an obvious isotope pattern, e. g. for chlorine, bromine, silicon or sulfur? The molecular ion shows all elements with stable isotopes in the compound. Is it possible to find out the maximum number of carbon atoms? This is always a problem with residue analysis as usually it is not possible to detect ^{13}C signal intensities with certainty. Also a noticeable absence of isotope signals, particularly with individual fragments, can be important for identifying the presence of phosphorus, fluorine, iodine, arsenic and other monoisotopic elements. Only the molecular or quasimolecular ions give complete information on isotopes.

5. Nitrogen rule
   Is nitrogen present? An uneven molecular mass indicates an uneven number of nitrogen atoms in the empirical formula.

6. Fragmentation pattern
   What information does the fragmentation pattern give? Are there pairs of fragments, the sum of which give the molecular weight? Which fragments could be formed from an α-cleavage? Here the use of a table giving details of fragmentation of molecular ions $(M-X)^+$ is advisable (see Table 3-12).

7. Key fragments
   What can be said about characteristic fragments in the lower mass region? Is there information on aromatic building blocks or ions formed through rearrangements (McLafferty, retro-Diels-Alder)? Here also tables with appropriate explanations are helpful (see Table 3-12).

8. Structure postulate
   Bringing together information rapidly leads to a rough interpretation, which initially gives a partial structure and finally it postulates the molecular structure. Two possibilities test this postulated structure: which fragmentation pattern would the proposed structure give? Is the proposed substance available as a reference and does this correspond to the unknown spectrum?

Within this interpretation scheme spectroscopic comparison through library searching is definitely placed at the beginning of the interpretation. Confirmation of a suggestion from a library search is achieved through an assessment using the above scheme. Interpretation has never finished simply with the print-out of the list of suggestions!

Steps to the Interpretation of Mass Spectra

1. Library search!
2. Only one substance?
3. Molecular ion?
4. Isotope pattern?
5. Nitrogen?
6. Fragmentation pattern?
7. Fragments?
8. Reference spectrum?

3.2.5.1 Isotope Patterns

For organic mass spectrometry only a few elements with noticeable isotope patterns are important, while in inorganic mass spectrometry there are many isotope patterns of metals, some of them very complex. From Table 3-7 it can be seen that the elements carbon, sulfur, chlorine, bromine and silicon consist of naturally occurring non-radioactive isotopes. The elements fluorine, phosphorus and iodine are among the few monoisotopic elements in the periodic table.

Table 3-7 Exact masses and natural isotope frequencies

Element	Isotope	Nominal mass[1] (g/mol)	Exact mass (g/mol)	Frequency[2] %	Factors for calculating the isotope intensity[3]	
					M+1	M+2
Hydrogen	^{1}H	1	1.007825	99.99		
	D, ^{2}H	2	2.014102	0.01		
Carbon	^{12}C	12	12.000000	98.9	1.1	0.006
	^{13}C	13	13.003354	1.1		
Nitrogen	^{14}N	14	14.003074	99.6	0.4	
	^{15}N	15	15.000108	0.4	0.4	
Oxygen	^{16}O	16	15.994915	99.76		
	^{17}O	17	16.999133	0.04	0.04	
	^{18}O	18	17.999160	0.20		0.20
Fluorine[4]	F	19	18.998405	100		
Silicon	^{28}Si	28	27.976927	92.2	5.1	
	^{29}Si	29	28.976491	4.7		3.4
	^{30}Si	30	29.973761	3.1		
Phosphorus	P	31	30.973763	100		
Sulfur	^{32}S	32	31.972074	95.02	0.8	
	^{33}S	33	32.971461	0.76		4.4
	^{34}S	34	33.976865	4.22		
Chlorine	^{35}Cl	35	34.968855	75.77		32.5
	^{37}Cl	37	36.965896	24.23		
Bromine	^{79}Br	79	78.918348	50.5		98.0
	^{81}Br	81	80.916344	49.5		
Iodine	I	127	126.904352	100		

[1] The calculation of the nominal mass of an empirical formula is carried out using the mass numbers of the most frequently occurring isotope e.g. Lindane $^{12}C_6\,^1H_6\,^{35}Cl_6$ M 288.

[2] The frequency of occurrence is a relative parameter. The isotope frequencies of an element add up to 100%.

[3] In the isotope pattern of the ion the intensity of the first mass peak (nominal mass) is assumed to be 100%. The intensities of the isotope peaks (satellites) M+1 and M2 are given by multiplying the factors with the number of atoms of an element in the ion.
Example: $C_{10}H_{22}$ M$^+$ 142   intensity m/z 143: 10 · 1.1 = 11%
C_6Cl_6 M$^+$ 282   intensity m/z 284: 6 · 32.5 = 195%
S_6 M$^+$ 192   intensity m/z 194: 6 · 4.4 = 26.4%

[4] Fluorine, sodium, aluminium, phosphorus, manganese, arsenic and iodine, for example, appear as monoisotopic elements in mass spectrometry.

If isotope signals appear in mass spectra there is the possibility that these elements can be recognised by their typical pattern and also the number of them in molecular and fragment ions can be determined. With carbon there are limitations to this procedure in residue analysis, because usually the quantity of substance available is too small for a sufficiently stable analysable signal to be obtained. In order to allow conclusions to be drawn on the maximum number of carbon atoms, a larger quantity of substance is necessary. For this the technique of individual mass registration (SIM, MID) with longer dwell times is particularly suitable for giving good ion statistics. In the evaluation in the case of carbon only the maximum number of carbon atoms (isotope intensity/1.1%) can be calculated, as contributions from other elements must be taken into account.

Some elements, in particular the halogens chlorine and bromine, which are contained in many active substances, plastics, and other technical products, can be recognised by the typical isotope patterns. These easily recognised patterns are shown in Figs. 3-30 to 3-37. The intensities shown are scaled down to a unit ion stream of the isotope pattern. The lowering of the specific response of the compound as a function of, for example, the degree of chlorination, is shown. The simple occurrence of the elements shown as a series is used as a reference in each case. The relative intensities within an isotope pattern are given as a percentage in the caption underneath the isotope lines with the mass contribution to the molecular weight based on the most frequent occurrence in each case. (Source: isoMass – PC program for calculation and analysis of isotope patterns by Ockels, SpectralService). What is noticeable for chlorine and bromine is the distance between the isotopes of two mass units. Provided that chlorine and bromine occur separately, the degree of chlorination or bromination can be easily determined by comparison of the variation in the intensities. For compounds which contain both chlorine and bromine the degree of substitution cannot be determined by comparison of the patterns alone. In these cases the high atomic weight of bromine is a help (for calculation of retention indices from the empirical formula see Section 3.2.2). In GC/MS coupling relatively high molecular weights (fragment ions) are detected through the presence of bromine in a molecule even at short retention times. In library searches mixed isotope patterns of chlorine and bromine are reliably recognised.

With sulfur the distance between the isotope peaks is also two mass units. However, when the proportion of sulfur in the molecule is low (e.g. in the case of phosphoric acid esters) it is difficult to be sure of the presence of sulfur (Fig. 3-36). Here a detailed investigation of the fragmentation is necessary. Higher sulfur contents (see also Fig. 3-164) give clear information. The combinations of sulfur and chlorine in which chlorine is clearly dominant are also shown (Fig. 3-37). Differences are extremely difficult to see with the naked eye and with computers only when measurements are obtainable with good ion statistics.

Silicon occurs very frequently in residue analysis. Silicones get into the analysis through derivatisation (silylation), partly through clean-up (joint grease), but more frequently through bleeding from the septum or the column (septa of autosampler bottles, silicone phases). The typical isotope pattern of all silicone masses can be recognised rapidly and excluded from further evaluation measures (Fig. 3-35).

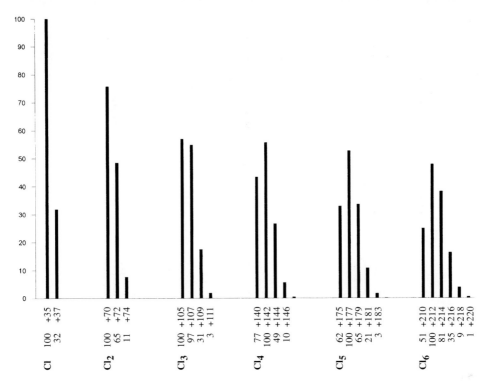

Fig. 3-30 Isotope pattern of chlorine (Cl to Cl$_6$)

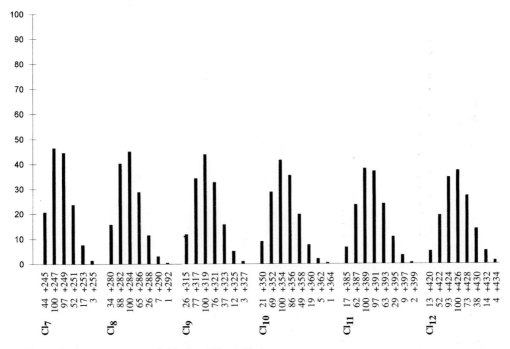

Fig. 3-31 Isotope pattern of chlorine (Cl$_7$ to Cl$_{12}$)

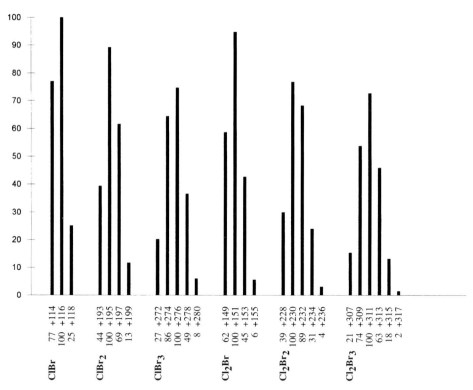

Fig. 3-32 Isotope pattern of chlorine/bromine (ClBr to Cl$_2$Br$_3$)

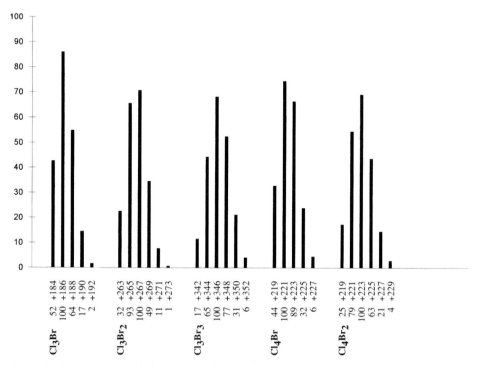

Fig. 3-33 Isotope pattern of chlorine/bromine (Cl$_3$Br to Cl$_4$Br$_2$)

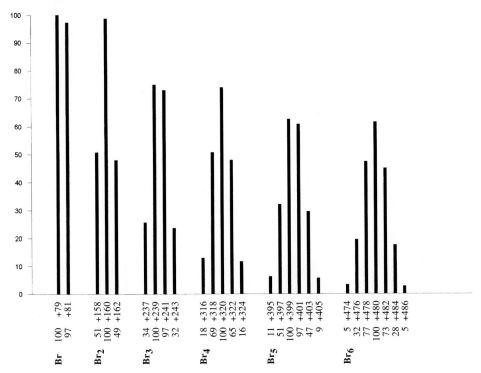

Fig. 3-34 Isotope pattern of bromine (Br to Br$_6$)

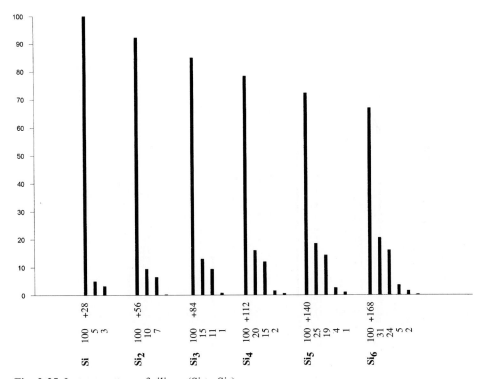

Fig. 3-35 Isotope pattern of silicon (Si to Si$_6$)

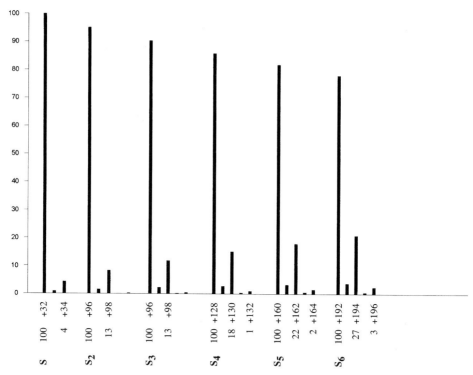

Fig. 3-36 Isotope pattern of sulfur (S to S$_6$)

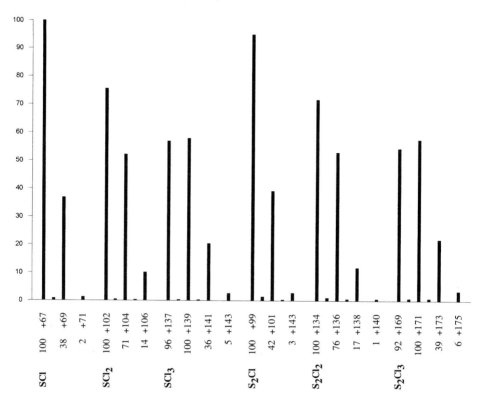

Fig. 3-37 Isotope pattern of sulfur/chlorine (SCl to S$_2$Cl$_3$)

3.2.5.2 Fragmentation and Rearrangement Reactions

The starting point for a fragmentation is the molecular ion (EI) or the quasimolecular ion (CI). A large number of reactions which follow the primary ionisation are described here. All the reactions follow the thermodynamic aim of achieving the most favourable energy balance possible. The basic mechanisms which are involved in the production of spectra of organic compounds will be discussed briefly here (Fig. 3-38). For more depth the references cited should be consulted (Budzikiewicz, Pretsch et al., Howe et al., McLafferty et al.).

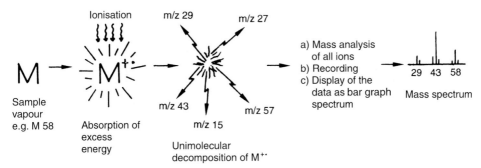

Fig. 3-38 Principle of mass spectrometry: formation of molecular ions as the starting point of a fragmentation (after Frigerio)

There are two possible mechanisms for the cleavage of carbon chains following ionisation, known as α-cleavage and formation of carbenium ions. The starting point in each case is the localisation of positive charge on electron-rich structures in the molecule.

α-Cleavage

α-cleavage takes place after ionisation by loss of one nonbonding electron from a heteroatom (e.g. in amines, ethers, ketones, see Tables 3-8 to 3-10) or on formation of an allylic or benzylic carbenium ion (from alkenes, alkylaromatics):

Amines

$$R_1\text{-}\overset{+\bullet}{\underset{R_2}{N}}\text{-}CH_2\text{-}R \longrightarrow R_1\text{-}\overset{+}{\underset{R_2}{N}}=CH_2 + R^\bullet$$

Table 3-8 Characteristic ions from the α-cleavage of amines

m/z	R_1	R_2
30	H	H
44	CH_3	H
58	CH_3	CH_3

Ethers

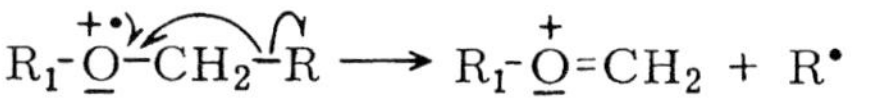

$R_1-\overset{+\bullet}{\underline{O}}-CH_2-R \longrightarrow R_1-\overset{+}{\underline{O}}=CH_2 + R^\bullet$

Table 3-9 Characteristic ions from the α-cleavage of ethers

m/z	R_1
31	H
45	CH_3
59	C_2H_5
73	C_3H_7
87	C_4H_9
etc.	

Ketones

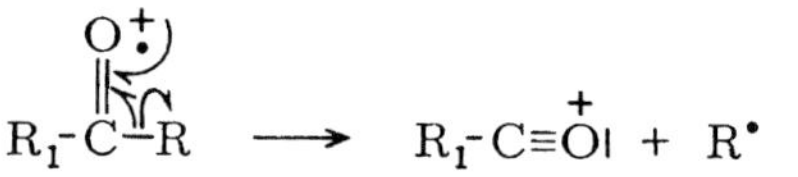

$R_1-\overset{\overset{O^{+\bullet}}{\|}}{C}-R \longrightarrow R_1-C\equiv\overset{+}{O}\text{I} + R^\bullet$

Table 3-10 Characteristic ions from the α-cleavage of ketones

m/z	R_1
29	H
43	CH_3
57	C_2H_5
71	C_3H_7
85	C_4H_9

Formation of Carbenium Ions

Fragmentation involving formation of carbenium ions takes place at a double bond in the case of aliphatic carbon chains (allylic carbenium ion) or at a branch. With alkylaromatics side chains are cleaved giving a benzylic carbenium ion (benzyl cleavage), which dominates as the tropylium ion m/z 91 in many spectra of aromatics.

Alkenes

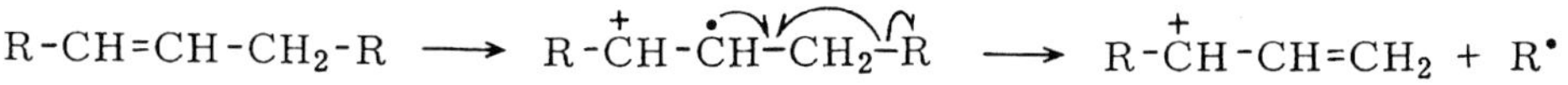

$R-CH=CH-CH_2-R \longrightarrow R-\overset{+}{C}H-CH-CH_2-R \longrightarrow R-\overset{+}{C}H-CH=CH_2 + R^\bullet$

Branched carbon chains

$R_3\overset{+}{C}-R \longrightarrow R_3C^+ + R^\bullet$

The formation of carbenium ions occurs preferentially at tertiary branches rather than secondary ones.

Alkylaromatics (benzyl cleavage)

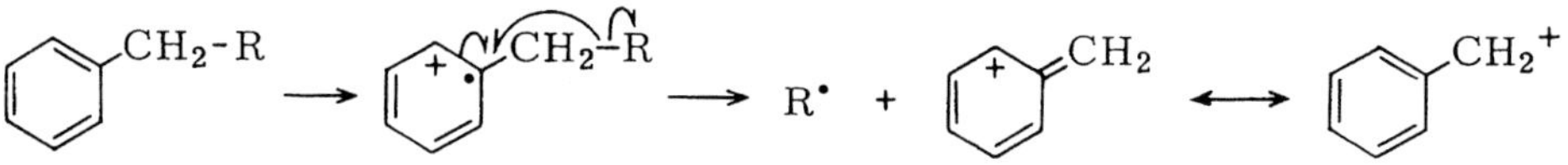

Loss of Neutral Particles

The elimination of stable neutral particles is a common fragmentation reaction. These include H_2O, CO, CO_2, NO, HCN, HCl, $RCOOH$, and alkenes (see also McLafferty rearrangement). These reactions can usually be recognised from the corresponding ions and mass differences in the spectra. Eliminations are particularly likely to occur when α-cleavage is impossible.

Alcohols (loss of water)

$$[C_nH_{2n+1}OH]^{+\cdot} \longrightarrow [C_nH_{2n}]^{+\cdot} + H_2O$$

Carbonyl Functions (loss of CO)

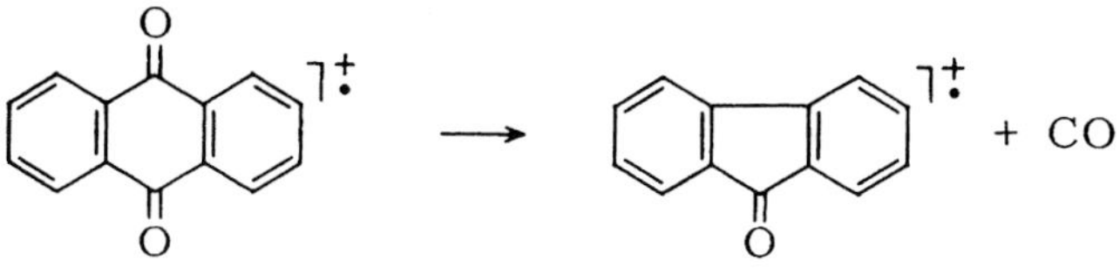

[structure: anthraquinone radical cation] $\longrightarrow$ [structure: fluorenone radical cation] $+ CO$

Heterocycles (loss of HCN)

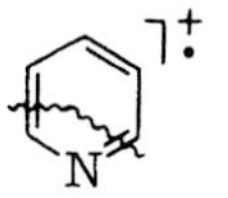

[structure: pyridine radical cation] $\longrightarrow$ $C_4H_4^{+\cdot} + HCN$

Retro-Diels-Alder (loss of alkenes)

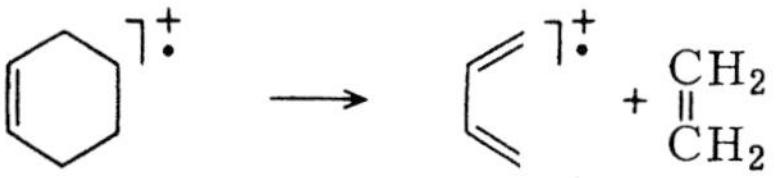

[structure: cyclohexene radical cation] $\longrightarrow$ [structure: butadiene radical cation] $+ \begin{array}{c} CH_2 \\ \| \\ CH_2 \end{array}$

The McLafferty Rearrangement

The McLafferty rearrangement involves the migration of an H-atom in a six-membered ring transition state. The following conditions must be fulfilled for the rearrangement to take place:

– The double bond C=X is C=C, C=O or C=N.
– There is a chain of three σ-bonds ending in a double bond.
– There is an H-atom in the γ-position relative to the double bond which can be abstracted by the element X of the double bond.

According to convention the McLafferty rearrangement is classified as the loss of neutral particles (alkene elimination with a positive charge remaining on the fragment formed from the double bond, see Table 3-11).

Table 3-11 Characteristic ions formed in the McLafferty rearrangement

m/z	R_1	found
44	H	in aldehydes
60	OH	in organic acids
74	O–CH_3	in methyl esters

Table 3-12 Mass correlations to explain cleavage reactions $(M - X)^+$ and key fragments X^+

m/z	Fragment X	$M^+ - X$	Explanations
12	C		
13	CH		
14	CH_2, N, N_2^{++}		
15	CH_3	$M^+ - 15$	Nonspecific, CH_3 at high intensity
16	O, NH_2, O_2^{++}, CH_4	$M^+ - 16$	Rarely CH_4, (but frequently $R^+ - CH_4$ in alkyl fragments), O from N-oxides and nitro compounds, NH_2 from anilines
17	OH, NH_3	$M^+ - 17$	Nonspecific O-indication, NH_3 from primary amines
18	H_2O, NH_4	$M^+ - 18$	Nonspecific O-indication, strong for many alcohols, some acids, ethers and lactones
19	H_3O, F	$M^+ - 19$	F-indication
20	HF, Ar^{++}, CH_2CN^{++}	$M^+ - 20$	F-indication
21	$C_2H_2O^{++}$ (rarely)		
22	CO_2^{++}		
23	Na (rarely)		
24	C_2		
25	C_2H	$M^+ - 25$	Rarely with a terminal $C\equiv CH$ group
26	C_2H_2, CN	$M^+ - 26$	From purely aromatic compounds, rarely from cyanides
27	C_2H_3, HCN	$M^+ - 27$	CN from cyanides, C_2H_3 from terminal vinyl groups and some ethyl esters
28	C_2H_4, N_2, CO	$M^+ - 28$	CO from aromatically bonded O, ethylene through RDA from cyclohexenes, by H-migration from alkyl groups, nonspecific from alicyclic compounds
29	C_2H_5, CHO	$M^+ - 29$	Aromatically bonded O, nonspecific with hydrocarbons
30	C_2H_6, H_2NCH_2, NO, CH_2O, BF (N-fragment)	$M^+ - 30$	CH_2O from cyclic ethers and aromatic methyl ethers, NO from nitro compounds and nitro esters
31	CH_3O, CH_2OH, CH_3NH_2, CF, (O-fragment)	$M^+ - 31$	Methyl esters, methyl ethers, alcohols
32	O_2, CH_3OH, S	$M^+ - 32$	Methyl esters, some sulfides and methyl ethers

Table 3-12 (continued)

m/z	Fragment X	$M^+ - X$	Explanations
33	CH_3OH_2, SH, CH_2F	$M^+ - 33$	SH nonspecific S-indication, $M^+ - 18-15$ nonspecific O-indication, strong with alcohols
34	SH_2, (S-fragment)	$M^+ - 34$	Nonspecific S-indication, strong with thiols
35	^{35}Cl, SH_3	$M^+ - 35$	Chlorides, nitrophenyl-compounds ($M^+ - 17 - 18$)
36	HCl, C_3	$M^+ - 36$	Chlorides
37	^{37}Cl, C_3H		
38	$H^{37}Cl$, C_3H_2		
39	C_3H_3	$M^+ - 39$	Weak with aromatic hydrocarbons
40	Ar, C_3H_4	$M^+ - 40$	Rarely with CH_2CN
41	C_3H_5, CH_3CN	$M^+ - 41$	C_3H_5 from alicyclic compounds, CH_3CN from aromatic N-methyl and o-C-methyl heterocycles
42	$CH_2=C=O$, C_3H_6, C_2H_4N	$M^+ - 42$	Nonspecific with aliphatic and alicyclic systems, strong through RDA from cyclohexenes, by rearrangement from α-, β-cyclohexenones, enol and enamine acetates: McLafferty product
43	CH_3CO, C_3H_7, C_2H_4N, CONH	$M^+ - 43$	Acetyl, propyl, aromatic methyl ethers ($M^+ - 15-28$), nonspecific with aliphatic and alicyclic systems
44	CO_2, CH_3NHCH_2, CH_2CHOH, (N-fragment)	$M^+ - 44$	CO_2 from acids, esters, butane from aliphatic hydrocarbons
45	C_2H_5O, HCS, (Sulfides)	$M^+ - 45$	Ethyl esters, ethyl ethers, lactones, acids, CO_2H from some esters; CH_3NHCH_3 from dimethylamines
46	C_2H_5OH, NO_2	$M^+ - 46$	Ethyl esters, ethyl ethers, rarely acids, nitro compounds, n-alkanols ($M^+ - 18-28$)
47	CH_3S, $C^{35}Cl$, $C_2H_5OH_2$, $CH(OH)_2$, (S-fragment)		
48	CH_3SH, $CH^{35}Cl$		
49	$C^{37}Cl$, $CH_2^{35}Cl$		
50	C_4H_2, $CH_3^{35}Cl$		
51	C_4H_3, (Aromatic fragment)		
52	C_4H_4, $CH_3^{37}Cl$, (Aromatic fragment)		
53	C_4H_5		
54	⬡, C_2H_4CN	$M^+ - 54$	Cyclohexene (RDA)
55	C_4H_7, C_2H_3CO	$M^+ - 55$	C_4H_7 from alicyclic systems and butyl esters
56	C_4H_8, C_2H_4CO	$M^+ - 56$	Nonspecific with alkanes and alicyclic systems
57	C_4H_9, C_2H_5CO, C_3H_2F	$M^+ - 57$	Nonspecific with alkanes and alicyclic systems

Table 3-12 (continued)

m/z	Fragment X	M⁺ − X	Explanations
58	CH_3COHCH_2, C_2H_5-$CHNH_2$, $C_2H_6NCH_2$	$M^+ - 58$	C_3H_6O from α-methylaldehydes and acetonides
59	C_2H_6COH, $C_2H_5OCH_2$, CO_2CH_3, CH_3CONH_2	$M^+ - 59$	Methyl esters
60	$CH_2CO_2H_2$, CH_2ONO	$M^+ - 60$	O-acetates ($M^+ - AcOH$), methyl esters ($M^+ - CH_3OH - CO$)
61	$CH_3CO_2H_2$, C_2H_4SH		
62	$HOCH_2CH_2OH$	$M^+ - 62$	Ethylene ketals
63	C_5H_3		
64	SO_2	$M^+ - 64$	SO_2 cleavage from sulfonic acids
65	C_5H_5		
67	furyl (furan radical)		
68	CH₂=CH-CH=CH-CH₂ , C_4H_4O, C_3H_6CN		
69	C_5H_9, C_3H_5CO, CF_3, C_3HO_2 (1,3-Dioxyaromatics)		
70	C_5H_{10}, pyrrolyl (N-H)		
71	C_5H_{11}, C_4H_7CO, furanyl-methyl		
72	$C_4H_{10}N$, $C_3N_7NHCH_2$, $C_2H_5COHCH_2$		
73	$CO_2C_2H_5$, $C_3H_7OCH_2$, $CH_2CO_2CH_3$, C_4H_8OH, (O-fragments)		
74	$CH_2=COHOCH_3$, $CH_3CH=COHOH$		
75	$C_2H_5CO_2H_2$, $C_2H_5SCH_2$, $CH_3OCHOCH_3$, (Dimethyl acetates)		
76	C_6H_4		
77	C_6H_5		
78	C_6H_6		
79	C_6H_7, ^{79}Br		
80	C_6H_8, $H^{79}Br$, pyridinyl, N-methylpyrrolyl-CH_2, CH_3S_2H		
81	C_6H_9, ^{81}Br, furyl-CH_2		
82	C_6H_{10}, $H^{81}Br$		
83	C_6H_{11}, C_4H_7CO		
84	piperidinyl, N-methylpyrrolidinyl		
85	C_6H_{13}, C_4H_9CO		
86	$C_3H_7COH=CH_2$		
87	$CO_2C_3H_7$, $CH_3CO_2C_2H_5$, $CH_2CH_2CO_2CH_3$, (O-fragments)		
88	$CH_2=COHOC_2H_5$, $CH_3CH=COHOCH_3$		
91	benzyl ($C_6H_5CH_2$), furyl-Cl , n-Alkyl chlorides		

Table 3-12 (continued)

92 [pyridine-CH$_2$], [cyclohexadiene=CH$_2$, -H, H]

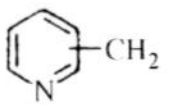

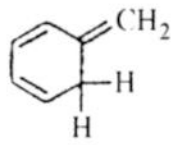

93 CH$_2^{79}$Br, [=NH, H, H structure]

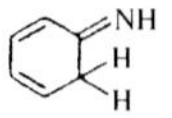

94 CH$_3^{79}$Br, [cyclohexadienone -H, H], [pyrrole-N(H)-CO]

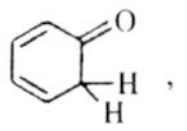

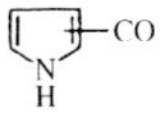

95 CH$_2^{81}$Br, [furan-CO]

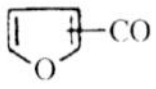

96 C$_5$H$_{10}$CN, CH$_3^{81}$Br

97 C$_7$H$_{13}$, [thiophene-CH$_2$]

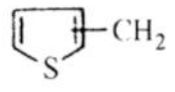

98 [piperidine N-CH$_2$]

99 C$_7$H$_{15}$, [O-C=O] (Ethylene ketals)

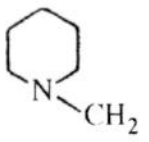

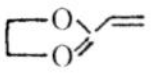

104 C$_2$H$_5$CHONO$_2$, [styrene], [methylenecyclohexadiene]

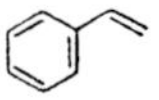

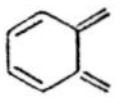

105 [Ph-CO], [Ph-N=N], [Ph-C$_2$H$_4$]

111 [thiophene-CO]

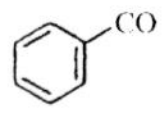

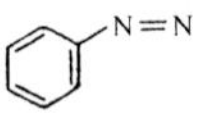

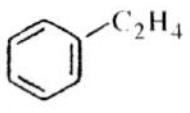

115 [indene-H]

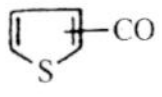

119 C$_2$F$_5$, [Ph-O-C(CH$_3$)$_2$], [Ph(CH$_3$)-O-CO]

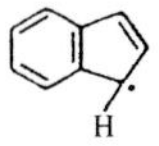

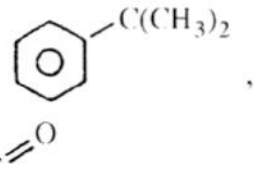

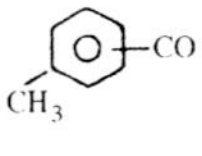

120 [benzofuranone C=O, O]

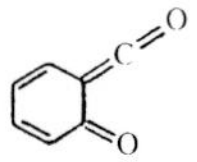

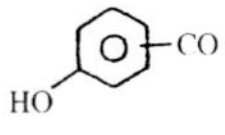

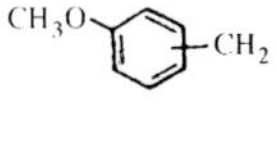

121 [Ph(HO)-O-CO], [CH$_3$O-Ph-CH$_2$]

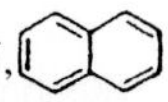

127 J

128 HJ, [naphthalene]

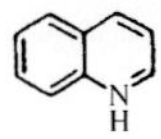

130 [quinoline-N-H]

131 C$_3$F$_5$

135 [cyclic-Br] (n-Alkyl bromides)

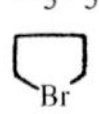

Table 3-12 (continued)

141	benzofuran-type structure with –CH₂
142	quinoline-type structure with –CH₂
149	phthalic acid structure (Phthalates)
152	biphenylene structure

3.2.6 Mass Spectroscopic Features of Selected Substance Classes

3.2.6.1 Volatile Halogenated Hydrocarbons

This group of compounds does not belong to a single class (Figs. 3-39 to 3-48). In a single analysis more than 60 aliphatic and aromatic compounds (Magic 60) can be determined by headspace GC/MS (static/purge and trap). A common feature is the appearance of chlorine and bromine isotope patterns in the mass spectra. With aliphatic compounds molecular ions do not always appear. With increasing molecular size the M^+ intensities decrease. Usually the loss of Cl (and also Br, F) as a radical from the molecular ion occurs. Fluorine can be recognised as HF from the difference of 20 u or as the CF fragment m/z 31, and bromine from signals with significantly higher masses but relatively short retention times in the GC. For detection it is a good idea to include the masses m/z 35/37 in the scan to guarantee ease of identification during spectroscopic comparison.

Aromatic halogenated hydrocarbons generally show an intense molecular ion. There is successive radical cleavage of chlorine. In the lower mass range the characteristic aromatic fragments appear with lower intensity.

3.2.6.2 Benzene/Toluene/Ethylbenzene/Xylenes (BTEX, Alkylaromatics)

Alkylaromatics form very stable molecular ions which can be detected with very high sensitivity (Figs. 3-49 to 3-57). The tropylium ion occurs at m/z 91 as the base peak, which is, for example, responsible for the uneven base peak in the toluene spectrum (M 92). The fragmentation of the aromatic skeleton leads to typical series of ions with m/z 38–40, 50–52, 63-67, 77–79 ('aromatic rubble'). Ethylbenzene and the xylenes cannot be differentiated from their spectra because they are isomers. In these cases the retention times of the components are more meaningful.

3.2 Substance Identification 267

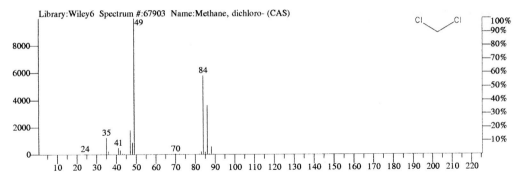

Fig. 3-39 Dichloromethane (R30) CH$_2$Cl$_2$, M: 84, CAS Reg. No.: 75-09-02

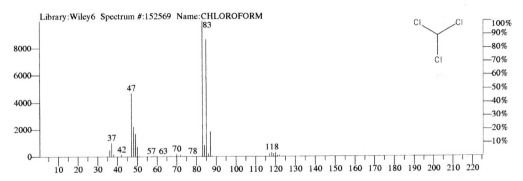

Fig. 3-40 Chloroform CHCl$_3$, M: 118, CAS Reg. No.: 67-66-3

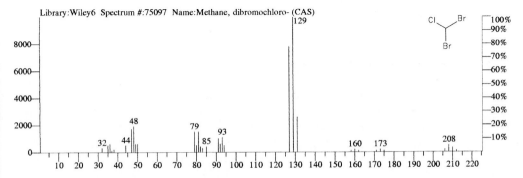

Fig. 3-41 Dibromochloromethane CHBr$_2$Cl, M: 206, CAS Reg. No.: 124-48-1

268   3 Evaluation of GC/MS Analyses

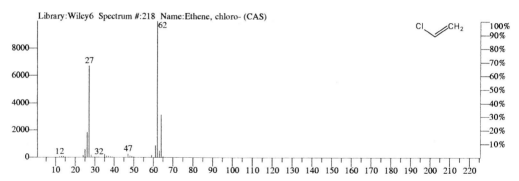

Fig. 3-42 Vinyl chloride C$_2$H$_3$Cl, M: 62, CAS Reg. No.: 75-01-4

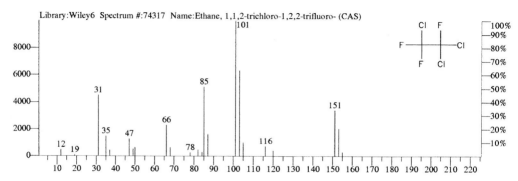

Fig. 3-43 1,1,2-Trifluoro-1,2,2-trichloroethane (R113) C$_2$Cl$_3$F$_3$, M: 186, CAS Reg. No.: 76-13-1

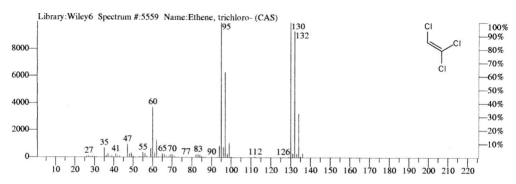

Fig. 3-44 Trichloroethylene C$_2$HCl$_3$, M: 130, CAS Reg. No.: 79-01-6

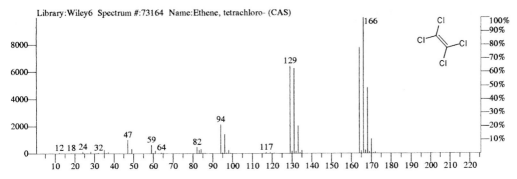

Fig. 3-45 Tetrachloroethylene (Per) C$_2$Cl$_4$, M: 164, CAS Reg. No.: 127-18-4

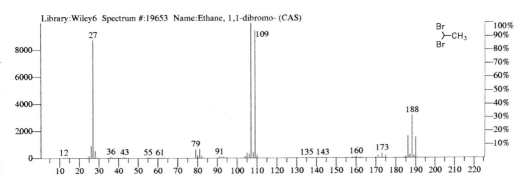

Fig. 3-46 1,1-Dibromoethane C$_2$H$_4$Br$_2$, M: 186, CAS Reg. No.: 557-91-5

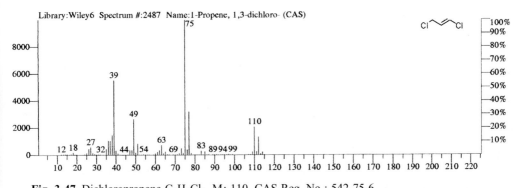

Fig. 3-47 Dichloropropene C$_3$H$_4$Cl$_2$, M: 110, CAS Reg. No.: 542-75-6

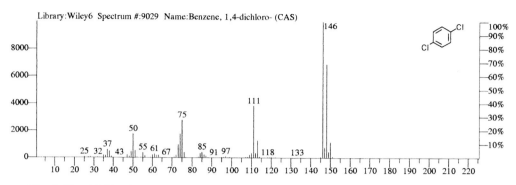

Fig. 3-48 p-Dichlorobenzene $C_6H_4Cl_2$, M: 146, CAS Reg. No.: 106-46-7

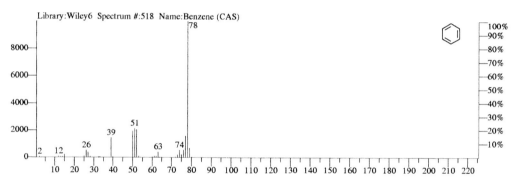

Fig. 3-49 Benzene C_6H_6, M: 78, CAS Reg. No.: 71-43-2

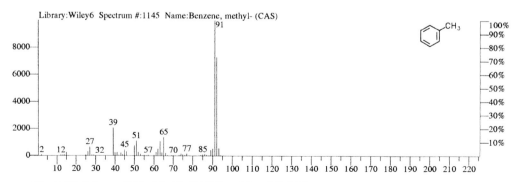

Fig. 3-50 Toluene C_7H_8, M: 92, CAS Reg. No.: 108-88-3

3.2 Substance Identification

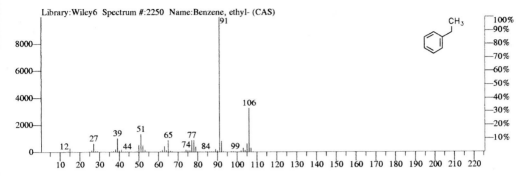

Fig. 3-51 Ethylbenzene C_8H_{10}, M: 106, CAS Reg. No.: 100-41-4

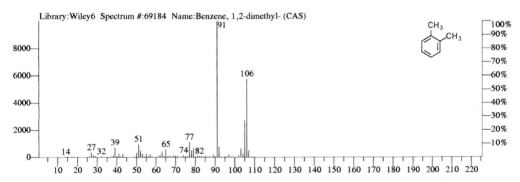

Fig. 3-52 o-Xylene C_8H_{10}, M: 106, CAS Reg. No.: 95-47-6

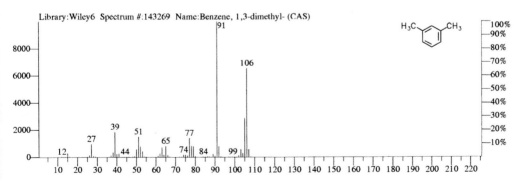

Fig. 3-53 m-Xylene C_8H_{10}, M: 106, CAS Reg. No.: 108-38-3

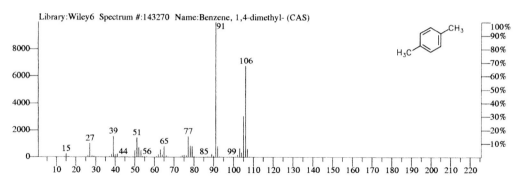

Fig. 3-54 p-Xylene C_8H_{10}, M: 106, CAS Reg. No.: 106-42-3

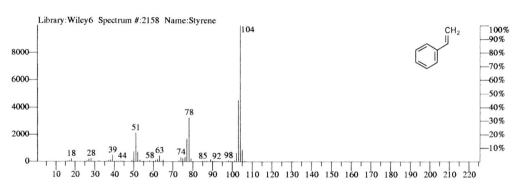

Fig. 3-55 Styrene C_8H_8, M: 104, CAS Reg. No.: 100-42-5

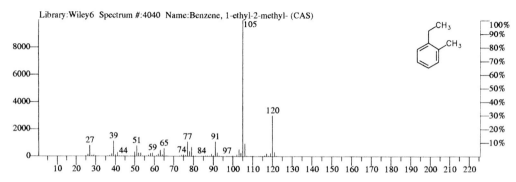

Fig. 3-56 1-Ethyl-2-methylbenzene C_9H_{12}, M: 120, CAS Reg. No.: 611-14-3

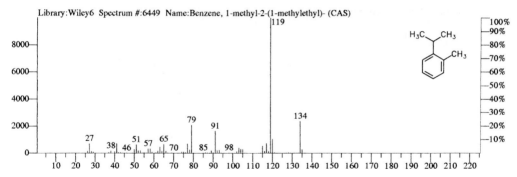

Fig. 3-57 1-Methyl-2-isopropylbenzene $C_{10}H_{14}$, M: 134, CAS Reg. No.: 1074-17-5

With alkyl side chains the problem of isomerism must be taken into account. For dimethylnaphthalene, for example, there are 10 isomers! Alkylaromatics fragment through benzyl cleavage. From a propyl side chain onwards benzyl cleavage can take place with H-transfer to the aromatic ring (to $C_8H_{10}^+$, m/z 106) or without transfer (to $C_8H_9^+$, m/z 105) depending on the steric or electronic conditions.

3.2.6.3 Polynuclear Aromatics

Polynuclear aromatics form very stable molecular ions (Figs. 3-58 to 3-66). They can be recognised easily from a 'half mass' signal caused by doubly charged molecular ions as m/2z = 1/2 m/z. Masses in the range m/z 100 to m/z 320 should be scanned for the analysis. In this range all polycondensed aromatics from naphthalene to coronene (including all 16 EPA components) can be determined and a possible matrix background of hydrocarbons with aliphatic character can be almost completely excluded from detection.

3.2.6.4 Phenols

In mass spectrometry phenolic substances are determined by their aromatic character (Figs. 3-67 to 3-72). Depending on the side chains, intense molecular ions and less intense fragments appear. In GC/MS phenols are usually chromatographed as their methyl esters or acetates. In residue analysis chlorinated and brominated phenols are the most important and can be recognised by their clear isotope patterns. The loss of CO (M-28) gives a less intense signal but is a clear indication of the presence of phenols. Halogenated phenols clearly show the loss of HCl (M-36) and HBr (M-80) in their spectra. With phenols isomers are also best recognised from their retention times rather than their mass spectra.

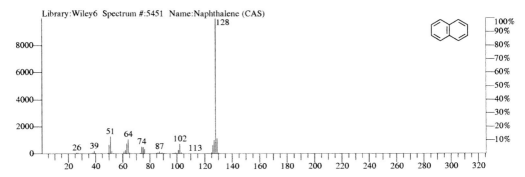

Fig. 3-58 Naphthalene $C_{10}H_8$, M: 128, CAS Reg. No.: 91-20-3

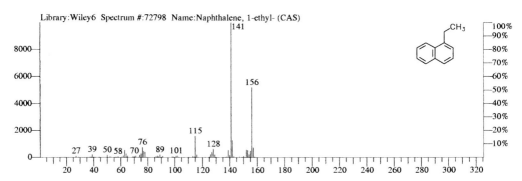

Fig. 3-59 1-Ethylnaphthalene $C_{12}H_{12}$, M: 156, CAS Reg. No.: 1127-76-0

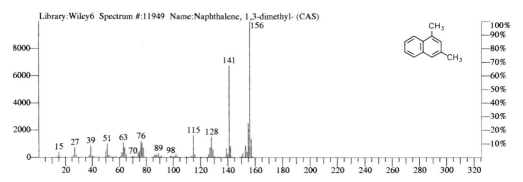

Fig. 3-60 1,3-Dimethylnaphthalene $C_{12}H_{12}$, M: 156, CAS Reg. No.: 575-41-7

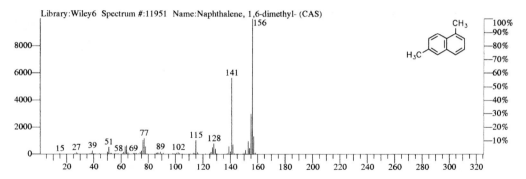

Fig. 3-61 1,6-Dimethylnaphthalene $C_{12}H_{12}$, M: 156, CAS Reg. No.: 575-43-9

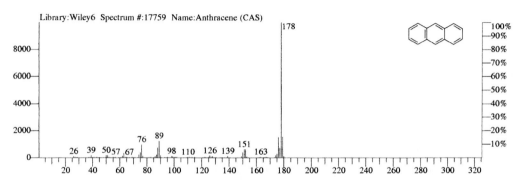

Fig. 3-62 Anthracene $C_{14}H_{10}$, M: 178, CAS Reg. No.: 120-12-7

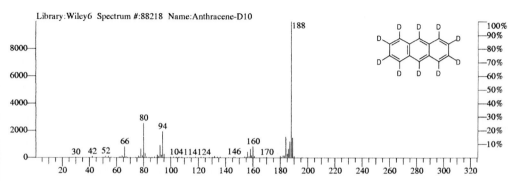

Fig. 3-63 Anthracene-d_{10} $C_{14}D_{10}$, M: 188, CAS Reg. No.: 1719-06-8

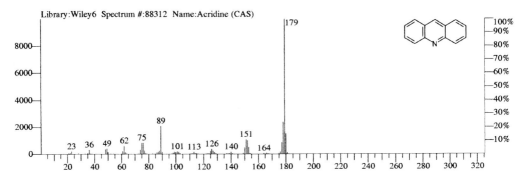

Fig. 3-64 Acridine $C_{13}H_9N$, M: 179, CAS Reg. No.: 260-94-6

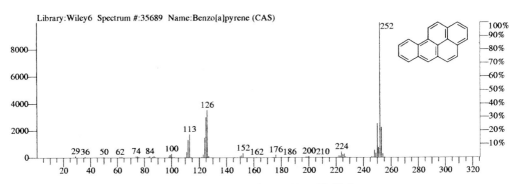

Fig. 3-65 Benzo[a]pyrene $C_{20}H_{12}$, M: 252, CAS Reg. No.: 50-32-8

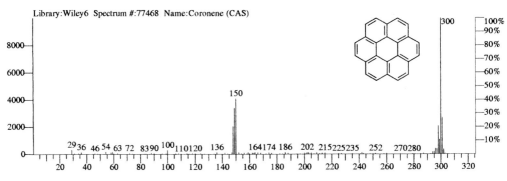

Fig. 3-66 Coronene $C_{24}H_{12}$, M: 300, CAS Reg. No.: 191-07-1

3.2 Substance Identification   277

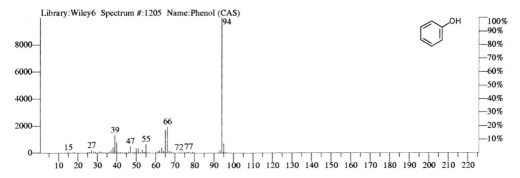

Fig. 3-67 Phenol C_6H_6O, M: 94, CAS Reg. No.: 108-95-2

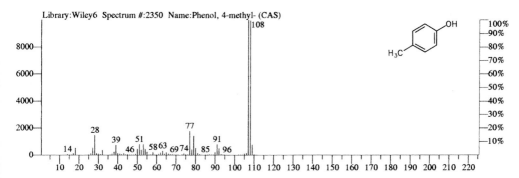

Fig. 3-68 p-Cresol C_7H_8O, M: 108, CAS Reg. No.: 106-44-5

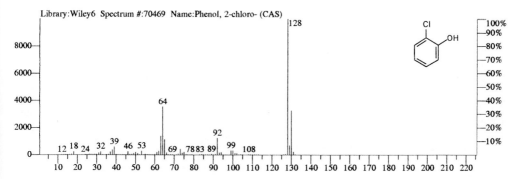

Fig. 3-69 o-Chlorophenol C_6H_5ClO, M: 128, CAS Reg. No.: 95-57-8

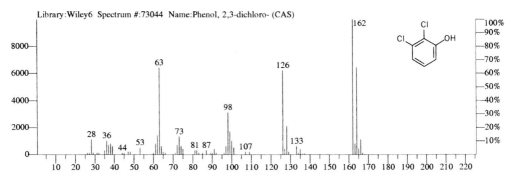

Fig. 3-70 2,3-Dichlorophenol $C_6H_4Cl_2O$, M: 162, CAS Reg. No.: 576-24-9

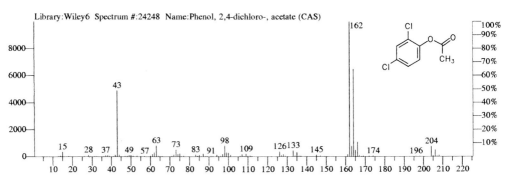

Fig. 3-71 2,4-Dichlorophenyl acetate $C_8H_6Cl_2O_2$, M: 204, CAS Reg. No.: 6341-97-5

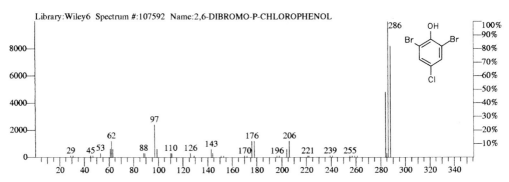

Fig. 3-72 2,6-Dibromo-4-chlorophenol $C_6H_3Br_2ClO$, M: 284

3.2.6.5 Plant Protection Agents

The character of the active substance forms the basis of the classification of these compounds (Figs. 3-73 to 3-104). In a collection of pesticide spectra there is therefore a wide variety of compound classes, which are covered to some extent by the other substance classes described here. Even when considering what appears to be a single group, such as phosphoric acid esters, it is virtually impossible to make any generalisations on fragmentation. Usually only stable compounds with aromatic character form intense molecular ions. In other cases molecular ions are of lower intensity and in trace analyses cannot be isolated from the matrix. For many plant protection agents (phosphoric acid esters, triazines, phenylureas etc), the use of chemical ionisation is advantageous for confirming identities or allowing selective detection.

Chlorinated Hydrocarbons

In this group of organochlorine pesticides there is a large number of different types of compound (Figs. 3-73 to 3-82). In the example of Lindane and HCB the difference between compounds with saturated and aromatic character can clearly be seen (molecular ion, fragmentation).

The high proportion of chlorine in these analytes leads to intense and characteristic isotope patterns. With the nonaromatic compounds (polycyclic polychlorinated alkanes made by Diels-Alder reactions, e.g. Dieldrin, Aldrin) spectra with large numbers of lines are formed through extensive fragmentation of the molecule. These compounds can be readily analysed using negative chemical ionisation, whereby the fragmentation is prevented.

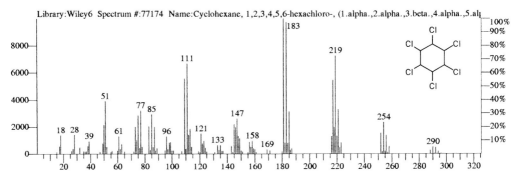

Fig. 3-73 Lindane $C_6H_6Cl_6$, M: 288, CAS Reg. No.: 58-89-9

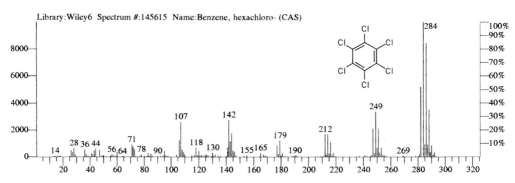

Fig. 3-74 Hexachlorobenzene (HCB) C_6Cl_6, M: 282, CAS Reg. No.: 118-74-1

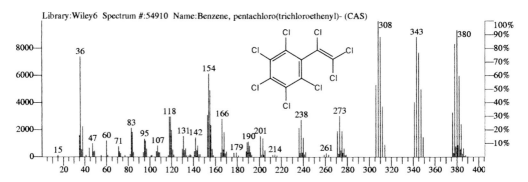

Fig. 3-75 Octachlorostyrene C_8Cl_8, M: 376, CAS Reg. No.: 29082-74-4

3.2 Substance Identification   281

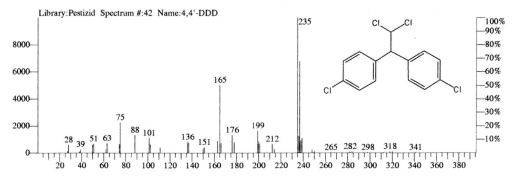

Fig. 3-76 4,4'-DDD $C_{14}H_{10}Cl_4$, M: 318, CAS Reg. No.: 72-54-8

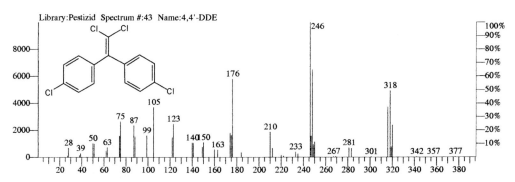

Fig. 3-77 4,4'-DDE $C_{14}H_8Cl_4$, M: 316, CAS Reg. No.: 72-55-9

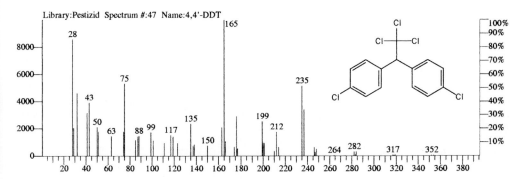

Fig. 3-78 4,4'-DDT $C_{14}H_9Cl_5$, M: 352, CAS Reg. No.: 50-29-3

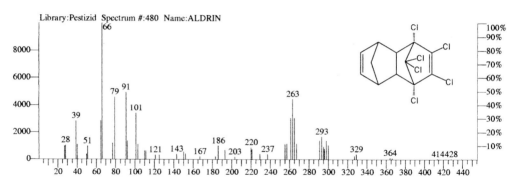

Fig. 3-79 Aldrin $C_{12}H_8Cl_6$, M: 362, CAS Reg. No.: 309-00-2

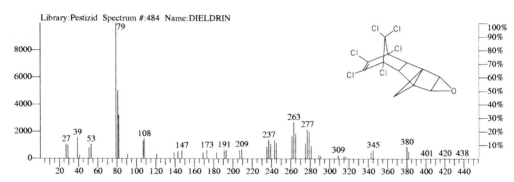

Fig. 3-80 Dieldrin $C_{12}H_8Cl_6O$, M: 378, CAS Reg. No.: 60-57-1

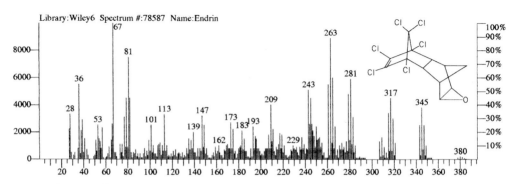

Fig. 3-81 Endrin $C_{12}H_8Cl_6O$, M: 378, CAS Reg. No.: 72-20-8

3.2 Substance Identification   283

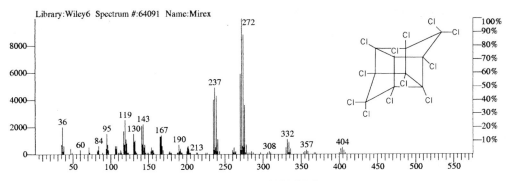

Fig. 3-82 Mirex $C_{10}Cl_{12}$, M: 540, CAS Reg. No.: 2385-85-5

Triazines

Triazine herbicides are substitution products of 1,3,5-triazines and thus belong to a single series of substances (Figs. 3-83 to 3-85). Hexazinon is also determined together with the triazine group in analysis (Fig. 3-86). Without exception the EI spectra of triazines show a large number of fragment ions and usually also contain the molecular ion with varying intensity. The high degree of fragmentation is responsible for the low specific response of triazines in trace analysis. All triazine analyses can be confirmed and quantified readily by positive chemical ionisation (e.g. with NH_3 as the reagent gas).

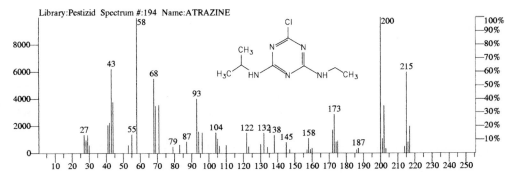

Fig. 3-83 Atrazine $C_8H_{14}ClN_5$, M: 215, CAS Reg. No.: 1912-24-9

284   3 Evaluation of GC/MS Analyses

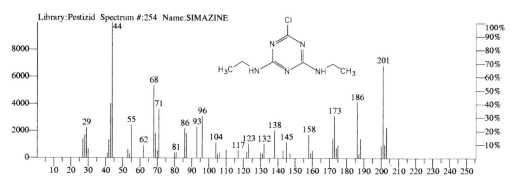

Fig. 3-84 Simazine $C_7H_{12}ClN_5$, M: 210, CAS Reg. No.: 122-34-9

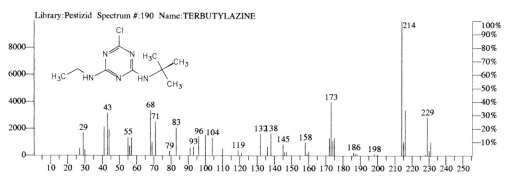

Fig. 3-85 Terbutylazine $C_9H_{16}ClN_5$, M: 229, CAS Reg. No.: 5915-41-3

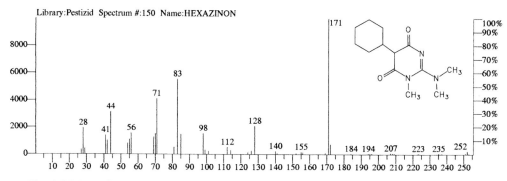

Fig. 3-86 Hexazinon $C_{12}H_{20}N_4O_2$, M: 252, CAS Reg. No.: 51235-04-2

Carbamates

The highly polar carbamate pesticides cannot always be analysed by GC/MS (Figs. 3-87 to 3-89). The low thermal stability leads to decomposition even in the injector. Substances with definite aromatic character, however, form stable intense molecular ions.

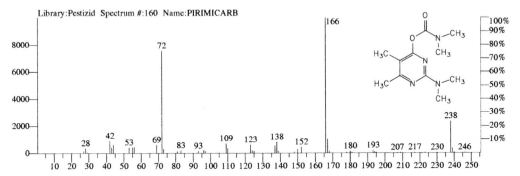

Fig. 3-87 Pirimicarb $C_{11}H_{18}N_4O_2$, M: 238, CAS Reg. No.: 23103-98-2

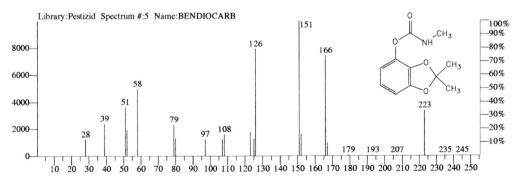

Fig. 3-88 Bendiocarb $C_{11}H_{13}NO_4$, M: 223, CAS Reg. No.: 22781-23-3

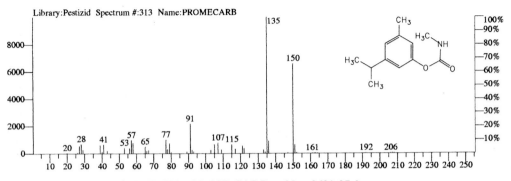

Fig. 3-89 Promecarb $C_{12}H_{17}NO_2$, M: 207, CAS Reg. No.: 2631-37-0

Phosphoric Acid Esters

This large group of pesticides does not exhibit uniform behaviour in mass spectrometry (Figs. 3-90 to 3-95). In trace analysis the detection of molecular ions is usually difficult, except in the case of aromatic compounds (e. g. Parathion). The high degree of fragmentation frequently extends into the area of matrix noise. Because of this it is more difficult to detect individual compounds, as in full scan analysis low starting masses must be used. Positive chemical ionisation is suitable for phosphoric acid esters because generally a strong CI reaction can be expected as a result of the large number of functional groups. The presence of the halogens Cl or Br can only be determined with certainty from the (quasi)molecular ions.

Phosphoric acid esters are also used as highly toxic chemical warfare agents (Tabun, Sarin and Soman, see Section 3.2.6.10).

The occurrence of fragments belonging to a particular group in the spectra of phosphoric acid esters has been intensively investigated (Table 3-13). Phosphoric acid esters are subdivided in the usual way as follows:

I   Dithiophosphoric acid esters $(RO)_2$-P(S)-S-Z
II Thionophosphoric acid esters $(RO)_2$-P(S)-O-Z
III Thiophosphoric acid esters $(RO)_2$-P(O)-S-Z
IV Phosphoric acid esters $(RO)_2$-P(O)-O-Z

Table 3-13 Fragments typical of various groups from phosphoric acid ester (PAE) pesticides (after Stan)

Group	R		m/z 93	m/z 97	m/z 109	m/z 121	m/z 125
Dithio-PAE	Ia	CH_3	+	–	–	–	+
	Ib	C_2H_5	+	+	–	+	+
Thiono-PAE	IIa	CH_3	+	–	+	–	+
	IIb	C_2H_5	+	+	+	+	+
Thiol-PAE	IIIa	CH_3	(+)	–	+	–	+
	IIIb	C_2H_5	–	+	+	+	–
PAE	IVa	CH_3	(+)	–	+	–	–
	IVb	C_2H_5	+/–	–	+	–	–

3.2 Substance Identification

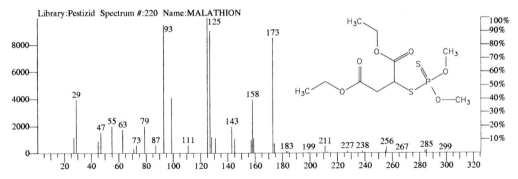

Fig. 3-90 Malathion $C_{10}H_{19}O_6PS_2$, M: 330, CAS Reg. No.: 121-75-5

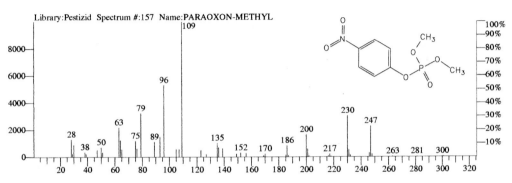

Fig. 3-91 Paraoxon-methyl $C_8H_{10}NO_6P$, M: 247, CAS Reg. No.: 950-35-6

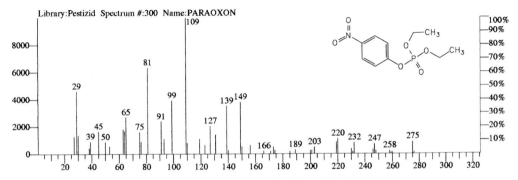

Fig. 3-92 Paraoxon(-ethyl) $C_{10}H_{14}NO_6P$, M: 275, CAS Reg. No.: 311-45-5

288   3 Evaluation of GC/MS Analyses

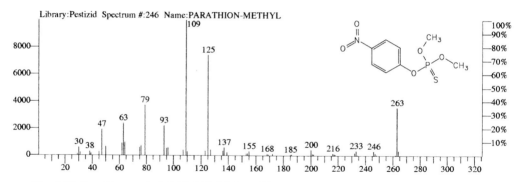

Fig. 3-93 Parathion-methyl $C_8H_{10}NO_5PS$, M: 263, CAS Reg. No.: 298-00-0

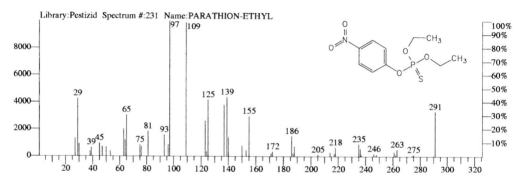

Fig. 3-94 Parathion-ethyl $C_{10}H_{14}NO_5PS$, M: 291, CAS Reg. No.: 56-38-2

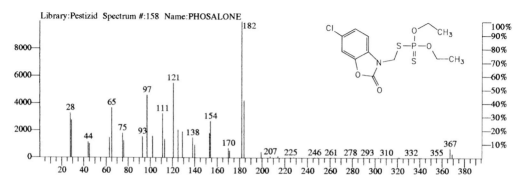

Fig. 3-95 Phosalone $C_{12}H_{15}ClNO_4PS_2$, M: 367, CAS Reg. No.: 2310-17-0

Phenylureas

GC/MS can only be used for the determination of phenylureas after derivatisation of the active substances. HPLC or HPLC/MS are currently the most suitable analytical methods because of the thermal lability and polarity of these compounds. A spectrum with few lines is frequently obtained in GC/MS analyses, which is dominated by the dimethylisocyanate ion m/z 72 which is specific to the group. The molecular ion region is of higher specificity but usually lower intensity (Figs. 3-96 to 3-101).

The phenylureas are rendered more suitable for GC by methylation of the azide hydrogen (e.g. with trimethylsulfonium hydroxide (TMSH) in the PTV injector after Färber). The mass spectra of the methyl derivatives correspond to those of the parent substances, except that the molecular ions are 14 masses higher with the same fragmentation pattern (Figs. 3-97, 3-99 and 3-101).

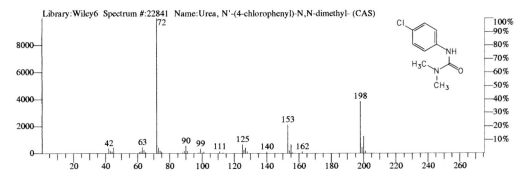

Fig. 3-96 Monuron $C_9H_{11}ClN_2O$, M: 198, CAS Reg. No.: 150-68-5

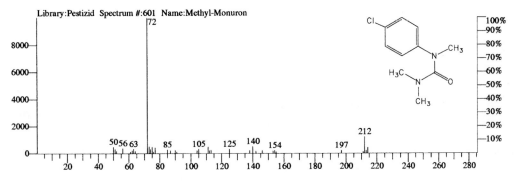

Fig. 3-97 Methyl-Monuron $C_{10}H_{13}ClN_2O$, M: 212

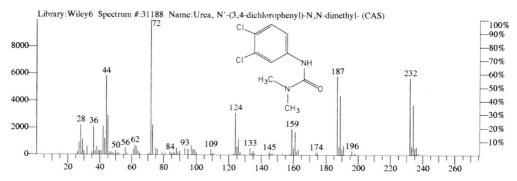

Fig. 3-98 Diuron $C_9H_{10}Cl_2N_2O$, M: 232, CAS Reg. No. 330-54-1

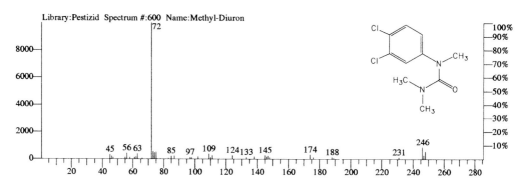

Fig. 3-99 Methyl-Diuron $C_{10}H_{12}Cl_2N_2O$, M: 246

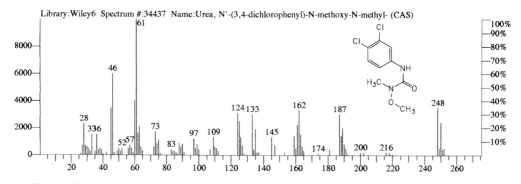

Fig. 3-100 Linuron $C_9H_{10}Cl_2N_2O_2$, M: 248, CAS Reg. No.: 330-55-2

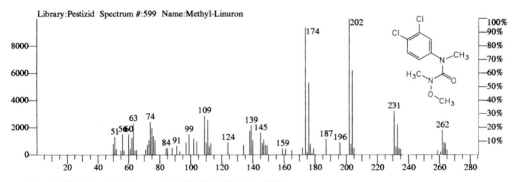

Fig. 3-101 Methyl-Linuron $C_{10}H_{12}Cl_2N_2O_2$, M: 262

Phenoxyalkylcarboxylic Acids

The free acids cannot be used with GC/MS in the case of trace analysis. They are determined using the methyl ester (Figs. 3-102 to 3-104). If the aromatic character predominates, intense molecular ions occur in the upper mass range. Increasing the length of the side chains significantly reduces the intensity of the molecular ion and leads to signals in the lower mass range. It should be noted that the presence of Cl or Br can only be determined with certainty from the (quasi)molecular ion. With EI the molecular ion fragments losing a Cl radical. Because of this the isotope signals of the fragments cannot be evaluated conclusively (see MCPB methyl ester). A final confirmation can be achieved through chemical ionisation.

292 3 Evaluation of GC/MS Analyses

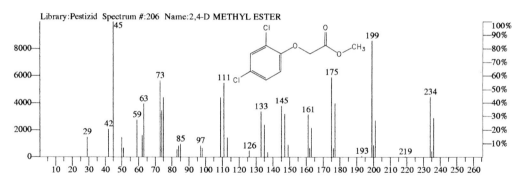

Fig. 3-102 2,4-D Methylester $C_9H_8Cl_2O_3$, M: 234, CAS Reg. No.: 1928-38-7

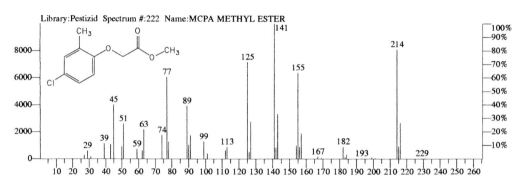

Fig. 3-103 MCPA Methylester $C_{10}H_{11}ClO_3$, M: 214, CAS Reg. No.: 2436-73-9

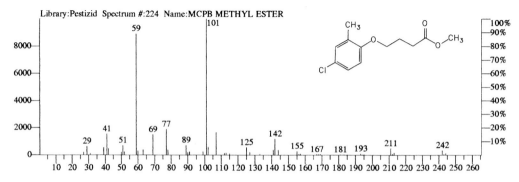

Fig. 3-104 MCPB Methylester $C_{12}H_{15}ClO_3$, M: 242, CAS Reg. No.: 57153-18-1

3.2.6.6 Polychlorinated Biphenyls (PCBs)

The spectra of polychlorinated biphenyls have similar features independent of the degree of chlorination (Figs. 3-105 to 3-114). As they are aromatic, their molecular ions are strongly pronounced. The degree of chlorination can clearly be determined from the isotope pattern. Fragmentation involves successive loss of Cl radicals and, in the lower mass range, degradation of the basic skeleton. For data acquisition the mass range above m/z 100 or 150 is required, so that detection of PCBs is usually possible above an accompanying matrix background. Individual isomers with a particular degree of chlorination have almost identical mass spectra. They can be differentiated on the basis of their retention times and therefore good gas chromatographic separation is a prerequisite for the determination of PCBs. The isomers 31 and 28 (nomenclature after Ballschmitter and Zell) are used as resolution criteria (Table 3-14). The spectra of compounds with different degrees of chlorination are shown in the following figures.

Table 3-14 PCB congeners for quantitation (after Ballschmitter)

PCB No.	Structure	
28	2,4,4′	Cl_3–PCB
52	2,2′,5,5′	Cl_4–PCB
101	2,2′,4,5,5′	Cl_5–PCB
118	2,3′,4,4′,5	Cl_5–PCB
138	2,2′,3,4,4′,5	Cl_6–PCB
153	2,2′,4,4′,5,5′	Cl_6–PCB
180	2,2′,3,4,4′,5,5′	Cl_7–PCB
209	2,2′,3,3′,4,4′,5,5′,6,6′	Cl_{10}–PCB [a]

[a] used as internal standard

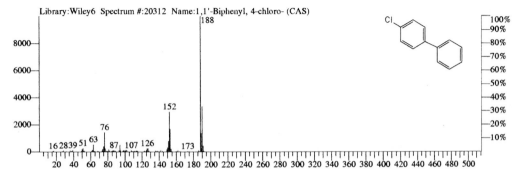

Fig. 3-105 Monochlorobiphenyl $C_{12}H_9Cl$, M: 188

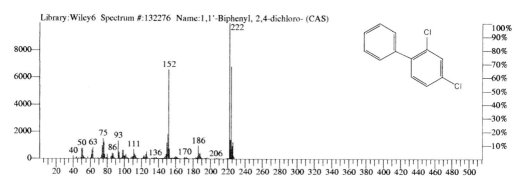

Fig. 3-106 Dichlorobiphenyl $C_{12}H_8Cl_2$, M: 222

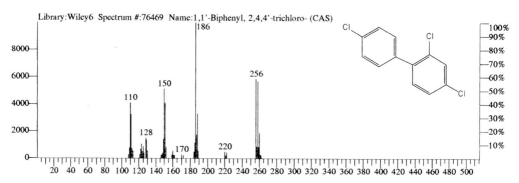

Fig. 3-107 Trichlorobiphenyl (e. g., PCB 28, 31) $C_{12}H_7Cl_3$, M: 256

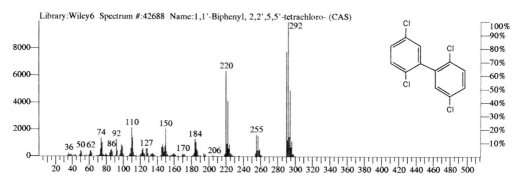

Fig. 3-108 Tetrachlorobiphenyl (e. g., PCB 52) $C_{12}H_6Cl_4$, M: 290

3.2 Substance Identification   295

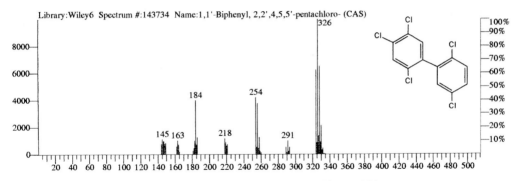

Fig. 3-109 Pentachlorobiphenyl (e.g., PCB 101, 118) $C_{12}H_5Cl_5$, M: 324

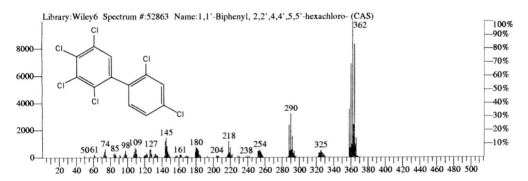

Fig. 3-110 Hexachlorobiphenyl (e.g., PCB 138, 153) $C_{12}H_4Cl_6$, M: 358

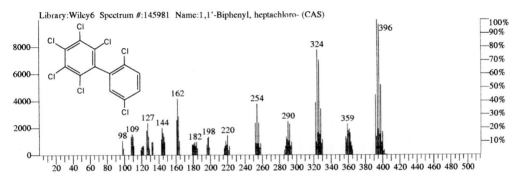

Fig. 3-111 Heptachlorobiphenyl (e.g., PCB 180) $C_{12}H_3Cl_7$, M: 392

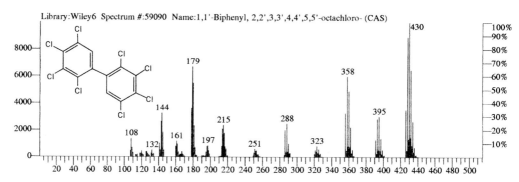

Fig. 3-112 Octachlorobiphenyl $C_{12}H_2Cl_8$, M: 426

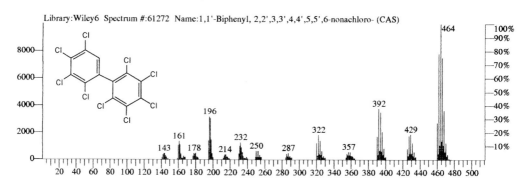

Fig. 3-113 Nonachlorobiphenyl $C_{12}HCl_9$, M: 460

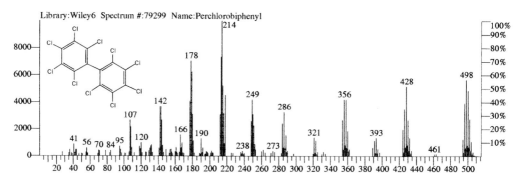

Fig. 3-114 Decachlorobiphenyl (PCB 209) $C_{12}Cl_{10}$, M: 494

3.2.6.9 Explosives

All explosives have a high proportion of oxygen in the form of nitro groups. GC/MS analysis of the nonaromatic compounds (Hexogen, Octogen, Nitropenta etc.) is problematic because of decomposition, even in the injector. However, aromatic nitro compounds and their metabolites (aromatic amines) can be detected with extremely high sensitivity because of their stability (Figs. 3-124 to 3-139). With EI molecular ions usually appear at low intensity because nitro compounds eliminate NO (M-30). o-Nitrotoluenes are an exception because a stable ion is formed after loss of an OH radical (M-17) as a result of the proximity of the two groups (ortho effect) (Figs. 3-124 and 3-125). If the scan is run including the mass m/z 30, the mass chromatogram can show the general elution of the nitro compounds. In this area chemical ionisation is useful for confirming and quantifying results. In particular water has proved to be a useful CI gas in ion trap systems for residue analysis of old munitions.

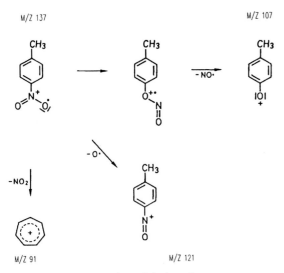

Fig. 3-124 Fragmentation of 4-nitrotoluene

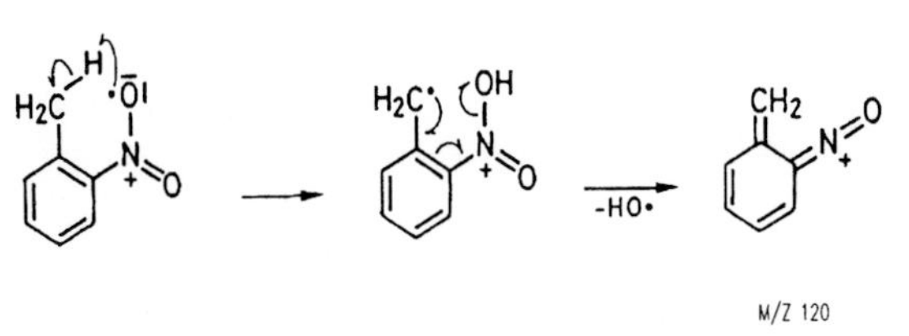

Fig. 3-125 Fragmentation of 2-nitrotoluene (ortho effect)

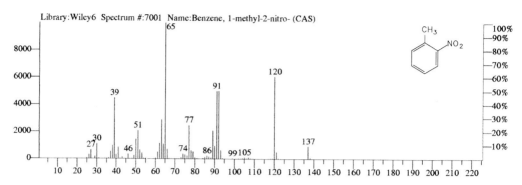

Fig. 3-126 2-Nitrotoluene $C_7H_7NO_2$, M: 137, CAS Reg. No.: 88-72-2

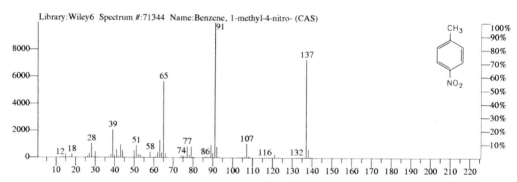

Fig. 3-127 4-Nitrotoluene $C_7H_7NO_2$, M: 137, CAS Reg. No.: 99-99-0

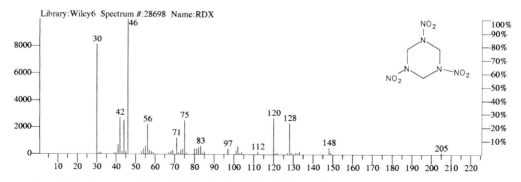

Fig. 3-128 Hexogen (RDX) $C_3H_6N_6O_6$, M: 222, CAS Reg. No.: 121-82-4

3.2 Substance Identification   303

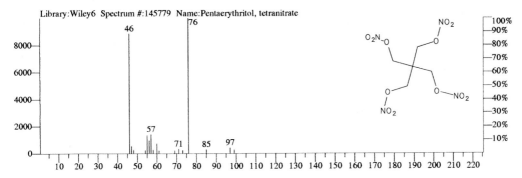

Fig. 3-129 Nitropenta (PETN) $C_5H_8N_4O_{12}$, M: 316, CAS Reg. No.: 78-11-5

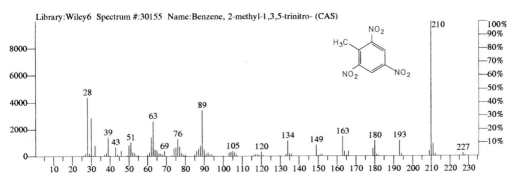

Fig. 3-130 Trinitrotoluene (TNT) $C_7H_5N_3O_6$, M: 227, CAS Reg. No.: 118-96-7

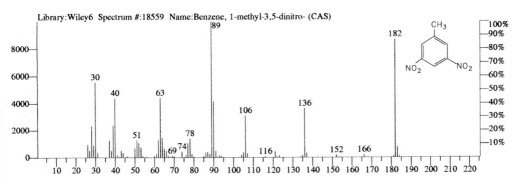

Fig. 3-131 3,5-Dinitrotoluene (3,5-DNT) $C_7H_6N_2O_4$, M: 182, CAS Reg. No.: 618-85-9

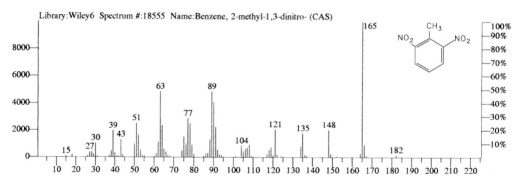

Fig. 3-132 2,6-Dinitrotoluene (2,6-DNT) $C_7H_6N_2O_4$, M: 182, CAS Reg. No.: 606-20-2

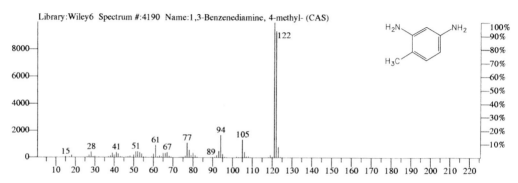

Fig. 3-133 2,4-Diaminotoluene (2,4-DAT) $C_7H_{10}N_2$, M: 122, CAS Reg. No.: 95-80-7
(Note: 2,4-DAT and 2,6-DAT form a critical pair during GC separation)

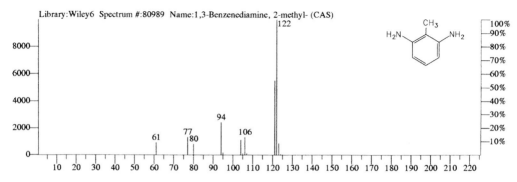

Fig. 3-134 2,6-Diaminotoluene (2,6-DAT) $C_7H_{10}N_2$, M: 122, CAS Reg. No.: 823-40-5

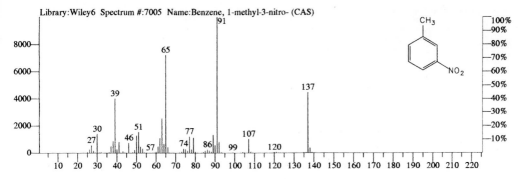

Fig. 3-135 (Mono-)3-nitrotoluene (3-MNT) C$_7$H$_7$NO$_2$, M: 137, CAS Reg. No.: 99-08-1

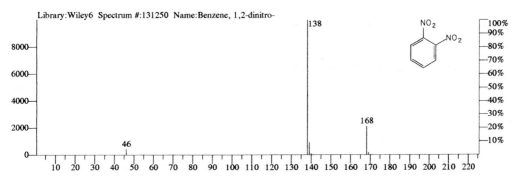

Fig. 3-136 1,2-Dinitrobenzene (1,2-DNB) C$_6$H$_4$N$_2$O$_4$, M: 168, CAS Reg. No.: 528-29-0

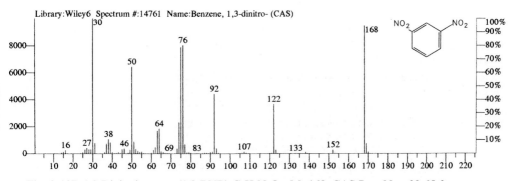

Fig. 3-137 1,3-Dinitrobenzene (1,3-DNB) C$_6$H$_4$N$_2$O$_4$, M: 168, CAS Reg. No.: 99-65-0

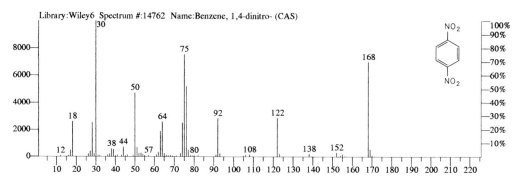

Fig. 3-138 1,4-Dinitrobenzene (1,4-DNB) $C_6H_4N_2O_4$, M: 168. CAS Reg. No.: 100-25-4

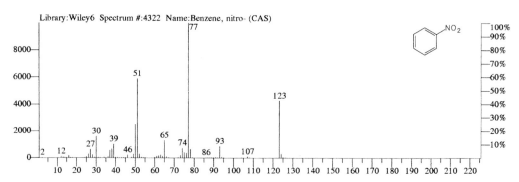

Fig. 3-139 (Mono-)nitrobenzene (MNB) $C_6H_5NO_2$, M: 123, CAS Reg. No.: 98-95-3

3.2.6.10 Chemical Warfare Agents

The identification of chemical warfare agents is important for testing disarmament measures and for checking disused military sites. Here also there is no single chemical class of substances. Volatile phosphoric acid esters and organoarsenic compounds belong to this group (Figs. 3-140 to 3-145). Chemical ionisation is the method of choice for identification and confirmation of identity.

3.2 Substance Identification   307

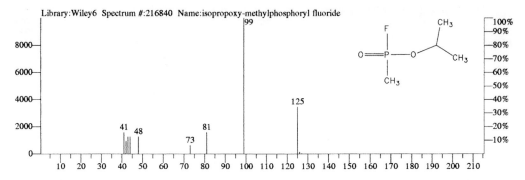

Fig. 3-140 Sarin $C_4H_{10}FO_2P$, M: 140, CAS Reg. No.: 107-44-8

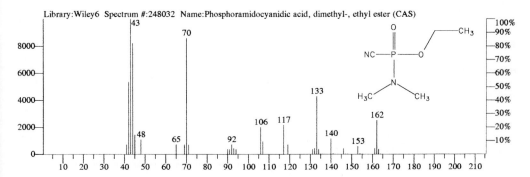

Fig. 3-141 Tabun $C_5H_{11}N_2O_2P$, M: 162, CAS Reg. No.: 77-81-6

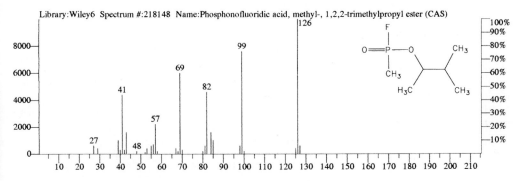

Fig. 3-142 Soman $C_7H_{16}FO_2P$, M: 182, CAS Reg. No.: 96-64-0

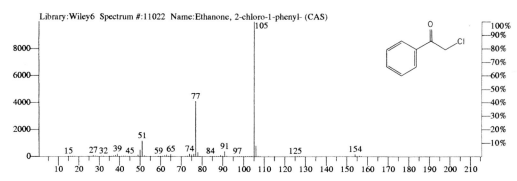

Fig. 3-143 Chloroacetophenone (CN) C_8H_7ClO, M: 154, CAS Reg. No.: 532-27-4

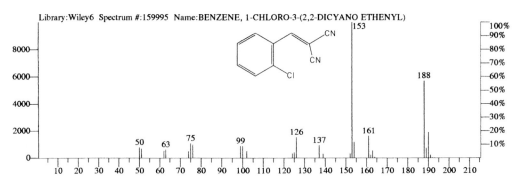

Fig. 3-144 o-Chlorobenzylidenemalnonitrile (CS) $C_{10}H_5ClN_2$, M: 188, CAS Reg. No.: 2698-41-1

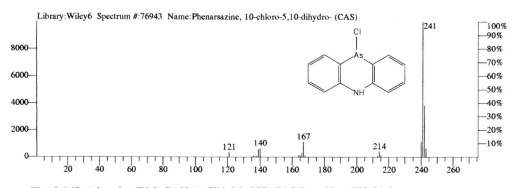

Fig. 3-145 Adamsite (DM) $C_{12}H_9AsClN$, M: 277, CAS Reg. No.: 578-94-9

3.2.6.11 Flameproofing Agents

Multiply brominated aromatic compounds, which can only be chromatographed in a few cases because of their high molecular weights, are used as flameproofing agents (Figs. 3-146 and 3-147). The molecular weights of the polybrominated biphenyls (PBB) and biphenyl ethers (PBBE), which are frequently used, have molecular weights of up to 1000 (decabromobiphenyl M 950). For GC/MS analysis pyrolysis is used with solid samples, which gives rise to readily identifiable and chromatographable products. Flameproofing agents have recently become of interest to analysts because burning materials containing PBB and PBBE can lead to formation of polybrominated dibenzo-dioxins and –furans. The spectra of PBB and PBBE flameproofing agents are characterised by the symmetrical isotope pattern of the bromine and the high stability of the aromatic molecular ion. Flameproofing agents based on brominated alkyl phosphates have a greater tendency to fragment.

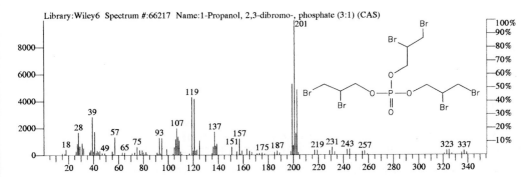

Fig. 3-146 Bromkal P67–6HP (Tris) $C_9H_{15}Br_6O_4P$, M: 692, CAS Reg. No.: 126-72-7

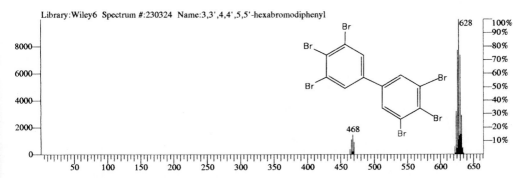

Fig. 3-147 Hexabromobiphenyl (HBB) $C_{12}H_4Br_6$, M: 622

3.3 Quantitation

Besides the identification of components of a mixture, the use of GC/MS systems to determine the proportions of the different components is important. The need to determine quantitatively an increasing number of components in complicated matrices in ever smaller concentrations makes the use of GC/MS systems in routine analysis appropriate for economic reasons. Gone are the days when only positive results from classical GC were confirmed by GC/MS. In many areas of residue analysis the development of routine multicomponent methods has only become possible through the selectivity of detection and the specificity of identification of GC/MS systems. The development of GC/MS data systems has therefore been successful in recent years, particularly in areas where the use of the mass spectrometric substance information coincides with the integration of substance peaks and thus increases the certainty of quantitation. Compared with chromatography data systems of GC and HPLC, there are therefore differences and additional possibilities which arise from the use of the mass spectrometer as the detector.

Scan Rate of the Chromatogram

Mass spectrometers do not continuously record the substance stream arriving in the detector (as for example with FID, ECD etc.). The chromatogram is comprised of a series of measurement points which are represented by mass spectra. The scan rate chosen by the user establishes the time interval between the data points. The maximum possible scan rate depends on the scanning velocity of the spectrometer. It is determined by the width of the mass range to be acquired and the necessity of achieving an analytical detection capacity of the instrument which is as high as possible. For routine measurements scan rates of 0.5 to 1.0 s/scan are usually chosen. Compared with a sharp concentration change in the slope of a GC peak, these scan rates are only slow.

In the integration of GC/MS chromatograms, the choice of scan rate is particularly important for the distribution of the peak area values. With the scan rate of 1s/scan, which is frequently used, the peak area of rapidly eluting components cannot be determined reliably as too few data points for the correct plotting of the GC peak are recorded (Fig. 3-148). The incorporation of the top of the peak in the calculation of height and area cannot be carried out correctly in these cases. Under certain circum-

Fig. 3-148 Comparison of an actual chromatogram with one reconstructed from data points at a low scan rate (after Chang)

stances small peaks next to sharp peaks can be lost in the plot. In the case of a distribution of area values in quantitative determinations, special attention should first be paid to the recording parameters before a possible cause is looked for in the detector itself.

To determine the optimal scan rate, the base width of the components eluting early should first be investigated. In practice it has been shown that the peak area of a symmetrical peak can be described well by ca. 10 measuring points. Therefore, the base width (elution time) of the most rapidly eluting component should be a measure of the scan rate in quantitative GC/MS analysis. There is no problem with base widths in the region of ca. 10 s, but volatile compounds elute with peak widths of 5 s or even lower in good chromatography. Here the limits of scanning instruments become clear, as faster scan rates (<0.5 s/scan) lead to a decrease in the dwell time per ion and thus to a significant loss in sensitivity. Individual mass recording (SIM/MID) which allows high sensitivity at high scan rates, or parallel detection of ions by ion trap or time-of-flight mass spectrometers are possible solutions.

3.3.1 Decision Limit

The question of when a substance can be said to be detected cannot be answered differently for quantitative GC/MS compared with all other chromatographic systems. The answer lies in the determination of the signal/noise ratio.

In the basic adjustment of mass spectrometers, unlike classical GC detectors, the zero point is adjusted correctly (electrometer zero) to ensure the exact plot of isotope patterns. For this the adjustment is chosen in such a way that minimum noise of the electronics is determined which can then be removed during data acquisition by software filters. Besides electrical noise, there is also chemical noise (matrix, column bleed, leaks etc.), particularly in trace analysis.

The decision as to whether a substance has been detected or not is usually assessed in the signal domain. Here it is established that the decision limit is such that the smallest detectable signal from the substance can be clearly differentiated from a blank value (critical value).

In the measurement of a substance-free sample (blank sample) an average signal is obtained which corresponds to a so-called blank value. Multiple measurements confirm this blank value statistically and give its standard deviation. With GC/MS this blank value determination is carried out in practice in the immediate vicinity to the substance peak, whereby the noise widths before and after the peak are taken into consideration. The average noise and the signal intensity can also be determined manually from the print-out. A substance can be said to be detected if the substance signal exceeds a certain multiple of the noise width (standard deviation). This value is chosen arbitrarily as 2, 3 or higher (depending on the laboratory or on the application) and is used as the deciding criterion in most routine evaluations of GC/MS data systems (Fig. 3-149).

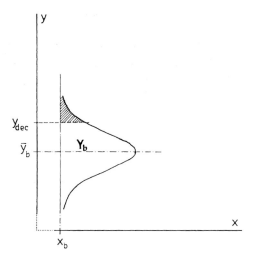

Fig. 3-149 Statistical interpretation of the decision limit
y_b = random sample of blank values
y_{dec} = decision limit
In a statistically defined system the decision limit, i.e. the smallest detectable signal which can be differentiated from a blank value, can only be obtained from multiple measurements of the blank value and the analysis at a given error probability (after Ebel)

3.3.2 Limit of Detection

How does the decision limit relate to the limit of detection (LOD)? Usually both terms are used synonymously, which is totally incorrect! The LOD of a method is never given in counts or parts of a scale but is given a dimension, such as pg/µl, ng/l, or ppb etc., in any case in the substance domain! The transfer from the signal domain of the decision limit into the substance domain of the LOD is effected using a valid calibration function (Fig. 3-150)! Nevertheless, the LOD can only be regarded as a qualitative parameter as here the uncertainty in the calibration function is not taken into account.

3.3.3 Limit of Quantitation

The limit of quantitation (LOQ) of a method is also given as a quantity of substance or concentration in the substance domain. This limit incorporates the calibration and thus also the uncertainty (error consideration) of the measurements. Unlike the LOD it is guaranteed statistically and gives the lower limiting concentration which can be unambiguously determined quantitatively. It can differ significantly from the blank value.

As a component can only be determined after it has been detected, the LOQ cannot be lower than the LOD. As there is a relative uncertainty in the result of ca. 100% at a

3.3 Quantitation 313

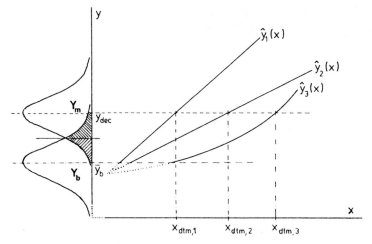

Fig. 3-150 Definition of the limit of detection (LOD) from the decision limit
The limit of detection (LOD) is defined as that quantity of substance, concentration or content X_{dtm} which is given using the calibration function from the smallest detectable signal y_{dec} (decision limit). Calibration functions with different sensitivities (slopes) lead to different limits of detection at the same signal height (after Ebel)

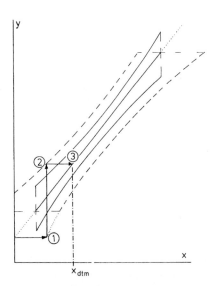

Fig. 3-151 Statistical definition of the limit of quantitation
1 Measured value at quantity of substance = 0 (signal domain)
2 Signal size which can be differentiated significantly taking the standard deviation into account
3 Limit of quantitation x_{dtm} from the calibration function
The limit of quantitation is the lower limiting concentration x_{dtm} at a fixed statistical error probability which can definitely be quantitatively determined and be differentiated significantly from zero concentration (after Ebel)

concentration of a substance corresponding to the LOD in an analysis sample, the LOQ must be correspondingly higher than the LOD, depending on the requirement.

3.3.4 Sensitivity

The term sensitivity, which is frequently used to describe the quality of a residue analysis, or the current state of a measuring instrument, is often used incorrectly as a synonym for the lowest possible LOD or LOQ. A sensitive analysis procedure, however, exhibits a large change in signal with a small change in substance concentration. The sensitivity of a procedure thus describes the increase a of a linear calibration function (see Section 3.3.5). At the same confidence interval of the measured points of a calibration function (Fig. 3-152), sensitive analysis procedures give a narrower confidence interval than less sensitive ones! The LOD is independent of the sensitivity of an analytical method.

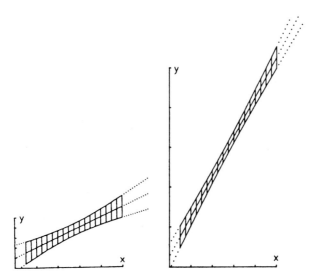

Fig. 3-152 Confidence intervals for calibration curves at the same distribution of the measured values but with different slopes (after Ebel)

3.3.5 The Calibration Function

For an analysis procedure that has been worked out, the calibration function is constructed from the measurement of known concentrations as, because of the dependence of the height of the signal on the operating parameters and the current state of the mass spectrometer, fixed response factors can only be used to a limited degree in GC/MS. The calibration function generally describes the dependence of the signal on the substance concentration. In the case of a linear dependence, the regression calculation gives a straight line for the calibration function, the equation for which contains the blank value a_0 and the sensitivity a:

$$f(x) = a_0 + a \cdot x \qquad (25)$$

The calibration function is defined exclusively within the working range given by the experimental calibration. GC/MS systems frequently achieve very low LODs so that, at a correspondingly dense collection of calibration points near the blank value, a nonlinear area is described. This area can be caused by unavoidable active sites (residual activities) in the system and 'swallows up' a small but constant quantity of substance. Such a calibration function tends to approach the x axis before reaching the origin.

In the upper concentration range the signal hardly increases at all with increasing substance concentration because of increasing saturation of the detector. The calibration function stops increasing and tends to be asymptotic (Fig. 3-153).

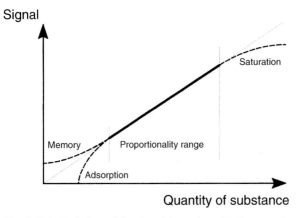

Fig. 3-153 Variation of the signal intensity with the quantity of substance

The best fit of the calibration curve to the measured points is determined by a regression calculation. The regression coefficient gives information on the quality of the curve fit. Linear regressions are not always suitable for the best fit in GC/MS analysis; quadratic regressions frequently give better results. Particularly in the area of trace analysis the type of fit within the calibration can change through the nonlinear effects described above. In such cases it is helpful to limit the regression calculation to that of a sample concentration lying close to the calibration level (local linearisation, next three to four data points).

If individual data systems do not allow this, a point to point calibration can be carried out instead. Important aspects regarding the optimisation of a calibration can be derived from the above:

– One-point calibrations have no statistical confirmation and can therefore only be used as an orientation.

- Multipoint calibrations must cover the expected concentration range. Extrapolation beyond the experimentally measured points is not allowed. (Calibration functions do not have to pass through the origin).
- Multiple measurements at an individual calibration level define the confidence interval which can be achieved.
- At the same number of calibration points those near the LOQ give an improvement in the fit (unlike the equidistant position of the calibration level.)

3.3.6 Quantitation and Standardisation

In order to determine the substance concentration in an unknown analysis sample, the peak areas of the sample are calculated using the calibration function and the results are given in terms of quantity or concentration. Many data systems also take into account the quantity weighed out and dilution or concentration steps in order to be able to give the concentration in the original sample.

For GC/MS the methods of external or internal standard calibration or standard addition are used as standardisation procedures.

External standardisation corresponds to the classical calibration procedure (Fig. 3-154). The substance to be determined is used to make a standard solution with a known concentration. Measurements are made on standard solutions of different concentrations (calibration steps, calibration levels). For calibration the peak areas determined are plotted against the concentrations of the different calibration levels.

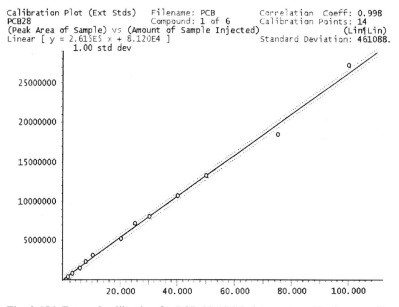

Fig. 3-154 External calibration for PCB 28 (GCQ data system, Finnigan, on-line thermodesorption GC/MS coupling, after Tschickard, Environmental Protection Office, Mainz)

With external standardisation, standard deviations of 5–10% are obtained in GC/MS, which occur even using highly reproducible injection techniques involving an autosampler. The causes of this distribution are injection errors and small changes in the mass spectrometer. As absolute values from different analysis runs are used for external standardisation, the calibration function and the sample measurement show all the effects which can cause a change in response between the mass spectrometric measurements. Various factors can contribute to this, e. g. slightly different ionisation efficiency through geometric changes in the filament, slightly varying transmission due to contamination of the lens systems with the matrix, or a contribution from the multiplier on signal production.

These factors can be compensated for by internal standardisation. The principle of the internal standard is based on the calculation of relative values which are determined within the same analysis. One or more additional substances are introduced as a fixed reference parameter, the concentration of which is kept constant in the standard solutions and is always added to the analysis sample at the same concentration. For the calculation the peak area values (or peak heights) of the substance being analysed relative to the peak area (or height) of the internal standard are used in the same analysis run. In this way all volume errors and variations in the function of the instrument are compensated for and quantitative determinations of the highest precision are achieved. Standard deviations of less than 5% can be achieved with internal standardisation.

The time at which the internal standard is added during the analysis depends on the analysis requirement. For example, the internal standard can be added to the sample at an early stage (surrogate standard) to simplify the clean-up. The addition of different standards at different individual stages of the clean-up allows the efficiency of individual clean-up steps to be monitored. Addition to the extract directly before the measurement can serve to confirm which instrument specification is required and to monitor the minimum signal/noise ratio necessary.

The choice of the internal standard is particularly important also in GC/MS. Basically the internal standard should behave as far as possible in the same way as the substance being analysed. Unlike classical GC detectors, the GC/MS procedure offers the unique possibility of using isotopically labelled, but nonradioactive analogues of the

---

Requirements of the Internal Standard in GC/MS Analysis:

- The internal standard chosen must be stable to clean-up and analysis and as inactive as possible.
- As far as possible the properties of the standard should be comparable to those of the analyte with regard to sample preparation and analysis; therefore isotopically labelled standards should ideally be used.
- The standard itself must not be present in the original sample.
- The retention behaviour in the GC should be adjusted so that elution occurs in the same section of the program (isothermal or heating ramp).
- The use of several standards allows them to be used as retention time standards.
- The retention behaviour of the internal standard should be adjusted to ensure that overlap with matrix peaks or other components to be determined is avoided and faultless integration is possible.
- The fragmentation behaviour (mass spectrometric response) should be comparable.
- The choice of the quantifying mass of the standard should exclude interference by the matrix or other components.

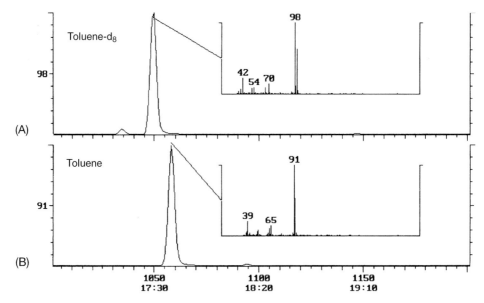

Fig. 3-155 Elution of toluene-d$_8$ as the internal standard using purge & trap GC/MS (mass chromatograms).
(A) Toluene-d$_8$, m/z 98
(B) Toluene m/z 91

substances being analysed. Deuterated standards are frequently used for this purpose as they fulfil the requirement of comparable behaviour during clean-up and analysis to the greatest degree. The extent of deuteration should always be sufficiently high for interference with the natural isotope intensities of the unlabelled substance to be excluded. The internal deuterated standard thus chosen can thus be detected selectively through its own mass trace and integrated (Fig 3-155).

In the clean-up of biological material the carrier effect can be exploited through the addition of an internal standard. The standard added at comparatively high concentrations can cover up active sites in the matrix and thus improve the ease of extraction of the substance being analysed. In chromatography deuterated standards have a slightly shorter retention time than the original analytes. This retention time difference, though small, is always visible in the mass chromatogram, which is used for selective integration of the individual components.

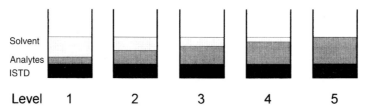

Fig. 3-156 Preparation of solutions for calibration with internal standards (c (ISTD) = constant, c (analytes) = variable, total volume = constant, fill with solvent, calculate concentrations on the basis of total volumes)

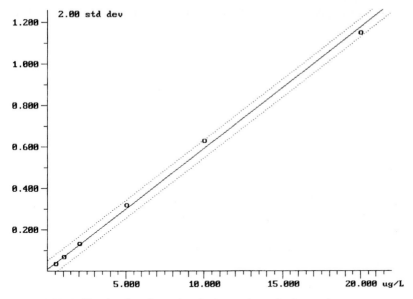

Fig. 3-157 Calibration function using the internal standard procedure.
Plot: area of the sample/area of the standard against quantity of sample

Table 3-15 List of volatile halogenated hydrocarbons found in a drinking water analysis using the ISTD procedure (I: internal standard, S: surrogate standard, A: analyte)

No.	Substance	Type	Scan #	Retention time	Me	Calculated conc.	Unit
1	Fluorobenzene IS1	I	680	17:00	BB	5.000	µg/L
2	1,2-Dichlorobenzene IS2	S	1050	26:15	BB	5.001	µg/L
3	Bromofluorobenzene SS1	S	943	23:34	BB	5.000	µg/L
9	Trichlorofluoromethane	A	424	10:36	BB	0.045	µg/L
11	Dichloromethane	A	522	13:03	BB	0.518	µg/L
17	Chloroform	A	630	15:45	BB	8.714	µg/L
18	1,1,1-Trichloroethane	A	641	16:01	BB	0.071	µg/L
21	Benzene	A	664	16:36	BB	0.056	µg/L
22	Carbon tetrachloride	A	651	16:17	BB	0.023	µg/L
24	Trichloroethylene	A	702	17:33	BV	0.060	µg/L
26	Dichlorobromomethane	A	732	18:18	BB	7.415	µg/L
29	Toluene	A	779	19:28	BB	0.167	µg/L
32	Dibromochloromethane	A	829	20:44	BB	4.860	µg/L
34	Tetrachloroethylene	A	813	20:19	BV	0.058	µg/L
35	Chlorobenzene	A	868	21:42	BB	0.032	µg/L
37	Ethylbenzene	A	874	21:51	BV	0.037	µg/L
38	meta, para-Xylene	A	881	22:02	VB	0.054	µg/L
39	Bromoform	A	923	23:05	BB	0.635	µg/L
45	Bromobenzene	A	954	23:51	BB	0.034	µg/L
51	1,2,4-Trimethylbenzene	A	998	24:57	BB	0.084	µg/L
55	Isopropyltoluene	A	1019	25:28	BB	0.057	µg/L
59	Naphthalene	A	1182	29:33	BB	0.155	µg/L
61	Hexachlorobutadiene	A	1176	29:24	BV	0.126	µg/L

In the preparation of standard solutions for calibration with internal standards, their concentration must be kept constant in all the calibration levels. Pipetting in separately the same volumes of the internal standard followed by different volumes of a mixed standard stock solution and then making up with solvent to a fixed volume in the sample vial has proved to be a good method (Fig. 3-156). The addition of the internal standard is now used in routine analysis by program control of the liquid autosampler. Keeping the mixed standard and the internal standard storage vessels separate simplifies many manual steps.

To calculate the results of analyses using internal standards, after integration of the standard chromatograms the areas of the analysis substance relative to the area of the internal standard are plotted against the quantity weighed out for the calibration (Fig. 3-157). The parameter determined relative to the internal standard is thus independent of deviations in the injection volume and possible variations in the performance of the detector, as all these influences affect the analyte and the internal standard to the same extent. To calculate the analysis results the ratio of the peak area for the analyte to that for the internal standard is determined and the concentration calculated using the calibration function (Table 3-15).

3.3.7 The Standard Addition Procedure

Matrix effects frequently lead to varying extraction yields. The headspace and purge and trap techniques in particular are affected by this type of problem. If measures for standardising the matrix are unsuccessful, the standard addition procedure can be used, as in, for example, atomic absorption (AAS) for the same reason. Here the calibration

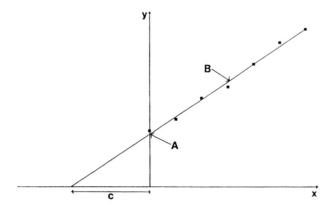

Fig. 3-158 Calibration using standard addition.
x axis: quantity of substance
y axis: signal intensity
(A) Intercept with the y axis: signal of the unaltered sample, quantity added = 0
(B) At this point the signal intensity has been doubled. If there is a direct dependence between concentration and signal, a quantity has been given at this point which corresponds to the content in the sample
(C) Content in the sample

is analogous to the external standardisation described above and involves addition of known quantities of the analyte to be determined. The calibration samples, however, are prepared with constant quantities of the analyte by addition of corresponding volumes of the standard solution. One sample is left as it is, i.e. no standard is added.

The analysis results are calculated by plotting the peak areas against the quantity added. The calibration function cuts the y axis at a height corresponding to the concentration of analyte in the sample, expressed as a peak area. The concentration in the sample can be read off by extrapolation of the calibration function to the point where it cuts the x axis (Fig. 3-158).

In spite of the good results obtained with this procedure, in practice there is one major disadvantage. A separate calibration must be carried out for every sample. On the other hand a calibration for a series of samples can be used with external or internal standards. A further criticism of the addition procedure arises from the statistical aspect. The linear

---

The Accuracy of Analytical Data

In spite of precise measuring procedures and careful evaluation of the data, the accuracy of analytical data is highly dependent on the sampling procedure, transportation to the laboratory and the sample preparation procedure (Fig. 3-159)! Even powerful instrumental analysis cannot correct errors that have already occurred.

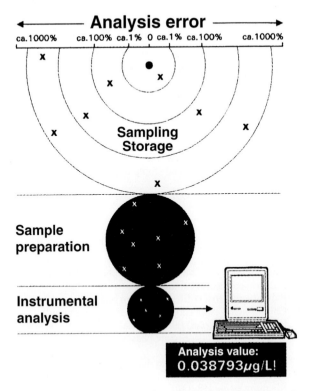

Fig. 3-159 Sources of error in analysis (after Hein/Kunze)

extrapolation of the calibration function is only carried out assuming its validity for this area also. A possible nonlinear deviation in this lower concentration range would give rise to considerable errors. To allow a better fit of the calibration function, a larger number of additions should be analysed. Nevertheless, the reservation still exists that the calibration curve must be extrapolated beyond the valid area for the values to be obtained.

3.4 Frequently Occurring Impurities

With increasing concentration during sample preparation and increasing sensitivity of GC/MS systems, the question of what are the necessary environmental conditions for trace analysis is becoming more important. Interfering signals from impurities are increasingly appearing (Table 3-16). These arise from 'cleaned' glass apparatus (e. g. rotary evaporators, pipettes), through contact with polymers e. g. cartridges from solid-phase extraction, septa of autosampler, solvent and sample bottles), from the solvents themselves (e. g. stabilisers) or even from the laboratory surroundings (e. g. dust, solvent vapours etc). There are also sources of interfering signals in the GC/MS system itself. These are often in the surroundings of the GC injector. They range from septum bleed or decomposing septa to impurities which have become deposited in the split vent and get into the measuring system on subsequent injections. The capillary column used

Table 3-16 Mass signals of frequently occurring contaminants

m/z values	Possible cause
149, 167, 279	Phthalate plasticisers, various derivatives
129, 185, 259, 329	Tri-n-butyl acetyl citrate plasticiser
99, 155, 211, 266	Tributyl phosphate plasticiser
91, 165, 198, 261, 368	Tricresyl phosphate plasticiser
108, 183, 262	Triphenylphosphine (synthesis by-product)
51, 77, 183, 201, 277	Triphenylphosphine oxide (synthesis by-product)
41, 55, 69, 83. ...	Hydrocarbon background (forepump oil, greasy fingers,
43, 57, 71, 85, ...	tap grease etc)
.. 99, 113, 127, 141, ...	
.. 285, 299, 313, 327, ...	
.. 339, 353, 367, 381, ...	
.. 407, 421, 435, 449, ...	
.. 409, 423, 437, 451, ...	
64, 96, 128, 160, 192, 224, 256	Sulfur (as S_8)
205, 220	Antioxidant 2,6-di-t-butyl-4-methylphenol (BHT, Ionol) and isomers (technical mixture)
115, 141, 168, 260, 354, 446	Poly(phenyl ether) (diffusion pump oil)
262, 296, 298	Chlorophenyl ether (impurity in diffusion pump oil)
43, 59, 73, 87, 89, 101, 103, 117, 133	Poly(ethylene glycol) (all Carbowax phases)
73, 147, 207, 221, 281, 355, 429	Silicone rubber (all silicone phases)
133, 207, 281, 355, 429	Silicone grease
233, 235	Rhenium oxide ReO_3^- (from the cathode in NCI)
217, 219	Rhenium oxide ReO_2^-
250, 252	Rhenium oxide $HReO_4^-$

3.4 Frequently Occurring Impurities

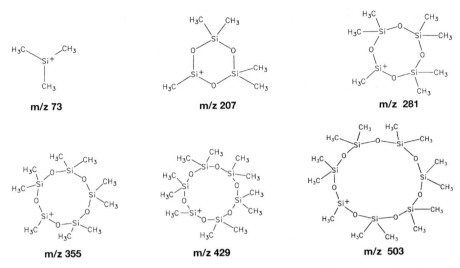

Fig. 3-160 Structures of the most important signals resulting from column bleed of silicone phases (after Spiteller)

is well known for sometimes high background noise caused by column bleed (silicone phases). As these typical mass signals constantly occur in trace analyses, the structures of the most frequently occurring ions are listed (Fig. 3-160). Sources of contamination in the mass spectrometer are also known. Among these are background signals arising from pump oil (fore pump, diffusion pump) and degassing products from sealing materials (e. g. Teflon) and ceramic parts after cleaning.

For trace analysis the carrier gas used (helium, hydrogen) should be of the highest possible purity from the beginning of installation of the instrument. Contamination from the gas supply tubes (e. g. cleaned up for ECD operations!) leads to ongoing interference and can only be removed at great expense. With central gas supply plants in particular gas purification (irreversible binding of organic contaminants to getter materials) directly at the entry to the GC/MS system is recommended for trace analysis. Leaks in the vacuum system, in the GC system and in the carrier gas supply always lead to secondary effects and should therefore be carefully eliminated. Under no circumstances should plastic tubing be used for the carrier gas supply to GC/MS systems (Table 3-17).

Table 3-17 Diffusion of oxygen through various line materials (air products)

Line material	Contamination by O_2 [ppm]
Copper	0
Stainless steel	0
Kel-F	0.6
Neoprene	6.9
Polyethylene (PE)	11
Teflon (PTFE)	13
Rubber	40

Note: measured in argon 6.0 at a line diameter of 6 mm and 1 m length with a flow rate of 5 l/h.

Some typical interfering components frequently occurring in residue analysis are listed together with their spectra (Figs. 3-161 to 3-175). They are listed in order of the mass of the base peak.

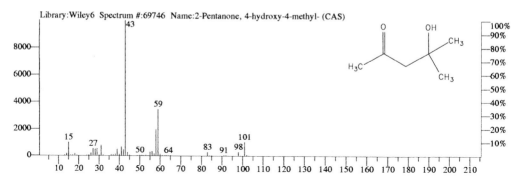

Fig. 3-161 Diacetone alcohol $C_6H_{12}O_2$, M: 116, CAS: 123-42-2
Occurrence: acetone dimer, forms from the solvent under basic conditions

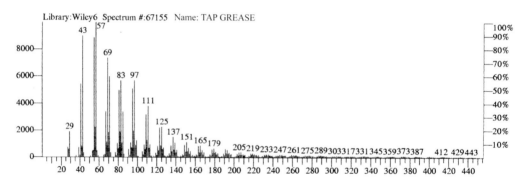

Fig. 3-162 Hydrocarbon background
Occurrence: greasy fingers as a result of maintenance work on the analyser, the ion source or after changing the column, background of forepump oil (rotatory slide valve pumps), joint grease

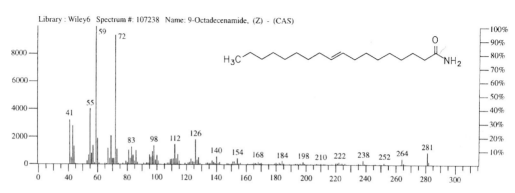

Fig. 3-163 Oleic acid amide
Occurrence: lubricant for plastic sheeting, erucamide occurs just as frequently

3.4 Frequently Occurring Impurities

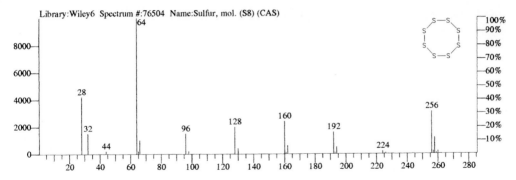

Fig. 3-164 Molecular sulfur S_8, M: 256, CAS: 10544-50-0
Occurrence: from soil samples (microbiological decomposition), impurities from rubber objects, and by-product in the synthesis of thio-compounds

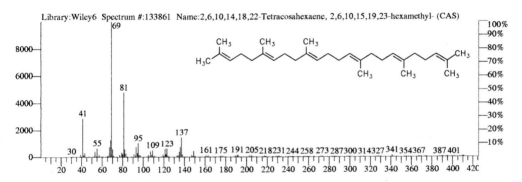

Fig. 3-165 Squalene $C_{30}H_{50}$, M: 410, CAS: 7683-64-9
Occurrence: stationary phase in gas chromatography

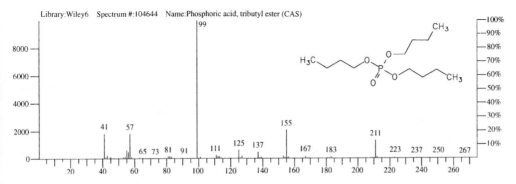

Fig. 3-166 Tributyl phosphate $C_{12}H_{27}O_4P$, M: 266, CAS: 126-73-8
Occurrence: widely used plasticiser

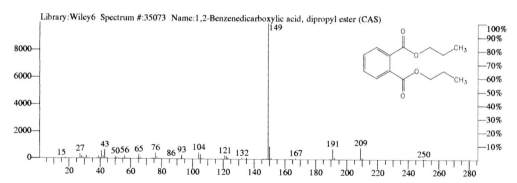

Fig. 3-167 Propyl phthalate $C_{14}H_{18}O_4$, M: 250, CAS: 131-16-8
Occurrence: plasticiser, widely used in many plastics (also SPE cartridges, bottle tops etc), typical mass fragment of all dialkyl phthalates m/z 149

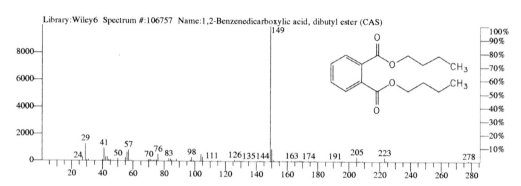

Fig. 3-168 Dibutyl phthalate $C_{16}H_{22}O_4$, M: 278, CAS: 84-74-2

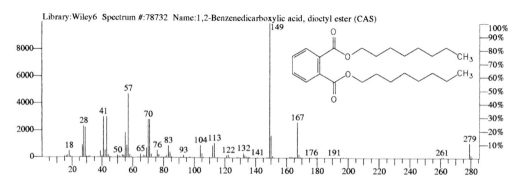

Fig. 3-169 Dioctyl phthalate $C_{24}H_{38}O_4$, M: 390, CAS: 117-84-0

3.4 Frequently Occurring Impurities 327

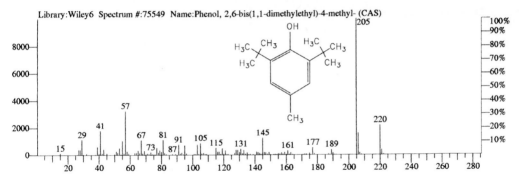

Fig. 3-170 Ionol (BHT) $C_{15}H_{24}O$, M: 220, CAS: 128-37-0
Occurrence: antioxidant in plastics, stabiliser (radical scavenger) for ethers, THF, dioxan, technical mixture of isomers

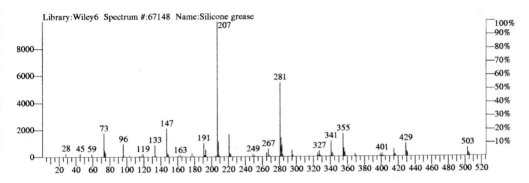

Fig. 3-171 Silicones, silicone grease
Occurrence: typical column bleed from many silicone phases, from septum caps of sample bottles, injectors etc (see also Fig. 3-160)

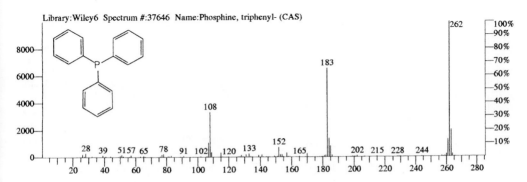

Fig. 3-172 Triphenylphosphine $C_{18}H_{15}P$, M: 262, CAS: 603-35-0
Occurrence: catalyst for syntheses

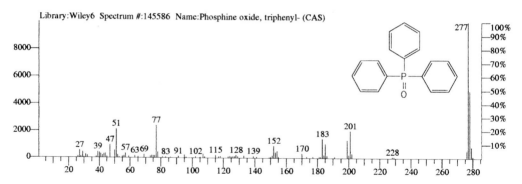

Fig. 3-173 Triphenylphosphine oxide $C_{18}H_{15}OP$, M: 278, CAS: 791-28-6
Occurrence: forms from the catalyst in syntheses (e.g. the Wittig reaction)

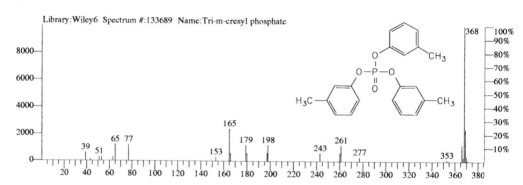

Fig. 3-174 Tri-m-cresyl phosphate $C_{21}H_{21}O_4P$, M: 368, CAS: 563-04-2
Occurrence: plasticiser (e.g. in PVC, nitrocellulose etc)

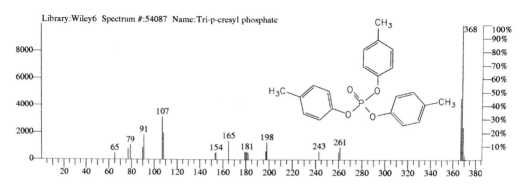

Fig. 3-175 Tri-p-cresyl phosphate $C_{21}H_{21}O_4P$, M: 368, CAS: 78-32-0
Occurrence: plasticiser (e.g. in PVC, nitrocellulose etc)

References for Chapter 3

Section 3.1.1

Bemgard, A., Colmsjö, A., Wrangskog, K., Prediction of Temperature-Programmed Retention Indexes for Polynuclear Aromatic Hydrocarbons in Gas Chromatography, Anal. Chem. 66 (1994), 4288–4294.

Ciccioli, P., Brancaleoni, E., Cecinato, A., Frattoni, M., A Method for the Selective Identification of Volatile Organic Compounds (VOC) in Air by HRGC-MS, In: 15th Int. Symp. Cap. Chromatogr., Riva del Garda, May 1993, P. Sandra (Ed.), Heidelberg: Huethig 1993, 1029–1042.

Deutsche Forschungsgemeinschaft, Gaschromatographische Retentionsindizes toxikologisch relevanter Verbindungen auf SE-30 oder OV-1, Mitteilung 1 der Kommission für Klinisch-toxikologische Analytik, Verlag Chemie: Weinheim 1982.

Hall, G. L., Whitehead, W. E., Mourer, C. R., Shibamoto, T., A New Gas Chromatographic Retention Index for Pesticides and Related Compounds, HRC & CC 9 (1986), 266–271.

Katritzky, A. L., Ignatchenko, E. S., Barcock, R. A., Lobanov, V. S., Karelson, M., Prediction of Gas Chromatographic Retention Times and Response Factors Using a General Quantitative StructureProperty Relationship Treatment, Anal. Chem. 66 (1994), 1799–1807.

Kostiainen, R., Nokelainen, S., Use of M-Series Retention Index Standards in the Identification of Trichothecenes by Electron Impact Mass Spectrometry, J. Chromatogr. 513 (1990), 31–37.

Lipinski, J., Stan, H.-J., Compilation of Retention Data for 270 Pesticides on Three Different Capillary Columns, In: 10th Int.Symp.Cap.Chromatogr., Riva del Garda, May 1989, P. Sandra (Ed.), Heidelberg: Huethig 1989, 597–611.

Manninen, A., et al., Gas Chromatographic Properties of the M-Series of Universal Retention Index Standards and their Application to Pesticide Analysis, J.Chromatogr. 394 (1987), 465–471.

Schomburg, G., Gaschromatographie, 2. Aufl., Weinheim: VCH 1987, 54–62.

Weber, E., Weber, R., Buch der Umweltanalytik, Band 4, Methodik und Applikationen in der Kapillargaschromatographie, GIT Verlag: Darmstadt 1992, 64–65.

Zenkevich, I. G., The Exhaustive Database for Gaschromatographic Retention Indizes of Low-Boiling Halogenhydrocarbons at PLOT Alumina Columns, In: 15th Int.Symp.Cap.Chromatogr., Riva del Garda, May 1993, P. Sandra (Ed.), Heidelberg: Huethig 1993, 181–186.

Section 3.2.3

Adams, R.P., Identification of Essential Oils by Ion Trap Mass Spectrometry, Academic Press: San Diego 1989.

Budzikiewicz, H., Massenspektrometrie, 4. Auflage, Wiley-VCH: Weinheim 1998.

Cairns, T., Siegmund, E., Jacobsen, R., Mass Spectral Data Compilation of Pesticides and Industrial Chemicals, Dpt. Health and Human Services, Food and Drug Administration, Office of Regulatory Affairs, Los Angeles CA. 1985.

Heller, C., Mass Spectral and GC Data of Drugs, Poisons, Pesticides, Pollutants and Their Metabolites, Toxichem 60, 3 (1993).

McLafferty, F. W., Stauffer, D. B., The Wiley/NBS Registry of Mass Spectral Data, Wiley & Sons: New York 1989.

Pfleger, K., Maurer, H. H., Weber, A., Mass Spectra and GC Data of Drugs, Poisons, Pesticides, Pollutants and Their Metabolites, Parts 1,2,3, 2. erw. Aufl., VCH: Weinheim 1992.

Section 3.2.4

Atwater, B. L., Stauffer, D. B., McLafferty, F. W., Peterson, D. W., Reliability Ranking and Scaling Improvements to the Probability Based Matching System for Unknown Mass Spectra, Anal. Chem. 57 (1985), 899–903.
Beynon, J. H., Brenton. A. G., Introduction to Mass Spectrometry. University Wales Press: Swansea 1982.
CLB-Redaktion, Die zwölfmillionste chemische Verbindung beschrieben, CLB 45 (1994), 215.
Davies, A. N., Mass Spectrometric Data Systems, Spectroscopy Europe 5 (1993), 34–38.
Fachinformationszentrum Chemie GmbH. MS Online Programm Mass-Lib-Retrieval Programm for Use with Mass Spectra Libraries, User Manual V 3.0, Fachinformationszentrum Chemie: Berlin 1988.
Henneberg, D., Weimann, B, Search for Identical and Similar Compounds in Mass Spectral Data Bases, Spectra 1984, 11–14.
Hübschmann, H. J., MS-Bibliothekssuche mittels INCOS und PBM, LABO Analytica 4 (1992), 102–118.
McLafferty, F. W., Turecek F., Interpretation of Mass Spectra, 4th. Ed., University Science Books: Mill Valley CA 1993.
Neudert, B., Bremser, W., Wagner, H., Multidimensional Computer Evaluation, Org. Mass Spectrom. 22 (1987) 321–329.
Palisade Corporation, *Benchtop/PBM Users Guide*, 1994.
Sokolow, S., Karnovsky, J., Gustafson, P., The Finnigan Library Search Program, Finnigan MAT Applikation No. 2, 1978.
Stauffer, D. B., McLafferty, F. W., Ellis, R. D., Peterson, D. W., Adding Forward Searching Capabilities to a Reverse Search Algorithm for Unknown Mass Spectra, Anal. Chem. 57 (1985), 771–773.
Stauffer, D. B., McLafferty, F. W., Ellis, R. D., Peterson, D. W., Probability-Based-Matching Algorithm with Forward Searching Capabilities for Matching Unknown Mass Spectra of Mixtures, Anal. Chem. 57 (1985), 1056–1060.
Stein, S. E., Estimating Probabilities of Correct Identification From Results of Mass Spectral Library Search, J. Am. Soc. Mass Spectrom. 5 (1994) 316–323.
Stein, S. E., Scott, D. R., Optimization and Testing of Mass Spectral Library Search Algorithms for Compound Identification, J. Am. Soc. Mass Spectrom. 5 (1994), 859–866.
Warr, W. A., Computer-Assisted Structure Elucidation-Part 1: Library Search and Spectral Data Collections, Anal. Chem. 65 (1993) 1045A–1050A.
Warr, W. A., Computer-Assisted Structure Elucidation – Part 2: Indirect Database Approaches and Established Systems, Anal. Chem. 65 (1993) 1087A–1095A.
Yang, L., Martin, M., Glazner, M., et al., Comparison of Three Benchtop Mass Spectrometers: Initial Development Phase of a Volatile Organic Analyzer for Space Station Freedom, presented at the ASMS, Tucson 1990.

Section 3.2.5

Budzikiewicz, H., Massenspektrometrie, 4. Auflage,Wiley-VCH: Weinheim 1998.
McLafferty, F. W., Turecek, F., Interpretation von Massenspektren. Heidelberg: Springer 1995.
Ockels, W., Hübschmann, H. J., Interpretation von Massenspektren I/II, Manuskript zum Fortbildungskursus der Axel Semrau GmbH, Sprockhövel 1994.

Section 3.2.6

Benz, W., Henneberg, D., Massenspektrometrie organischer Verbindungen, Akad. Verlagsgesellschaft: Frankfurt 1969.
Budzikiewicz, H., Massenspektrometrie, 4. Auflage, Wiley-VCH: Weinheim 1998.

Färber, H., Schöler, F., Gaschromatographische Bestimmung von Harnstoffherbiziden in Wasser nach Methylierung mit Trimethylaniliniumhydroxid oder Trimethylsulfoniumhydroxid, Vom Wasser 77 (1991,), 249–262.
Howe, I., et al., Mass Spectrometry, McGraw-Hill: New York 1981.
McLafferty, F. W., Mass Spectrometric Analysis – Molecular Rearrangements, Anal. Chem. 31 (1959), 82–87.
McLafferty, F. W., Stauffer, D. B., The Wiley/NBS Registry of Mass Spectral Data, 5th Ed., Wiley-Interscience: New York.
Ockels, W., Pestizid-Bibliothek, Axel Semrau GmbH: Sprockhövel 1993.
Ockels, W., Hübschmann, H. J., Interpretation von Massenspektren, Fortbildungskurs, Axel Semrau GmbH: Sprockhövel 1993.
Ockels, W., Hübschmann, H. J., Interpretation von Massenspektren II – Substanzidentifizierung in der Rückstandsanalytik, Fortbildungskurs, Axel Semrau GmbH: Sprockhövel 1994.
Pretsch, E., et al., Tabellen zur Strukturaufklärung organischer Verbindungen mit spektrometrischen Methoden, 3. Auflage, 1. korr. Nachdruck, Springer: Berlin 1990.
Schröder, E., Massenspektrometrie, Springer: Berlin 1991.
Seibl, J., Massenspektroskopie, Akad. Verlagsgesellschaft: Frankfurt 1970.
Stan, H.-J., Abraham, B., Jung, J., Kellert, M., Steinland, K., Nachweis von Organophosphorinsecticiden durch Gas-Chromatographie-Massenspektroskopie, Fres. Z. Anal. Chem. 287 (1977), 271–285.
Stan, H.-J., Kellner, G., Analysis of Organophosphoric Pesticide Residues in Food by GC/MS Using Positive and Negative Chemical Ionisation, In: Baltes, W., Czodik-Eysenberg, P. B., Pfannhauser, W. (Eds.), Recent Development in Food Analysis, Proceedings of the First European Conference on Food Chemistry (EURO FOOD CHEM I), Vienna 17–20 February 1981, Verlag Chemie: Weinheim 1981, 183–189.

Section 3.3.6

Chang, C., Parallel Mass Spectrometry for High Performance GC and LC Detection, Int. Laboratory 5 (1985) 58–68.
Doerffel, K., Statistik in der analytischen Chemie, VCH: Weinheim 1984.
Ebel, S., Dorner, W., Jahrbuch Chemielabor 1987, VCH: Weinheim 1987.
Ebel, S., Kamm, U., Fres. Z. Anal.Chem. 316 (1983) 382–385.
Funk, W., Dammann, V., Donnevert, G., Qualitätssicherung in der Analytischen Chemie, VCH: Weinheim 1992.
Guichon, G., Guillemin, C. L., Quantitative Gas Chromatography, Elsevier: Amsterdam, Oxford, New York, Tokyo 1988.
Hein, H., Kunze, W., Umweltanalytik mit Spektrometrie und Chromatographie, VCH: Weinheim 1994.
Meyer, V. R., Richtigkeit bei der Peakflächenbestimmung, GIT Fachz. Lab. 1 (1994) 4–5.
Miller, J. N., The Method of Standard Additions, Spectroscopy Europe 1992, 4/6, 26–27.
Montag, A., Beitrag zur Ermittlung der Nachweis- und Bestimmungsgrenze analytischer Meßverfahren, Fres. Z.Anal.Chem. 1982 (312), 96–100.
Nachweis-, Erfassungs- und Bestimmungsgrenze, DIN 32 645, Beuth-Verlag, Berlin.
Naes, T., Isakson, T., The Importance of Outlier Detection in Spectroscopy, Spectroscopy Europe 1992, 4/4, 32–33.
Vogelsang, J., Höchstmengenüberschreitung und Streuung der Analysenwerte bei der Rückstandsanalyse von Pestiziden in Trinkwasser, Lebensmittelchem. 1990 (44), 7.

Section 3.4

Spiteller, M., Spiteller, G., Massenspektrensammlung von Lösemitteln, Verunreinigungen, Säulenbelegmaterialien und einfachen aliphatischen Verbindungen, Wien: Springer 1973.

4 Applications

The applications given here have been chosen in order to describe typical areas of use of GC/MS, such as air, water, soil, foodstuffs, the environment, waste materials, drugs and pharmaceutical products. Special emphasis has been placed on current and reproducible examples which give an idea of what is going on in routine laboratories. The selection cannot be totally representative of the use of modern GC/MS, but shows the main areas into which the methodology has spread and will continue to do so. In addition, in special areas of application, such as the analysis of gases or aromas, isotope-specific measuring procedures and other powerful examples of the use of GC/MS are described.

Most of the applications described are compiled from the references cited and documented with various print-outs and lists. The analysis conditions are described in full to allow adaptation of the methods. If any of the methods have been published, the sources are given for each section. References to other directly related literature are also given.

4.1 Air Analysis According to EPA Method TO-14

Key words: air, SUMMA canister, thermodesorption, volatile halogenated hydrocarbons, cryofocusing, water removal, thick film column

The EPA (US Environmental Protection Agency) describes a process for sampling and analysis of volatile organic compounds (VOCs) in the atmosphere. It is based on the collection of air samples in passivated stainless steel canisters (SUMMA canisters). The organic components are then separated by GC and determined using conventional GC detectors or by mass spectrometry (Fig. 4-1). The use of mass spectrometers allows the direct positive detection of individual components.

For an SIM analysis the mass spectrometer is programmed in such a way that a certain number of compounds in a defined retention time range are detected. These SIM ranges are switched periodically so that a list of target compounds can be worked through. In the full scan mode the mass spectrometer works as a universal detector during data acquisition.

A cryoconcentrator with a 3-way valve system is used for concentration. The sample can be let in by two routes without having to alter the screw joints on the tubing. Usually the inlet is via a mass flow regulator to a cryofocusing unit. The direct measurement of the sample volume provides very precise data. A Nafion drier is used to dry the air.

334 4 Applications

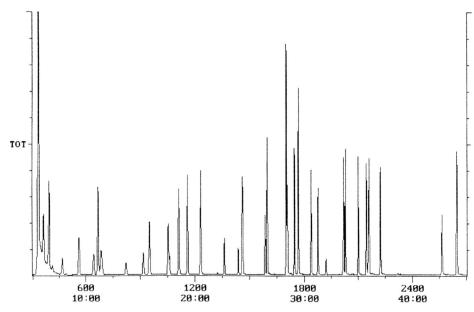

Fig. 4-1 A typical chromatogram for a 1000 ml sample of the 10 ppbv standard

---

SUMMA Canisters

The analysis of volatile organic compounds is frequently carried out by adsorption on to suitable materials. For this Tenax is mainly used (see Section 2.1.4). The limits of this adsorption method lie in the adsorption efficiency, which is dependent upon the compound, the breakthrough of the sample at higher air concentrations, the impossibility of multiple measurements on a sample and the possible formation of artefacts. Stainless steel canisters, whose inner surfaces have been passivated by the SUMMA process, do not exhibit these limitations. This passivation process involves polishing the inner surface and then applying a Cr/Ni oxide layer. Containers treated in this way have been used successfully for the collection and storage of air samples. Purification and handling of these canisters and the sampling apparatus must be carried out carefully, however, because of possible contamination problems.

Another means of sample injection is the loop injection. The sample is drawn through a 5 ml sample loop directly into the cryoconcentrator. This method is suitable for highly concentrated samples, as only a small quantity of sample is required. In this case the drier is avoided.

The Nafion drier is a system for removal of water from the air sample, which uses a semipermeable membrane. Nonpolar compounds pass through the membrane while polar ones, such as water, are held by the membrane and diffuse outwards. The outer side of the membrane is dried by a clean air stream and the water thus separated is removed from the system. The Nafion drier is recommended by the TO-14 method to prevent blockage of the cryofocusing unit by the formation of ice crystals.

In this application the GC is coupled to the mass spectrometer by an open split interface. A restrictor limits the carrier gas flow. Open coupling was chosen because the sensitivity of the ion trap GC/MS makes the concentration of large quantities of air superfluous. Open coupling dilutes the moisture, which may be contained in the sample, to an acceptable level so that cryofocusing can be used without additional drying.

Analysis Conditions

Open split interface:	SGE type GMCC/90 (SGE), mounted in the transfer line to the MS, restrictor 0.05 mm, adjusted to 2.5% transmission
GC/MS system:	Finnigan MAGNUM Mode: EI, 35–300 u Scan rate: 1 s/scan
GC separation:	J&W DB-5, 60 m × 0.25 mm × 1 µm. The thick film columns DB-1 or DB-5 guarantee chromatographic separation even at start temperatures just above room temperature if cryofocusing is used. Helium 1 ml/min.
Program:	Start: 35 °C, 6 min Ramp: 8 °C/min Final temp.: 200 °C
Concentrator:	Grasby Nutech model 3550A cryoconcentrator with 354A cryofocusing unit, Nafion drier
Autosampler:	Nutech 3600 16-position sampler
Sample injection:	The sample is drawn out of the SUMMA canister through the Nafion drier and reaches the cryoconcentrator cooled to –160 °C. It is then heated rapidly to transfer the sample to the cryofocusing unit of the GC (Fig. 4-2). At –190 °C the sample is focused in a fused silica column of 0.53 mm ID and then heated to 150 °C for injection.
Calibration:	A 6-point calibration based on 1 litre samples from a SUMMA canister was carried out. The concentration range for the calibration was between 0.1 ppbv and 20 ppbv (Fig. 4-3).

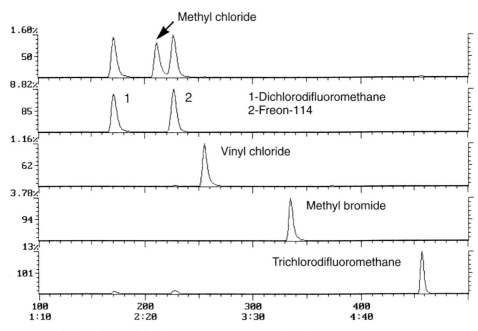

Fig. 4-2 Elution of gases following injection using cryofocusing

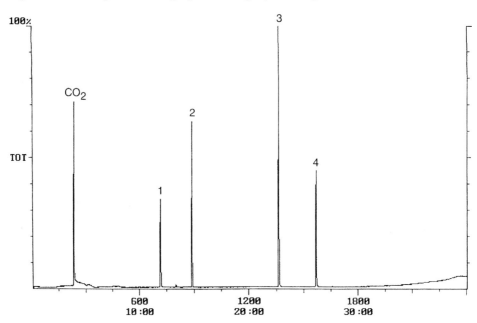

Fig. 4-3 Continuous flushing of the gas lines ensures contamination-free analyses
Internal standards used here:
(1) Bromochloromethane
(2) 1,4-Difluorobenzene
(3) Chlorobenzene- d_5
(4) Bromofluorobenzene

Limit of Detection

For the EPA method TO-14 (Figs. 4-4 and 4-5) a limit of detection of 0.1 ppbv is required. This requirement is achieved even at the high split rate of the interface.

Results

With this method data acquisition in the full scan mode allows mass spectra to be run with subsequent identification through library searching even at the required limit of detection of 0.1 ppbv. Figures 4-6 and 4-7 show examples of comparability of the spectra and the identification of dichlorobenzene at 20.0 ppbv and 0.1 ppbv. With the procedure described both the compounds required according to TO-14 and other unexpected components in the critical concentration range can be identified.

Compound	LOD	Compound	LOD
Dichlorodifluoromethane	0.01 ppbv	trans-1,3-Dichloropropene	0.01 ppbv
Methyl chloride	0.01 ppbv	Toluene	0.01 ppbv
Freon-114	0.02 ppbv	1,1,2-Trichloroethane	0.01 ppbv
Vinyl chloride	0.01 ppbv	1,2-Dibromoethane	0.01 ppbv
Methyl bromide	0.01 ppbv	Tetrachloroethylene	0.01 ppbv
Ethyl chloride	0.08 ppbv	Chlorobenzene	0.01 ppbv
Trichlorfluoromethane	0.01 ppbv	Ethylbenzene	0.01 ppbv
1,1-Dichloroethylene	0.02 ppbv	m/p-Xylenes	0.01 ppbv
Dichloromethane	0.01 ppbv	Styrene	0.02 ppbv
3-Chloropropane	0.02 ppbv	o-Xylene	0.01 ppbv
Freon-113	0.01 ppbv	1,1,2,2-Tetrachloroethane	0.09 ppbv
1,1-Dichloroethane	0.01 ppbv	4-Ethyltoluene	0.02 ppbv
cis-1,2-Dichloroethylene	0.01 ppbv	1,3,5-Trimethylbenzene	0.02 ppbv
Chloroform	0.01 ppbv	1,2,4-Trimethylbenzene	0.01 ppbv
1,1,1-Trichlorethane	0.01 ppbv	1,3-Dichlorbenzene	0.01 ppbv
1,2-Dichloroethane	0.01 ppbv	Benzyl chloride	0.08 ppbv
Benzene	0.01 ppbv	1,4-Dichlorobenzene	0.01 ppbv
Carbon tetrachloride	0.01 ppbv	1,2-Dichlorobenzene	0.01 ppbv
1,2-Dichloropropane	0.02 ppbv	1,2,4-Trichlorobenzene	0.01 ppbv
Trichloroethylene	0.02 ppbv	Hexachlorobutadiene	0.02 ppbv
cis-1,3-Dichloropropene	0.01 ppbv		

Fig. 4-4 Limits of detection for compounds used in the EPA method TO-14

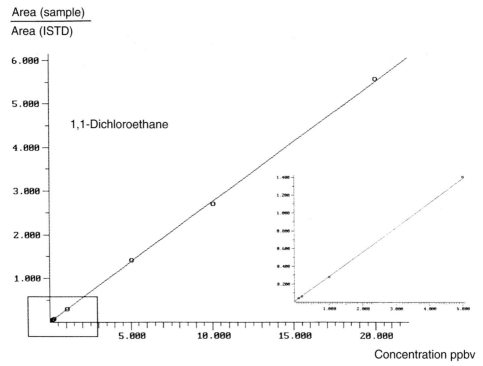

Fig. 4-5 Six point calibration from 0.1 to 20 ppbv. The lower region is shown magnified. ppbv = parts per billion in volume.

4.1 Air Analysis According to EPA Method TO-14 339

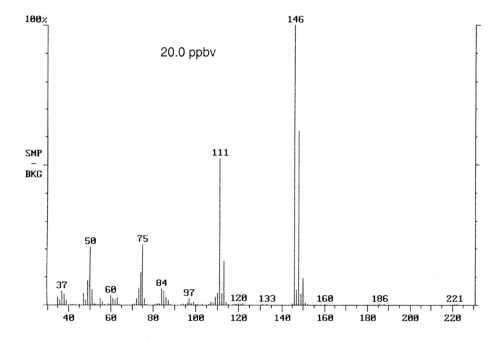

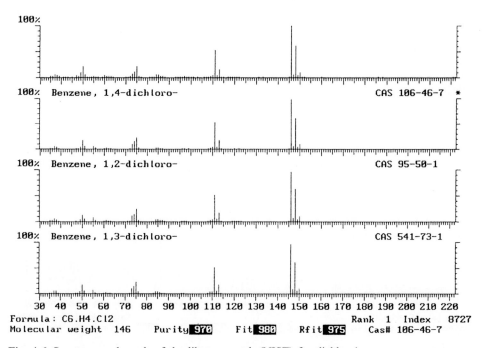

Fig. 4-6 Spectrum and result of the library search (NIST) for dichlorobenzene at a concentration of 20 ppbv.

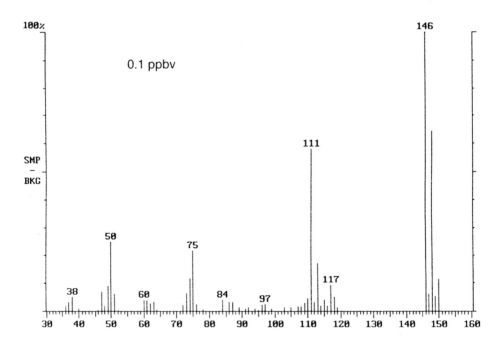
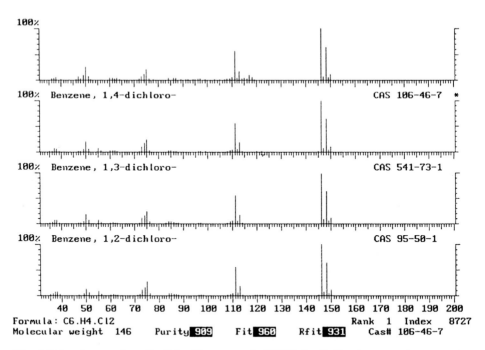

Fig. 4-7 Spectrum and result of the library search (NIST) for dichlorobenzene at a concentration of 0.1 ppbv.

4.2 BTEX using Headspace GC/MS

Key words: headspace, pressure balanced injection, BTEX, reproducibility, sensitivity, automatic evaluation

The determination of BTEX components using the static headspace technique, for example for soil and water samples, is now a standard procedure designed for a high sample throughput. The reproducibility and stability of the procedure are of particular interest with regard to GC/MS coupling. The data given here demonstrate the capabilities of the system. Computer-based automatic evaluation is necessary for carrying out long series of analyses. By using GC/MS coupling the selectivity of the mass spectrometer can help to identify the analytes on the basis of the mass spectra in the evaluation of the analyses.

Analysis Conditions

Static headspace: Perkin Elmer HS40 headspace sampler, 2 ml sample volume (blank value-free water fortified with 1 µg/l BTEX), incubation at 290 °C, 45 min (without shaker), sampling interval 0.05 min (sample injection)

GC/MS system: Finnigan MAGNUM
Injector: PTV/2
250 °C isotherm
Split 10 ml/min
Column: SGE HT8, 25 m × 0.25 mm × 0.25 µm
Program: Start: 40 °C for 2 min
Ramp: 10 °C/min
Final temp.: 200 °C

Mass spectrometer: Scan mode: EI, 33–240 u
Scan rate: 0.75 s/scan

Results

Figure 4-8 shows the total ion chromatogram of the mass traces for benzene m/z 78, toluene m/z 91 and the isomers of xylene and ethylbenzene m/z 91, 106. A peak caused by CS_2 can be recognised from the mass trace m/z 78. Presumably CS_2 gets into the analysis because this is a product emitted from the butyl rubber septums of the headspace bottles. The isomers meta- and para-xylene elute at the same time from the column chosen. This concentration of 2 µg/l with an excellent signal to noise ratio of ca. 900:1 (Fig. 4-9) can be detected, from which a limit of detection of less than 0.01 µg/l can be expected.

A calibration file is used for detection and evaluation of the peaks. Each substance entry (Fig. 4-10) contains mass spectra, retention times and other parameters for identification of the BTEX components (see Section 3.3). The automatic evaluation compares the mass spectra at a given retention time in a search window and only carries out the peak integration after positive identification. The peak areas determined are entered

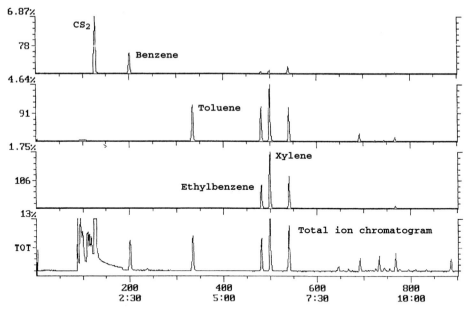

Fig. 4-8 Typical chromatogram from the headspace GC/MS analysis of BTEX components.
Below: total ion current chromatogram (33–240 u)
Above: mass chromatogram for benzene (m/z 78), toluene (m/z 91) and xylene with ethylbenzene (m/z 91, 106)

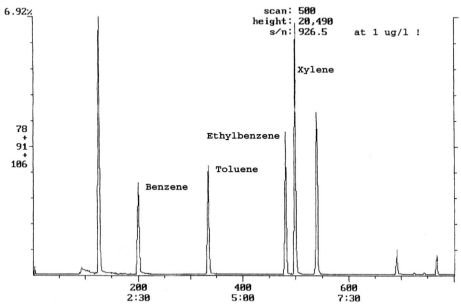

Fig. 4-9 Mass chromatogram for selective masses with the signal/noise ratio for the m,p-xylene peak

4.2 BTEX using Headspace GC/MS

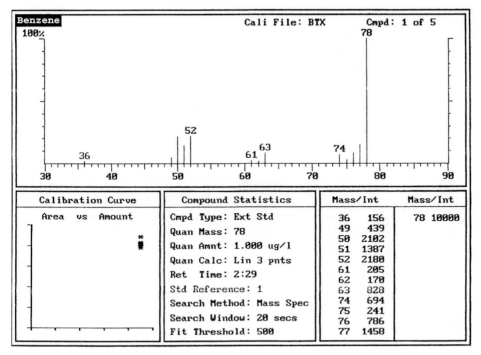

Fig. 4-10 Entry in the calibration file: benzene

Table 4-1 Limits of quantitation and average recoveries for the halogenated hydrocarbons and BTEX aromatics investigated

Compound	Limit of quantitation [µg/l]	Recovery [%]
Trifluoromethane	0.01	97.5
1,1-Dichloroethylene	0.01	98.2
Dichloromethane	0.01	101.2
1,2-Dichloroethane	0.01	94.6
1,1-Dichloroethane	0.01	93.8
Chloroform	0.01	102.7
1,1,1-Trichloroethane	0.01	97.6
Carbon tetrachloride	0.01	94.5
Trichloroethane	0.01	101.6
Bromodichloromethane	0.01	95.7
1,2-Dichloropropane	0.01	95.3
1,1,2-Trichloroethylene	0.01	94.8
Dibromochloroethane	0.01	91.8
Tetrachloroethylene	0.01	98.7
Chlorobenzene	0.05	87.4
Bromoform	0.01	89.9
1,1,2,2-Tetrachloroethane	0.02	90.8
Benzene	0.01	99.4
Toluene	0.01	96.7
Ethylbenzene	0.01	101.7
m/p-Xylene	0.01	95.3
o-Xylene	0.01	96.3

```
Quantitation Report     Quanfile: BTX-12              Quan Entries: 5
Comment: BTX 1ug/l      HS40: 90 grdC, 45', 0.05' sampl.
Sorted via: Entry Number ↑                            (S) = Standard
```

Cal	Name of Compound	S	Scan#	R Time	Me	Calc Amt(A)	Units
1	Benzene	E	201	2:31	BB	1.001	ug/l
2	Toluene	E	334	4:10	BB	1.001	ug/l
3	Xylene-1	E	481	6:01	BV	1.000	ug/l
4	Xylene-2	E	499	6:14	BB	1.000	ug/l
5	Ethylbenzene	E	540	6:45	BB	1.001	ug/l

```
External Standards Report    Filename: BTX           Quan Mass: 78
Benzene                      Compound: 1 of 5        Cali Pnts: 8
Sorted via: Entry Order ↑    Units: ug/l             (S) = Standard
```

No	Ent Date	Time	Pk Area	Ret T.	Inj Amount	Ar/Amount
1	Aug-24-92	11:09	27,700	2:29	1.000	27700.000
2	Aug-24-92	11:10	27,771	2:29	1.000	27771.000
3	Aug-24-92	11:10	28,425	2:29	1.000	28425.000
4	Aug-24-92	11:11	28,630	2:30	1.000	28630.000
5	Aug-24-92	11:11	29,281	2:30	1.000	29281.000
6	Aug-24-92	11:11	31,354	2:30	1.000	31354.000
7	Aug-24-92	11:16	29,455	2:30	1.000	29455.000
8	Aug-24-92	11:20	28,425	2:29	1.000	28425.000

Fig. 4-11 Entry giving automatically determined peak areas in the quantitation list from eight analyses

into the file and are converted into concentration values if a calibration is present. In the present case the absolute peak areas from eight consecutive samples were determined and evaluated (Fig. 4-11). The values show very good agreement. The average deviation from the mean for all components is ca. 3% (e.g. benzene). Here the variances of the whole procedure from sample preparation to mass spectrometric detection are included because internal standardisation has not been carried out.

Using the same parameters the analysis of volatile halogenated hydrocarbons and BTEX can be carried out together.

Reference: A. Mülle, AS Applikationslabor, Sprockhövel.

4.3 Simultaneous Determination of Volatile Halogenated Hydrocarbons and BTEX

Key words: volatile halogenated hydrocarbons, BTEX, gas from landfill sites, seepage water, purge & trap, Tenax, internal standard

The analysis of seepage water (Fig. 4-12) and the condensate residue from the incineration of gas from landfill sites (Fig. 4-13) are two examples of the effective simultaneous control of the volatile halogenated hydrocarbon and BTEX concentrations. The purge and trap technique was chosen here so that as wide a spectrum of unknown components as possible could be determined together. The complex elution sequence can be worked out easily using GC/MS by means of selective mass signals. Toluene-d_8 is used as the internal standard for quantitation.

Analysis Conditions

Concentrator:	Tekmar purge & trap system LSC 2000	
	Trap:	Tenax
	Sample volume:	5 ml
	Sample temp.:	40 °C
	Standby:	30 °C
	Sample preheat:	2.50 min
	Purge duration:	10 min
	Purge flow:	40 ml/min
	Desorb preheat:	175 °C
	Desorb temp.:	180 °C
	Desorb time:	4 min
	MCM Desorb:	5 °C
	Bake:	10 min at 200 °C
	MCM bake:	90 °C
	BGB:	OFF
	Mount:	60 °C
	Valve:	220 °C
	Transfer line:	200 °C
	Connection to GC:	LSC 2000, inserted into the carrier gas supply of the injector
GC/MS system:	Finnigan	
Injector:	PTV cold injection system	
Injector program:	isotherm, 200 °C	
	split OPEN, ca. 20 ml/min	
Column:	Restek Rtx-624, 60 m × 0.32 mm × 1.8 µm	
Temp. program:	Start:	40 °C, 5 min
	Program:	15 °C/min
	Final temp.:	200 °C, 9.5 min
Mass spectrometer:	Scan mode:	EI, 45–220 u
	Scan rate:	1 s/scan

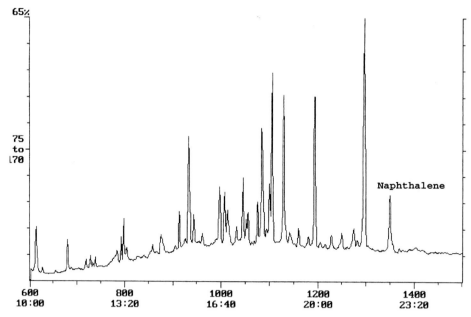

Fig. 4-12 Purge & trap GC/MS chromatogram for the analysis of a condensate from a block heating power station run on gas from landfill sites

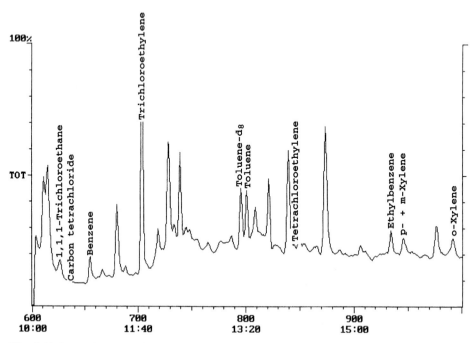

Fig. 4-13 Purge & trap GC/MS chromatogram for the analysis of a seepage water sample from a landfill site (section)

4.3 Simultaneous Determination of Volatile Halogenated Hydrocarbons and BTEX

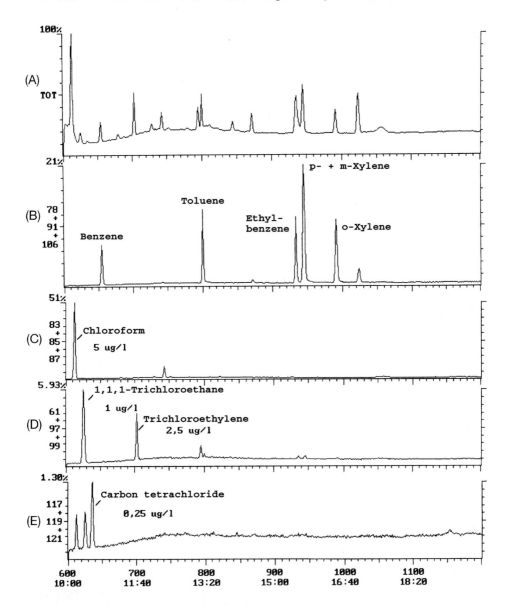

Fig. 4-14 Analysis of a standard volatile halogenated hydrocarbon/BTEX mixture
(A) Total ion current (for conditions see text)
(B) BTEX components (m/z 78, 91, 106)
(C) Chloroform (m/z 83, 85, 87)
(D) 1,1,1-Trichloroethane (m/z 61, 97, 99)
(E) Carbon tetrachloride (m/z 117, 119, 121)
(F) Trichloroethylene (m/z 95, 130, 132)
(G) Bromodichloromethane (m/z 83, 85, 129)
(H) Tetrachloroethylene (m/z 166)
(I) Dibromochloromethane (m/z 79, 127, 129)
(J) Bromoform (m/z 173)

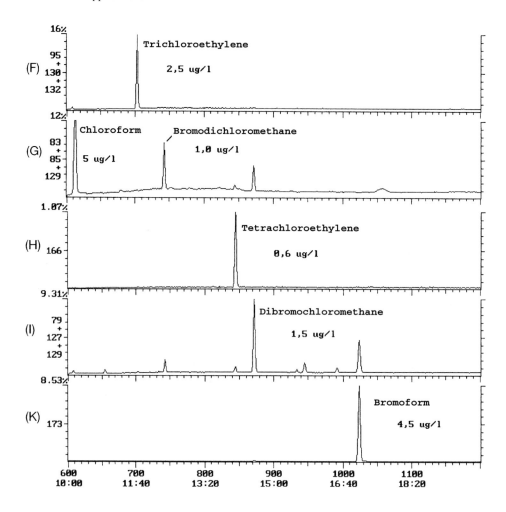

Fig. 4-14 (continued)

Results

The elution sequence of the volatile halogenated hydrocarbons compared to the BTEX aromatics is shown in Fig. 4-14 in the form of mass chromatograms using TIC in a standard analysis. The internal standard, toluene-d_8 (1 µg/l) elutes immediately before the toluene peak (see TIC). The column used has very high thermal stability until the end of the temperature program. This is shown by a completely flat base line. Benzene elutes here between tetrachloroethane and trichloroethylene, ortho-xylene, as the last BTEX component, between dibromochloromethane and bromoform. In the concentration range of around 1 µg/l volatile halogenated hydrocarbons and BTEX are determined together in an analysis with a good signal/noise ratio.

Reference: H. G. Ostrop, AGR mbH, Chemische Untersuchung, Gelsenkirchen.

4.4 Static Headspace Analysis of Volatile Priority Pollutants

Key words: headspace, VOCs, water, soil, quantitation, internal standard

Many analytical techniques have been used for the quantitation of volatile organic compounds (VOC) in water and soil, including liquid-liquid microextraction (LLME), solid phase microextraction (SPME) and purge & trap (P/T). Automated static headspace analysis offers the advantages of simplicity and robustness especially when large sample throughput is required. A typical application of the static headspace method described here is the analysis of surface and waste water, see Fig. 4-15, other techniques such as purge & trap GC/MS are better designed for ultra-trace analysis of drinking water. The samples were transferred to 20 ml headspace vials, together with an internal standard (e.g. toluene-d_8). "Salting-out" was achieved by saturating the sample with sodium sulfate. Using the sample agitation feature, headspace equilibrium is reached very quickly, allowing all sampling operations to take place during the GC run time.

The GC was operated in accordance with the "whole column trapping technique" using a standard medium bore capillary column with a 1 μm stationary phase and liquid CO_2 as coolant.

The analytes were trapped in the column inlet at subambient temperature, without the need for a dedicated cold trapping device. Trapping of the VOCs helped to maintain optimum chromatographic efficiency by focusing the analytes at the column inlet. High sensitivity was achieved by injecting a relatively high volume of headspace using a low split flow. The injecting speed was controlled in order to prevent the injector from overflowing. The MS was operated in scanning electron impact (EI) mode, allowing acquisition of full mass spectra and thereby enabling both targeting analysis and identification of unknowns during a single analysis (Fig. 4-16). At 0.5 ug/l, all compounds were identified by automated library searching, using the NIST 98 library (Fig. 4-17).

The quantitation was based on an internal standard method, using an isotopically labelled analogue of toluene. A surrogate standard was also added to the samples in order to control the long-term spectral balance stability and to check the MS tuning criteria.

Analysis Conditions

GC/MS system:	Finnigan Trace MS system
Injector:	split/splitless injector
	Injector temperature: 200 °C
	split mode
	constant flow at 3.0 ml/min
Autosampler:	CE Instruments HS2000
	Headspace sample volume: 1 ml
	Incubation: 40 °C, 10 min
GC column:	J&W Scientific DB-1, 30 m × 0,32 mm × 1,0 μm
Temperature program:	−40 °C at start
	10° C/min to 260 °C
Transfer line:	direct coupling to MS ion source
	260° C constant

Mass spectrometer: EI mode
                     ion source at 200 °C
MS conditions: Scan cycle time: 0.4 s
                     Scan range: 45–270 u

Results

The technique of static headspace GC/MS offers significant benefits for laboratories tasked with running VOC analyses. The virtual elimination of sample carry-over, thanks to the programmable temperature cleaning cycle of the syringe and needle heaters in the headspace autosampler, obviates the need for running blank samples between specimen samples.

The wide linear dynamic range of the mass spectrometer, even running in full scan mode, permits target compounds to be accurately quantified over a concentration range of at least 3 decades. Full scan operation enables unknown peaks to be automatically detected and identified. Even after a period of several weeks of unattended operation, highly sensitive and reproducible results can be demonstrated.

The precision study, based on 72 replicate injections of the low standard (0.5 µg/l), represents analyses acquired over 3 weeks of continuous operation (see Table 4-2). Such high precision is normally associated with quadrupole mass spectrometers running in the selected ion monitoring (SIM) mode; yet these results were obtained in full scan mode. The R-square results show how linear the system is; the figures were derived from multiple calibration curves spanning 3 orders of magnitude (0.1 to 100 µg/l) and accumulated over 3 days (Fig. 4-18). The calculated limits of detection (LOD) for each target compound lie in the low ppb range, making this technique suitable for analyses of VOCs in water or other materials like soil or sediments (Fig. 4-19).

The system yields very high stability and sensitivity. The % RSDs and limits of detection were based on analyses performed at 0.5 µg/l (n = 72, over 3 weeks of continuous operation). The limits of detection computed according to LOD = 3SD. R-square values were obtained as described in the text.

Reference: Marc Termonia, DCMS Laboratories, Brussels

4.4 Static Headspace Analysis of Volatile Priority Pollutants

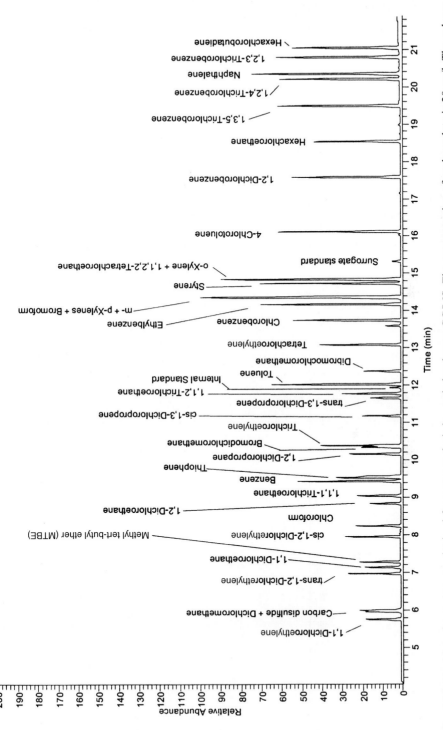

Fig. 4-15 Total ion chromatogram obtained for water analysis by static headspace GC/MS. The concentration of each analyte is 20 µg/l. The column (DB1, 30 m × 0.32 mm × 1 µm) is programmed from −40 °C to 260 °C at 10 °C/min. 1 ml of headspace is injected in the split mode.

352   4 Applications

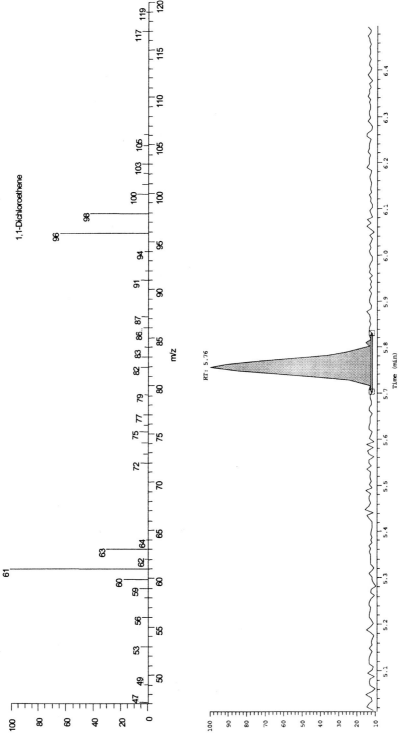

Fig. 4-16 Mass spectrum and chromatogram obtained for 1,1-dichloroethylene at 0.5 µg/l

4.4 Static Headspace Analysis of Volatile Priority Pollutants

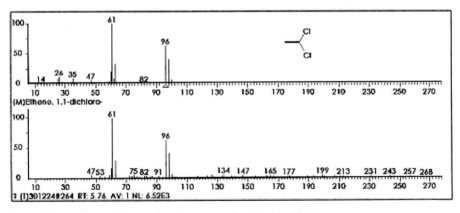

Fig. 4-17 Automated library searching using the NIST 98 library

Table 4-2 Substance list, LODs are given for static headspace method in water

Substance name	% RSD	R-square	LOD, µg/l
1,1-Dichloromethane	0.552	0.9998	0.010
Carbon disulfide	0.945	0.9990	0.013
Dichloromethane	1.150	0.9992	0.018
trans-1,2-Dichloroethane	0.698	0.9997	0.012
MTBE	0.697	0.9993	0.013
cis-1,2-Dichloroethane	1.318	0.9963	0.022
Chloroform	0.578	0.9996	0.013
1,2-Dichloroethane	0.362	0.9989	0.011
1,1,1-Trichloroethane	0.747	0.9998	0.014
Benzene	0.568	0.9994	0.016
Thiophene	0.793	0.9992	0.012
Carbon tetrachloride	0.385	0.9996	0.008
Bromodichloromethane	0.397	0.9995	0.010
Trichloroethane	0.585	0.9994	0.008
cis-1,3-Dichloropropene	0.909	0.9933	0.019
trans-1,3-Dichloropropene	1.017	0.9883	0.026
1,1,2-Trichloroethane	0.772	0.9989	0.015
Toluene	0.409	0.9997	0.016
Dibromochloromethane	0.225	0.9994	0.006
Tetrachloroethane	0.432	0.9997	0.006
Chlorobenzene	0.792	0.9990	0.007
Ethylbenzene	0.365	0.9996	0.011
Bromoform	0.397	0.9987	0.013
m- + p-Xylenes	0.219	0.9999	0.011
Styrene	0.833	0.9998	0.013
o-Xylene	0.193	0.9993	0.005
1,1,2,2-Tetrachloroethane	0.663	0.9983	0.013
4-Chlorotoluene	0.178	0.9996	0.007
1,2-Dichlorobenzene	0.605	0.9996	0.006
Hexachloroethane	0.496	0.9996	0.007
1,3,5-Trichlorobenzene	2.032	0.9936	0.021
1,2,4-Trichlorobenzene	2.550	0.9912	0.026
Naphthalene	1.216	0.9950	0.016
1,2,3-Trichlorobenzene	2.825	0.9933	0.027
Hexachlorobutadiene	1.569	0.9933	0.020

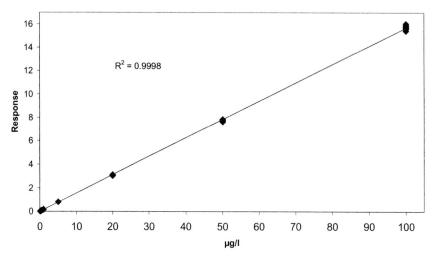

Fig. 4-18 Calibration graph obtained for 1,1-dichloroethylene over the 0.1 to 100 µg/l concentration range. The graph is based on nine calibrations (consecutive calibrations were repeated over three days).

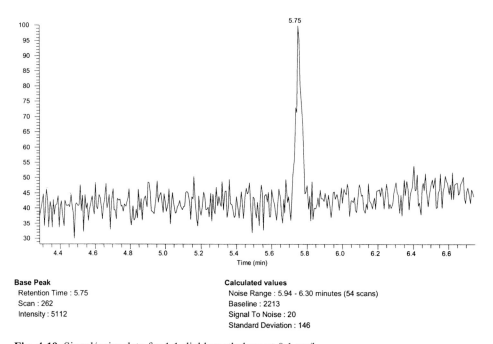

Base Peak
Retention Time : 5.75
Scan : 262
Intensity : 5112

Calculated values
Noise Range : 5.94 - 6.30 minutes (54 scans)
Baseline : 2213
Signal To Noise : 20
Standard Deviation : 146

Fig. 4-19 Signal/noise data for 1,1-dichloroethylene at 0.1 µg/l

4.5 MAGIC 60 – Analysis of Volatile Organic Compounds

Key words: purge & trap, water, soil, volatile halogenated hydrocarbons, BTEX, drinking water, EPA, internal standards

The method described here for identification and determination of volatile organic compounds in water and soil is taken from the EPA methods 8260, 8240 and 524.2. These approximately 60 substances are referred to by the term MAGIC 60. The method also allows the determination of compounds in solid materials, such as soil samples, and is generally applicable for a broad spectrum of organic compounds which have sufficiently high volatility and low solubility in water for them to be determined effectively by the purge and trap procedure. Vinyl chloride can also be determined with certainty by this method (Figs. 4-20 and 4-21).

Gastight syringes (5 ml, 25 ml) with open/shut valves have been shown to be useful for the preparation of water samples. The syringe is carefully filled with the sample from the plunger, the plunger inserted, and the volume taken up adjusted. The standard solutions are added through the valve with a µl syringe. The syringe is then connected to the purge and trap apparatus via the open/shut valves and the prepared sample transferred to a needle sparger or frit sparger vessel.

The analysis of soil samples is similar to that of water, but, depending on the expected concentration of volatile halogenated hydrocarbons in the sample, a different preparation

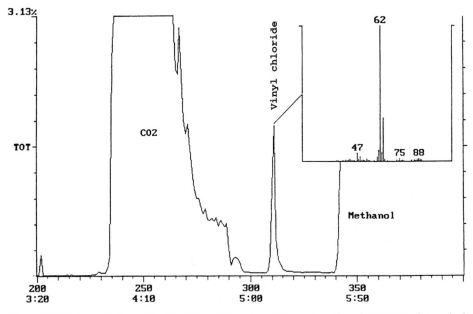

Fig. 4-20 Elution of vinyl chloride (50 µg/l) between CO_2 and methanol (DB-624, for analysis parameters see text). The methanol peak can be avoided by using poly(ethylene glycol) as the solvent for the preparation of standard solutions (Fig. 4-21)

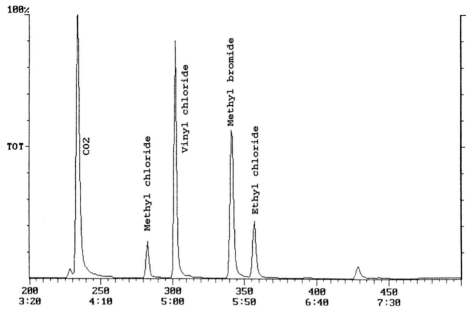

Fig. 4-21 Elution of vinyl chloride (50 µg/l) from a DB-VRX column (for analysis parameters see text) with poly(ethylene glycol) as the solvent

procedure is chosen. For samples where a concentration of less than 1mg/kg is expected, 5 g of the sample are placed directly in a 5 ml needle sparger and treated with 5 ml reagent water. Solutions of the internal standards are added with a µl syringe. At higher concentrations 4 g of the sample are weighed into a 10 ml vessel which can be closed with a Teflon-coated septum (e.g. headspace vial), and are treated with 9.9 ml methanol and 0.1 ml of the surrogate standard solution. After the solid phase has settled, 5–100 µl of the methanol phase are taken up, depending on the concentration, in a prepared 5 ml (25 ml) syringe together with reagent water. After addition of the internal standard, the sample is injected into the purge and trap apparatus (needle sparger).

Analysis Conditions

Sample material:	water or soil
Concentrator:	Tekmar purge & trap system LSC 2000
Trap:	Supelco Vocarb 3000, room temperature
Autosampler:	Tekmar ALS 2016 with 25 ml frit sparge glass vessels for water samples and needle sparge vessels for soil samples
Purge & trap parameters:	Purge gas: 40 ml/min
	Pre-purge time: 0 min
	MCM: 0 °C/90 °C
	Sample temp.: room temperature for water and soil
	Purge time: 12 min
	Dry purge: 0 min
	Desorb preheat: 255 °C for Vocarb 3000

	Desorption: 260 °C, 4 min Bake mode: 260 °C, 20 min Bake gas bypass: ON after 120 s Valves: 110 °C Transfer line: 110 °C
GC/MS system:	Finnigan MAGNUM
Injector:	PTV-cold injection system, 200 °C isotherm, direct connection (insertion) of the purge & trap concentrator into the carrier gas supply line of the PTV
Column:	J&W DB-VRX, 60 m × 0.32 mm × 1.8 µm DB-624 or Rtx-624, 60 m × 0.32 mm × 1.8 µm
Program:	Start: 40 °C, 5 min Program 1: 7 °C/min to 180 °C Program 2: 15 °C/min to 220 °C Final temp.: 220 °C, 5 min
Mass spectrometer:	Scan mode: EI, 33–260 u Scan rate: 1 s/scan
Calibration:	Internal/surrogate standards: toluene-d_8, fluorobenzene, 4-bromofluorobenzene, 1,2-dichloroethane-d_4 Solvent for all standard solutions: methanol
Cleaning the purge & trap unit:	In the case of contamination cleaning the concentrator can be necessary. Clean purge vessels filled with water are used and the following short program is put into operation: Purge time: 5 min Desorption: 1 min Bake mode: 5 min Valves, transfer line: 200 °C

All other parameters remain unchanged.

Results

The search and identification of individual analytes in the chromatograms (Fig. 4-22) are carried out automatically on the basis of reference data, such as the retention time and mass spectrum, which are stored in a calibration file (Fig. 4-23). For quantitation, a calibration with eight steps from 0.1 µg/l to 40 µg/l is constructed (Fig. 4-24). The purge and trap procedure gives very good linearity over this range. A standard analysis of the compounds listed in Table 4-3 in water is shown in Fig. 4-25.

Reference: T. Egert, Dr. J. Kurz, Institut Fresenius, Tanusstein.

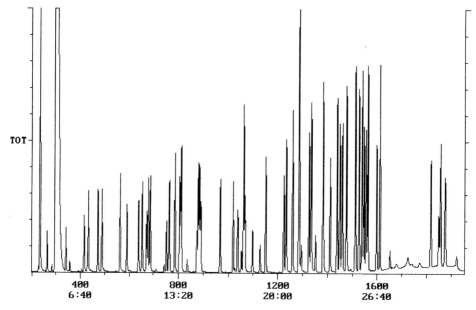

Fig. 4-22 MAGIC 60: standard chromatogram of a soil sample.
All analytes: 100 µg/kg, internal standard: 10 µg/kg (for analysis parameters see text)

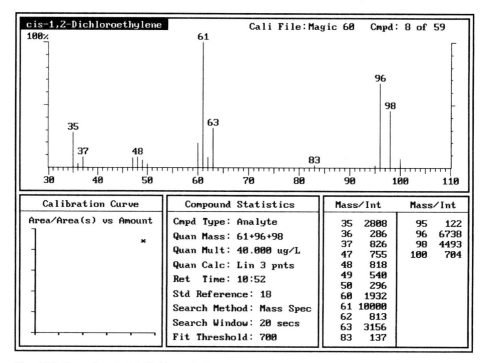

Fig. 4-23 Entry for cis-1,2-dichloroethylene from the calibration file

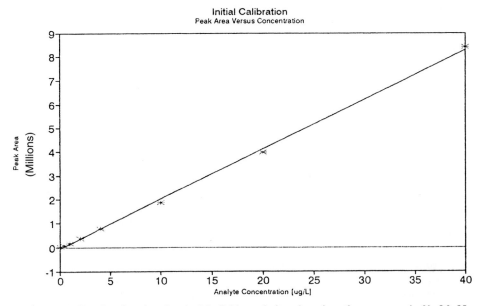

Fig. 4-24 Calibration function for cis-1,2-dichloroethylene based on the masses m/z 61, 96, 98. Correlation 0.9991, relative standard deviation: 4.96%

Table 4-3 MAGIC 60 substance list with details of quantifying masses, CAS numbers, limits of detection (MDL, minimum detection limit) and limits of quantitation (MQL, minimum quantitation limit), arranged alphabetically (for the method, see text)

Compound	m/z	CAS no.	MDL [µg/l]	MQL [µg/l]
Benzene	77, 78	71-43-2	0.05	0.1
Bromobenzene	77, 156, 158	108-86-1	0.1	0.2
Bromochloromethane	49, 128, 130	74-97-5	0.05	0.1
Bromodichloromethane	83, 85, 127	75-27-4	0.05	0.1
Methyl bromide	94, 96	74-96-4	n.b.	n.b.
Bromoform	173, 175, 252	95-25-2	0.05	0.1
n-Butylbenzene	91, 134	104-51-8	0.05	0.1
sec-Butylbenzene	105, 134	135-98-8	0.05	0.1
t-Butylbenzene	91, 119	98-06-6	0.05	0.1
Chlorobenzene	77, 112, 114	108-90-7	0.1	0.2
Ethyl chloride	64, 66	75-00-3	n.b.	n.b.
Methyl chloride	50, 52	74-87-3	n.b.	n.b.
Chloroform	83, 85	67-66-3	0.05	0.1
2-Chlorotoluene	91, 126	95-49-8	0.05	0.1
4-Chlorotoluene	91, 126	106-43-4	0.05	0.1
Dibromochloromethane	127, 129	124-48-1	0.05	0.1
1,2-Dibromo-3-chloropropane	75, 155, 157	96-12-8	0.25	0.5
Dibromomethane	93, 95, 174	74-95-3	0.1	0.2
1,2-Dibromoethane	107, 109	106-93-4	0.1	0.2
1,2-Dichlorobenzene	111, 146	95-50-1	0.15	0.2

Table 4-3 (continued)

Compound	m/z	CAS no.	MDL [µg/l]	MQL [µg/l]
1,3-Dichlorobenzene	111, 146	541-73-1	0.1	0.2
1,4-Dichlorobenzene	111, 146	106-46-7	0.1	0.2
1,1-Dichloroethane	63, 112	75-34-3	0.05	0.1
1,2-Dichloroethane	62, 98	107-06-2	0.1	0.2
1,1-Dichloroethylene	61, 63, 96	75-35-4	0.05	0.1
cis-1,2-Dichloroethylene	61, 96, 98	156-59-4	0.05	0.1
trans-1,2-Dichloroethylene	61, 96, 98	156-60-5	0.05	0.1
Dichlorodifluoromethane	85, 87	75-71-8	0.1	0.2
Dichloromethane	49, 84, 86	75-09-2	0.5	0.5
1,2-Dichloropropane	63, 112	78-87-5	0.1	0.2
1,3-Dichloropropane	76, 78	142-28-9	0.05	0.1
2,2-Dichloropropane	77, 97	590-20-7	0.05	0.1
1,1-Dichloropropylene	75, 110, 112	563-68-6	0.05	0.1
cis-1,3-Dichloropropylene	75, 110, 112	10061-01-5	0.05	0.1
trans-1,3-Dichloropropylene	75, 110, 112	10061-01-5	0.05	0.1
Ethylbenzene	91, 106	100-41-4	0.05	0.1
Hexachlorobutadiene	225, 260	87-68-3	0.15	0.2
Isopropylbenzene	105, 120	48-82-8	0.1	0.2
4-Isopropyltoluene	91, 119, 134	99-87-6	0.05	0.1
Naphthalene	128	91-20-3	0.5	0.5
Styrene	78, 104	100-42-5	0.1	0.2
1,1,1,2-Tetrachloroethane	131, 133	620-30-6	0.1	0.2
1,1,2,2-Tetrachloroethane	83, 85, 131	79-34-5	0.1	0.2
Tetrachloroethylene	129, 166, 168	127-18-4	0.1	0.2
Carbon tetrachloride	117, 119	56-23-5	0.05	0.1
Toluene	91, 92	108-88-3	*	*
1,2,4-Trichlorobenzene	180, 182	120-82-1	0.3	0.3
1,2,3-Trichlorobenzene	180, 182	87-61-6	0.4	0.4
1,1,1-Trichloroethane	61, 97, 99	71-55-6	0.05	0.1
1,1,2-Trichloroethane	83, 85, 97	79-020-5	0.1	0.2
Trichloroethylene	95, 130, 132	79-01-6	0.05	0.1
Trichlorofluoromethane	101, 103	75-69-4	0.05	0.1
1,2,3-Trichloropropane	75, 77	96-18-4	0.1	0.2
1,2,4-Trimethylbenzene	105, 120	95-63-6	0.1	0.2
1,3,5-Trimethylbenzene	105, 120	108-67-8	0.1	0.2
Vinyl chloride	62, 64	75-01-4	n.b.	n.b.
m-Xylene	91, 106	108-38-3	0.1	0.2
o-Xylene	91, 106	95-47-6	0.1	0.2
p-Xylene	91, 106	95-47-6	0.1	0.2

Internal standards:

4-Bromofluorobenzene	95, 174	
1-Chloro-2-bromopropane	77, 79	
1,2-Dichlorobenzene-d_4	115, 150, 152	
1,2-Dichloroethane-d_4	65, 102	
Fluorobenzene	77, 96	
Toluene-d_8	70, 98, 100	

4.5 MAGIC 60 – Analysis of Volatile Organic Compounds

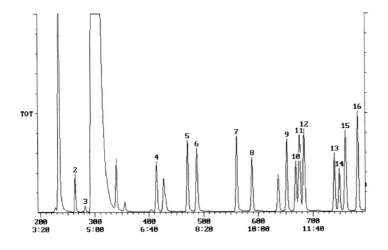

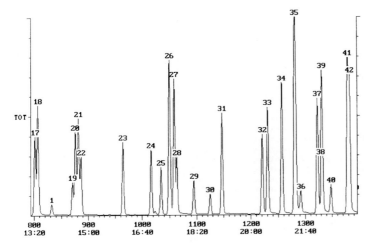

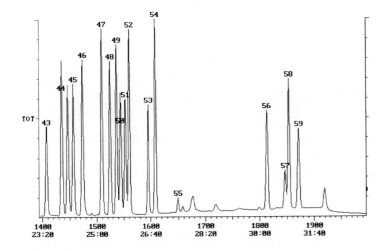

Fig. 4-25

1	Benzene	Active	I	13:52
2	Fluorobenzene	Active	A	4:24
3	Dichlorodifluoromethane	Active	A	4:42
4	Chloromethane	Active	A	6:54
5	Trichlorofluoromethane	Active	A	7:50
6	1,1-Dichloroethylene	Active	A	8:07
7	Methylene chloride	Active	A	9:20
8	trans-1,2-Dichloroethylene	Active	A	9:48
9	1,1-Dichloroethane	Active	A	10:52
10	cis-1,2-Dichloroethylene	Active	A	11:09
11	Bromochloromethane	Active	A	11:15
12	Chloroform	Active	A	11:24
13	2,2-Dichloropropane	Active	A	12:20
14	1,2-Dichloroethane-d4	Active	A	12:29
15	1,2-Dichloroethane	Active	A	12:40
16	1,1,1-Trichloroethane	Active	A	13:02
17	1,1-Dichloropropylene	Active	A	13:22
18	Carbon tetrachloride	Active	A	13:27
19	Dibromomethane	Active	A	14:31
20	1,2-Dichloropropane	Active	A	14:36
21	Trichloroethylene	Active	A	14:42
22	Bromodichloromethane	Active	A	14:47
23	cis-1,3-Dichloropropylene	Active	A	16:04
24	trans-1,3-Dichloropropyle	Active	A	16:56
25	1,1,2-Trichloroethane	Active	A	17:14
26	Toluene-d8	Active	A	17:30
27	Toluene	Active	A	17:38
28	1,3-Dichloropropane	Active	A	17:44
29	Dibromochloromethane	Active	A	18:15
30	1,2-Dibromoethane	Active	A	18:44
31	Tetrachloroethylene	Active	A	19:07
32	1,1,1,2-Tetrachloroethane	Active	A	20:21
33	Chlorobenzene	Active	A	20:30
34	Ethylbenzene	Active	A	20:56
35	m,p-Xylene	Active	A	21:21
36	Bromoform	Active	A	21:32
37	Styrene	Active	A	22:02
38	1,1,2,2-Tetrachloroethane	Active	A	22:09
39	o-Xylene	Active	A	22:11
40	1,2,3-Trichloropropane	Active	A	22:28
41	Isopropylbenzene	Active	A	22:58
42	4-Bromofluorobenzene	Active	A	23:02
43	Bromobenzene	Active	A	23:27
44	2-Chlorotoluene	Active	A	23:55
45	4-Chlorotoluene	Active	A	24:16
46	1,3,5-Trimethylbenzene	Active	A	24:34
47	tert.-Butylbenzene	Active	A	25:09
48	1,2,4-Trimethylbenzene	Active	A	25:24
49	sec.-Butylbenzene	Active	A	25:37
50	1,3-Dichlorobenzene	Active	A	25:44
51	1,4-Dichlorobenzene	Active	A	25:53
52	4-Isopropyltoluene	Active	A	26:00
53	1,2-Dichlorobenzene	Active	A	26:35
54	n-Butylbenzene	Active	A	26:48
55	1,2-Dibromo-3-Chloropropa	Active	A	27:29
56	1,2,4-Trichlorobenzene	Active	A	30:13
57	Naphthalene	Active	A	30:46
58	Hexachlorobutadiene	Active	A	30:53
59	1,2,3-Trichlorobenzene	Active	A	31:12

Fig. 4-25 MAGIC 60: Standard chromatogram for a water sample.
All analytes: 20 µg/l, internal standard: 2 µg/l (for components see above, for analysis parameters see text)

4.6 Vinyl Chloride in Drinking Water

Key words: vinyl chloride, purge & trap, Vocarb 4000, PLOT column, quantitation

For the sensitive and selective determination of vinyl chloride, chromatographic separation from other volatile components is necessary in special cases. Instead of a packed column, a fused silica column coated with Porapak was used. This type column can be used both with cryofocusing and also for direct coupling with the mass spectrometer.

Analysis Conditions

Sample material:	drinking water	
Concentrator:	Tekmar purge & trap system LSC 2000 with AquaTek 50 sample dispenser	
	25 ml sample volume, 70 s sample transfer	
Trap:	Supelco Vocarb 4000 at room temperature	
Purge & trap parameters:	Purge gas:	40 ml/min
	Mount:	40 °C
	Pre-purge time:	1.5 min
	Sample temp.:	35 °C
	Purge time:	10 min
	Dry purge:	3 min
	MCM:	0 °C/90 °C
	Desorb preheat:	245 °C
	Desorption:	250 °C, 4 min
	Bake mode:	280 °C, 5 min
	Bake gas bypass:	OFF
	Valves:	110 °C
	Transfer line:	120 °C
GC/MS system:	Finnigan ITS 40	
Injector:	Tekmar cryofocusing -30 °C	
Column:	Chrompack PoraPlot Q, 25 m × 0.32 mm × 10 µm, cf. 2.5 m particle trap	
Program:	Start:	40 °C, 5 min
	Program 1:	10 °C/min to 100 °C
	Program 2:	20 °C/min to 250 °C
	Final temp.:	250 °C, 1.5 min
Mass spectrometer:	Scan mode:	EI, 55–249 u
	Scan rate:	0.5 s/scan
Calibration:	external standardisation	

Results

The PLOT column used (porous layer open tubular) exhibits considerable retention of chlorinated volatile components. For vinyl chloride the retention time is ca. 12:30 min with very good separation from the other volatile halogenated hydrocarbons. The mass

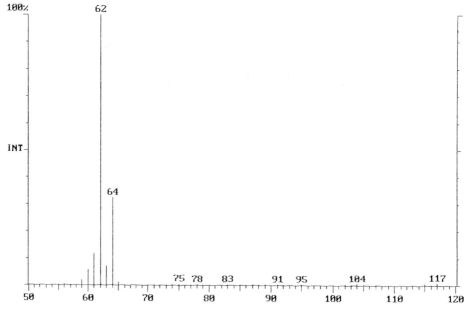

Fig. 4-26 Mass spectrum of vinyl chloride in a standard run, 1 µg/l

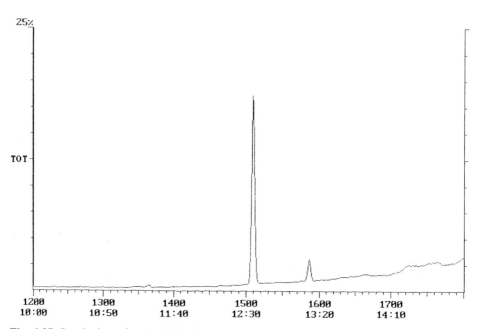

Fig. 4-27 Standard run for vinyl chloride on PoraPlot Q, 1 µg/l

4.6 Vinyl Chloride in Drinking Water

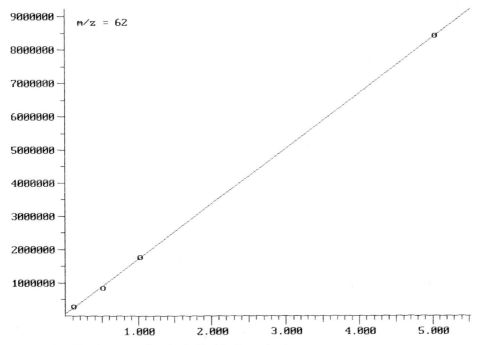

Fig. 4-28 Calibration curve for vinyl chloride determination

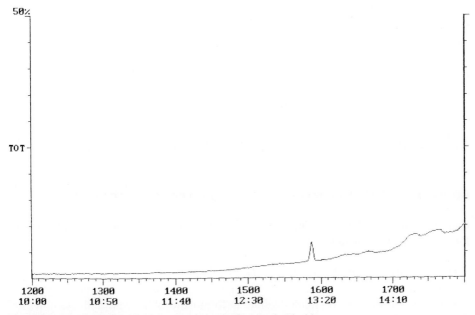

Fig. 4-29 Analysis of a drinking water sample for vinyl chloride

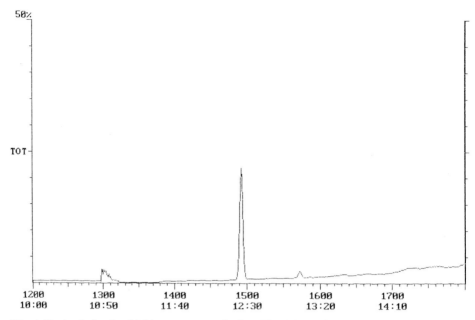

Fig. 4-30 Analysis of a drinking water sample from Fig. 4-29 spiked with 1 µg/l vinyl chloride

spectrometric evaluation of vinyl chloride uses the masses m/z 62 and 64 (Fig. 4-26). The chromatogram of a standard with a concentration of 1 µg/l is shown in Fig. 4-27. The calibration for the determination of vinyl chloride in drinking water covers the concentration range 0.1 µg/l to 5.0 µg/l (Fig. 4-28). Figures 4-29 and 4-30 show chromatograms of a drinking water sample with and without addition of standard. The drinking water sample shows no signal for vinyl chloride (<0.1 µg/l).

Reference: Dr. C. Schlett, B. Pfeifer, Gelsenwasser AG, Gelsenkirchen.

4.7 Chloral Hydrate in Surface Water

Key words: chloral hydrate, hydrolysis, chloroform, purge & trap, Vocarb 4000

Chloral hydrate can be determined analytically by quantitative conversion into chloroform in an alkaline medium:

$$CCl_3CH(OH)_2 + NaOH \rightarrow CHCl_3 + HCOONa + H_2O$$

For the determination of chloral hydrate in formulations and sprays, the samples are diluted appropriately and adjusted to pH 13.5 with 10 M NaOH. The reaction can take place within 5 min in the sample vessel of the purge and trap autosampler at 110 °C in the drying oven.

Analysis Conditions

Sample material:	aqueous pesticide sprays, surface water	
Concentrator:	Tekmar purge & trap system LSC 2000 with AquaTek 50 sample dispenser	
	25 ml sample volume, 60 s sample transfer	
Trap:	Supelco Vocarb 4000, room temperature	
Purge & trap parameters:	Purge gas:	40 ml/min
	Mount:	40 °C
	Pre-purge time:	0 min
	Sample temp.:	35 °C
	Purge time:	10 min
	Dry purge:	3 min
	MCM:	0 °C/90 °C
	Desorb preheat:	245 °C
	Desorption:	250 °C, 4 min
	Bake mode:	280 °C, 5 min
	Bake gas bypass:	OFF
	Valves:	100 °C
	Transfer line:	200 °C
GC/MS system:	Finnigan ITS 40	
Injector:	Tekmar cryofocusing –30 °C	
Column:	Restek Rtx-1701, 30 m × 0.25 mm × 0.1 µm	
Program:	Start:	60 °C, 4 min
	Program 1:	5 °C/min to 150 °C
	Program 2:	20 °C/min to 280 °C
	Final temp.:	280 °C, 3.5 min
Mass spectrometer:	Scan mode:	EI, 60–99 u
	Scan rate:	0.5 s/scan
Calibration:	external standardisation	

Results

The chromatograms show a chloroform standard in Fig. 4-31 and in Fig. 4-32 a standard run with 50 µg/l chloral hydrate, determined as chloroform. The chromatogram of a real sample with a concentration of ca. 50 µg/l is shown in Fig. 4-33. The sample also contains large quantities of toluene.

The standard solutions cannot necessarily be stored. A considerable decrease in concentration is observed even while standing in the autosampler.

Reference: Dr. C. Schlett, B. Pfeifer, Gelsenwasser AG, Gelsenkirchen.

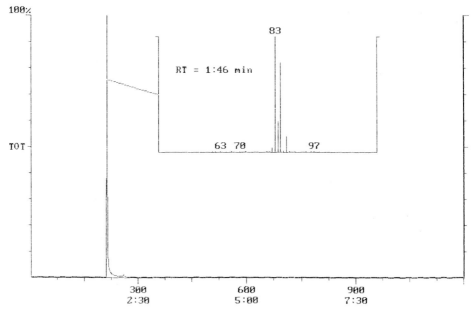

Fig. 4-31 Standard run for chloroform

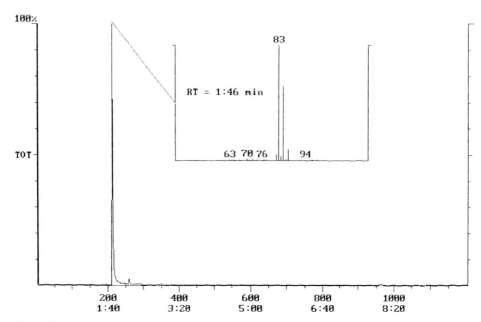

Fig. 4-32 Standard run for chloral hydrate, 50 µg/l

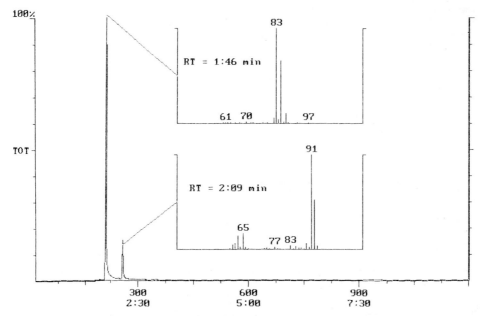

Fig. 4-33 Actual water sample with chloral hydrate (ca. 50 µg/l) and toluene

4.8 Field Analysis of Soil Air

Key words: volatile halogenated hydrocarbons, BTEX, adsorption, thermodesorption, soil air, mobile analysis, library search

Gaseous samples (soil air, gases from landfill sites, exhaust gases, air in enclosed spaces) are analysed for volatile halogenated and aromatic solvents by adsorption of the components in thermodesorption tubes. Because of the high affinity for volatile halogenated hydrocarbons and the low adsorption of water vapour, Tenax has proved particularly suitable for this purpose. Sampling can be actively effected using a pump, which, for example, can be connected to a purposely constructed bore column, and the volume flow can be controlled. Passive collection by diffusion can also be used.

A thermodesorption tube filled with Tenax, which should be fitted with a glass frit and which is closed with a silanised glass wool plug and conditioned, is used for sample preparation.

The mobile GC/MS system SpectraTrak 620 provides all the necessary equipment for active sampling, volume flow measurement and analysis, whereby the adsorption/desorption tube remains in the same piece of equipment. After sampling and analysis an evaluation is carried out automatically according to the requirements of the user.

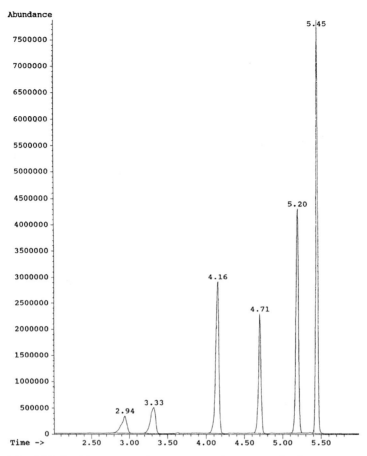

Fig. 4-34 Total ion current chromatogram of a gas sample after concentration and desorption

Analysis Conditions

Sample collection:	0.5 l soil air
GC/MS system:	VIKING SpectraTrak 620 with HP-MSD 5971 A
Adsorbent:	ca. 100 mg Tenax
Desorption:	200 °C, 1 min
Column:	Restek Rtx-5, 20 m × 0.18 mm × 0.40 µm
Carrier gas:	helium, 1.5 bar (20 psi)
Split:	10 ml/min
Program:	Start: 50 °C
	Ramp: 15 °C/min
	Final temp.: 200 °C
Mass spectrometer:	Scan mode: EI, full scan, 45–250 u
	Scan rate: 0.5 s/scan

```
                    Viking SpectraTrak 600
                    Library Search Report
Disk File Name:     C:\CHEMPC\DATA\SCONGCMS
Operator:
Date Acquired:      22 Sep 93    1:56 pm
Cycle:              SCONGCMS
Run Type:
Data File:
Sample Name:        0.5 L Volumen auf tenax midtune @14psi
Misc Info:          Tenax only approx 100mg @15C/min

DA Method:          C:\CHEMPC\METHODS\DEFAULT.M
Integrator:         ChemStation
Integration Events: AutoIntegrate
Search Libraries:   C:\DATABASE\NBS75K.L           Minimum Quality:   0
Unknown Spectrum:   Apex minus start of peak


 Pk#   RT   Area%        Library/ID              Ref#     CAS#      Qual

  1   2.94   4.01  C:\DATABASE\NBS75K.L
                   Benzene                       62628  000071-43-2  91
                   Benzene                       62627  000071-43-2  91
                   Benzene                       62626  000071-43-2  91

  2   3.33   5.68  C:\DATABASE\NBS75K.L
                   Trichloroethylene              5300  000079-01-6  97
                   Trichloroethylene             65175  000079-01-6  97
                   Trichloroethylene             65176  000079-01-6  94

  3   4.16  20.68  C:\DATABASE\NBS75K.L
                   Toluene                       63030  000108-88-3  94
                   Toluene                         965  000108-88-3  91
                   Toluene                       63029  000108-88-3  87

  4   4.71  12.65  C:\DATABASE\NBS75K.L
                   Tetrachloroethylene           67757  000127-18-4  97
                   Tetrachloroethylene           13222  000127-18-4  97
                   Tetrachloroethylene           67759  000127-18-4  96

  5   5.20  24.74  C:\DATABASE\NBS75K.L
                   Benzene, chloro-              63935  000108-90-7  94
                   Benzene, chloro-              63934  000108-90-7  91
                   Benzene, chloro-               2491  000108-90-7  87

  6   5.45  32.25  C:\DATABASE\NBS75K.L
                   p-Xylene                      63701  000106-42-3  91
                   p-Xylene                      63702  000106-42-3  87
                   Benzene, 1,3-dimethyl-         2027  000108-38-3  74
```

Fig. 4-35 Library search report after searching the NBS library

372   4 Applications

Results

Figure 4-34 shows the chromatogram of a synthetic gas sample, which contains ca. 10 ppm of each of the six components. They are identified using a library search with the NIST library containing ca. 75 000 spectra (Fig. 4-35). The search results are printed out individually for visual examination of the matching of the spectra (Fig. 4-36).

In the field a rapid on-line analysis can be incorporated, which allows reliable identification of organic components in gases with an automatic recording device.

Reference: J. Niebel, Axel Semrau GmbH & Co, Applikationslabor, Sprockhövel.

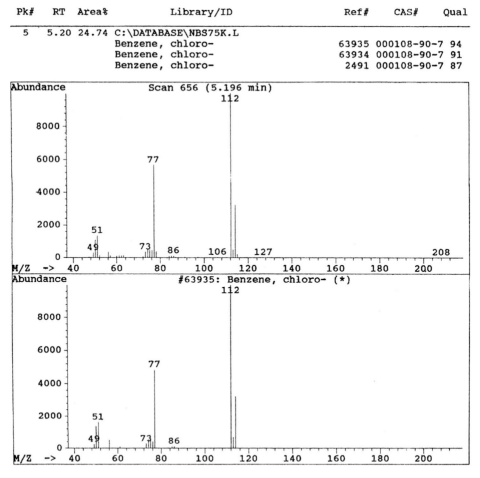

Fig. 4-36 Spectroscopic comparison for chlorobenzene

4.9 Field Analysis of Solvent Contamination of Soil

Key words: volatile halogenated hydrocarbons, BTEX, thermodesorption, Tenax, soil, mobile analysis, library search, PBM

The thermodesorption device built into the mobile GC/MS system SpectraTrak 620 is used for the field analysis of soil samples for volatile halogenated and aromatic solvents. In order for a soil sample to behave like a charged thermodesorption tube, the corresponding tube is filled with ca. 2 cm Tenax which is separated from the soil filling by a plug of silanised glass wool (2 cm soil filling corresponds to ca. 200 mg soil).

The analysis method developed for field testing requires the user to first purge the desorption tube with carrier gas in the cold to drive out the remaining air. The volatile components are then driven out of the soil by rapid heating and retarded in the Tenax plug in order to be injected on to the GC column in the same heating cycle. At the end of the analysis the results are qualitatively and quantitatively evaluated automatically using the NBS spectral library and prepared calibrations.

Analysis Conditions

Sample collection: ca. 200 mg soil

Mobile GC/MS system: VIKING SpectraTrak 620 with HP-MSD 5971A
Adsorbent: Tenax
Desorption: helium cold flow, 1 min, 200 °C, 1 min
Column: Restek Rtx-5, 20 m × 0.18 mm × 0.40 µm
Carrier gas: helium, 1.5 bar (20 psi)
Split: 50 ml/min
Program: Start: 50 °C, 2 min
Ramp: 10 °C/min to 250 °C

Mass spectrometer: Scan mode: full scan, 45–180 u
Scan rate: 0.5 s/scan

Results

The sample tube filled with Tenax and the soil sample is shown in Fig. 4-37. Figure 4-38 shows the TIC chromatogram of a soil sample which was prepared with a mixture of solvents at a concentration of 1 ppm. In spite of the long process of desorption from the soil, the combination with Tenax as the inserted adsorbent gives an excellent peak shape using split injection on to a column with small internal diameter but high film thickness. For field analysis the procedure provides rapid on-line sample injection with reliable separation of the components and high sensitivity and is easily quantified. Qualitative and quantitative evaluation take place automatically at the end of data acquisition (Figs. 4-39 to 4-41).

Reference: J. Niebel, Axel Semrau GmbH & Co, Applikationslabor, Sprockhövel.

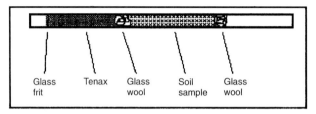

Fig. 4-37 Charging the sample tube with Tenax and soil sample

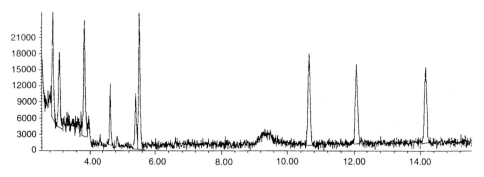

Fig. 4-38 Soil with ca. 1 ppm solvent mixture (volatile halogenated hydrocarbons and BTEX), total ion current chromatogram

Extract of the ion 91 characteristic of toluene

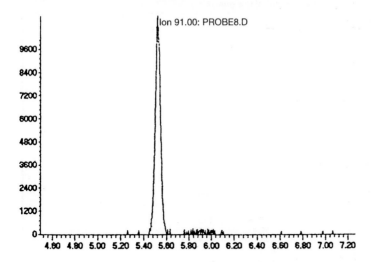

The corresponding mass spectrum of toluene

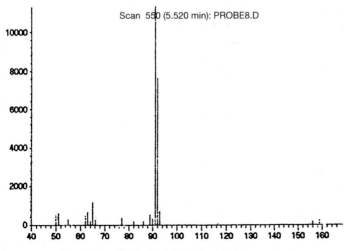

Fig. 4-39 Mass trace and spectrum of toluene from a soil sample

Library search report (NBS library) drawn up automatically
after the end of the analysis

```
           Viking SpectraTrak 600
           Library Search Report

Disk File Name:    C:\CHEMPC\DATA\SCONGCMS\EARTH.R\PROBE8.D
Operator:          joerg Niebel
Date Acquired:     17 Aug 93  5:07 pm
Cycle:             SCONGCMS
Run Type:          EARTH.R
Data File:         PROBE8.D
Sample Name:       tenax mit erde 0.2gr. & 1 ul spike
Misc Info:         1min 200°C

DA Method:         C:\CHEMPC\METHODS\DEFAULT.M
Integrator:        ChemStation
Integration Events: AutoIntegrate
Search Libraries:  C:\DATABASE\NBS54K.L         Minimum Quality:  60
Unknown Spectrum:  Apex minus start of peak


Pk#  RT   Area%    Library/ID                    Ref#   CAS#         Qual

1   2.89  10.60  C:\DATABASE\NBS54K.L
                 Ethene, chloro-                 132    000075-01-4   78    Comment: Wrong result,
                 Cyclobutane, 1,3-dichloro-      3624   055887-82-6   43    substance is Dichloromethane
                 Cyclobutane, 1,2-dichloro-      3626   017437-39-7   43

2   3.10  9.62   C:\DATABASE\NBS54K.L
                 1,3-Benzenediol, 2-methyl-      3671   000608-25-3   78
                 1,3-Hexadien-5-yne              381    010420-90-3   72
                 2,4-Hexadiyne                   379    002809-69-0   72
                 Comment: True search result see rank 4

3   3.84  13.50  C:\DATABASE\NBS54K.L
                 Methane, bromodichloro-         10961  000075-27-4   90
                 Methane, trichloro-             3053   000067-66-3   59
                 Hexanoic acid, 2-methyl-3-oxo-, 13422  029304-40-3   53

4   4.64  5.27   C:\DATABASE\NBS54K.L
                 1-Propene, 1,3-dichloro-, (Z)-  2031   010061-01-5   90
                 1-Propene, 1,2-dichloro-        2028   000563-54-2   90
                 1-Propene, 1,1-dichloro-        2030   000563-58-6   86

5   5.41  5.94   C:\DATABASE\NBS54K.L
                 1-Propene, 1,3-dichloro , (Z)   2031   010001-01-5   78
                 Propane, 1,1,1,2-tetrachloro-   14932  000812-03-3   78
                 1-Propene, 1,2-dichloro-        2028   000563-54-2   78

6   5.52  15.67  C:\DATABASE\NBS54K.L
                 Toluene                         891    000108-88-3   86
                 Benzeneacetamide, N-methyl-     8230   006830-82-6   59
                 Tricyclo[3.2.2.02,4]non-        23383  062249-53-0   59

7   10.66 13.95  C:\DATABASE\NBS54K.L
                 Benzene, ethyl-                 1829   000100-41-4   64
                 Thiocyanic acid, phenylmethyl este 8187 003012-37-1  43
                 1,3-Cyclopentadiene, 5-(1-methylet 1833 002175-91-9  38

8   12.07 13.05  C:\DATABASE\NBS54K.L
                 No matches found                              Comment: Bromoform (CHBr₃) is not in NBS library

9   14.16 12.39  C:\DATABASE\NBS54K.L
                 Ethane, 1,1,2,2-tetrachloro-    11850  000079-34-5   64
                 Ethane, 1,1-dichloro-2,2-difluoro- 5234 000471-43-2  40
                 Ethane, 2,2-dichloro-1,1,1-trifluo 8689 000306-83-2  40


           Wed Aug 18 13:12:40 1993
```

Fig. 4-40 Library search report (NBS library) drawn up automatically after the end of data acquisition

Automatic quantitation, using a three point calibration by the ESTD method

```
         Viking SpectraTrak 600
         Quantitation Report
         btex

Calibration Last Updated: Tue Aug 17 15:28:24 1993
External Standard Report: Summary
_____

Disk File Name: C:\CHEMPC\DATA\SCONGCMS\EARTH.R\PROBE8.D
Operator:      joerg Niebel
Date Acquired: 17 Aug 93  5:07 pm
Cycle:         SCONGCMS
Run Type:      EARTH.R
Data File:     PROBE8.D
Sample name:   Tenax with 0.2 g soil & 1 ul spike
Misc Info:     1min 200°C

DA Method:     C:\CHEMPC\METHODS\EARTH.M
Integrator:    ChemStation
_____

Pk # Compound          Ret Time   Amount     Peak Type

 1 Benzene              3.09      1.1 mg/kg
 2 Toluene              5.52      0.9 mg/kg
 3 Ethylbenzene        10.65      0.7 mg/kg

         Wed Aug 18 13:49:54 1993
```

Fig. 4-41 Automatic quantitation of the BTEX components using an external three point calibration

4.10 Residual Monomers and Polymerisation Additives

Key words: purge & trap, polymers, residual monomers, additives

The purge and trap procedure is suitable for concentrating volatile components from polymers or materials in which plastics have been processed. The analysis of a group of these types of compounds is shown below. For the purge and trap analysis, the aqueous migration solutions of the corresponding incubation experiments are used.

Analysis Conditions

Concentrator:	Tekmar purge & trap system LSC 2000, needle sparge glass vessel, 5 ml	
Trap:	Supelco Vocarb 3000, cooled to −20 °C with TurboCool (liq. CO_2)	
Purge & trap parameters:	Purge gas:	40 ml/min
	Pre-purge time:	0 min
	MCM:	0 °C/90 °C
	Sample temp.:	room temperature
	Purge time:	12 min
	Dry purge:	0 min
	Desorb preheat:	255 °C with Vocarb 3000
	Desorption:	260 °C, 4 min
	Bake mode:	260 °C, 20 min

378   4 Applications

	Bake gas bypass:	ON after 120 s
	Valves:	110 °C
	Transfer line:	110 °C
GC/MS system:	Finnigan	
Injector:	PTV cold injection system 200 °C isotherm, split 30 ml/min direct connection (insertion) of the purge & trap concentrator into the carrier gas line of the PTV	
Column:	J&W DB-VRX, 60 m × 0.32 mm × 1.8 µm	
Program:	Start:	40 °C, 5 min
	Program 1:	7 °C/min to 180 °C
	Program 2:	15 °C/min to 220 °C
	Final temp.:	220 °C, 5 min
Mass spectrometer:	Scan mode:	EI, 33–260 u
	Scan rate:	1 s/scan

Results

Figure 4-42 shows the chromatogram of 26 selected monomers and additives. The components are assigned according to the entry number in the calibration file and from the retention time (Fig. 4-43). α-Bromostyrene is given as a typical entry. The standard mixture was injected in polyethylene glycol, which itself does not appear in the analysis. The first, not recorded, peak in the chromatogram is CO_2 from the sample vial. The concentrations of the substances (Fig. 4-44) are significantly higher than in the analysis of volatile halogenated hydrocarbons and BTEX in this application in the standard. To

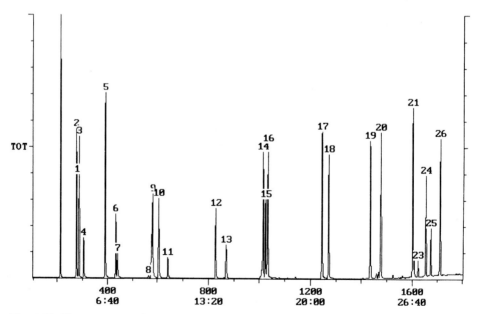

Fig. 4-42 Chromatogram of selected monomers and polymerisation additives (for analysis parameters see text; for peak list see Fig. 4-44)

4.10 Residual Monomers and Polymerisation Additives

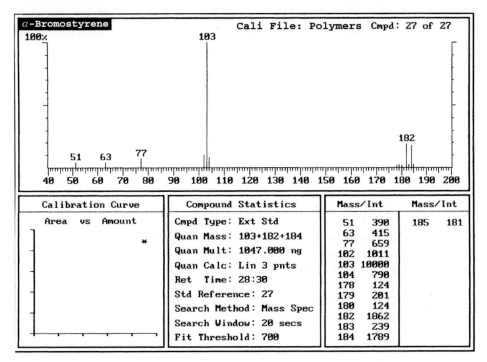

Fig. 4-43 Entry for α-bromostyrene in the calibration file

Cal	Name of Compound	S	Scan #	R Time	Me	Calc Amt(A)	Units
1	Vinyl chloride	E	276	4:36	BB	1811.001	ng
2	Acetaldehyde	E	275	4:35	BV	10010.001	ng
3	1,3-Butadiene	E	285	4:45	MM	1539.001	ng
4	Ethylene oxide	E	384	5:04	BB	5330.001	ng
5	Propylene oxide	E	391	6:31	BB	10129.001	ng
6	Vinylidene chloride	E	433	7:13	BB	1020.001	ng
7	Acrylonitrile	E	440	7:20	VB	1989.000	ng
8	Vinyl acetate	E	565	9:25	BB	1055.001	ng
9	Methyl vinyl ketone	E	576	9:36	MM	9558.001	ng
10	Butyraldehyde	E	581	9:41	BB	9830.001	ng
11	Methacrylonitrile	E	605	10:05	BB	1997.001	ng
12	Methyl acrylate	E	640	10:40	BB	1009.001	ng
13	Ethyl acrylate	E	828	13:48	BB	1911.001	ng
14	Epichlorohydrin	E	870	14:30	BB	9670.001	ng
15	1,7-Octadiene	E	1014	16:54	BV	981.001	ng
16	Ethyl methacrylate	E	1024	17:04	BB	1006.000	ng
17	1-Octene	E	1032	17:12	VB	1006.001	ng
18	Butyl acrylate	E	1245	20:45	BB	2023.000	ng
19	Styrene	E	1271	21:11	BB	1009.000	ng
20	α-Methylstyrene	E	1434	23:54	BB	1025.001	ng
21	p-Methylstyrene	E	1476	24:36	VB	987.001	ng
22	p-Chlorostyrene	E	1602	26:42	BB	986.001	ng
23	Ethylvinylbenzene isomer-1	E	1623	27:03	BB	221.000	ng
24	Ethylvinylbenzene isomer-2	E	1623	27:03	BB	221.000	ng
25	Divinylbenzene isomer-1	E	1655	27:35	BB	221.001	ng
26	Divinylbenzene isomer-2	E	1674	27:54	BB	221.001	ng
27	α-Bromostyrene	E	1711	28:31	BB	1047.001	ng

Fig. 4-44 Quantitation report with the components used in the standard analysis (the substance list refers to the peaks shown in Fig. 4-42)

make the capacity suitable for the column and mass spectrometer used, the split ratio is varied according to the requirements. The analysis parameters are comparable with those of the volatile halogenated hydrocarbon/BTEX analysis (see Section 4.5).

Reference: T. Egert, Dr J. Kurz, Institut Fresenius, Taunusstein.

4.11 Field Analysis of Odour Emissions

Key words: mobile analysis, odour contamination, thermodesorption, acetic acid, library comparison

Complaints by inhabitants of new housing very near a composting plant led to investigations concerning the nature and extent of odour emissions. The measurements were carried out using a mobile GC/MS system when the emissions occurred. The aim of the screening analysis was to characterise the individual potential sources of an odour by means of patterns from the total ion current chromatograms used as fingerprints and thus to determine the actual sources.

Analysis Conditions

Sample collection: Supelco thermodesorption tubes Carbotrap 370, 800 ml air in each case

Mobile GC/MS system: VIKING SpectraTrak 600 with HP-MSD 5971A
Injector: thermal desorption at 250 °C
Column: J&W DB-5, 30 m × 0.32 mm × 0.25 μm
Mass spectrometer: Scan mode: EI, 10–250 u
Scan rate: 1 s/scan

Results

During the measurements a definite acid odour was established on a field path 150 m away (Fig. 4-45). The library comparison gave acetic acid as the main component (Figs. 4-45 and 4-46).

The measurement in the green cuttings delivery area gave a chromatogram which is characterised by the elution of hydrocarbons and terpenes in the middle retention time range (Fig. 4-47). The measurement at the exit of the biofilter, from which the air was discharged into the environment, showed a readily comparable peak pattern (Fig. 4-48). According to the chromatogram, the major causes of the odour were traced back to the rotting vegetable matter with high acetic acid emissions and the open storage of green cuttings with characteristic terpene emissions.

Reference: Dr. Vorholz, ITU Ingenieurgemeinschaft technischer Umweltschutz, Berlin

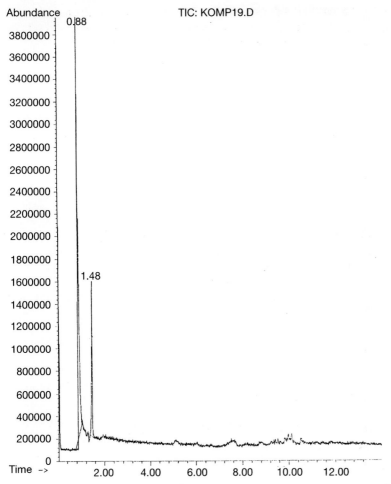

Fig. 4-45 Total ion current chromatogram of a pollution measurement at a distance of 150 m. The main component elutes with a retention time of 1.48 min.

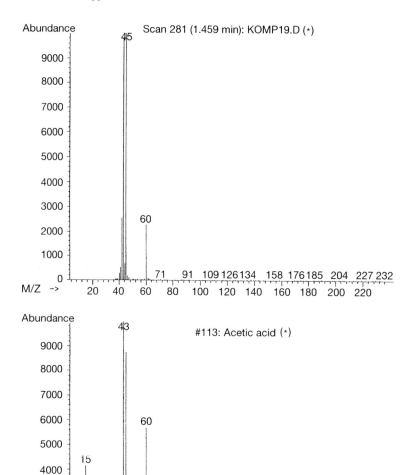

Fig. 4-46 The main component from Fig. 4-45 is identified as acetic acid by a PBM comparison with the NBS library.
Above: spectrum from the pollution measurement
Below: suggestion from the NBS library

4.11 Field Analysis of Odour Emissions

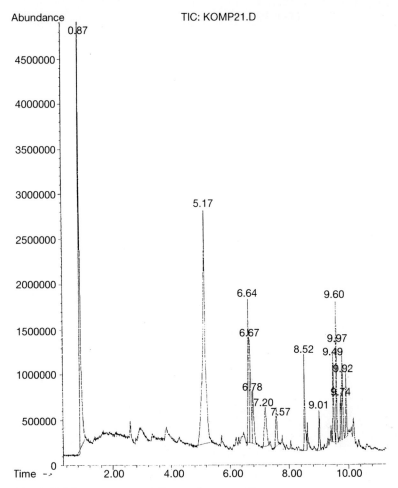

Fig. 4-47 Measurement on delivery of green cuttings

384   4 Applications

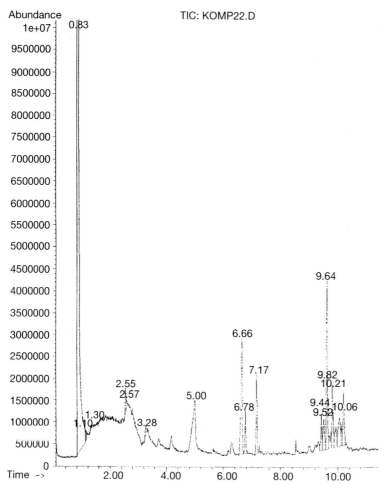

Fig. 4-48 Measurement at the exit of a biofilter in the area of the green cuttings delivery

4.12 Geosmin and Methylisoborneol in Drinking Water

Key words: odour contamination, purge & trap, polarity, full scan sensitivity, quantitation

Most published complaints about the quality of drinking water relate to odour and taste. The investigation of 2-methylisoborneol (2-MIB) and Geosmin therefore demands particular attention to analytical detail in the investigation of water quality. Both compounds are formed by micro-organisms and are responsible for a somewhat musty component of the odour.

Up to now the closed loop stripping technique requiring ca. 3.5 h per sample has been used for sample clean-up. By using the purge and trap technique in combination with a storage ion trap mass spectrometer, the required limits of detection of 10 ppt in ca. 30 min analysis time can be achieved.

Analysis Conditions

Concentrator:	Tekmar purge & trap system LSC 2000	
Trap:	Tenax/silica gel/SP2100	
Purge & trap parameters:	Sample volume:	20 ml
	Sample preheat:	3 min at 80 °C
	Purge:	12 min, 45 ml/min, 80 °C
	Desorption:	4 min, 200 °C
GC/MS system:	Finnigan	
Injector:	split injector, split ca. 10:1	
Column:	J&W DB-5-MS, 30 m × 0.25 mm × 0.25 µm	
Mass spectrometer:	Scan mode:	EI, full scan, 70–140 u
	Scan rate:	1 s/scan

Results

The mass spectra and structures of 2-MIB and Geosmin are shown in Figs. 4-49 and 4-50. Both spectra were run at a dilution of 2.5 ppt (50 pg/20 ml) (Fig. 4-51). Both compounds are polar because of the alcoholic OH group, so that the use of purge and trap is better than that of the static headspace procedure. By comparison of the direct injection of a solution of 100 pg per substance (Fig. 4-52) with a purge and trap procedure with this solution (Fig. 4-53), a purge efficiency for 2-MIB of 19% and for Geosmin of 84% were found. These values are affected considerably by the sample temperature chosen during the purge phase because of the polar character of the substances.

For both compounds calibration using external standardisation gives good linearity of the procedure, particularly in the region of the limit of quantitation (Figs. 4-54 and 4-55).

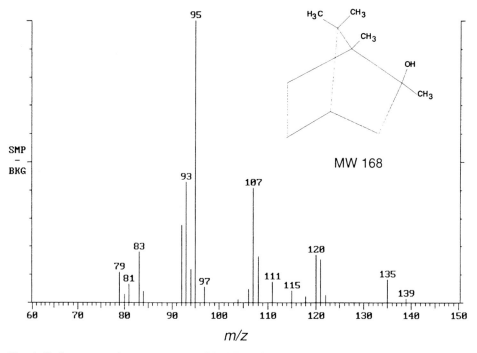

Fig. 4-49 Structure and mass spectrum of 2-MIB at 2.5 ppt

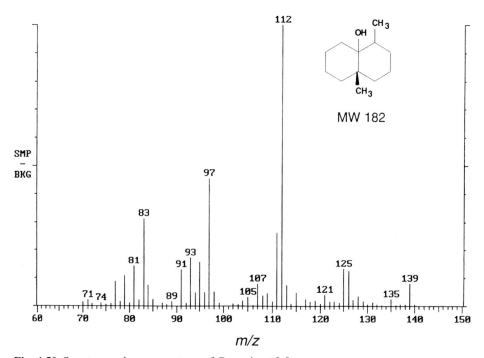

Fig. 4-50 Structure and mass spectrum of Geosmin at 2.5 ppt

4.12 Geosmin and Methylisoborneol in Drinking Water

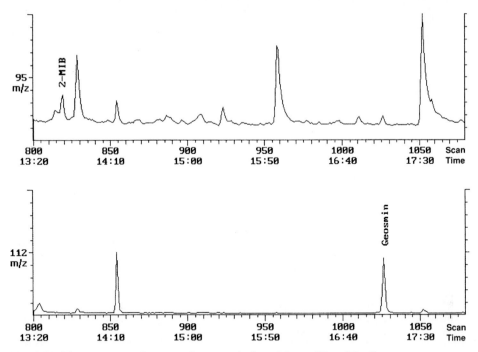

Fig. 4-51 Chromatogram of a purge & trap analysis at 2.5 ppt (50 pg/20 ml)

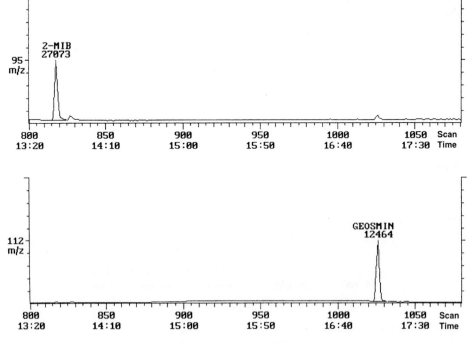

Fig. 4-52 Direct injection of 100 pg portions 2-MIB and Geosmin

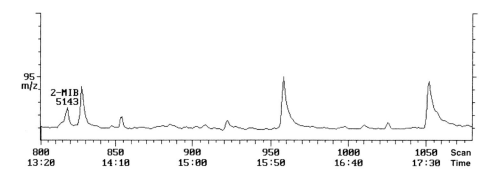
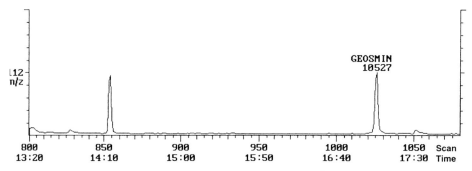

Fig. 4-53 Purge & trap analysis of 100 pg samples 2-MIB and Geosmin

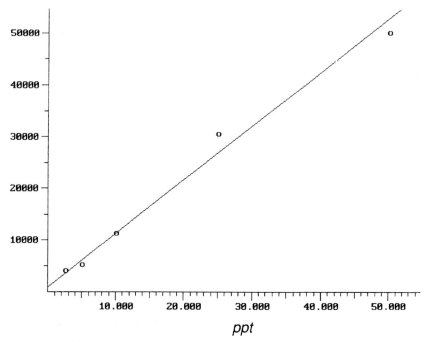

Fig. 4-54 Calibration for 2-MIB in the region 2.5 to 50 ppt

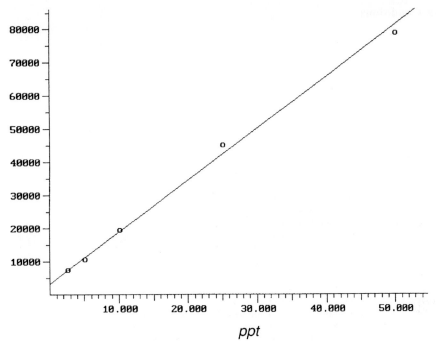

Fig. 4-55 Calibration for Geosmin in the region 2.5 to 50 ppt

4.13 Substituted Phenols in Drinking Water

Key words: odour contamination, drinking water, chlorophenols, bromophenols, alkylphenols, acetylation, chemical ionisation

Even at very low concentrations, substituted phenols lead to intense and unpleasant odours in drinking water. The various phenolic components can be formed from humic substances during chlorination of drinking water. The analysis procedures described up to now cannot guarantee reproducible and above all sensitive analyses. By combination of a special sample preparation procedure, known as extractive enrichment, with a sensitive mass spectrometer, phenols can be detected down to 1 ng/l in drinking, untreated and other water samples.

Analysis Conditions

Sample preparation: water sample 5000 ml (pH 11), addition of internal standard 2,4,6-trichlorophenol-$^{13}C_6$
+ 20 ml acetic anhydride, 15 min
+ 100 ml methanol
then glass wool filtration

SPE extraction: 1 g C_{18}-material
+ 6 ml methanol
+ 6 ml water
sample flow 1000 ml/h
dry under N_2 (ca. 500 ml/min)
elution with 2 × 1 ml acetone
concentration to 0.5 ml

GC/MS system: Finnigan ITS 40
Injector: PTV cold injection system
Injector program: Start: 60 °C, 1 min
Program: 20 °C/s
Final temp.: 250 °C
Column: Rtx-1701, 30 m × 0.25 mm × 0.1 μm
Temp. program: Start: 60 °C, 1 min
Program 1: 5 °C/min to 150 °C
Program 2: 20 °C/min to 280 °C
Final temp.: 280 °C, 4.5 min

Mass spectrometer: Scan mode: EI/CI, 50–350 u
Scan rate: 1 s/scan

Results

The method described allows the determination of 29 substituted phenols (Table 4-4, Fig. 4-56). To work out the analyses the individual components are isolated from the total ion current by selective mass chromatograms (Table 4-5, Figs. 4-57 and 4-58). By use of the ion trap mass spectrometer, the phenolic compounds contained in a drinking water sample can be identified from their mass spectra even at the very low limits of quantitation (Fig. 4-59). In the EI spectra the molecular ions do not appear or give very weak signals. Confirmation of molecular weights can be achieved using chemical ionisation with methane or methanol (Fig. 4-60).

Reference: Dr. C. Schlett, B. Pfeifer, Gelsenwasser AG, Gelsenkirchen.

Table 4-4 Composition of the standard from Fig. 4-56

No.	Phenol, analysed as the acetate	No.	Phenol, analysed as the acetate
1	2,6-Dimethylphenol	16	4-Chloro-3-methylphenol
2	2-Chlorophenol	17	2,4-Dichlorophenol
3	2-Ethylphenol	18	3,5-Dichlorophenol
4	3-Chlorophenol	19	2,3-Dichlorophenol
5	2,5-Dimethylphenol	20	3,4-Dichlorophenol
6	4-Chlorophenol	21	2,4,6-Trichlorophenol
7	2,4-Dimethylphenol	22	2,3,6-Trichlorophenol
8	3-Ethylphenol	23	2,3,5-Trichlorophenol
9	3,5-Dimethylphenol	24	2,4,5-Trichlorophenol
10	2,3-Dimethylphenol	25	2,6-Dibromophenol
11	3,4-Dimethylphenol	26	2,4-Dibromophenol
12	2-Chloro-5-methylphenol	27	2,3,4-Trichlorophenol
13	4-Chloro-5-methylphenol	28	2,4,6-Tribromophenol
14	2,6-Dichlorophenol	29	4,6-Dichlororesorcinol
15	4-Bromophenol		Internal Standard: 2,4,6-trichlorophenol-$^{13}C_6$

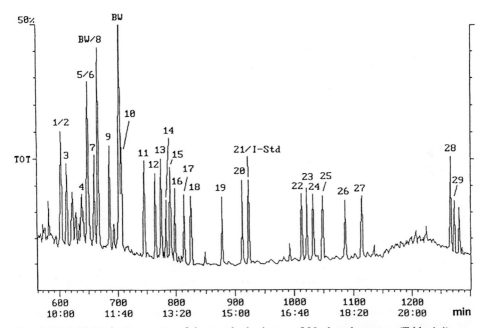

Fig. 4-56 GC/MS chromatogram of the standard mixture of 29 phenyl acetates (Table 4-4)

Table 4-5 Specific masses of the phenyl acetates (EI)

Phenol, analysed as the acetate	m/z values
Ethylphenols	107, 122
Dimethylphenols	107, 122
Chloromethylphenols	107, 142
Chlorophenols	128, 130
Dichlorophenols	162, 164
Trichlorophenols	196, 198
Dichlororesorcinol	178, 180
Bromophenols	172, 174
Dibromophenols	250, 252
Tribromophenols	330, 332
Internal standard 2,4,6-trichlorophenol-$^{13}C_6$	202

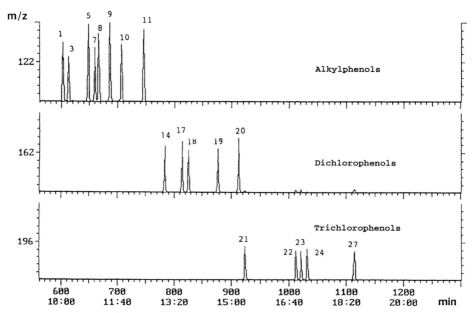

Fig. 4-57 Mass chromatogram of alkylphenols (m/z 122), dichlorophenols (m/z 162) and trichlorophenols (m/z 196)

4.13 Substituted Phenols in Drinking Water 393

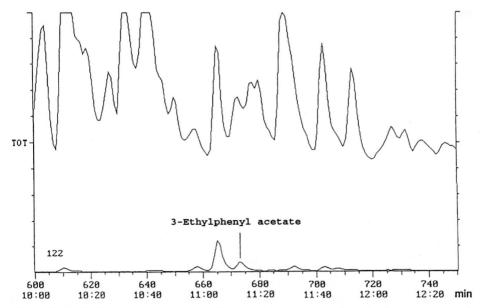

Fig. 4-58 Evaluation of a water sample with a total ion current and a selective mass trace m/z 122 for alkylphenyl acetates

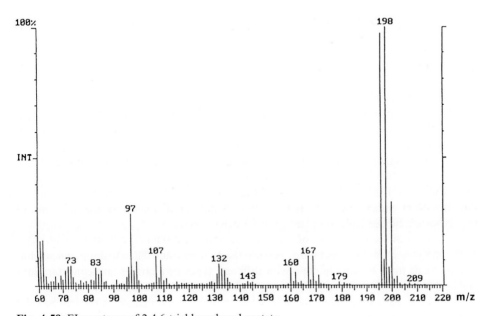

Fig. 4-59 EI spectrum of 2,4,6-trichlorophenyl acetate

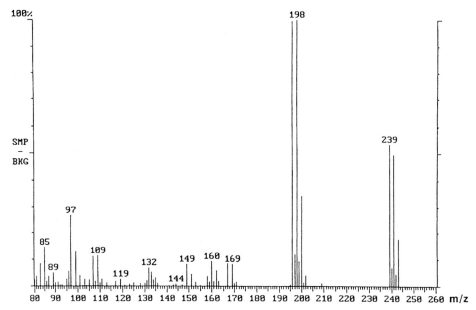

Fig. 4-60 CI spectrum of 2,4,6-trichlorophenyl acetate

4.14 Pesticides in Tea

Key words: SFE, matrix contamination, ECD, complete mass spectra, INCOS library search, pesticide library, chemical ionisation

The maximum permissible concentrations of plant protection agents are regulated by various laws and regulations. Checking demands the adherence to extremely low threshold values, which requires very powerful analysis methods. An increase in the number of active substances and reaction products requires very reliable multicomponent methods for effective monitoring purposes. The Pesticide Handbook lists ca. 800 active substances in the general section on currently used pesticides alone. In analysis of pesticide residues the use of GC/MS instruments has been the necessary reference procedure from the beginning because of their high information output. Today the use of GC/MS is indispensable in routine testing for pesticides. Screening analyses involving detection with ECD, NPD and MS-SIM require confirmation of the detection of the active substances with complete EI or CI mass spectra.

Because of the large number of samples and active substances, automatic evaluation of the analyses in routine operations is particularly important. The clean-up and separation methods must be able to cope with different types of matrices and concentrations and be stable for the necessary sensitivity with large series of samples.

Extracts of tea leaves have particularly high matrix contamination. Classical Soxhlet extraction and also SFE are used as clean-up procedures. Extracts from SFE can be used directly with GC/MS. As tea comes from various different plantation areas, the spectrum of expected pesticides is unlimited. In practice different groups of active substance are present, including those which are prohibited in some countries.

The identification of individual active substances in the evaluation of analyses is based on retention times in the normal way. However, changing matrix conditions lead to deviations. In many cases confirmation by GC/MS analysis using library spectroscopic comparison is necessary. The active substance spectra in tea analyses contain clear matrix signals or signals of co-eluting substances because of their low concentrations, which are sometimes in the lower pg region. These effects can only be partially compensated for, even by background subtraction of the spectra. Chemical ionisation allows higher selectivity in the determination of pesticide active substances. This, in turn, allows additional confirmation in the analysis of tea extracts.

Analysis Conditions

SFE extraction:	Suprex PrepMaster with AccuTrap	
	Fluid:	CO_2
	Modifier:	methanol
	Static extraction:	10 min, 400 bar, 80 °C
	Dynamic extraction:	2 ml/min, 30 ml, 400 bar, 2% MeOH, 80 °C
	Extract isolation:	Cryotrap, +5 °C
		Desorption at 40 °C
		with 1 ml methanol, 0.5 ml/min
GC/MS system:	Finnigan MAGNUM	
Injector:	PTV cold injection system, splitless	
Injector program:	Start:	60 °C, 0.5 min
	Program:	250 °C/min to 280 °C
	Final temp.:	280 °C, 26 min
Column:	SGE HT5, 30 m × 0.22 mm × 0.1 µm	
	Helium, 1 bar	
Temp. program:	Start:	60 °C, 1 min
	Program:	8 °C/min to 270 °C
	Final temp.:	270 °C, 15 min
Mass spectrometer:	Scan mode:	EI, full scan, 80–370 u
		CI, 100–500 u, reagent gas methanol
	Scan rate:	1 s/scan

Results

Figure 4-61 shows a typical chromatogram of a tea extract after SFE clean-up, which was taken with an ECD as the detector. The labelled peak was shown to be Ethion. To confirm the identity in the GC/MS analysis the NIST, Wiley and pesticides spectral libraries and the collection of pesticide CI spectra are available. The INCOS search procedure was used. In order to assess the efficiency of the library search with matrix-containing samples, a mixture of 19 pesticides including Ethion was added to a typical tea extract in the relevant concentrations.

In the total ion current of the GC/MS the components are eluted close together representing most of the contents of the tea leaves. Even in the more specific CI analysis more than 60 components are localised by the peakfinder in an approximately six minute long section of the chromatogram at a high elution temperature (Fig. 4-62). The selective mass chromatograms of the individual components are used to evaluate the data and the signal/noise ratio is determined.

Ethion is shown selectively on the mass trace of the strongest fragment m/z 231(Fig. 4-63). The fragment m/z 153 clearly overlaps with other compounds. The spectral comparison with the NIST library leads to Ethion as the first and Terbufos as the second suggestion (Fig. 4-64). The compound is identified as Ethion from consideration of the retention index (Ethion 2256, Terbufos 1774).

A very difficult elution situation is seen for Heptachlor in Fig. 4-65. The peak overlaps with several other substance peaks. The spectrum cannot be completely cleaned up by subtraction of the background. Detailed investigation of this co-elution situation shows, however, that the specific fragments of Heptachlor are present and maximises them at the same time (Fig. 4-65). A spectroscopic comparison using the INCOS search

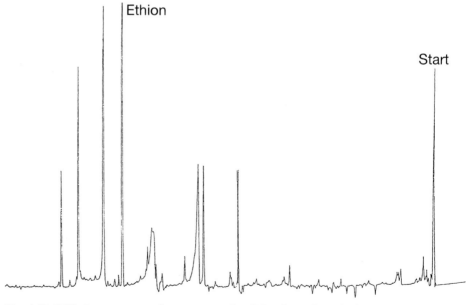

Fig. 4-61 ECD chromatogram of a tea extract after SFE. The peak marked is shown to be Ethion

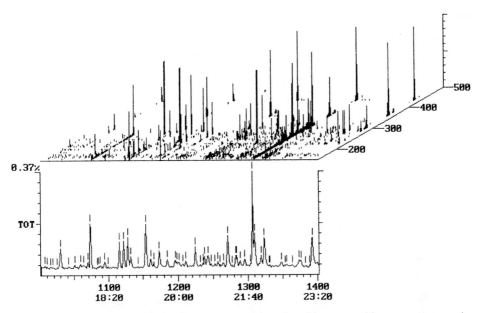

Fig. 4-62 CI (methanol) analysis of a tea extract. More than 60 protonatable components are localised in a section corresponding to ca. 6 min with the peakfinder.

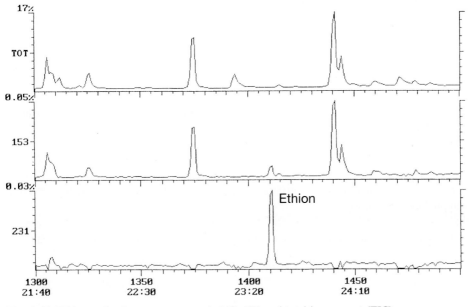

Fig. 4-63 Ethion: selective mass traces m/z 153, 231 and total ion current (TIC)

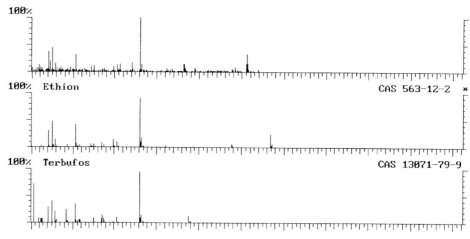

Fig. 4-64 INCOS search in the NIST library gives the suggestions Ethion and Terbufos

with the pesticide library gives the search results Mirex, Heptachlor and Chlordecon with FIT values above 750, in spite of the highly contaminated spectrum (Fig. 4-66). The very low RFIT values are a result of the overlapping signals. Here also the retention index is a very useful tool for identification (Mirex 2618, Heptachlor 1912, Chlordecon 2330).

Alachlor was recovered in higher concentrations in the sample than were originally fortified. The two dominant fragments of Alachlor show the active substance in the mass chromatograms very selectively (Fig. 4-67). A search using the NIST library gives the search results Alachlor and Butachlor. It is easily possible to differentiate between them on the basis of their fragmentation patterns (Fig. 4-68) and this is confirmed by the results of the CI analysis (Fig. 4-69).

The elution of Endosulfan is an example of a frequently occurring situation whereby a trace component is overlapped by an intense peak (Fig. 4-70). While in this case the evaluation of the EI analysis only gives an indication because of the fragmentation of the Endosulfan, the active substance can be identified with certainty by chemical ionisation (Fig. 4-71). The quasimolecular ion m/z 405 formed by protonation shows Endosulfan on the mass trace under a high peak in the total ion current. Comparison of the CI library of pesticides gives a corresponding suggestion for the CI spectrum (Fig. 4-72).

All 19 pesticides in the fortified tea sample were able to be identified using EI and CI data with the SFE GC/MS method described (Table 4-6). For samples of this type containing large quantities of matrix, data acquisition of EI and CI chromatograms of a sample have proved useful.

Reference: P. Brand, J. Niebel, Axel Semrau GmbH & Co. Applikationslabor, Sprockhövel.

4.14 Pesticides in Tea   399

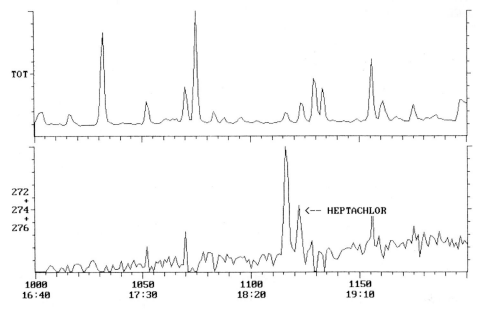

Fig. 4-65 Heptachlor: co-elution of matrix components

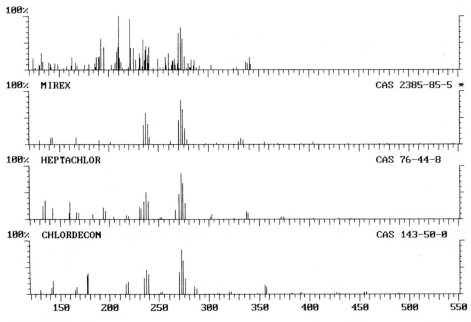

Fig. 4-66 INCOS search in the pesticide library with the suggestions Mirex, Heptachlor and Chlordecon

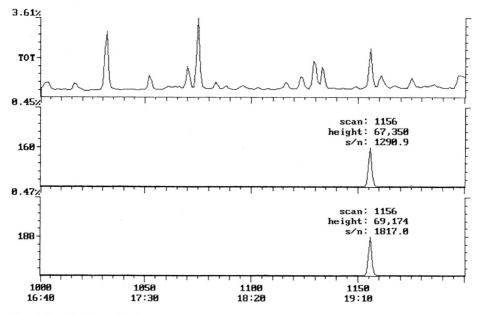

Fig. 4-67 Alachlor: selective mass traces m/z 160, 188 and total ion current (TIC)

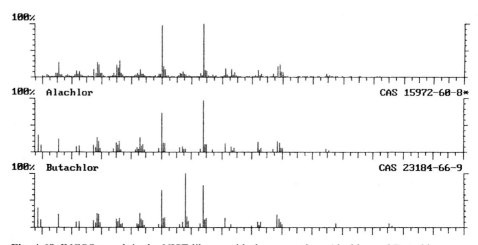

Fig. 4-68 INCOS search in the NIST library with the suggestions Alachlor and Butachlor

4.14 Pesticides in Tea

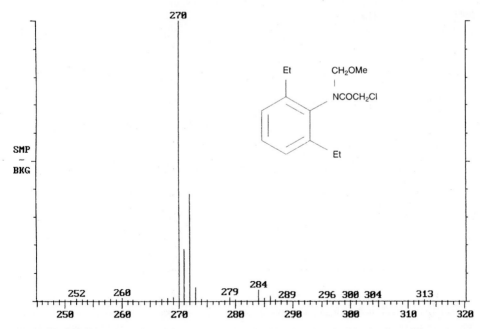

Fig. 4-69 Confirmation of the identification of Alachlor by chemical ionisation with methanol: m/z 270 as a protonated molecular ion, isotope pattern of a Cl atom

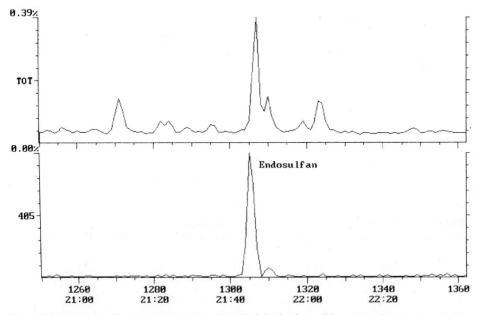

Fig. 4-70 Analysis of a tea extract using chemical ionisation with methanol leads to selective detection of Endosulfan (M 404) on the mass trace m/z 405

402   4 Applications

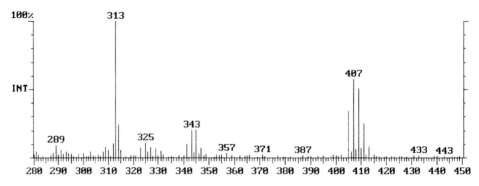

Fig. 4-71 CI spectrum (methanol) of the Endosulfan peak from Fig. 4-70

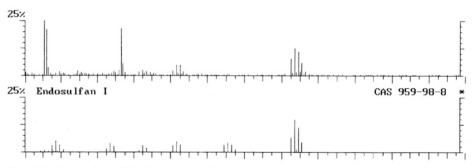

Fig. 4-72 Comparison of the CI spectrum from Fig. 4-71 with a collection of CI pesticide spectra

Table 4-6 Active substance concentrations in the fortified tea sample

Active substance	Concentration [pg/µl]	Active substance	Concentration [pg/µl]
Alachlor	22	HCB	7
o,p-DDE	22	α-HCH	11
p,p-DDE	22	β-HCH	7
p,p-DDD	22	Heptachlor	11
o,p-DDT	29	Lindane	11
p,p-DDT	22	Quintozen	11
Dicofol	40	S 421	35
Dieldrin	26	Tecnacen	17
α-Endosulfan	11	Tetradifon	30
Ethion	28		

4.15 GC/MS/MS Target Compound Analysis of Pesticide Residues in Difficult Matrices

Key words: pesticides, residue analysis, selectivity, MS/MS, confirmation, quantitation

The analysis of biological materials, such as plant and animal matter, for pesticide residue compounds using GC/MS has traditionally been difficult due to the very complex sample matrix. Approaches such as ECD and NPD detection, SIM, large volume injection or more sample concentration has been used for increasing the sensitivity of the method. But in matrices such as these, selectivity, another trace level requirement, is often the limiting factor. Simply injecting more sample necessarily means injecting more sample matrix and chemical background. The SIM approach, while enhancing the sensitivity at the expense of full scan mass spectral information, increases the monitored signal of ions common to both the background and the analyte within the retention time window. Real improvements in trace level complex matrix analyses depend on improving the detectivity (S/N) of the method by improving both the sensitivity and the selectivity.

The use of MS/MS in GC detection eliminates the matrix that causes the difficulty when using single stage MS. Thus, any background ions that may have been at the fragment ion m/z values are removed. The resultant product ion spectrum at the retention time is in presence and abundance due entirely to the m/z value of the selected pesticide precursor ion and not to the chemical background. Through this process the chemical background and matrix have been eliminated and low level detection is enhanced.

The samples used for the assay were fish tissue and river water from an industrial region. Aliquots of the homogenised fish tissue (typically 10 g) are treated with 100 g anhydrous sodium sulfate and extracted three times with 100 ml methylene chloride. The extract portions are combined, concentrated and purified using gel chromatography. Extracts are further cleaned using either a deactivated alumina/silica gel column or a Maxi Clean Florisil PR Cartridge (Alltech) and solvent elution. River water samples have been filtered and preconcentrated using C_{18}-SPE cartridges with standard methods described elsewhere.

For data acquisition multiple MS/MS scan events are time programmed as known from SIM acquisitions throughout the chromatographic run. Individual scan events are automatically set at the appropriate times so that as different target analytes elute, the conditions necessary for their optimal determination are activated. Precursor ions are usually chosen using the known SIM masses. MS/MS fragmentation is achieved using initial standard values with little individual response optimisation of the activation voltage.

Analysis Conditions

GC/MS system:	Finnigan GCQ plus (Polaris)
Injector:	split/splitless injector
	4 mm ID deactivated glass liner
	(for organochlorine pesticides with a 1 cm glass wool plug 3 cm from top)
	splitless injection, split valve open after 1 min at 50 ml/min
	injector temperature 300 °C
Autosampler:	CE Instruments AS2000
	Sample volume: 1 µl
	Air plug: 2 µl
	Hold time before injection: 2 s, after injection 0 s
GC column:	J&W Scientific DB-XLB, 30 m × 0,25 mm × 0,25 µm
Flow:	constant flow at 1.5 ml/min (ca. 40 cm/s)
Temp. program:	60 °C for 1 min
	40 °C/min to 150 °C, 1 min
	5 °C/min to 300° C, 4.75 min
Transfer line:	direct coupling to MS ion source
	300 °C constant
Mass spectrometer:	EI mode
	ion source at 200 °C
MS/MS conditions:	organochlorine pesticides, see Table 4-7
	NP-pesticides, see Table 4-8

Table 4-7 MS/MS acquisition parameters for organochlorine pesticides

Compound Name	Acquisition Mode	Precursor Ion m/z	Excitation Voltage V	q Value	Product Scan Range m/z
4,4'-DDD	SRM MS/MS	235	0.80	0.225	40–440
4,4'-DDE	SRM MS/MS	318	0.80	0.225	40–440
4,4'-DDT	SRM MS/MS	235	0.80	0.225	40–440
Alachlor	SRM MS/MS	188	0.80	0.225	40–440
Aldrin	SRM MS/MS	293	0.80	0.225	40–440
α-BHG	SRM MS/MS	181	0.80	0.225	40–440
α-Chlordane	SRM MS/MS	373	0.80	0.225	40–440
β-BHC	SRM MS/MS	181	0.80	0.225	40–440
cis-Nonachlor	SRM MS/MS	407	0.80	0.225	40–440
Dieldrin	SRM MS/MS	279	0.80	0.225	40–440
Endosulfan I	SRM MS/MS	339	0.60	0.225	40–440
Endosulfan II	SRM MS/MS	339	0.60	0.225	40–440
Endosulfan sulfate	SRM MS/MS	387	0.65	0.225	40–440
Endrin	SRM MS/MS	281	0.80	0.225	40–440
Endrin aldehyde	SRM MS/MS	345	0.65	0.225	40–440
γ-BHC (Lindane)	SRM MS/MS	181	0.80	0.225	40–440
γ-Chlordane	SRM MS/MS	373	0.80	0.225	40–440
Heptachlor	SRM MS/MS	272	0.80	0.225	40–440
Heptachlor epoxide	SRM MS/MS	353	0.70	0.225	40–440
Methoxychlor	SRM MS/MS	227	0.90	0.225	40–440

4.15 GC/MS/MS Target Compound Analysis of Pesticide Residues

Table 4-8 MS/MS acquisition parameters for nitrogen/phosphorus pesticides

Compound Name	Retention Time min:s	Acquisition Mode	Precursor Ion m/z	Excitation Voltage V	q Value	Product Scan Range m/z
Dichlorvos	9:10	SRM MS/MS	185.00	1.20	0.33	70–200
Mevinphos	11:18	SRM MS/MS	192.00	0.80	0.33	100–200
Heptenophos	13:31	SRM MS/MS	215.00	1.10	0.33	80–220
Propoxur	13:84	SRM MS/MS	152.00	1.00	0.33	60–155
Demeton-S-methyl	13:99	SRM MS/MS	142.00	0.90	0.33	50–150
Ethoprophos	14:17	SRM MS/MS	200.00	0.90	0.33	100–210
Desethylatrazin	14:65	SRM MS/MS	172.00	1.10	0.33	80–180
Bendiocarb	14:86	SRM MS/MS	166.00	0.80	0.33	80–170
Tebutan	14:95	SRM MS/MS	190.00	0.90	0.33	80–200
Dimethoate	15:73	SRM MS/MS	125.00	0.90	0.33	50–130
Simazine	15:96	SRM MS/MS	201.00	0.90	0.33	100–210
Atrazine	16:14	SRM MS/MS	200.00	1.30	0.33	100–210
Terbumeton	16:35	SRM MS/MS	169.00	1.00	0.33	70–175
Terbutylazine	16:62	SRM MS/MS	214.00	1.30	0.33	100–220
Dimpylate	17:03	SRM MS/MS	304.00	0.90	0.33	130–310
Terbazil	17:29	SRM MS/MS	161.00	1.20	0.33	80–170
Etrimfos	17:55	SRM MS/MS	292.00	0.90	0.33	150–300
Pirimicarb	17:95	SRM MS/MS	238.00	0.80	0.33	100–240
Desmethryn	18:23	SRM MS/MS	213.00	0.95	0.33	100–220
Ametryn	18:98	SRM MS/MS	227.00	0.90	0.33	100–230
Prometryn	19:10	SRM MS/MS	241.00	0.90	0.33	150–250
Terbutryn	19:56	SRM MS/MS	185.00	1.00	0.33	150–310
Pirimiphosmethyl	19:71	SRM MS/MS	305.00	0.75	0.33	60–190
Cyanazine	20:48	SRM MS/MS	225.00	1.10	0.33	60–230
Penconazole	21:62	SRM MS/MS	248.00	1.10	0.33	130–250
Triadimenol	22:05	SRM MS/MS	128.00	1.00	0.33	60–130
Methidathion	22:46	SRM MS/MS	145.00	0.80	0.33	60–150
Fluazifob-buthyl	24:63	SRM MS/MS	383.00	1.00	0.33	150–390
Oxadixyl	25:41	SRM MS/MS	163.00	0.80	0.33	60–170
Benalaxyl	26:19	SRM MS/MS	266.00	0.90	0.33	85–270
Hexazinon	26:95	SRM MS/MS	171.00	1.20	0.33	80–180
Azinphosmethyl	29:61	SRM MS/MS	132.00	1.10	0.33	60–140
Fenarimol	30:55	SRM MS/MS	139.00	1.00	0.33	60–145
Pyrazophos	30:83	SRM MS/MS	265.00	0.90	0.33	100–280

Results

Several commercially available mixes of pesticide compounds were used for calibration of the system by serial dilution and external standard techniques. A least squares regression was applied to curve fit the calibration data. Calibration curves were linear from below 0.01 ng/µl to 10 ng/µl (organochlorine pesticides) with correlation coefficients from 0.994 to 0.999.

The first example illustrates in Figs. 4-73 to 4-75 the determination of two Chlordane isomers. As can be seen in Fig. 4-73 the single stage technique, whether monitoring the

406   4 Applications

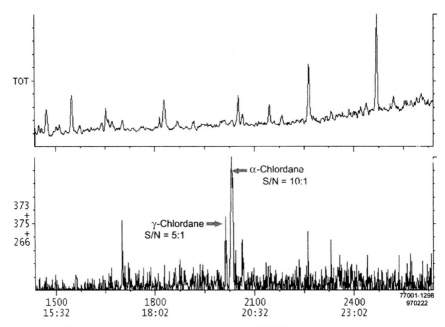

Fig. 4-73 Single stage MS analysis of fish extract 998 F2

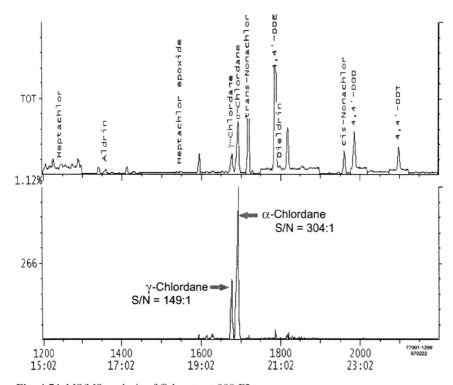

Fig. 4-74 MS/MS analysis of fish extract 998 F2

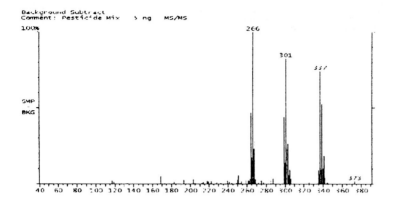

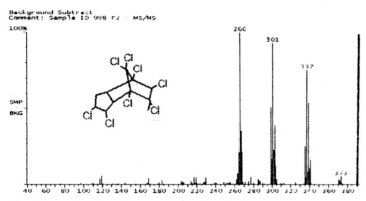

Fig. 4-75 Confirmation of α-Chlordane (the top spectrum represents the MS/MS library spectrum for reference)

total ion current (TIC) or three characteristic masses (m/z 373+375+266), can barely distinguish between the analyte signal and the complex matrix at this concentration level (550 ppb α-Chlordane in fish). MS/MS eliminates the matrix difficulties and allows much clearer detection of the analyte (Fig. 4-74) with an additional full product ion mass spectrum for confirmation (Fig. 4-75). Other compounds detected and confirmed are labelled. An example of a calibration curve is given in Fig. 4-76 for α-Chlordane showing a correlation factor of 0.998 over seven calibration points.

Another example shows the determination of Captan (Fig. 4-77) representative of nitrogen and phosphorus pesticides in difficult matrices. The single stage GC/MS analysis of a river water sample shows a susceptible peak at retention time 16.12 min indicating the occurrence of Captan deterioration (Fig. 4-78). The corresponding mass spectrum is dominated by chemical background ions (Fig. 4-79). The specific ions of Captan can be found within at lower levels. Mass m/z 149 is not diagnostic as there may be a contribution from ubiquitously occurring phthalate plasticisers. The high ion intensity of m/z 299 cannot be explained from the molecular ion of Captan in this case.

Using m/z 149 as the precursor ion for the MS/MS determination, the unambiguous confirmation of Captan in the sample at high S/N (signal/noise) levels is achieved (Fig.

408   4 Applications

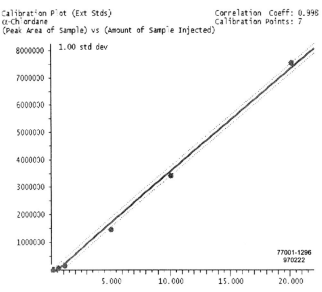

Fig. 4-76 α-Chlordane calibration curve (corr. coefficient 0.998)

4-80). The resulting product ion spectrum gives clear evidence of the structure-specific ions of Captan as proof (Fig. 4-81).

The retention and acquisition parameters of common nitrogen- and phosphorus-containing pesticides (phosphoric acid esters, atrazines, etc.) are listed in Table 4-7. It is recommended that the excitation parameters are used initially with equal standard values (0.9 V), which may be slightly adjusted in the case of response optimisation. The quality of the product ion spectra is not affected (Fig. 4-82). A typical standard run in MS/MS acquisition mode is shown in Figs. 4-83 and 4-84. Even at the low 10 pg/ul level the compounds are detected with high S/N values and full MS/MS product spectra for confirmation (Fig. 4-82).

The use of MS/MS techniques can provide a very reliable and indeed essential means of analysis in the determination of target compounds in complex matrices. The improved selectivity of the analysis from the MS/MS process, along with the inherent sensitivity of the ion trap mass spectrometer can often result in an order of magnitude or more of improvement in the signal to noise performance with related dramatic improvements in limits of detection, over single stage full scan or SIM methodologies.

Reference: J. Gummersbach, ThermoQuest Applications Laboratory, Egelsbach, Germany

4.15 GC/MS/MS Target Compound Analysis of Pesticide Residues

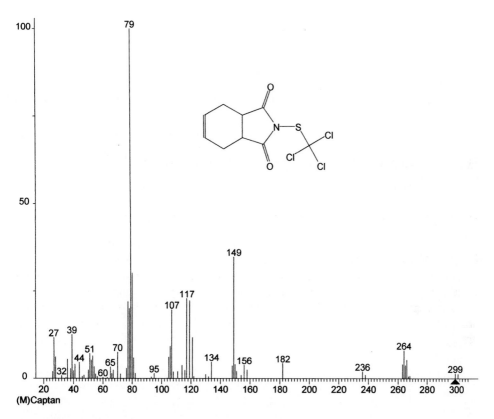

Fig. 4-77 Captan – EI Spectrum with formula (M 299)

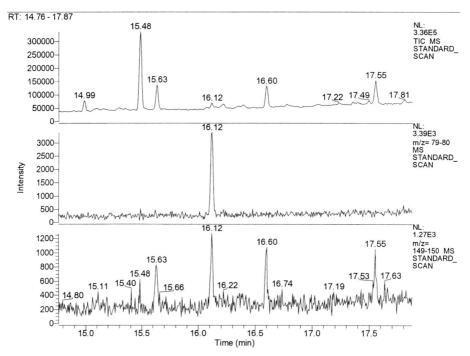

Fig. 4-78 Single stage MS analysis of river water sample with Captan mass chromatograms m/z 79, 149

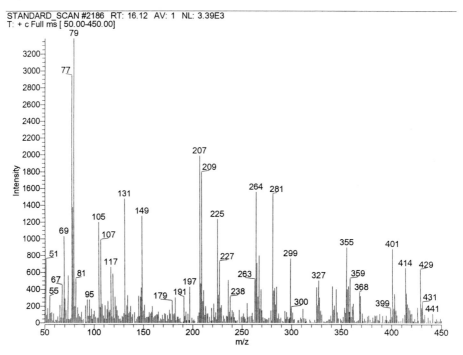

Fig. 4-79 Mass spectrum of GC peak eluting 16.12 min – Captan? Specific ions m/z 79, 107, 149 are within chemical background (m/z 149 may result from phthalate plasticiser)

4.15 GC/MS/MS Target Compound Analysis of Pesticide Residues 411

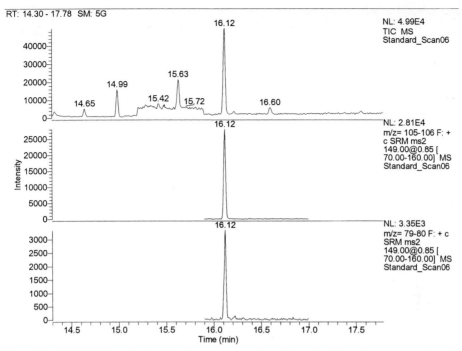

Fig. 4-80 GC/MS/MS analysis of a river water sample using m/z 149 as the precursor ion with Captan product ion chromatograms m/z 105, 79

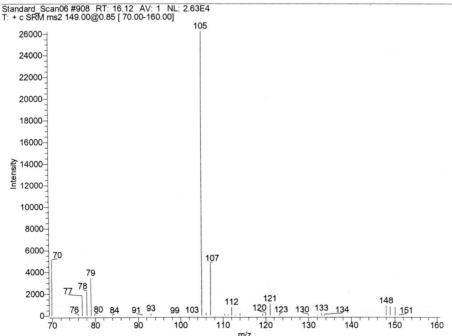

Fig. 4-81 MS/MS Spectrum of GC peak eluting at 16.12 min (m/z 149 as precursor ion) – Confirmation for Captan with structure specific ions m/z 79, 105, 107 free of chemical background

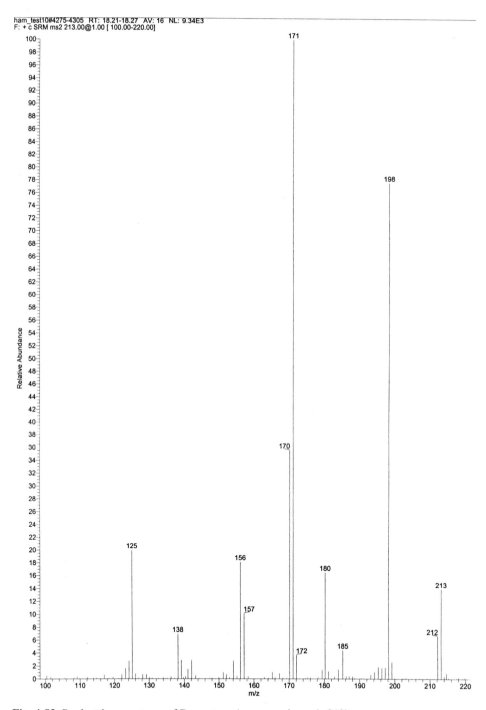

Fig. 4-82 Product ion spectrum of Desmetryn (precursor ion m/z 213)

4.15 GC/MS/MS Target Compound Analysis of Pesticide Residues

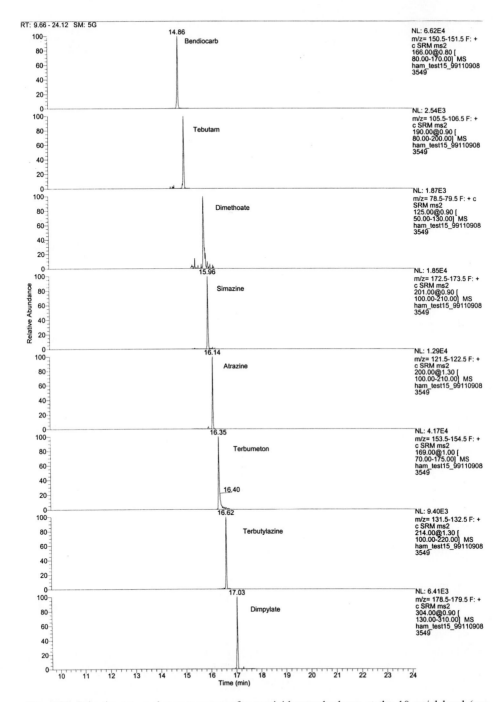

Fig. 4-83 Selective mass chromatograms of a pesticide standard run at the 10 pg/ul level (see Table 4-7, part 1)

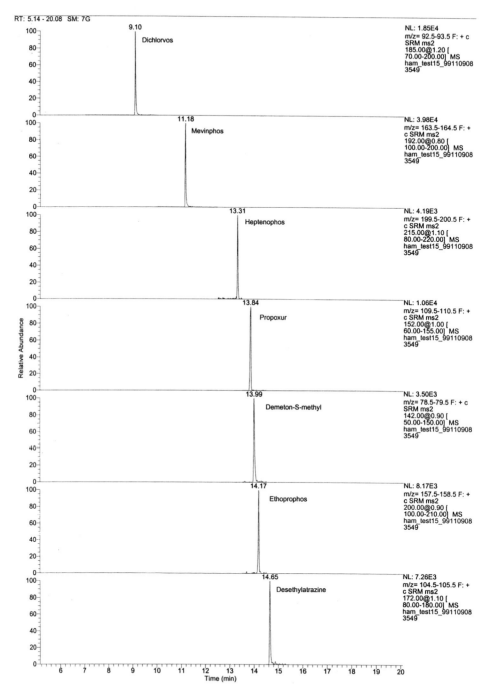

Fig. 4-84 Selective mass chromatograms of a pesticide standard run at the 10 pg/ul level (see Table 4-7, part 2)

4.16 Triazine Herbicides in Drinking and Untreated Water

Key words: herbicides, triazines, SPE, threshold values, drinking water, internal standard, automatic evaluation

The European ordinance on drinking water prescribes threshold values for the concentration of individual pesticide active substances of 0.1 µg/l and, based on the total of all the active substances, of 0.5 µg/l. Triazine herbicides have become recognised because the active substances Atrazine and Simazine, which are mainly used in the cultivation of maize, have significantly exceeded the threshold values. NPD is used as the standard detector for triazine herbicides in GC. The false identification of, for example, Propazine as tris(2-chloroethyl) phosphate (a flameproofing agent from paints and varnishes) cannot be established with NPD through the retention time comparison alone. The confirmation of positive identification is therefore fundamentally necessary.

Analysis Conditions

Sample preparation: 200 ml water are drawn through a purified and conditioned RP_{18} solid phase cartridge (SPE) at pH 6–8 and eluted with 2 × 2 ml acetone. A florisil column can be connected after the cartridge for purification.

GC/MS system: Finnigan MAGNUM
Injector: PTV cold injection system, splitless injection
Injector program: Start: 40 °C
 Program: 100 °C/min
 Final temp.: 280 °C

Guard column: empty column, deactivated, 5 m × 0.53 mm
Column: Permabond SE-54, 25 m × 0.25 mm × 0.50 µm
Temp. program: Start 40 °C, 2 min
 Program 1: 16 °C/min to 150 °C
 Program 2: 10 °C/min to 200 °C
 Program 3: 8 °C/min to 275 °C
 Final temp.: 275 °C, 30 min

Mass spectrometer: Scan mode: EI, full scan, 60–400 u
 Scan rate: 0.5 s/scan

Calibration: The internal standard, Dipropetryn, is added to SPE sample preparation and hexadecanoic acid nitrile (HDN) after the extraction; calibration is carried out over the whole procedure including SPE.

Results

The list of active substances covered by the analysis method described is given in Fig. 4-85 in order of the retention times. Naphthalene-d_8, Dipropetryn and hexadecanoic acid nitrile (HDN) are used as internal standards. Bromazil is given with two entries and different quantifying masses in order to cover possible interference with peak integration by matrix effects. There is an entry in the calibration file for each substance (Fig. 4-86). The file contains the retention time of the substance, the mass spectrum and the selective quantitation mass for automatic evaluation. Calculation of concentrations is carried out using the appropriate calibration function (Fig. 4-87).

The section of a standard chromatogram shows the elution sequence of the active substances with good component separation (Fig. 4-88). Overlapping peaks, such as Desisopropylatrazine (4) and Trifluralin (5) are differentiated completely because of their differing mass signals. The elution sequence of the substances Simazine, Atrazine, Propazine and Terbutylazine is shown together with their mass spectra in Fig. 4-89.

To work out the analysis data the most recent calibration file is used (see Section 3.3). A spectroscopic comparison is first carried out in a time window around a given retention time. If an active substance is recognised from the spectrum, the peak is integrated on the mass trace of a selective fragment and the concentration is calculated using the calibration function. With automatic evaluation informative documentation of the procedures is necessary. A typical print-out of the analysis documentation records the current conditions (Fig. 4-90). On the right-hand side of the figure the spectroscopic comparison between the two most recent analyses is shown: above is the substance spectrum of the analysis being evaluated (sample) and below is the substance spectrum of the last testing of the calibration (standard enhanced). For spectra actually measured a difference is calculated which can indicate an unexpected interference in the chromatogram. On the left-hand side of the figure the active substance and the calculated concentration are given. At the bottom the substance peaks of the active substance and the standard used are shown in the selective mass chromatogram. If other substances elute,

1	Naphthalene-d_8	Active	I	9:29
2	Dipropetryn	Active	I	18:14
3	HDN	Active	I	18:37
4	Desisopropylatrazine	Active	A	14:31
5	Trifluralin	Active	A	14:37
6	Desethylatrazine	Active	A	14:39
7	Desethylterbutylazine	Active	A	14:52
8	Simazine	Active	A	15:37
9	Atrazine	Active	A	15:42
10	Propazine	Active	A	15:47
11	Terbutylazine	Active	A	16:01
12	Sebutylazine	Active	A	16:43
13	Vinclozoline	Active	A	17:21
14	Bromazil (205)	Active	A	18:05
15	Bromazil (207)	Active	A	18:05
16	Metolachlor	Active	A	18:21
17	Cyanazine	Active	A	18:25
18	Pendimethalin	Active	A	19:09
19	Metazachlor	Active	A	19:14
20	Hexazinon	Active	A	22:49

Fig. 4-85 List of herbicides which are checked using automatic evaluation

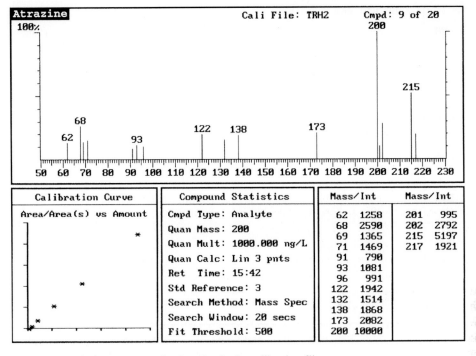

Fig. 4-86 Analysis parameters for Atrazine in the calibration file

which are not given in the list of active substances searched for, the mass spectra of these additionally appearing substances are extracted by the evaluation routine and identified by library comparison (Fig. 4-91). With this automatic evaluation a manual control can also be carried out if required.

The triazine herbicides can also be effectively analysed by chemical ionisation. Their EI spectra have a particularly large number of lines because of extensive fragmentation (Fig. 4-92). The distribution of the ion current among many fragments limits the detectability among the background present in matrix samples. Chemical ionisation prevents extensive fragmentation and gives the quasimolecular ion $(M+H)^+$ as the only intense signal in the CI spectrum (Fig. 4-93). As the ion current is concentrated on a single mass in the CI spectrum, a much higher signal/noise ratio can be achieved. In the case of Simazine an increase from S/N 171 in the EI mode to S/N 1034 in the CI mode is possible. At the same time the selectivity of the detection increases, as shown in the mass chromatograms of the quasimolecular ion m/z 202 (Fig. 4-93) compared with the EI spectrum with m/z 201 (Fig. 4-92).

Reference: H. Korpien, M. Krill, Dr. K. Post, Stadtwerke Frankfurt am Main, Betriebszweig Wasser, Frankfurt am Main

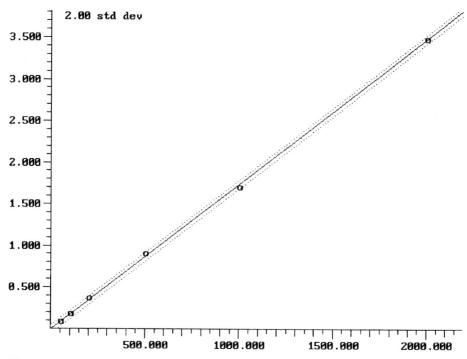

Fig. 4-87 Linear calibration function for Atrazine from 50 ng/l to 2000 ng/l

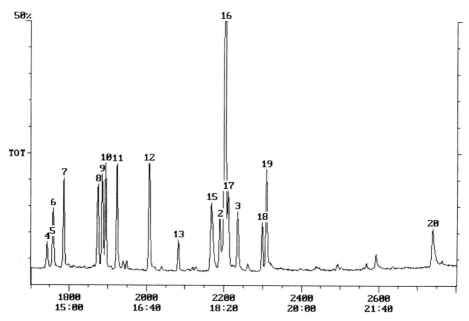

Fig. 4-88 Typical chromatogram of the standard showing the active substance number from the calibration file (Fig. 4-85)

4.16 Triazine Herbicides in Drinking and Untreated Water

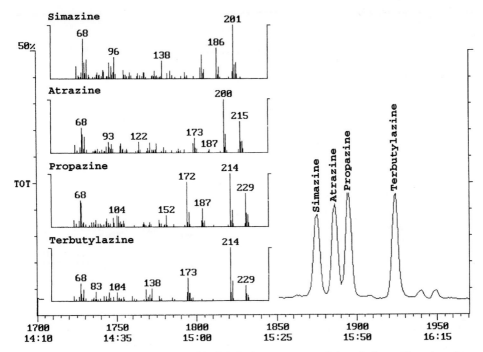

Fig. 4-89 Part of a chromatogram with the elution sequence of the triazines: Simazine, Atrazine, Propazine, Terbutylazine

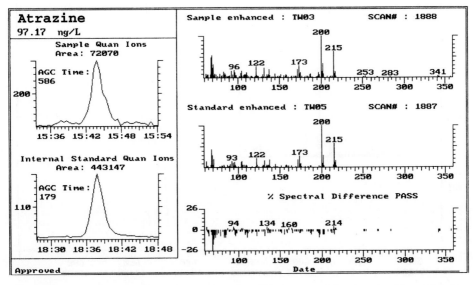

Fig. 4-90 Documentation of evaluation using the substance report for Atrazine (for explanation see text)

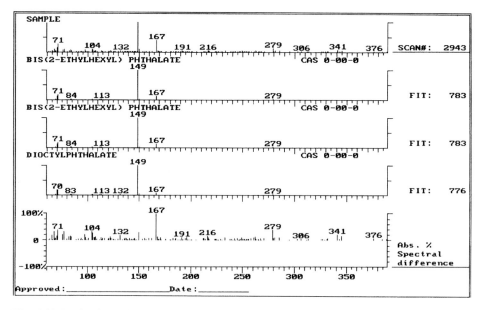

Fig. 4-91 Evaluation of additional substances appearing in an analysis by library comparison (UAR, unknown analytical response)

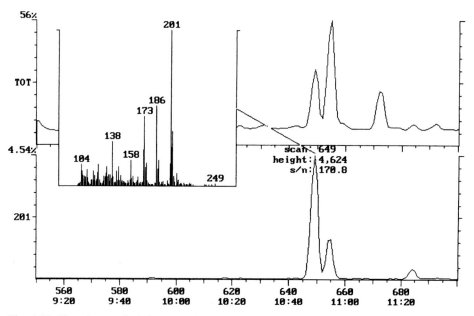

Fig. 4-92 Simazine: analysis in EI mode. Signal/noise ratio 171:1 on the mass trace of the molecular ion m/z 201

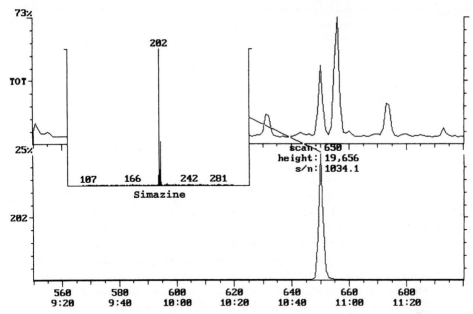

Fig. 4-93 Simazine: analysis in CI mode (isobutane). Signal/noise ratio 1034:1 on the mass trace of the quasimolecular ion m/z 202

4.17 Nitrophenol Herbicides in Water

Key words: herbicides, nitrophenols, rotational perforator, methylation

The determination of nitrophenols increased in importance after this substance class was suspected of being one of the causes of the death of forests. Nitrophenols and their derivatives are determined using GC-NPD or GC/MS. The analytes are first extracted with a pentane/ether mixture and then methylated with diazomethane. The recoveries lie in the range 70–100%. The limit of detection of the procedure is 0.25 µg/l.

Analysis Conditions

Sample preparation: liquid/liquid extraction in Brodesser low density rotation perforator (Fig. 4-94), methylation with diazomethane.

Gas chromatograph: Carlo Erba Mega
Injector: on-column, 7.5 µl injection volume
Guard column: retention gap, length 5 m, ID 0.32 mm, deactivated
Column: J&W DB-5, 30 m × 0.25 mm × 0.25 µm
Helium 2.5 ml/min, 1.2 bar

Program: Start: 70 °C, 1 min
                 Program 1:   10 °C/min to 130 °C
                 Program 2:   20 °C/min to 270 °C
                 Final temp.:   270 °C, 10 min

Mass spectrometer: Finnigan ITD 800 with direct coupling
                     Scan mode: EI, full scan, 50–400 u
                     Scan rate: 1 s/scan

Results

Bromoxynil, Ioxynil, 2,6-dimethyl-4-nitrophenol and DNOC can be assigned unambiguously from their characteristic molecular ions (Fig. 4-95). The evaluation of the chromatograms is carried out using mass chromatograms of selective ion signals (Fig. 4-96).

The procedure described is suitable for detecting and quantifying nitrophenols in aqueous media. Excess of the derivatisation agent leads to interference in the form of peak broadening arising from precipitation in the retention gap. While nitrophenol detection is unaffected by interference by background, contaminants formed by side reactions of the sample contents are detected in the GC/MS system when operating in the full scan mode.

Reference: Prof. H. F. Schöler, Institut für Sedimentforschung, Heidelberg University.

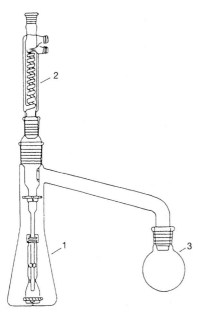

Fig. 4-94 Ludwig light phase rotation perforator for liquid/liquid extraction with specific light solvents.
(1) Extraction vessel
(2) Condenser
(3) Solvent storage

4.17 Nitrophenol Herbicides in Water 423

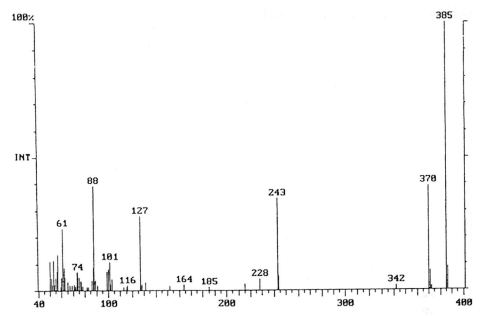

Fig. 4-95 Mass spectrum of Ioxynil methyl ether

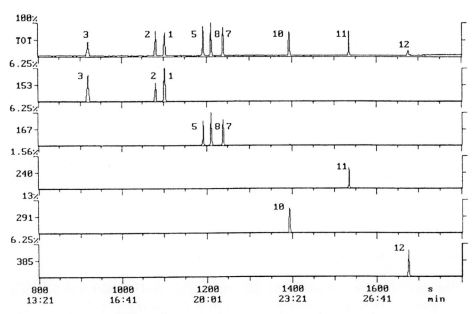

Fig. 4-96 Chromatogram of nitrophenols (as methyl ethers) in the total ion current and selective mass chromatogram (identification see Table 4-9)

Table 4-9 List of nitrophenols investigated (as methyl ethers)

No.	Nitrophenol derivative	Molecular mass
1	4-Nitrophenol, methyl ether	153
2	2-Nitrophenol, methyl ether	153
3	3-Nitrophenol, methyl ether	153
4	2-Methyl-3-nitrophenol, methyl ether	167
5	3-Methyl-4-nitrophenol, methyl ether	167
6	4-Methyl-3-nitrophenol, methyl ether	167
7	5-Methyl-2-nitrophenol, methyl ether	167
8	4-Methyl-2-nitrophenol, methyl ether	167
9	Dinoseb, methyl ether	240
10	Dinoterb, methyl ether	240
11	Bromoxynil, methyl ether	291
12	Ioxynil, methyl ether	385
13	2,4-Dinitrophenol, methyl ether	198
14	3,4-Dinitrophenol, methyl ether	198
15	2,5-Dinitrophenol, methyl ether	198
16	2,6-Dinitrophenol, methyl ether	198
17	4,6- Dinitro-o-cresol, methyl ether	212
18	2,6- Dimethyl-4-nitrophenol, methyl ether	181

4.18 Dinitrophenol Herbicides in Water

Key words: herbicides, dinitrophenols, drinking water, surface water, SPE, methylation, chemical ionisation

This method was drawn up for the determination of 2,4-dinitrophenol herbicides in drinking and untreated water, taking into account the threshold values for drinking water control. Dinitrophenols and their derivatives are particularly important in drinking water treatment because of the toxicity of their metabolites (aminonitro- and diaminophenols). The following belong to this group of substances (Table 4-10):

Type A (free phenol): Dinoseb, Dinoterb, DNOC
Type B (esters): Dinoseb acetate, Dinoterb acetate, Binapacryl, Dinobuton

The active substances of type A are determined as the methyl ethers after treatment with diazomethane in the cartridge of the solid phase extraction before elution.

Analysis Conditions

Sample preparation: adsorption on glass cartridges with Carbopack B (120–400 µm, 250 mg, Supelco), conditioning with 10 ml methanol/acetic acid 1:1, 10 ml ethyl acetate, 2 ml methanol, 10 ml doubly distilled water,
addition of 0.9 ml conc. HCl to 250 ml water sample, adsorption with ca. 4 ml/min flow rate, then drying with N_2
derivatisation with 2 ml diazomethane in t-butyl methyl ether directly on the SPE cartridge (!), 30 min reaction time,
elution with 5 ml ethyl acetate, concentration under N_2 to 100 µl final volume.

GC/MS system: Finnigan MAGNUM with autosampler CTC A 200 S
Injector: PTV cold injection system,
Injector program: Start: 80 °C
Program: 300 °C/min to 250 °C
Column: J&W DB-5, 30 m × 0.25 mm × 0.25 µm
Program: Start: 70 °C
Program: 12 °C/min to 250 °C

Mass spectrometer: Scan mode: EI, full scan, 50–450 u
CI, full scan, 50–600 u, reagent gas methanol
Scan rate: 1 s/scan

Table 4-10 Structures of 2,4-dinitrophenol herbicides

Compound	R1	R2
DNOC	-OH	-methyl
Dinoterb	-OH	-t-butyl
Dinoseb	-OH	-s-butyl
Dinoseb acetate	-acetate	-s-butyl
Dinoterb acetate	-acetate	-t-butyl
Dinobuton	-carbonate	-s-butyl
Binapacryl	-acrylate	-s-butyl
2,4-Dinitrophenol	-OH	-H

Results

Because of the acidity of the compounds of type A, derivatisation with diazomethane to the methyl ethers is recommended for GC separation (Figs. 4-97 and 4-98). Dinoseb and Dinoterb are differentiated by means of their EI spectra (Figs. 4-99 and 4-100). If these compounds are analysed in the CI mode, the two substances must be differentiated using the retention time, as the CI mass spectra of both herbicides show the $(M+H)^+$ ion with m/z 255 (Figs. 4-101 and 4-102).

In chemical ionisation Dinoseb and Dinoterb acetates and Dinobuton show the protonated phenol fragments as the base peak. Dinobuton also shows the fragment m/z 177

(Table 4-11). Binapacryl does not form a quasimolecular ion with methanol as the reagent gas. In the analysis of compounds of type B Dinoseb and Dinoterb are formed as hydrolysis products and under certain circumstances can only be quantified as a total parameter (free phenols and esters). Hydrolysis can also occur during the analysis of compounds of type A.

Reference: F. Heimlich, H. Mayer, Dr. J. Nolte, ISAS Institut für Spektrochemie und angewandte Spektroskopie, Dortmund.
The results of this work were presented as a poster to the 8th IUPAC Congress of Pesticide Chemistry, Washington DC, 4–9th July 1994.

Table 4-11 Specific ions for EI/CI analyses of dinitrophenol herbicides (R_t retention time, M molecular mass)

Substance	R_t	M	EI [m/z (%)]	CI [m/z (%)]
DNOC, methyl ether	10:04	212	182(100), 165(70), 212(20)	213(100)
Dinoterb, methyl ether	11:25	254	239(100), 209(55), 254(20)	255(100)
Dinoseb, methyl ether	11:34	254	225(100), 195(70), 254(20)	255(100)
Dinoseb acetate	12:10	282	89(100), 211(55), 240(40)	241(100)
Dinoterb acetate	12:19	282	225(100), 177(50), 240(20)	241(100)
Dinobuton	13:33	326	211(100), 163(30), 205(60)	241(100), 177(20)
Binapacryl	15:07	322	83(100)	83(100)

4.18 Dinitrophenol Herbicides in Water

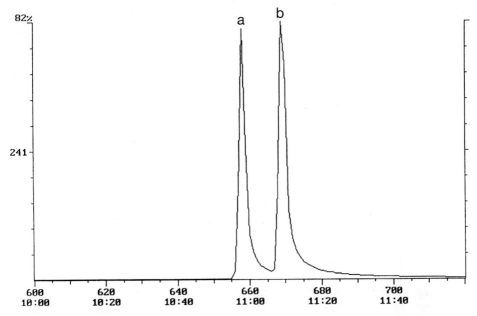

Fig. 4-97 The analysis of Dinoterb (a) and Dinoseb (b) shows severe peak tailing because of the acidity of the compounds

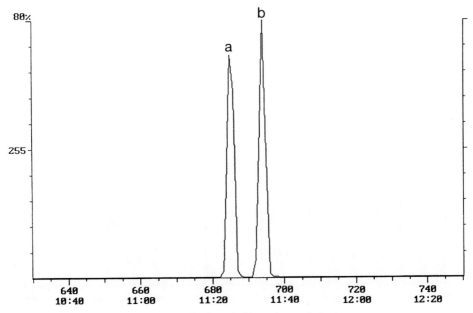

Fig. 4-98 Analysis of Dinoterb (a) and Dinoseb (b) as the methyl ethers

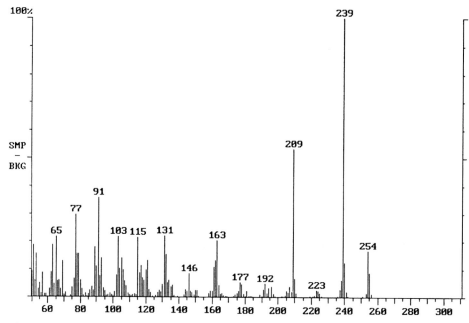

Fig. 4-99 EI spectrum of Dinoterb methyl ether

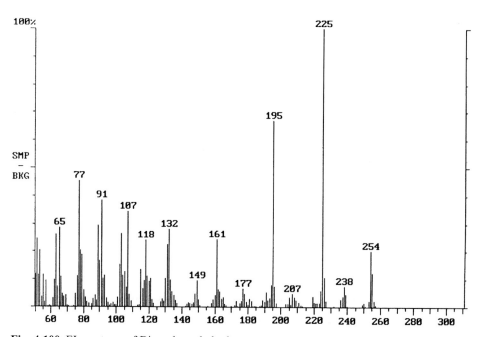

Fig. 4-100 EI spectrum of Dinoseb methyl ether

4.18 Dinitrophenol Herbicides in Water

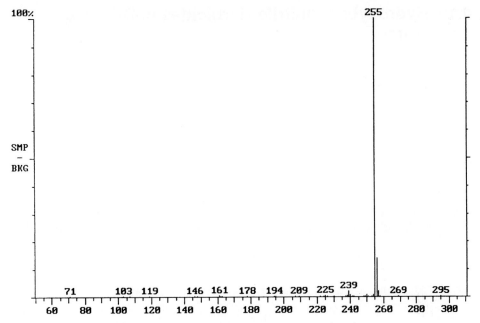

Fig. 4-101 CI spectrum of Dinoterb methyl ether, reagent gas methanol

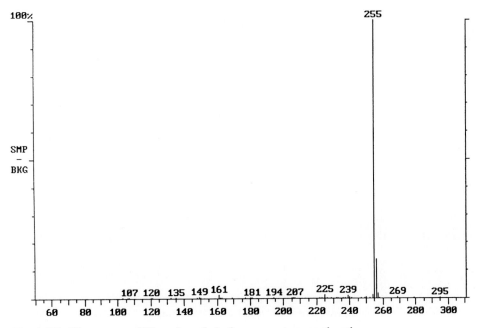

Fig. 4-102 CI spectrum of Dinoseb methyl ether, reagent gas methanol

4.19 Hydroxybenzonitrile Herbicides in Drinking Water

Key words: herbicides, dihalogenated hydroxybenzonitrile, drinking water, SPE, methylation, chemical ionisation

The method of determination of dihalogenated hydroxybenzonitrile herbicides was drawn up on the basis of the threshold values for the control of drinking water. During the development of a sensitive analysis procedure, the highly polar nature of these active substances in particular had to be taken into account. Quantitative desorption from the solid phase of the enrichment procedure (SPE) is made more difficult by their hydrophilic properties. Therefore in this procedure methylation is carried out on the SPE cartridge. The substances are determined as their methyl ethers after reaction with diazomethane (Table 4-12).

Table 4-12 Structures of the dihalogenated hydroxybenzonitriles and Mecoprop

Compound	Structure
Bromoxynil octyl ester Bromoxynil	NC–⟨Br,Br⟩–O–C(=O)(CH$_2$)$_6$CH$_3$ → NC–⟨Br,Br⟩–OH
Ioxynil	NC–⟨I,I⟩–OH
Chloroxynil	NC–⟨Cl,Cl⟩–OH
Mecoprop	Cl–⟨CH$_3$⟩–O–C(CH$_3$)(H)–C(=O)OH

Analysis Conditions

Sample preparation: adsorption on glass cartridges with Carbopack B (120–400 µm, 250 mg, Supelco), conditioning with 10 ml ethanol/acetic acid 1:1, 10 ml ethyl acetate, 2 ml ethanol, 2 ml doubly distilled water,
addition of 0.3 ml conc. HCl to 250 ml water sample, adsorption with ca. 4 ml/min flow rate, then drying with N_2,
derivatisation with 2 ml diazomethane in t-butyl methyl ether directly on the SPE cartridge (!), 30 min reaction time,
elution with 5 ml ethyl acetate, concentration under N_2 to 100 µl final volume.

GC/MS system: Finnigan MAGNUM with autosampler CTC A200S
Injector: PTV cold injection system,
Injector program: Start: 80 °C
Program: 300 °C/min to 250 °C
Column: J&W DB-5, 30 m × 0.25 mm × 0.25 µm
Program: Start: 70 °C
Program: 12 °C/min to 250 °C
Transfer line: 220 °C

Mass spectrometer: Scan mode: EI, full scan, 50–450 u
CI, full scan, 100–600 u, reagent gas methanol
Scan rate: 1 s/scan

Results

The herbicides are adsorbed well on the solid phase material. The methylation with diazomethane on the SPE cartridge is finished after ca. 30 min reaction time. The recoveries determined are 75% for Bromoxynil, 80% for Ioxynil, and 50% for Chloroxynil. For Mecoprop as the representative of the phenoxyalkylcarboxylic acids 95% recovery is achieved. Mass spectrometric detection using chemical ionisation with methanol as the reagent gas is better than EI detection (Fig. 4-103, Table 4-13). The limits of detection for Chloroxynil (Fig. 4-104) and Mecoprop (Fig. 4-105) are 10 ng/l and for Bromoxynil (Fig. 4-106) and Ioxynil (Fig. 4-107) 3 ng/l at a signal/noise ratio of 5. When using the method it should be noted that Bromoxynil can also appear as the degradation product of Bromoxynil alkylcarboxylic acid esters and of Bromfen oxime.

Reference: B. Grass, Dr. J. Nolte, ISAS Institut für Spektrochemie und angewandte Spektroskopie, Dortmund.
The results of this work were presented as a poster to the 24th International Symposium on Environmental Analytical Chemistry, Ottawa 16–19th May 1994.

432   4 Applications

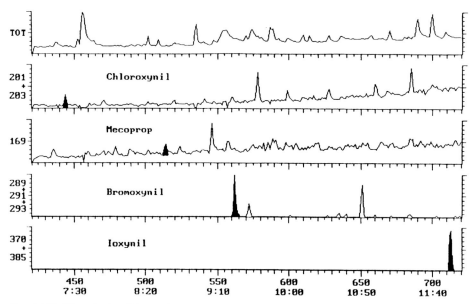

Fig. 4-103 EI analysis of the methyl ethers in a spiked water sample (10 ng/l)

Table 4-13 Specific ions for the EI/CI analysis of hydroxybenzonitrile herbicides and Mecoprop (R_t retention time, M molecular mass)

Substance	R_t	M	EI [m/z (%)]	CI [m/z (%)]
Chloroxynil methyl ether	7:23	201	201(100), 203(65)	202(100), 204(70)
Mecoprop methyl ester	8:33	228	228(100), 169(95), 107(90)	169(100), 229(35)
Bromoxynil methyl ether	9:22	291	291(100), 289(40), 293(45)	292(100), 290(50), 294(50)
Ioxynil methyl ether	11:52	385	385(100), 370(20)	386(100)

4.19 Hydroxybenzonitrile Herbicides in Drinking Water

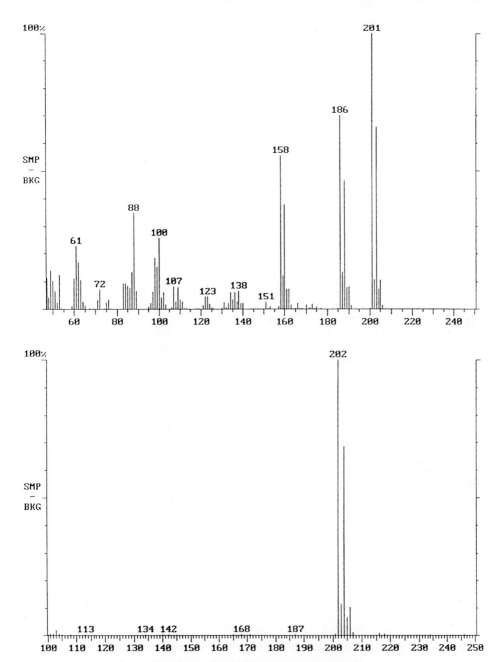

Fig. 4-104 Chloroxynil methyl ether (above: EI, below: methanol CI)

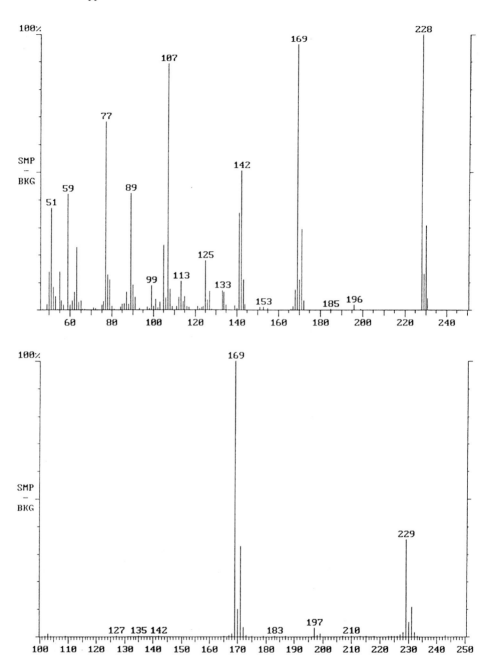

Fig. 4-105 Mecoprop methyl ether (above: EI, below: methanol CI)

4.19 Hydroxybenzonitrile Herbicides in Drinking Water

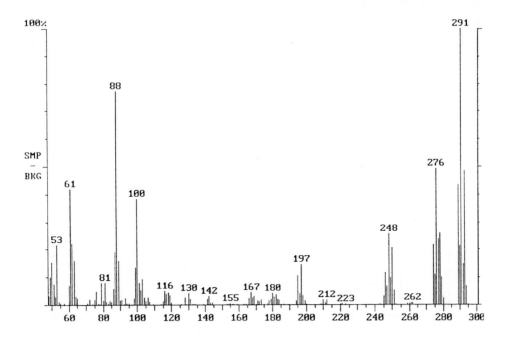

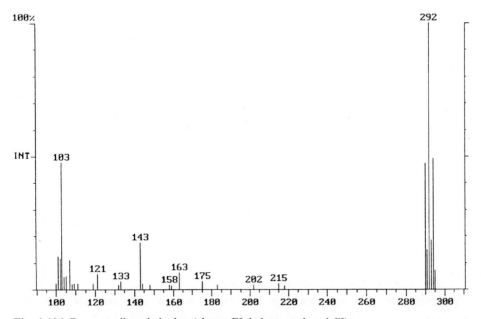

Fig. 4-106 Bromoxynil methyl ether (above: EI, below: methanol CI)

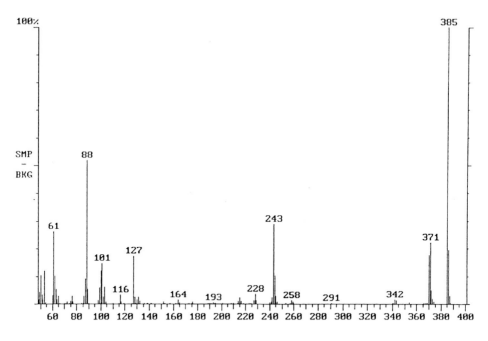
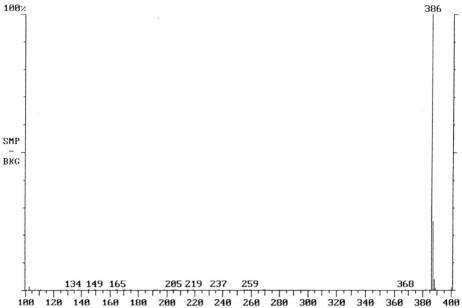

Fig. 4-107 Ioxynil methyl ether (above: EI, below: methanol CI)

4.20 Determination of Phenoxyalkylcarboxylic Acids in Water Samples

Key words: herbicides, water, residue analysis, SPE extractions, quantitation, automation

Phenoxyalkylcarboxylic acids are used as selective herbicides, in particular for combating dicotyledonous weeds in cereal crops after the seeds have come up. This class of active substances has been used for many years, for example with the well known post-emergence herbicides 2,4-D and 2,4,6-T. These substances are characterised by their high stability (in the dark) and water solubility. Because of their high polarity effective sample preparation and complete derivatisation of the extracts is very important in the development and execution of a multicomponent procedure.

The sample preparation is done via automated solid phase extraction (for aqueous samples). Extraction is carried out using RP-C_{18}-cartridges (e.g. Baker) using automatic clean-up with an Auto-Trace workstation. Up to six samples can be processed at the same time.

The RP-C_{18}-columns are conditioned as follows:

3 ml hexane
3 ml acetone
3 ml methanol
2 × 5 ml distilled water

The water sample (1 l) is adjusted to pH 2 with sulfuric acid and an internal standard (10 ml, 2,4-D-d_6, 10 ng/µl in methanol) is added. The sample thus prepared is applied to the SPE cartridge (6 ml/min). After enrichment the cartridge is rinsed with 3 ml distilled water, dried for 90 min in a stream of nitrogen (20 ml/min) and then eluted with 4 ml acetone (0.6 ml/min).

The extracts are derivatised by methylation with diazomethane. The acetone extract is concentrated to ca. 150 µl with nitrogen and ca. 1 ml diazomethane added until a yellow coloration remains. The sealed solution is kept for ca. 1 hour in the dark at room temperature. The solution is then concentrated to a volume of ca. 150 µl and made up to exactly 1.0 ml with n-hexane.

Analysis Conditions

GC/MC-system: Finnigan GC Q

Gas Chromatography:
Injector: 250 °C
Column: J&W DB5MS, 30 m × 0.25 mm × 0.25 µm
Temperature program: 50 °C, 1 min
                    8 °/min to 300 °C
                    300 °C, 5 min
Carrier gas: helium 5.5
Rate: 40 cm/s

Pressure programming: constant flow
MS coupling: direct
Transfer line: 275 °C

Mass Spectrometry:
Ionisation: EI, 70 eV
Mode: full scan
Mass range: 50–400 u
Scan rate: 0.5 s
Target value: 150
Threshold: 80 counts
High mass adjust: 85 %
Ion trap offset: 10 V
Tuning: autotune parameter

Calibration:
Standard: internal standards method
                        Standardisation during the whole procedure using 2,4-D-d$_3$
                        Addition of 10 µl standard solution to sample before clean-up corresponds to 100 ng/l ISTD
Calibration steps: 50, 100, 250, 500, 750 ng/l
Regression: linear over all the calibration steps

To draw up the calibration the dilutions given above are prepared and measured (Table 4-14).

Table 4-14 List of components (as methyl ethers) in the stock solution for calibration

No.	Active substance	Molecular formula	M	Rt	m/z (quan)	ISTD
1	2,4-D-d$_6$, ISTD	$C_3D_6H_8Cl_2O_3$	240	18:02	205	1
2	Dicamba	$C_9H_8Cl_2O_3$	234	16:14	188+203+234	1
3	Mecoprop	$C_{11}H_{13}ClO_3$	228	16:43	142+169+228	1
4	MCPA	$C_{10}H_{11}ClO_3$	214	17:01	141+155+214	1
5	Dichlorprop	$C_{10}H_{10}Cl_2O_3$	248	17:42	162+189+248	1
6	Bromoxynil	$C_8H_5Br_2NO$	289	18:01	248+276+291	1
7	2,4-D	$C_9H_8Cl_2O_3$	234	18:04	175+199+234	1
8	Triclopyr	$C_8H_6Cl_3NO_3$	269	18:53	212+210+269	1
9	Fenoprop	$C_{10}H_9Cl_3O_3$	282	19:44	196+223+282	1
10	2,4,5-T	$C_9H_7Cl_3O_3$	268	20:10	209+233+268	1
11	MCPB	$C_{12}H_{15}ClO_3$	242	20:13	101+107+242	1
12	Fluroxypyr	$C_8H_7Cl_2FN_2O_3$	268	20:55	211+209+268	1
13	2,4-DB	$C_{11}H_{12}Cl_2O_3$	262	21:07	101+162+231	1
14	Bentazon	$C_{11}H_{14}N_2O_3S$	254	21:32	175+212+254	1
15	Ioxynil	$C_8H_5I_2NO$	385	21:47	243+370+385	1
16	Haloxyfop	$C_{16}H_{13}ClF_3NO_4$	375	24:19	288+316+375	1

Results

A standard chromatogram (Fig. 4-108, step 500 ng/l) is used to build up the substance library (Fig. 4-109). This data file contains all the information required for the reliable automatic identification of the derivatised active substances in real samples and the quantitation of the components found. This file is known as the calibration file because

4.20 Determination of Phenoxyalkylcarboxylic Acids in Water Samples

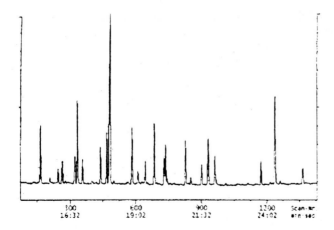

Fig. 4-108 Chromatogram of the standard solution (step 500 ng/l)

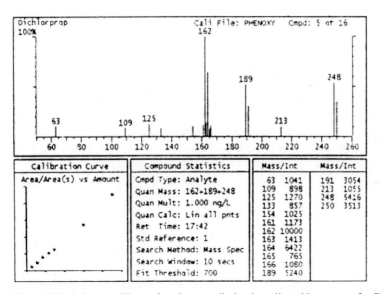

Fig. 4-109 Substance library for phenoxyalkylcarboxylic acids, extract for Dichlorprop

it contains, besides many fixed parameters for each component, the current retention time and quantitative calibration. Update programs allow adaptation to changing facts and allow the file to be used at any time after it was created. Additional active substances are integrated into the file as required.

To identify individual components the mass spectrum and the retention times of the compounds are used. Spectroscopic comparison is made using a retention time window on the basis of the full scan analysis and gives an FIT value as a comparison criterion. Integration of the peaks using the given selective quantitation masses is only carried out for active substances which have been detected. These selective quantitation masses are chosen using the methods used in SIM analysis. Unlike SIM analysis, however, in this procedure full scan data are obtained which allow spectroscopic comparison to confirm the identity of an active substance. In this way a greater reliability in the evaluation of the analysis is achieved. Peaks with a signal/noise ratio below a given value are rejected.

The GC peaks are integrated on the selective individual mass traces (mass chromatograms). Both the intensities of individual masses and the sum of masses and mass ranges can be used (Fig. 4-110). Peak integration establishing the beginning and end of the peak and the base line can be carried out automatically or manually as desired. Evaluation of the data is carried out using both peak area and peak height.

The calibration function can be plotted as a table or a curve after regression calculations (Fig. 4-111). This gives the function equation with the correlation coefficient and standard deviations.

Evaluation of Analyses

The chromatograms of analysis samples are evaluated using the calibration file (Fig. 4-112). Basically all the compounds in the chromatogram entered in the calibration file are searched for on the basis of the mass spectrum corresponding to a particular retention time. Provided that identification is positive (the criteria FIT value and signal to noise ratio are fulfilled), peak areas and heights are determined. Evaluation takes place in one process for all components (multicomponent analysis, target compound analysis).

For every series of samples at least two fortified water samples are analysed in addition in order to be able to critically assess the results.

More than 100 samples have been analysed with both internal and external standards. Some of the samples were also analysed using another GC/MS system (Finnigan TSQ 700). The results were comparable within the normal limits of accuracy.

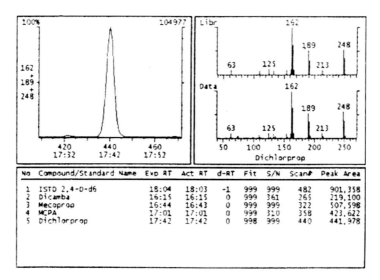

Fig. 4-110 Integration of the Dichlorprop peak on selective ion traces (here a sum) after checking the identity

4.20 Determination of Phenoxyalkylcarboxylic Acids in Water Samples

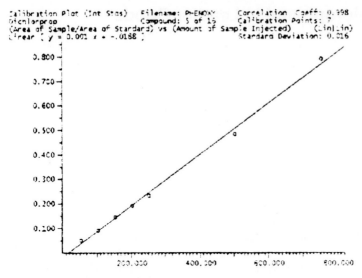

Fig. 4-111 Calibration function for Dichlorprop over a range of measurements from 50 to 750 ng/l (correlation factor 0.998)

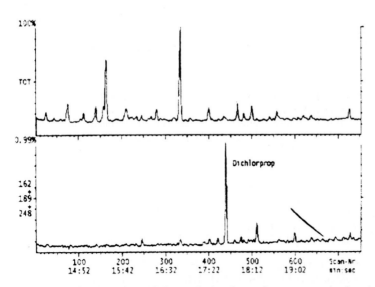

Fig. 4-112 Chromatogram of the analysis of a well water sample for phenoxyalkylcarboxylic acids. Bentazon was identified.

The process described consisting of automatic solid phase extraction, methylation with diazomethane and measurement of the extracts with the GCQ-GC/MS system has proved to be a reliable and rapid procedure for analysing phenoxyalkylcarboxylic acids in water samples.

Reference: F. Kriegsmann, Landesamt für Wasserwirtschaft, Munich

4.21 Pentachlorophenol in Leather Goods

Key words: pentachlorophenol (PCP), consumer articles, trimethylaniline hydroxide, TMAH

The discussion as to whether pentachlorophenol (PCP) used as a wood impregnating agent is harmful to health and the knowledge and that PCP can contribute to the pollution of the environment by dioxins led to the complete disappearance of the substance from wood protection agents some years ago. Legal requirements have made limitations to the use of the substance much more stringent. According to the legal situation it is basically prohibited to manufacture, bring into circulation or use PCP, its salts and compounds, preparations and PCP residues or products treated with it. Failure to comply with the legal requirements is a criminal offence.

The following threshold values were laid down: for preparations containing PCP as an impurity 0.01% and for products which are contaminated with PCP as a result of some treatment 5 mg/kg. Only that part of the product which is actually treated may be used to establish the concentration.

The concentration of PCP in leather goods is determined after extraction of the ground up material with toluene and subsequent methylation (described here) or acetylation.

Analysis Conditions

Sample preparation:	1 g ground up leather is treated with 10 ml toluene (possible ultrasonic treatment), 10 min
Derivatisation:	50 µl of the toluene extract are treated with 50 µl trimethylaniline hydroxide (TMAH) and shaken. Methylation occurs immediately.
Gas chromatograph:	Carlo-Erba Vega 6130
Injector:	splitless injection, 0.5 µl
Column:	J&W DB-5, 30 m × 0.32 mm × 0.1 µm Helium, 1 bar
Program:	Start: 55 °C, 1 min Program: 10 °C/min Final temp.: 250 °C, 20 min
Mass spectrometer:	Finnigan ITD 800 Scan mode: EI, full scan, 100–350 u Scan rate: 1 s/scan
Calibration:	external standardisation

Results

Pentachlorophenol is detected and determined as pentachloromethoxybenzene (M 278), which can be recognised with certainty because of the intense chlorine isotope pattern of the molecular ion and an almost equally strong fragmentation caused by methyl cleavage (Fig. 4-113).

The external calibration has been carried out from 0.1 mg/kg as the lowest value. The calibration function of the procedure shows it to have very good linearity (Fig. 4-114).

Reference: L. Matter, Dr. D. Schenker, Chemisches- und Lebensmittel-Untersuchungsamt der Stadt Duisburg, Duisburg.

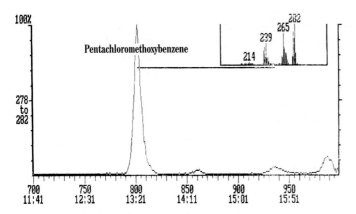

Fig. 4-113 PCP determination: mass chromatogram of pentachloromethoxybenzene

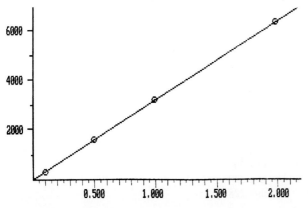

Fig. 4-114 Calibration using an external standard for PCP determined as the methyl ether

4.22 Polycyclic Aromatic Hydrocarbons

Key words: PAHs, hazard estimation, military waste, environment, soil, water, SFE, deuterated standards

Polycyclic aromatic hydrocarbons (PAHs) are a group of harmful organic substances which have come to be of public interest on account of the hazard estimation of sites where there are old waste deposits. On classical industrial sites, but also on disused military sites, considerable quantities of PAHs were discharged into the environment through the burning of fossil fuels. Their current ubiquitous occurrence has been traced to the incomplete combustion of organic substances and, in particular, to traffic emissions.

More than 100 polycyclic aromatic hydrocarbons have so far been detected. They always occur in complex mixtures and never as individual substances. There is relatively little confirmed information about their actual toxicity. About 10 substances are now classified as highly carcinogenic (Fig. 4-115), benzo(a)pyrene being regarded as the most dangerous of the PAHs because of its high toxicity.

During routine investigations by HPLC with diode array detection, overlapping substances in noticeably high concentrations keep on appearing, which are not contained in the list of PAHs (Table 4-15). The samples can be investigated in detail using GC/MS analysis, which is sufficiently sensitive for the necessary trace detection and at the same time gives the mass spectra of these additional substances.

Table 4-15 List of polyaromatic hydrocarbons determined by EPA method 8270. The components listed here are analysed routinely using GC/MS within the scope of the list analysis.

Substance	Molecular ion M^+
Naphthalene	m/z 128
Acenaphthylene	m/z 152
Acenaphthene	m/z 154
Fluorene	m/z 166
Phenanthrene	m/z 178
Anthracene	m/z 178
Fluoranthene	m/z 202
Pyrene	m/z 202
Benz(a)anthracene	m/z 228
Chrysene	m/z 228
Benzo(b)fluoranthracene	m/z 252
Benzo(k)fluoranthracene	m/z 252
Benzo(a)pyrene	m/z 252
Dibenz(a,h)anthracene	m/z 278
Indeno(123-cd)pyrene	m/z 276
Benzo(ghi)perylene	m/z 276

4.22 Polycyclic Aromatic Hydrocarbons 445

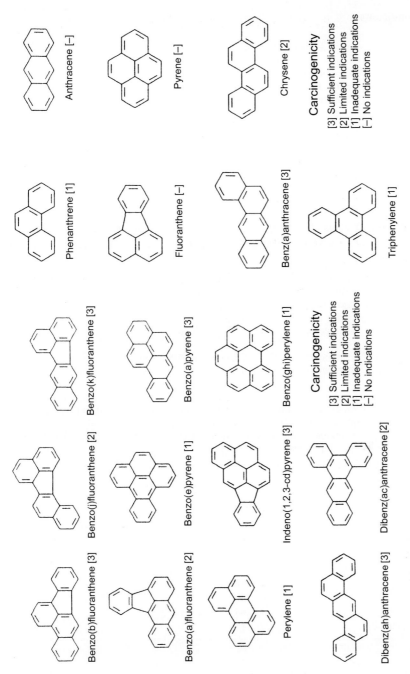

Fig. 4-115 Structural formulae and carcinogenicity of PAHs (after Speer)

Analysis Conditions

Sample material:	soil and other solid materials	
Sample preparation: SFE extraction:	extraction by SFE Suprex Prepmaster with AccuTrap	
	Pressure:	400 bar
	Temp.:	80–150 °C
	Flow:	2.5 ml
	Modifier:	2% methanol
	Static extraction:	15 min
	Dynamic extraction:	ca. 10 × the container volume
	Restrictor:	variable, 150 °C
	Cold trap:	+5 °C, cooled with technical grade CO_2
	Elution medium:	methanol
GC/MS system:	Finnigan ITS 40	
Injector:	PTV cold injection system	
	Start:	60 °C
	Ramp:	20 °C/s
	Final temp.:	320 °C, 10 min.
Column:	SGE HT5, 25 m × 0.22 mm × 0.1 µm	
Program:	Start:	60 °C
	Ramp:	12 °C/min.
	Final temp.:	320 °C, 7.5 min.
Mass spectrometer:	Scan mode:	EI, full scan, 100–300 u
	Scan rate:	1 s/scan
Calibration:	The quantitation of the PAHs is carried out using internal standards. A group of deuterated PAHs (e. g. Promochem, deuterated PAH standard solution, Table 4-16) are added to the sample clean-up and are assigned as standards during the evaluation of the individual PAHs according to their retention times.	

Table 4-16 List of internal standards for PAH analysis using GC/MS

Substance	Molecular ions M^+
1,4-Dichlorobenzene-d_4	m/z 150
Naphthalene-d_8	m/z 136
Acenaphthene-d_{10}	m/z 166
Phenanthrene-d_{10}	m/z 188
Chrysene-d_{12}	m/z 240
Perylene-d_{12}	m/z 264
Benzo(ghi)perylene-d_{12}	m/z 288

The conditions chosen here for sample preparation and analysis correspond to the parameters which are also used for the determination of nitroaromatics and PCBs. Through the use of an ion trap GC/MS the three classes of substance PAHs, PCBs (see Section 4.25) and nitroaromatics (see Section 4.28) can be detected in parallel in one analysis run.

Results

In Fig. 4-116 the total ion current is shown over the whole elution range from naphthalene to benzo(ghi)perylene as the typical chromatogram of a soil sample. For the evaluation the selective mass chromatograms of the individual components are used instead of the complex plot of the total ion current. Figure 4-117 shows the typical elution sequence for the mass trace m/z 252 with benzo(b,k)fluoranthene, benzo(e)pyrene and benzo(a)pyrene under the conditions stated. The concentrations of the PAHs in the sample are calculated automatically from the above list using a quantitation database (see Section 3.3).

The identification of other components present in the sample is first carried out manually by comparison of the spectra with the NIST library. The mass chromatogram m/z 168 (Fig. 4-118) from the analysis shown in Fig. 4-116 shows two intense signals, which can be explained by the library comparison. Figure 4-119 shows the spectrum of the main component and Fig. 4-120 the extract from the NIST library.

A waste water sample was investigated first for PAHs (Fig. 4-121). However, chlorinated compounds in clearly higher concentrations were detected in the sample by GC/MS (Fig. 4-122): trichlorobenzene (Fig. 4-123), tetrachloro- (Fig. 4-124) and pentachlorobiphenyls (Fig. 4-125), which were identified by spectroscopic comparison.

Reference: P. Lönz, TERRACHEM Essen GmbH, Essen.

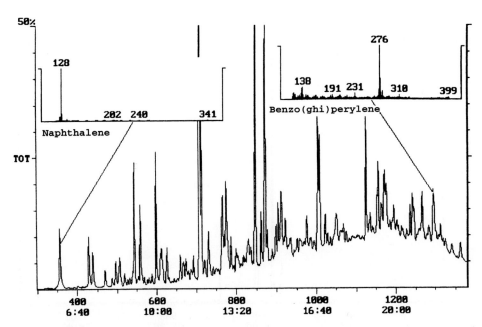

Fig. 4-116 Chromatogram (total ion current) of a typical soil extract with the spectra of naphthalene and benzo(ghi)perylene

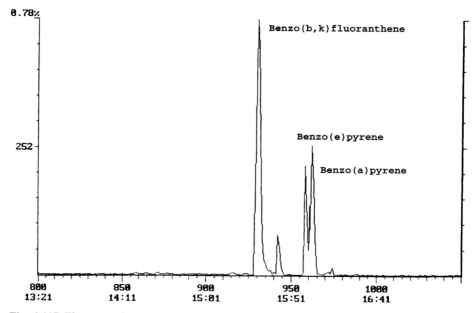

Fig. 4-117 The mass chromatogram for m/z 252 shows the separation of benzo(e)pyrene and benzo(a)pyrene (for the analysis parameters see text)

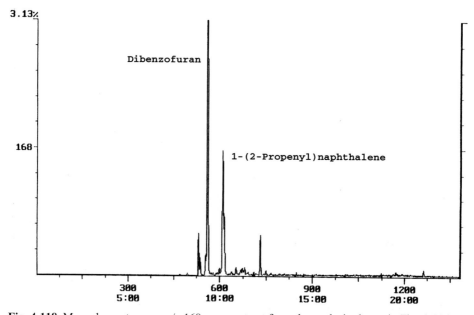

Fig. 4-118 Mass chromatogram m/z 168 as an extract from the analysis shown in Fig. 4-116

4.22 Polycyclic Aromatic Hydrocarbons   449

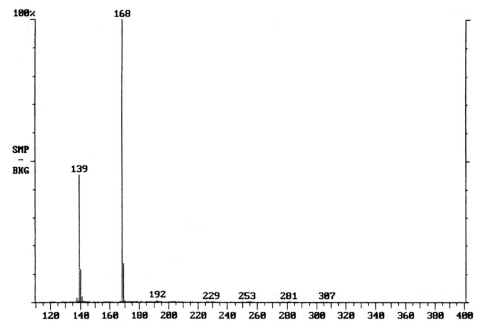

Fig. 4-119 Mass spectrum of the main component after subtraction of the background

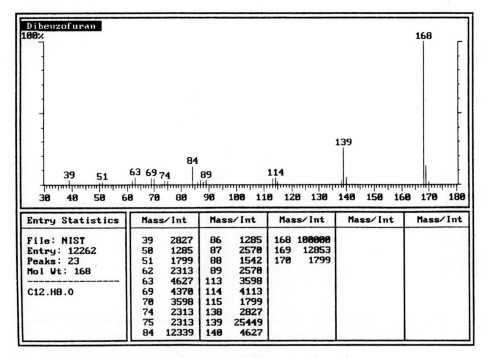

Fig. 4-120 Comparison spectrum from the NIST library for dibenzofuran

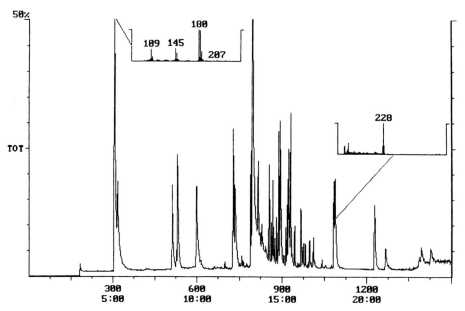

Fig. 4-121 Chromatogram of a waste water sample: the investigation for PAHs (e.g. m/z 228 chrysene/benz(a)anthracene) gave significantly higher concentrations of chlorinated substances, such as trichlorobenzene m/z 180/182

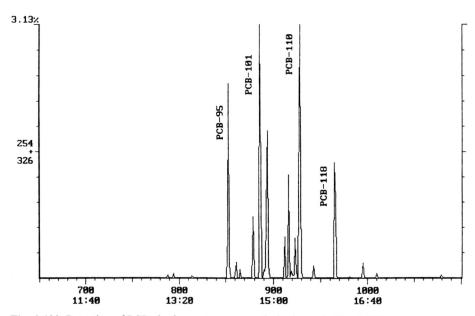

Fig. 4-122 Detection of PCBs in the waste water analysis shown in Fig. 4-121

4.22 Polycyclic Aromatic Hydrocarbons

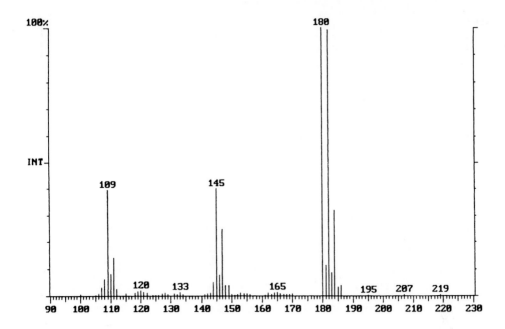

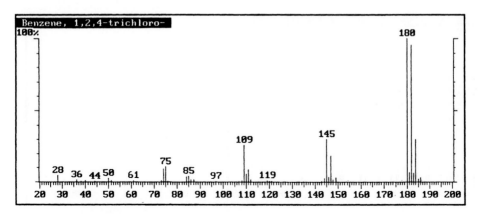

Fig. 4-123 NIST spectrum of 1,2,4-trichlorobenzene

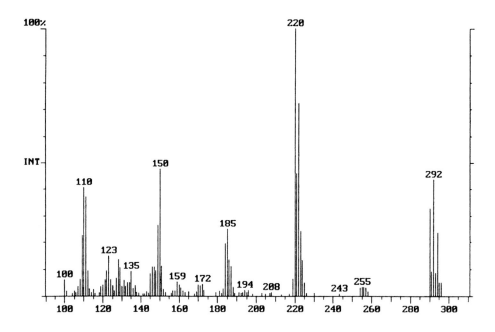
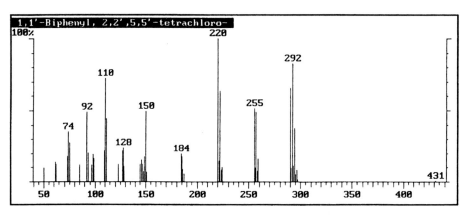

Fig. 4-124 NIST spectrum of tetrachlorobiphenyl

4.22 Polycyclic Aromatic Hydrocarbons   453

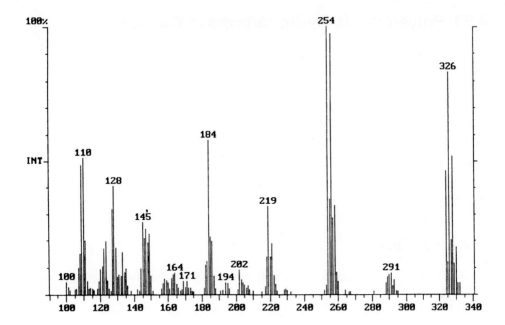

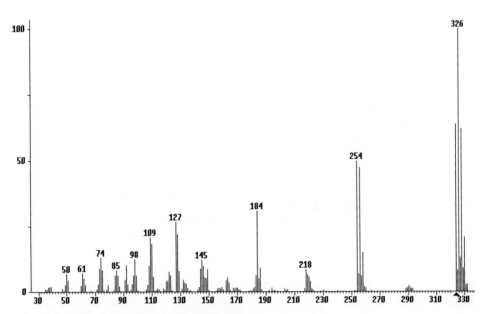

Fig. 4-125 NIST spectrum of pentachlorobiphenyl

454   4 Applications

4.23 Polyaromatic Hydrocarbons in City Air

Key words: PAHs, suspended particulates, Soxhlet, phenols, biphenyls, isomers, library comparison

One important cause of the ubiquitous distribution of PAHs is the domestic fire burning fossil fuels. During the heating period of 1990/91 suspended particulate fractions in the old town area of Halle (Saale, Germany) were obtained and evaluated. A particularly high exposure of PAHs was expected in the location chosen because of the high density of housing where the heating was exclusively with brown coal in unmodernised stoves and chimneys. An additional contribution was expected from inner-city traffic.

Analysis Conditions

Sample collection: Koch parallel impactor (Fraunhofer Gesellschaft Hanover), glass wool filter
Suction rate: 30 l/min
Charge: 49.6 mg dust from 345 m^3 air

Sample preparation: Soxhlet extraction with 180 ml toluene,
SPE purification with benzenesulfonic acid/silica gel and florisil column, taken up in hexane

GC/MS system: Finnigan ITS 40
Injector: SPI injector
Start: 100 °C
Ramp: 300 °C/min
Final temp.: 300 °C, 30 min.
Column: J&W DB-5, 30 m × 0.25 mm × 0.25 µm
Helium, pre-pressure 10 psi
Start: 100 °C for 3 min
Program 1: 10 °C/min to 200 °C
Program 2: 5 °C/min
Final temp.: 310 °C

Mass spectrometer: Scan mode: EI, full scan, 100–520 u
The scan range was extended to mass 520 to be able to determine PCBs which might be present.
Scan rate: 1 s/scan

Results

Starting with naphthalene a whole range of PAHs up to high molecular weight components with m/z 252, such as benzo(b,k)fluoranthene, benzo(e,a)pyrene, appeared in the samples of suspended particulates (Fig. 4-126). Figure 4-127 shows as part of the analysis the elution of fluoranthene and pyrene.

Besides the PAHs a series of related classes of compound were detected, among them alkylaromatics (Fig. 4-128), also in high concentrations, which were identified by

4.23 Polyaromatic Hydrocarbons in City Air   455

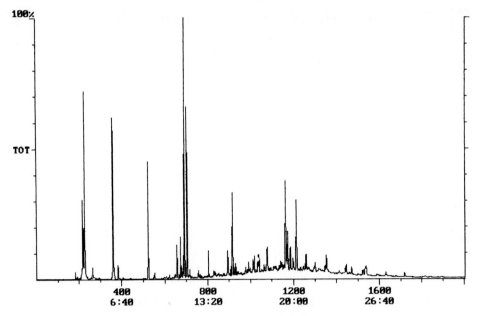

Fig. 4-126 Analysis of filter eluate: total ion current of the whole chromatogram

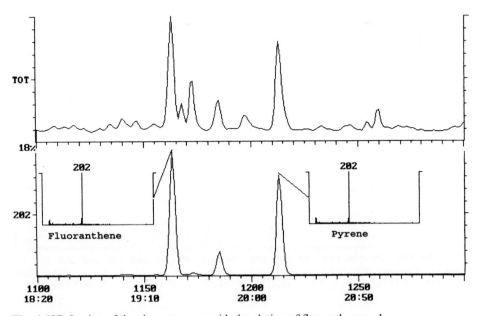

Fig. 4-127 Section of the chromatogram with the elution of fluoranthene and pyrene

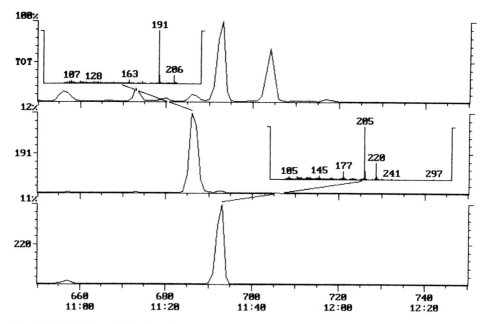

Fig. 4-128 Elution of alkylphenols

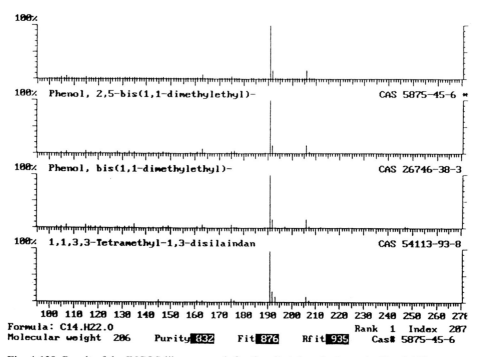

Fig. 4-129 Result of the INCOS library search for the alkylphenols shown in Fig. 4-128. Di-t-butylphenols

4.23 Polyaromatic Hydrocarbons in City Air   457

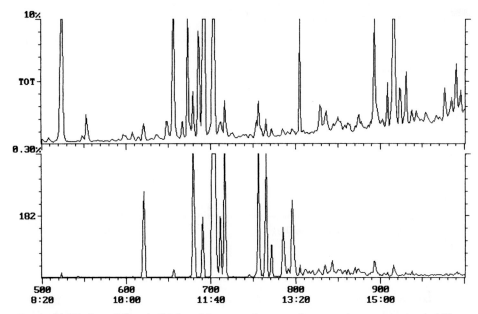

Fig. 4-130 Elution of dimethylbiphenyl isomers, shown as the mass chromatogram m/z 182

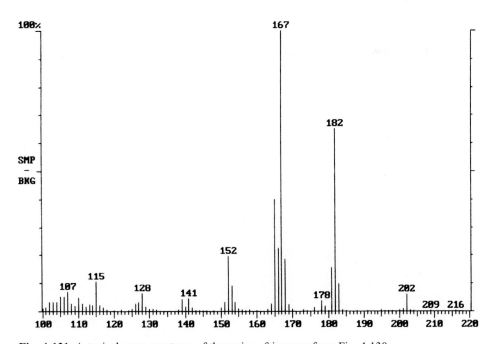

Fig. 4-131 A typical mass spectrum of the series of isomers from Fig. 4-130

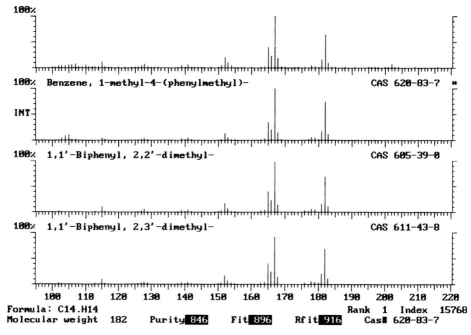

Fig. 4-132 Assignment of the spectrum as that of a dimethylbiphenyl isomer

a library comparison (Fig. 4-129). Identification of the isomers can only be carried out by comparison of retention times with standards. Thus in the retention time range of 10–15 min various dimethylbiphenyl isomers are eluted (Fig. 4-130). Because of the existence of isomers, the library search can only identify the substance class (Figs. 4-131 and 4-132). PCBs cannot be detected in air samples.

In the example of polycyclic aromatics the limits of mass spectrometry and the use of the coupling procedure for identification are clear. The isomers can indeed be clearly differentiated from other homologues, but they can only be identified by comparison of the retention times with those of standard substances.

4.24 Routine Analysis of 24 PAHs in Water and Soil

Key words: PAHs, SPE, Soxhlet, chemical ionisation, water, soil, dynamic region, deuterated standards

The process given here for the routine determination of PAHs is particularly suitable for samples which contain a strongly aliphatic hydrocarbon matrix, besides PAHs. Chemical ionisation with water is used as the ionisation procedure. This makes the undesired background selectively transparent and a higher response is achieved for the PAHs. Compared with EI analysis (Fig. 4-133) the response of compounds in the CI mode is about the same (Fig. 4-134). The signal/noise ratio with CI using water for compounds with a significant proton affinity exceeds that with EI.

Analysis Conditions

Sample preparation:	Water: 500 ml sample, solid phase extraction (SPE) on C_{18}-cartridges; liquid/liquid extraction with cyclohexane; concentration to 0.2 ml. Soil: 10 g sample with 10 g sodium sulfate, Soxhlet extraction with cyclohexane, 6 h, concentration to 1 ml. The standard mixture of 6 deuterated compounds is added to the extraction mixture.
GC/MS system:	Finnigan ITS 40, Finnigan ITD 800 with Siemens GC Sichromat 1
Injector program:	Start: 40 °C Program: rapidly to 320 °C Final temp.: 320 °C, 15 min.
Column:	Restek XTI-5, 30 m × 0.32 mm × 0.25 µm Helium, 2 bar
Temp. program:	Start: 40 °C, 1 min Program 1: 15 °C/min to 340 °C Final temp.: 340 °C
Mass spectrometer:	Scan mode: CI, full scan, 100–300 u, reagent gas water Scan rate: 1 s/scan
Calibration:	using deuterated internal standards (Table 4-17)

Results

The polycyclic aromatic hydrocarbons listed in Table 4-18 were investigated, of which 16 were analysed using the EPA method. A calibration file with the CI mass spectrum, retention time and calibration function are used for searching and identification of the compounds (Fig. 4-135). The specific mass trace $(M+H)^+$ is used to successfully identify the components, in order to check the minimum signal/noise ratio required and to determine the peak areas. Quantitation is then carried out with reference to the internal standard and the calibration function (Fig. 4-136).

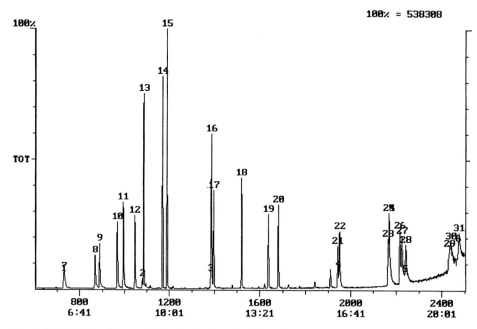

Fig. 4-133 PAH routine analysis: standard run in EI mode.
Internal standards: 0.5 ng/μl; PAHs: 5 ng/μl

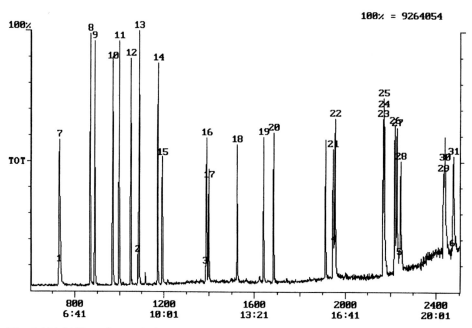

Fig. 4-134 PAH routine analysis: standard run in CI mode (reagent gas water).
Internal standards: 0.5 ng/μl; PAHs: 5 ng/μl

Table 4-17 Internal standards (dissolved in 80% dichloromethane-d_2 (99.6%) and 20% benzene-d_6 (99.6%), Promochem)

No.[1]	Compound	M[2]
0	1,4-Dichlorobenzene-d_4	150
1	Naphthalene-d_8	136
2	Acenaphthene-d_{10}	166
3	Phenanthrene-d_{10}	188
4	Chrysene-d_{12}	240
5	Perylene-d_{12}	264
6	Benzo(ghi)perylene-d_{12}	288

[1] Entry number in the calibration file
[2] The quasimolecular ions $(M+H)^+$ are formed in CI with water

Table 4-18 Polyaromatic hydrocarbon calibration substances (solution in toluene, certified by NIST, Promochem)

No.[1]	Compound	ISTD No.[2]	M[3]
7	Naphthalene*	1	128
8	2-Methylnaphthalene	1	142
9	1-Methylnaphthalene	1	142
10	Biphenyl	2	154
11	2,6-Dimethylnaphthalene	2	156
12	Acenaphthylene*	2	152
13	Acenaphthene*	2	154
14	2,3,5-Trimethylnaphthalene	3	170
15	Fluorene*	3	166
16	Phenanthrene*	3	178
17	Anthracene*	3	178
18	1-Methylphenanthrene	4	192
19	Fluoranthene*	4	202
20	Pyrene*	4	202
21	Benz(a)anthracene*	4	228
22	Chrysene*	4	228
23	Benzo(b)fluoranthene*	5	252
24	Benzo(k)fluoranthene*	5	252
25	Benzo(e)pyrene	5	252
26	Benzo(a)pyrene*	5	252
27	Perylene	5	252
28	Indeno(123-cd)pyrene*	6	276
29	Dibenz(a,h)anthracene*	6	278
30	Benzo(ghi)perylene*	6	276

* PAHs according to EPA (16 components)
[1] Entry number in the calibration file
[2] Assignment of the PAHs to the internal standard from Table 4–17
[3] The quasimolecular ions $(M+H)^+$ are formed through CI with water

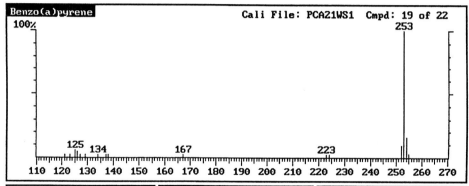

Fig. 4-135 Benzo(a)pyrene entry (CI spectrum) in the calibration file

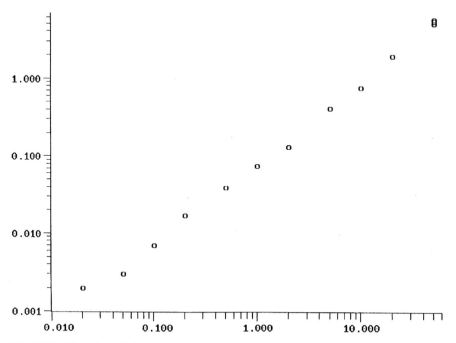

Fig. 4-136 Linearity of the water CI procedure for benzo(a)pyrene (correlation coefficient 0.998)

The limits of detection (S/N = 3) of the procedure are between 2 ng/l for volatile PAHs and 10 ng/l for higher-boiling ones with CI using water. The absolute limit of detection of the CI method is between 2 and 10 pg/injection at a linear range greater than 10^4.

Reference: A. Landrock, H. Merten, H. Richter, Institut Fresenius Sachsen, Dresden.

4.25 Polychlorinated Biphenyls in Milk and Milk Products

Key words: PCBs, milk, GPC, matrix, signal/noise ratio, INCOS library search

Polychlorinated biphenyls (PCBs) have been widely used in, for example, transformers, condensers, hydraulic oils, protective coatings and building materials, because of their physical properties. The toxic activity of PCBs mainly gives rise to liver damage. The hazard they pose to the environment lies in their high persistence and enrichment via the food chain. PCBs are recovered as lipid-soluble residues in plant and animal foodstuffs in widely varying concentrations. Milk and milk products are also affected and are therefore subject to continuous control.

To prepare a sample 0.5 g milk fat is applied to a gel permeation column (GPC, filling biobeads SX 3, 200–400 mesh) and then eluted with acetone/cyclohexane (1+3). The PCB fraction collected is evaporated nearly to dryness on a rotary evaporator and dissolved in 2 ml trimethylpentane. The solution is then further purified by treating with concentrated sulfuric acid to oxidise hydrocarbons. The extract is treated with 2 ml sulfuric acid, shaken for 10 min on a mixer and then centrifuged. The organic phase is removed and transferred to the sample vessels of the GC autosampler.

Unlike the analysis of industrial products, in the analysis of PCBs from biological material selective metabolisation of individual substances (congeners) must be taken into consideration, which can lead to an alteration in the original PCB sample. Of the 209 possible PCB congeners about 60 compounds are recovered in industrial mixtures. The qualitative and quantitative determination of the individual compounds is very time-consuming. In DIN 51 527 T1 six compounds are cited which can be determined analytically in order to calculate the total PCB content if the PCB sample justifies this.

Analysis Conditions

GC/MS system:	Finnigan Magnum ion trap system,
Injector:	PTV cold injection system
	total sample transfer, 2 µl injection volume
Injector program:	Start: 50 °C, 1.5 min
	Program: 240 °C/min to 300 °C
	Final temp.: 300 °C, 16 min
Column:	SGE HT5, 25 m × 0.25 mm × 0.1 µm
Temp. program:	Start: 60 °C, 1 min
	Program 1: 20 °C/min to 160 °C
	Program 2: 5 °C/min to 310 °C
	Final temp.: 310 °C, 5 min
Transfer line:	260 °C
Mass spectrometer:	Scan mode: EI, full scan, 140–520 u
	Scan rate: 1 s/scan

Results

The quantitative evaluation of standard analyses and real samples in the lowest concentration range is shown below (Figs. 4-137 to 4-141). In the analyses of standards PCBs of all degrees of chlorination are detected with good signal/noise ratios and complete spectra which can be confirmed by library comparison at a dilution of 5 pg/µl.

Evaluation of real samples shows in particular the presence of PCB congeners of higher degrees of chlorination in milk. While Cl_3- and Cl_4-PCBs are not detectable (Fig. 4-142), the various isomers in the C_5- to C_8-PCBs are detected through their mass spectra (Fig. 4-143) and confirmed by comparison with the NIST library (Figs. 4-144 to 4-146).

Reference: M. Wislicenus, Staatliche Milchwirtschaftliche Lehr- und Forschungsanstalt, Wangen/Allgäu.
J. Niebel, Axel Semrau GmbH & Co, Applikationslabor, Sprockhövel, Germany.

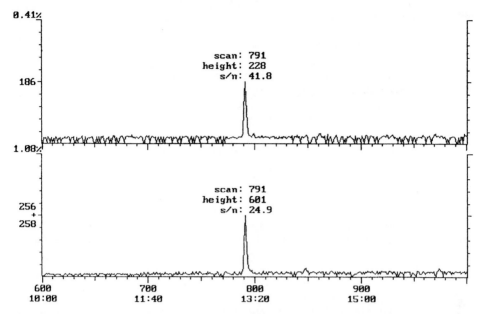

Fig. 4-137 PCB 33 (Cl$_3$-PCB), 5 pg/µl, detection on the mass traces m/z 186 (S/N 40:1) and m/z 256 + 258 (S/N 25:1)

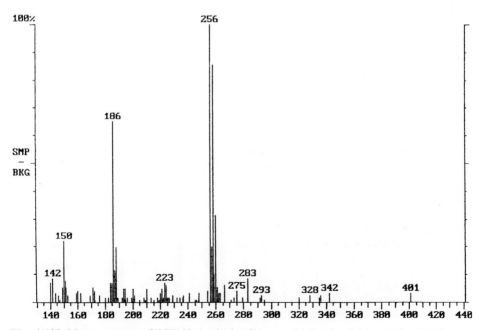

Fig. 4-138 Mass spectrum of PCB 33 (analysis of the standard at 5 pg/µl from Fig. 4-137)

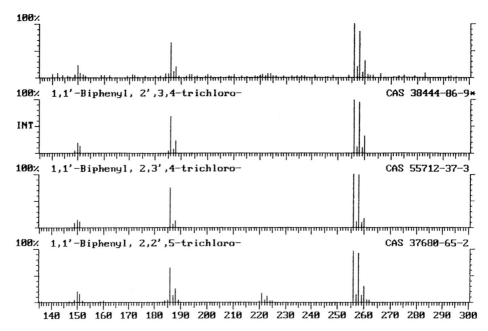

Fig. 4-139 The result of the INCOS library search gives an FIT value of 968 (standard analysis 5 pg/µl, spectrum from Fig. 4-138)

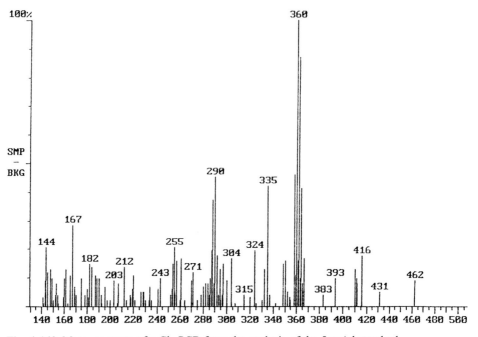

Fig. 4-140 Mass spectrum of a Cl_6-PCB from the analysis of the 5 pg/µl standard

4.25 Polychlorinated Biphenyls in Milk and Milk Products

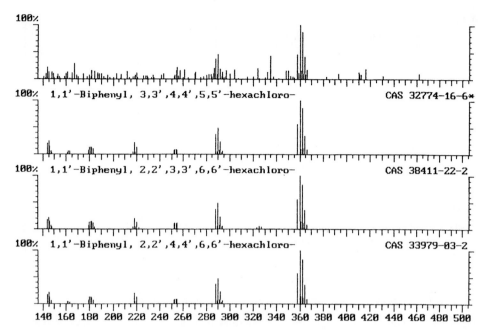

Fig. 4-141 The result of the INCOS library search gives an FIT value of 808 (standard analysis 5 pg/µl, spectrum from Fig. 4-140)

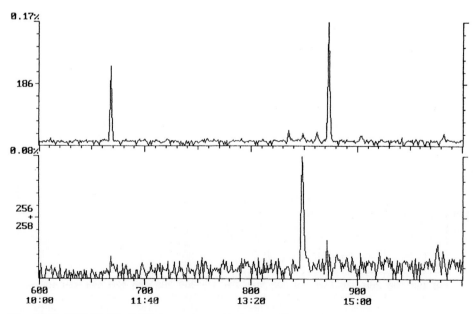

Fig. 4-142 PCB-33 (Cl$_3$-PCB) is not detectable in the milk sample (Fig. 4-137, scan 791)

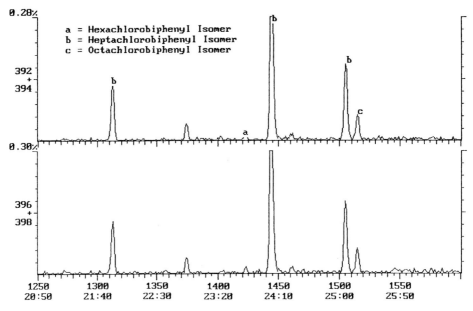

Fig. 4-143 Elution of hexa-, hepta- and octachloro-PCBs in a milk sample, the mass chromatograms of the Cl$_7$-PCB molecular ion cluster are shown

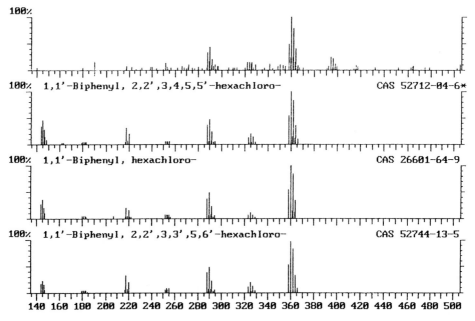

Fig. 4-144 Confirmation of a hexachlorobiphenyl by means of an INCOS library search (scan # 1422 from Fig. 4-143)

4.25 Polychlorinated Biphenyls in Milk and Milk Products

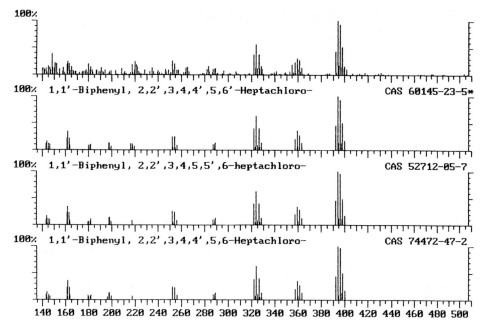

Fig. 4-145 Confirmation of a heptachlorobiphenyl by means of an INCOS library search (scan # 1312 from Fig. 4-143)

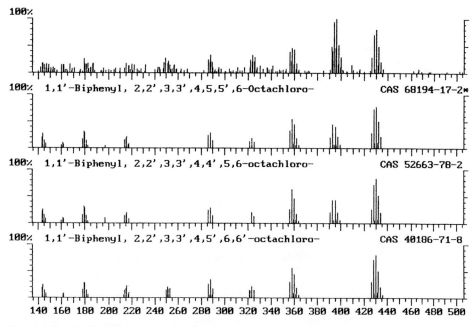

Fig. 4-146 Confirmation of an octachlorobiphenyl by means of an INCOS library search (scan # 1515 from Fig. 4-143)

4.26 Polychlorinated Biphenyls in Indoor Air

Key words: PCBs, sealants, air sampling, thermodesorption, Tenax, quantitation

Polychlorinated biphenyls (PCBs) were used in the period 1960 to 1980 as flameproofing plasticisers in sealants. The proportion used in polysulfide sealants was up to 30%.

Guidelines for the use of PCBs indoors were laid down by national authorities. For precautionary reasons these bodies suggested that PCB contamination below 300 ng/m^3 in the air could be tolerated. Within the range 300 to 3000 ng/m^3 it was recommended that the source of the contamination should be traced and removed. At concentrations above 3000 ng/m^3 measures should be introduced without question (e.g. a ban on the use of these rooms).

To sample PCBs from indoor air high volume samplers and polyurethane foam cartridges or porous polymers are generally used, which are later cleaned up in the laboratory according to the recommendations of national authorities. For the PCB determination described here a procedure was worked out which allows air sampling to be carried out on adsorption materials using an automatic thermodesorber without further sample preparation. An ion trap GC/MS was used as the detector.

The determination of all PCB congeners is not feasible. Because of this an overall determination according to DIN 51 527 part 1 (polychlorinated biphenyls in waste oil) was recommended. According to DIN 51 527 the calculation is based on the six congeners 28, 52, 101, 138, 153 and 180. The national authorities chose the congeners 28, 52, 101 and 153 for the determination of PCBs in air. To determine the overall concentration, the concentrations of these congeners are added and multiplied by the factor 6.

For air sampling the steel tubes of the thermodesorber are used packed with ca. 50 mg Tenax GR (35–16 mesh). The adsorbent filling is fixed with silanised glass wool and metal springs. In order to be able to determine the concentrations in the zero range of 50 ng/m^3, a sample volume of more than 200 l is necessary. Sampling thus takes ca. 3 h at a flow rate of 2 l/min. Du Pont P4000 personal air samplers were used.

Analysis Conditions

Sample collection: active, 3 h, 2 l/min
adsorbent Tenax GR

Thermodesorber: Perkin Elmer ATD 400
Desorption temp.: 350 °C
Desorption time: 30 min
Cold trap filling: 20 mg Tenax GR (backflush)
Cold trap temp.: Low: 30 °C
High: 350 °C, 5 min
Desorption flow: 80 ml/min
Input split: none
Output split: 30 ml/min
Transfer line: 225 °C
Valve temp.: 225 °C

Gas chromatograph:	Perkin Elmer model 8700
Column:	SGE HT-5, 30 m × 0.23 mm × 0.25 µm
Carrier gas:	helium, 125 kPa
Program:	Start: 60 °C, 1 min
	Program 1: 8 °C/min to 180 °C
	Program 2: 12 °C/min to 300 °C
	Final temp.: 300 °C, 10 min
Mass spectrometer:	Finnigan ITD 800, direct coupling
	Scan mode: EI, full scan, 100–400 u
	Scan rate: 0.5 s/scan
	Transfer line: 300 °C

Results

Figures 4-147 and 4-148 show a total ion current chromatogram and an individual mass chromatogram resulting from an air sample taken directly next to a sealant. The total calculated PCB concentration was ca. 7000 ng/m^3. An absolute calibration of 5 ng corresponds to a total PCB concentration of 600 ng/m^3 for 200 l of air. Figure 4-149 shows the calibration function for the PCB congener 52 from 1 to 50 ng per sample tube. The desorption of the PCB congeners from Tenax GR is readily reproducible with high transfer rates for the four congeners used for quantitation (Fig. 4-150). Using multiple desorption the transfer rates of the congeners 153, 138 and 180 can be increased further. However, hexa- and heptachloro-PCBs are insignificant because of their low vapour pressures. The thermodesorption system used has been shown to be extremely powerful and inert for PCB analysis so that even the use of PCB 209 (Decachlorobiphenyl) as internal standard is possible.

Reference: M. Tschickardt, Landesamt für Umweltschutz und Gewerbeaufsicht, Mainz.

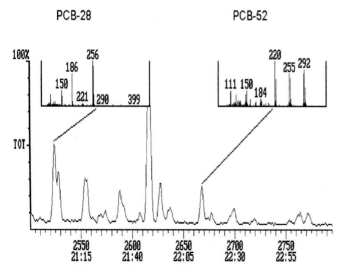

Fig. 4-147 Total ion current chromatogram of an air sample near the sealant

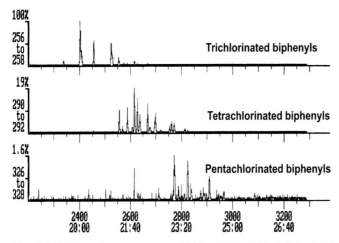

Fig. 4-148 Mass chromatograms of PCBs (3Cl-, 4Cl-, 5Cl-PCBs) in indoor air

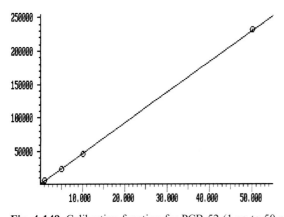

Fig. 4-149 Calibration function for PCB-52 (1 ng to 50 ng per tube)

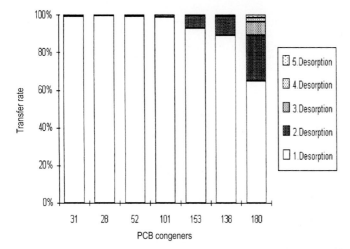

Fig. 4-150 Multiple desorption of 50 ng samples of PCB congeners from Tenax-GR

4.27 Screening for Dioxins and Furans

Key words: dioxins, furans, screening, resolution, LRMS, HRMS, isotope pattern, sensitivity, confirmation

The identification and quantitation of polychlorinated dibenzodioxins and dibenzofurans in environmental samples presents a challenge to the analyst even today. Because of their high chronic toxicity the isomers with 2,3,7,8-Cl substitution (Fig. 4-151) are the subject of the investigation. The evaluation of the concentrations determined is carried out taking into account the international toxicity equivalents (Table 4-19). Brominated compounds (inclusion of eight 2,3,7,8-brominated dioxins in the list of prohibited chemicals), polychlorinated xanthenes (PCXE), the xanthones (PCXO) and methylfurans, which have become known, for example, from bleaching processes in paper manufacture, are also increasingly included in dioxin analysis. The matrices involved in the analyses are extremely variable and range from water, soil and air to sewage sludge, fly ash, compost, waste and plant and animal tissues. Sample clean-up (Figs. 4-152 and 4-153) is particularly important, the aim being to concentrate the chlorinated dioxins/furans with the most effective possible purification of the extract. SFE (see Section 2.1.2) is particularly effective as the reduction of the sample preparation time allows a more rapid response in dioxin analysis.

The analysis of the extracts takes place as a stepwise screening procedure according to Fig. 4-154 by GC/MS with unit mass resolution (LRMS, low resolution MS). The use of the ion trap GC/MS allows the detection of the total Cl isotope cluster of molecular ions and thus allows for a high certainty of identification. It depends on the task involved whether the analysis is carried out in the full scan mode or by individual mass

7 of 75 PCDDs:

2,3,7,8-Tetra CDD
1,2,3,7,8-Penta CDD
1,2,3,4,7,8-Hexa CDD
1,2,3,6,7,8-Hexa CDD
1,2,3,7,8,9-Hexa CDD
1,2,3,4,6,7,8-Hepta CDD
Octa CDD

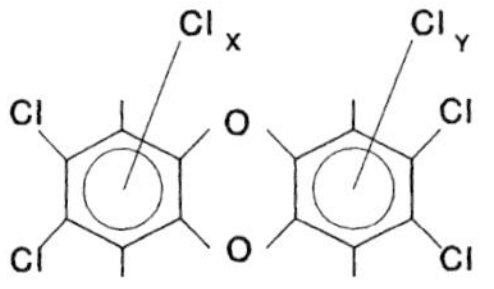

10 of 135 PCDFs:

2,3,7,8-Tetra CDF
1,2,3,7,8-Penta CDF
2,3,4,7,8-Penta CDF
1,2,3,4,7,8-Hexa CDF
1,2,3,6,7,8-Hexa CDF
1,2,3,7,8,9-Hexa CDF
2,3,4,6,7,8-Hexa CDF
1,2,3,4,6,7,8-Hepta CDF
1,2,3,4,7,8,9-Hepta CDF
Octa CDF

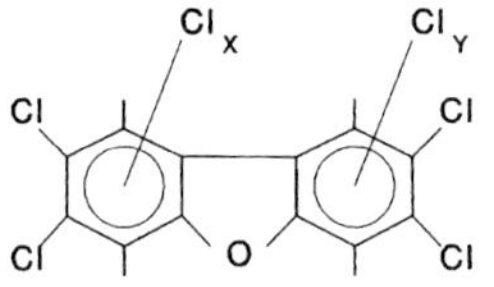

X, Y = 0 to 2

Fig. 4-151 Polychlorinated dibenzo-dioxins and dibenzofurans with 2,3,7,8-Cl substitution

Table 4-19 Toxicity equivalents of PCDD/PCDF isomers

PCDD/F	BGA	NATO/CCMS
2,3,7,8-TCDD	1.0	1.0
1,2,3,7,8-PCDD	0.1	0.5
1,2,3,4,7,8-HxCDD	0.1	0.1
1,2,3,6,7,8-HxCDD	0.1	0.1
1,2,3,7,8,9-HxCDD	0.1	0.1
1,2,3,4,6,7,8-HpCDD	0.01	0.01
OCDD	0.001	0.001
2,3,7,8-TCDF	0.1	0.1
1,2,3,7,8-PCDF	0.1	0.05
2,3,4,7,8-PCDF	0.1	0.5
1,2,3,4,7,8-HxCDF	0.1	0.1
1,2,3,6,7,8-HxCDF	0.1	0.1
1,2,3,7,8,9-HxCDF	0.1	0.1
2,3,4,6,7,8-HxCDF	0.1	0.1
1,2,3,4,6,7,8-HpCDF	0.01	0.01
1,2,3,4,7,8,9-HpCDF	0.01	0.01
OCDF	0.001	0.001
Other TCDD	0.01	0
Other PCDD	0.01	0
Other HxCDD	0.01	0
Other HpCDD	0.001	0
Other TCDF	0.01	0
Other PCDF	0.01	0
Other HxCDF	0.01	0
Other HpCDF	0.001	0

4.27 Screening for Dioxins and Furans

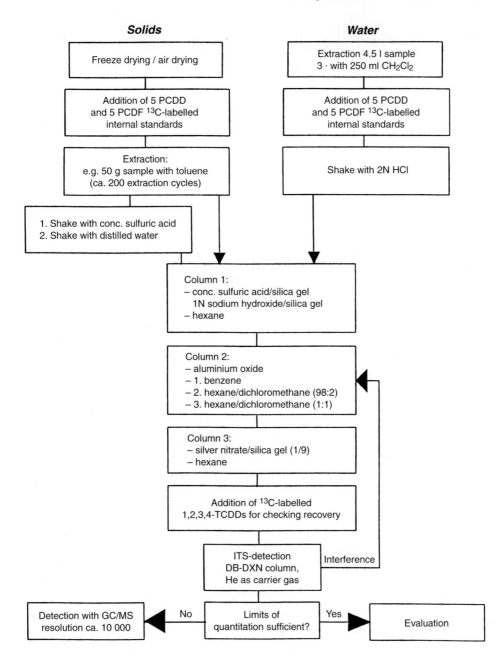

Fig. 4-152 Separation process for sample clean-up in PCDD/PCDF analysis

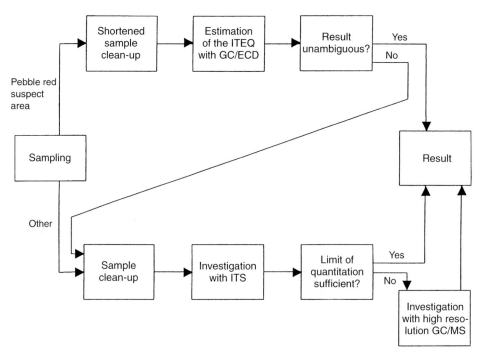

Fig. 4-153 Flow diagram showing the different steps of PCDD/PCDF analysis

recording (SIM, MID). While it is necessary to run mass spectra which are as full as possible when analysing samples of unknown composition, target compound analysis (TCA) investigating certain individual compounds is also possible in the SIM mode. In the case of individual mass recording and where the limits of quantitation in the full scan mode are not sufficient, the determination and confirmation by high resolution GC/MS with a double focusing magnetic sector mass spectrometer are necessary. The resolution necessary for dioxin determination is 10 000. In order to separate xanthenes from dioxins a still higher resolution must be achieved. The mass difference between TCDDs (m/z 319.8965) and TCXEs (m/z 319.9138) requires a resolution of over 18 500 to avoid false positive results. Electron impact ionisation is used for dioxin analysis. Chemical ionisation with detection of negative ions gives a low specific response because of the lower Cl content in the molecules of TCDDs and TCDFs.

```
MID   Time Windows
 #   Start   Measure   End         Cycletime
 1   14:00    2:50    16:50 min    0.55 sec
 2   16:50    3:00    19:50 min    0.57 sec
 3   19:50    3:00    22:50 min    0.60 sec
 4   22:50    2:50    25:40 min    0.65 sec
 5   25:50    3:00    28:50 min    0.70 sec
```

```
Window 1    mass     F int  gr  time(ms)
 # 1       303.9016     1   1    46.42
   2       305.8987     1   1    46.42
   3       313.9839  1 20   1     2.73
   4       315.9419     1   1    46.42
   5       317.9389     1   1    46.42
   6       319.8965     1   1    46.42
   7       321.8937     1   1    46.42
   8       331.9368     1   1    46.42
   9       333.9338     1   1    46.42
  10       351.9807  c 20   1     2.73
```

```
Window 2    mass     F int  gr  time(ms)
 # 1       313.9839  1 20   1     2.73
   2       339.8598     1   1    49.15
   3       341.8569     1   1    49.15
   4       351.9000     1   1    49.15
   5       353.8970     1   1    49.15
   6       355.8547     1   1    49.15
   7       357.8518     1   1    49.15
   8       363.9807  c 20   1     2.73
   9       367.8949     1   1    49.15
  10       369.8919     1   1    49.15
```

```
Window 3    mass     F int  gr  time(ms)
 # 1       363.9807  1 20   1     2.73
   2       373.8208     1   1    51.88
   3       375.8179     1   1    51.88
   4       385.8610     1   1    51.88
   5       387.8580     1   1    51.88
   6       389.8157     1   1    51.88
   7       391.8128     1   1    51.88
   8       401.8559     1   1    51.88
   9       403.8529     1   1    51.88
  10       413.9775  c 20   1     2.73
```

```
Window 4    mass     F int  gr  time(ms)
 # 1       407.7818     1   1    58.71
   2       409.7789     1   1    58.71
   3       413.9775  1 20   1     2.73
   4       419.8220     1   1    58.71
   5       421.8190     1   1    58.71
   6       423.7767     1   1    58.71
   7       425.7738     1   1    58.71
   8       435.8169     1   1    58.71
   9       437.8140     1   1    58.71
  10       463.9743  c 20   1     2.73
```

```
Window 5    mass     F int  gr  time(ms)
 # 1       425.9775  1 20   1     2.73
   2       441.7428     1   1    64.17
   3       443.7399     1   1    64.17
   4       453.7830     1   1    64.17
   5       455.7801     1   1    64.17
   6       457.7377     1   1    64.17
   7       459.7348     1   1    64.17
   8       463.9743  c 20   1     2.73
   9       469.7779     1   1    64.17
  10       471.7750     1   1    64.17
```

Fig. 4-154 Determination of PCDD/PCDFs by high resolution mass spectrometry: Retention time window and mass traces with exact masses of the congeners arriving for determination in 5 time windows (l lock mass, reference mass for high resolution measurement, c calibration mass, control of mass calibration, Finnigan MAT 95)

Analysis Conditions

Gas chromatography:
Column: DB 5-DXN, 60 m × 0.32 mm × 0.25 μm
Injector: PTV cold injection system
80 °C, 1 min
250 °C/min to 300 °C
300 °C, 30 min

LRMS analysis:
GC/MS system: Finnigan ITS 40 ion trap GC/MS system
Scan mode: EI, mass range 310–360 u and 370–480 u as the molecular ion ranges of the PCDD/PCDF isomers (Table 4-20)
Scan rate: 0.5 s/scan

HRMS analysis:
GC/MS system: Finnigan 8400 (double focusing magnetic sector instrument)
Resolution: 10 000
Scan mode: EI with individual mass registration MID, mass traces (Fig. 4-154)

Table 4-20 The most intense molecular ions of the PCDD/PCDF isomers (LRMS)

Substance	M^+ ions in the Cl cluster					
TCDD	320	322	324	326		
PCDD	354	356	358	360		
HxCDD	388	390	392	394	396	
HpCDD	422	424	426	428	430	
OCDD	456	458	460	462	464	466
TCDF	304	306	308	310		
PCDF	338	340	342	344		
HxCDF	372	374	376	378	380	
HpCDF	406	408	410	412	414	
OCDF	440	442	444	446	448	450

Results

With the determination of the complete molecular ion cluster of the PCDDs/PCDFs (Figs. 4-155 and 4-156) unambiguous identification of the individual components is possible using low resolution GC/MS analysis. Further confirmation by high resolution MS is not necessary here, as the limit of quantitation in the lower pg range is sufficient for the present sample.

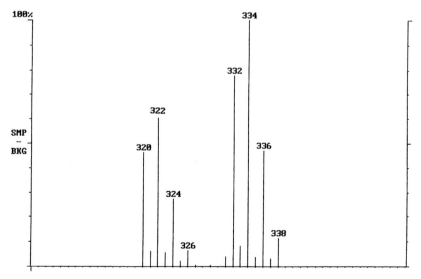

Fig. 4-155 Molecular ion clusters of the precursor TCDD (m/z 320, 322, 324, 326) and the internal standard added $^{13}C_{12}$-TCDD (m/z 332, 334, 336, 338)

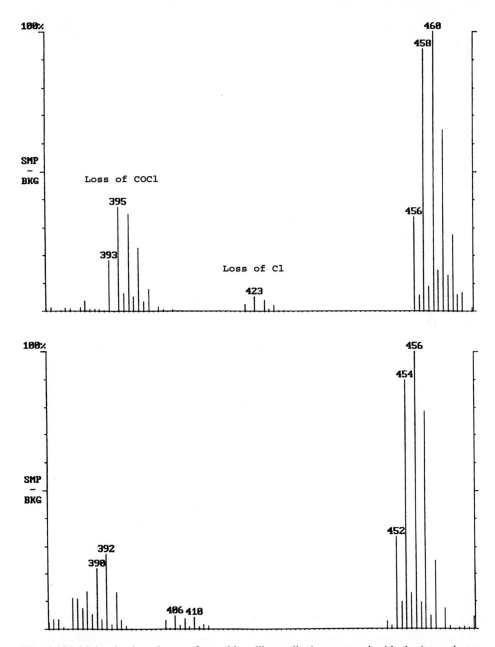

Fig. 4-156 Molecular ion cluster of octachlorodibenzodioxin compared with the internal standard $^{13}C_{12}$-OCDF added

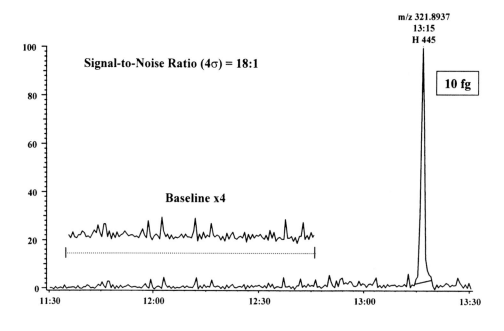
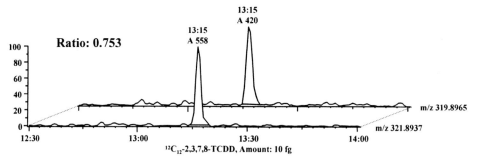
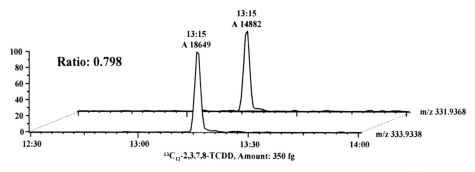

Fig. 4-157 Sensitivity measurement of 10 fg 2,3,7,8-TCDD coinjected with 350 fg $^{13}C_{12}$-TCDD standard. A signal/noise ratio of larger than 15:1 (4 σ) is achieved. The measured isotope ratios are within the required ±15% limits (theoretical value is 0.774) (Finnigan MAT)

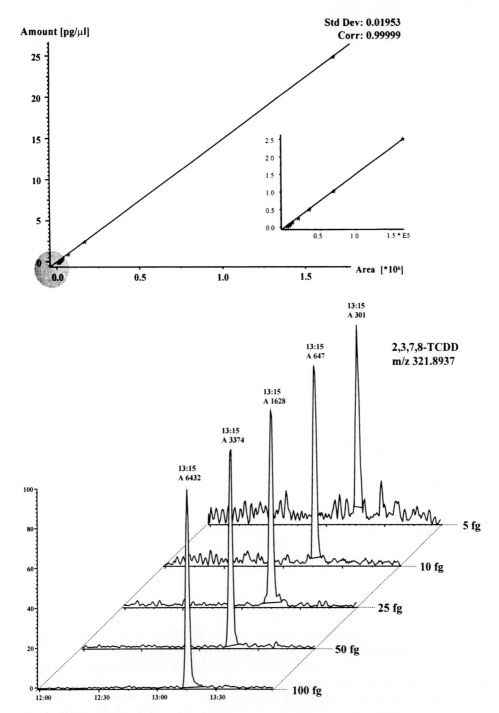

Fig. 4-158 Linearity test for 2,3,7,8-TCDD ranging from 5 fg to 25 pg. The grey spot is the part of the linearity curve shown in the inset and in the mass chromatograms of m/z 321.8937 (Finnigan)

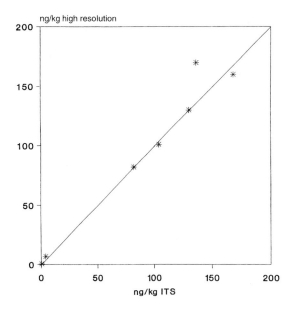

Fig. 4-159 Comparison of the LRMS (ion trap) and HRMS (high resolution) methods.
* ITEQ

Quality Assurance Measures in Dioxin Analysis
(according to Landesumweltamt (regional environmental office) NRW, Germany, FG 413, 1994)

- Addition of ten ^{13}C-labelled PCDD/F standards with 2,3,7,8-Cl substitution as internal standards (one substance per group of isomers).
- Carrying out careful extract preparation to remove interfering substances
- Regular testing of blank samples
- Regular testing of control samples
- Not analysing samples with strongly differing levels of contamination at the same time
- Parallel testing of diverse samples using low resolution (ion trap GC/MS) and high resolution mass spectrometry (double focusing magnetic sector instrument) by different analysts
- Double determinations
- Determination of the recovery of the internal ^{13}C standard ($<50\%$)
- Plausibility control by experienced analysts
- Participation in external round robin tests
- Complete avoidance of plastics during sample clean-up (also no Teflon)

If the sensitivity of this LRMS method is too low, detection is carried out by HRMS using a magnetic sector instrument with a resolution of 10 000. The considerably higher specificity of high resolution for dioxin analysis is shown in Figs. 4-157 and 4-158. The limits of detection for high resolution mass spectrometry are in the lower fg range. Comparison of the low and high resolution methods shows good agreement from the lowest to the highest measuring ranges (Fig. 4-159).

References: P. Bachhausen, Landesumweltamt NRW, Düsseldorf.
Dr. H. Münster, Finnigan GmbH, Applikationslabor, Bremen.

4.28 Analysis of Military Waste

Key words: military waste, nitroaromatics, TNT, degradation products, carcinogenicity, Soxhlet, choice of column, ECD, PID, chemical ionisation, PAHs

In the past little attention was paid to the ecological consequences of rearmament and disarmament. Only recently has the problem of military waste become a topic in residue analysis. The combined consequences of two world wars have not been considered for a long time and current political changes are leading to new types of contamination which will require analytical solutions within the framework of demilitarisation. This overall area will also therefore increase in importance in the future in the field of analysis.

There is no absolute definition of military waste. The term encompasses disused military sites where munitions or chemical weapons were manufactured, processed, stored or deposited. Soil and ground water are contaminated by the substances concerned and also by synthesis products, by-products, degradation products or by problematic fuels. Some of these compounds are liable to explode giving off toxic products. Explosives (nitroaromatics) and chemical warfare agents are of particular importance. The large quantity of water required for the manufacture of nitroaromatics led to the setting up of production plants in areas with large water supplies which are often still also used for public consumption. Many substances which are recovered today as metabolites or stabilisers are carcinogenic.

Table 4-21 By-products and degradation products of TNT with details of the carcinogenic potential

Compound	Assessment	Compound	Assessment
2-Nitrotoluene		2,6-Diaminotoluene	
3-Nitrotoluene		2,4,6-Triaminotoluene	
4-Nitrotoluene		2-Amino-6-nitrotoluene	
2,3-Dinitrotoluene	III A2	2-Amino-4-nitrotoluene	
2,4-Dinitrotoluene	III A2	6-Amino-2,4-dinitrotoluene	
2,6-Dinitrotoluene	III A2	4-Amino-2,6-dinitrotoluene	
3,4-Dinitrotoluene	III A2	2-Amino-4,6-dinitrotoluene	
2,4,6-Trinitrotoluene		3-Nitrobiphenyl	
2,4,5-Trinitrotoluene	III B	4-Nitrobiphenyl	III A2
2,3,4-Trinitrotoluene		3-Nitrobiphenyl	
1,2-Dinitrobenzene	III B	4-Nitrobiphenyl	III A2
1,3-Dinitrobenzene	III B	2,2'-Dinitrobiphenyl	
1,4-Dinitrobenzene	III B	2-Nitronaphthalene	III A2
2,3-Diaminotoluene		4-Aminobiphenyl	III A1
2,4-Diaminotoluene		1,3-Dinitronaphthalene	III B

The substances are assessed according to the ordinance on hazardous materials and the MAK value list:
III A1 = definitely carcinogenic in humans
III A2 = definitely carcinogenic in animals
III B = suspicion of potential carcinogenicity

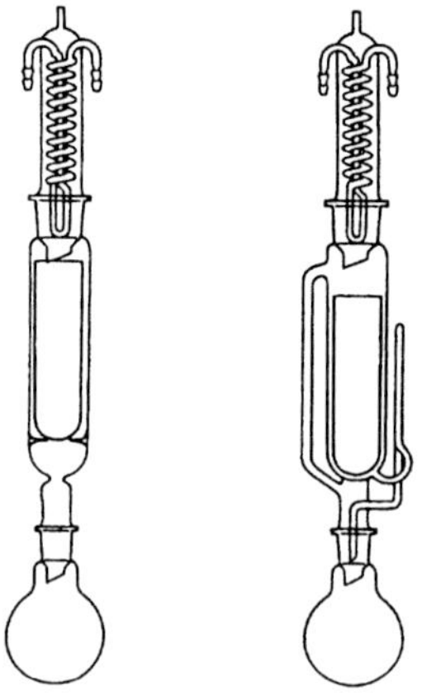

Compound/ extraction procedure	Methanol Soxhlet	Methanol flow extraction	Diethyl ether extraction
2-Nitrotoluene	85%	100%	91,7%
3-Nitrotoluene	82%	95%	90%
4-Nitrotoluene	73%	90%	85%
2,6-Dinitrotoluene	91%	95%	95%
2,4-Dinitrotoluene	72%	85%	90%
2,3-Dinitrotoluene	74%	90%	90%
3,4-Dinitrotoluene	34%	70%	80%
2,4,6-Trinitrotoluene	6%	70%	80%
1,3-Dinitrobenzene	66%	70%	85%
1,4-Dinitrobenzene	1%	70%	85%
4-Amino-2,6-dinitrotoluene	47%	45%	90%
2-Amino-4,6-dinitrotoluene	41%	45%	90%
3-Nitrobiphenyl	96%	95%	90%
2,2-Dinitrobiphenyl	92%	95%	95%
1,3-Dinitronaphthalene	56%	80%	95%
2,3-Diaminotoluene	–	35%	75%
2,4-Diaminotoluene	–	45%	80%
2,6-Diaminotoluene	–	50%	80%
2-Amino-4-nitrotoluene	–	65%	80%
2-Amino-6-nitrotoluene	–	65%	80%
2,6-Diamino-4-nitrotoluene	–	45%	80%

– Not determined

Fig. 4-160 Comparison of liquid extraction procedures for the analysis of nitroaromatics

Table 4-22 Detectors for the analysis of nitroaromatics and their metabolites

Detector	Advantages	Disadvantages
FID	High linearity	Low selectivity, Low sensitivity
ECD	Good sensitivity	Low linearity, Contamination when using highly concentrated samples, low sensitivity for substances with less than 2 nitro groups, correspondingly difficult detection of metabolites
NPD	Good selectivity for nitro compounds and metabolites	Low sensitivity
PID	The same response for nitro and amino compounds, field tests	Low sensitivity
ELD	N-specific, high linearity, simple calibration	Low sensitivity
GC/MS-EI	High sensitivity, good identification even of other accompanying substances	–
GC/MS-CI	High selectivity, additional confirmation from the molecular mass, PCI for nitroaromatics and metabolites, NCI only for nitro compounds	–

4.28 Analysis of Military Waste   485

Besides HPLC for routine control, GC with ECD and PID has been the normal procedure for residue analysis up till now. GC/MS is increasing in importance as the number of substances in the trace region as well as in high concentrations which need to be determined is increasing. As the spectrum of expected contaminants is usually difficult to estimate, the mass spectrometer is indispensable as the universal and specific detector.

The analysis of nitroaromatics and their metabolites is of particular importance in the assessment of military waste. 14 by-products and degradation products of 2,4,6-TNT (trinitrotoluene) alone are definitely classified or suspected as carcinogens (Table 4-21).

The analysis of TNT degradation products involves Soxhlet extraction or SFE for the clean-up of soil extracts, whereby the differing polarity of the substances to be extracted needs to be taken into consideration when choosing the extraction agent (Fig. 4-160). Nitroaromatics can be separated by HPLC or GC. GC/MS coupling is now the method of choice for determining them together with their metabolites. Classical GC detection using NPD, PID or ECD has limitations (Fig. 4-161 and Table 4-22). ECD is frequently used and exhibits high sensitivity for all compounds with two or more nitro groups. However, it can therefore only be used for the detection of certain metabolites (Fig. 4-162).

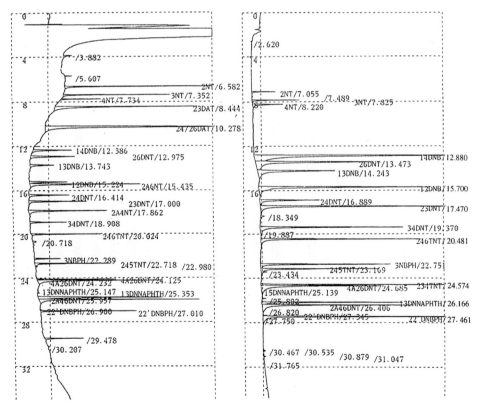

Fig. 4-161 Comparison of PID and ECD detectors in the analysis of nitroaromatics and their metabolites (column HT8, 400 pg/component, for conditions see text)

Analysis Conditions

Sample: soil samples

Sample preparation: continuous flow extraction with methanol or diethyl ether

GC/MS systems: (1) HP MSD 5971A, EI analysis
(2) Finnigan Magnum, EI/CI analyses

Injectors: (1) HP split injector, isotherm 270 °C, split operation
(2) PTV cold injection system, split operation
Start: 40 °C, 1 min
Program: 250 °C/min to 300 °C
Final temp.: 300 °C, 15 min

Columns: (1) Restek Rtx-200, 30 m × 0.25 mm × 0.25 μm
(2) SGE HT8, 25 m × 0.22 mm × 0.25 μm
Helium, 2 bar

Mass spectrometer: Scan mode: (1) EI, full scan, 29–250 u
(2) EI/CI, full scan, 100–400 u, reagent gas water
Scan rate: (1) and (2) 1s/scan in each case

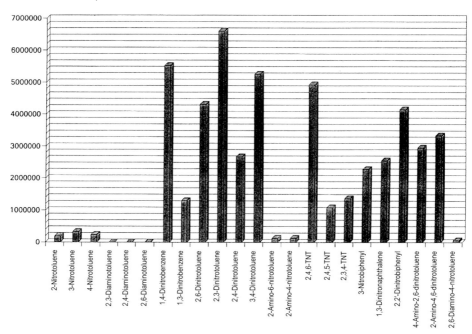

Fig. 4-162 Differing responses of nitroaromatics and their metabolites on ECD (400 pg samples)

4.28 Analysis of Military Waste

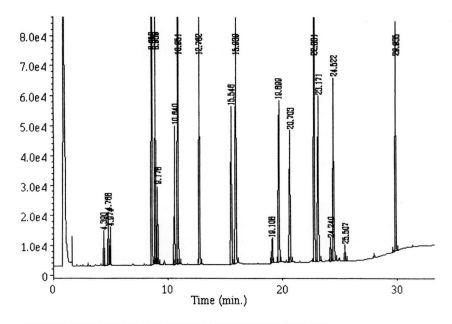

```
============================================
                 External Standard Report
============================================

Sample Name        : 2 ppm Mixture            Injection Number: 1
Instrument Method: NTH2_25.MTH

     4.390     22076    BB   0.027   1    2.000  2MNT
     4.768     30355    BB   0.025   1    2.000  3MNT
     4.974     17798    BV   0.024   1    2.000  4MNT
     8.658    265792    BB   0.042   1    2.000  14DNB
     8.959    256766    BB   0.042   1    2.000  26DNT
     9.178     78678    BB   0.046   1    2.000  13DNB
    10.640    151300    BB   0.050   1    2.000  24DNT
    10.951    374679    BB   0.047   1    2.000  23DNT
    12.762    297095    BB   0.054   1    2.000  34DNT
    15.546    213138    BB   0.063   1    2.000  3-NO2biphenyl
    15.939    331069    BB   0.059   1    2.000  246TNT
    19.108     30392    BB   0.057   1    2.000  245TNT
    19.699    213808    BB   0.062   1    2.000  234TNT
    20.703    188891    BB   0.067   1    2.000  1.3-DiNO2naphthalene
    22.831    396779    BB   0.065   1    2.000  2.2'-DiNO2biphenyl
    23.171    220906    BB   0.062   1    2.000  4NH2-2.6-DiNO2toluene
    24.522    251604    BB   0.064   1    2.000  2NH2-4.6-DiNO2toluene
    29.935    220999    BB   0.045   1    2.000  2.4-DiNO2Diphenylamine
```

Fig. 4-163 Separation of nitroaromatics on a Restek Rtx-1701 phase (60 m × 0.32 mm × 0.25 µm, H_2, HP 5890 GC/ECD, 80 °C, 2 min, 25 °C/min to 130 °C/min, 4 °C/min to 220 °C, 10 °C/min to 260 °C, 4.5 min)

Results

An analysis of nitroaromatics with ECD as the detector and hydrogen as the carrier gas is shown in Fig. 4-163. This separation is designed to be rapid and does not take metabolites into consideration. Rapid separation requires the appropriate choice of GC column. The critical pair 2,4- and 2,6-diaminotoluene can be separated well on the Rtx-200 column (trifluoropropylmethylsilicone phase) (Fig. 4-164), which is not possible with columns of the same length containing pure methylsilicone or carborane phases (HT8). As the isomers cannot be differentiated by mass spectrometry, gas chromatographic separation is a prerequisite for separate determination (Fig. 4-165). Other compounds from TNT analysis which can sometimes co-eluate with them can be differentiated by mass spectrometry.

The EI spectra of nitroaromatics frequently show an intense fragment at m/z 30 which results from rearrangement and fragmentation of the nitro group. The molecular ion signals are small and can therefore be completely undetectable in matrix-rich samples. With chemical ionisation all the substances involved form intense (quasi)molecular ions which can be detected with high S/N values because of their higher masses (Fig. 4-166). A series of typical CI spectra measured using water as the reagent gas are shown in Fig. 4-167 for various nitro- and amino-substitutions. The outstanding selectivity of chemical ionisation allows the determination of the equally important concentrations of polycondensed aromatics (PAHs) in the same analysis. As the PAHs are very accessible by CI, it is only necessary to assign the corresponding $(M+H)^+$ masses in order to be able to determine their concentrations (Fig. 4-168). The evaluation of the analysis is thus only directed to a further group of compounds. A second analysis run on the same sample is no longer necessary.

Reference: J. Niebel, Axel Semrau GmbH & Co, Applikationslabor, Sprockhövel.

4.28 Analysis of Military Waste

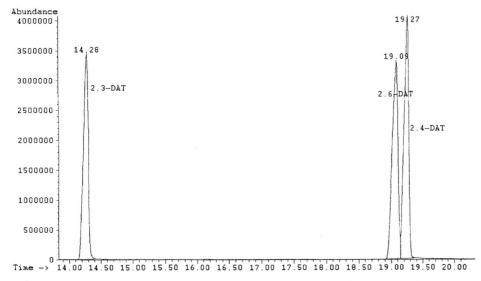

Fig. 4-164 Chromatographic separation of the critical pair 2,4- and 2,6-diaminotoluene on the Rtx-200 capillary column

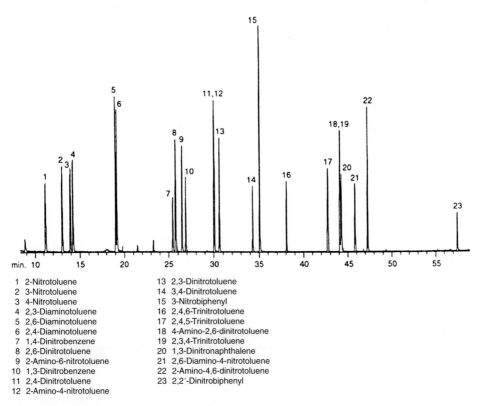

1 2-Nitrotoluene
2 3-Nitrotoluene
3 4-Nitrotoluene
4 2,3-Diaminotoluene
5 2,6-Diaminotoluene
6 2,4-Diaminotoluene
7 1,4-Dinitrobenzene
8 2,6-Dinitrotoluene
9 2-Amino-6-nitrotoluene
10 1,3-Dinitrobenzene
11 2,4-Dinitrotoluene
12 2-Amino-4-nitrotoluene
13 2,3-Dinitrotoluene
14 3,4-Dinitrotoluene
15 3-Nitrobiphenyl
16 2,4,6-Trinitrotoluene
17 2,4,5-Trinitrotoluene
18 4-Amino-2,6-dinitrotoluene
19 2,3,4-Trinitrotoluene
20 1,3-Dinitronaphthalene
21 2,6-Diamino-4-nitrotoluene
22 2-Amino-4,6-dinitrotoluene
23 2,2'-Dinitrobiphenyl

Fig. 4-165 Complete chromatographic separation of nitroromatics and metabolites on the Rtx-200 capillary column

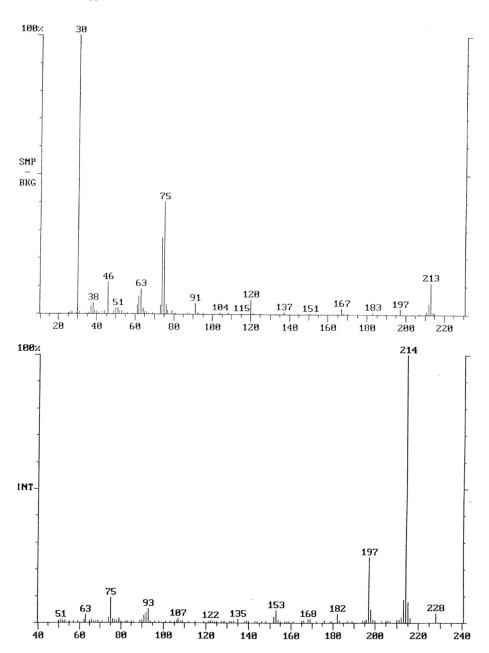

Fig. 4-166 Mass spectra of trinitrobenzenes
Above: EI spectrum with extensive fragmentation down to m/z 30 (NO$^+$) from the nitro group
Below: CI spectrum (methane) with (M+H)$^+$ as the base peak

4.28 Analysis of Military Waste

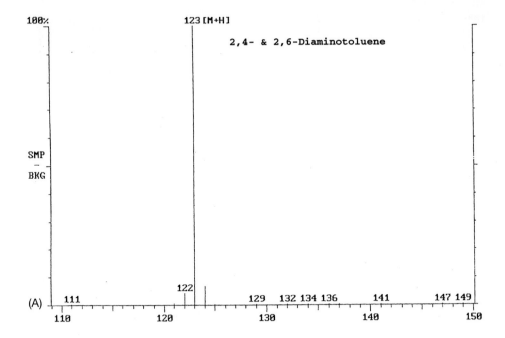

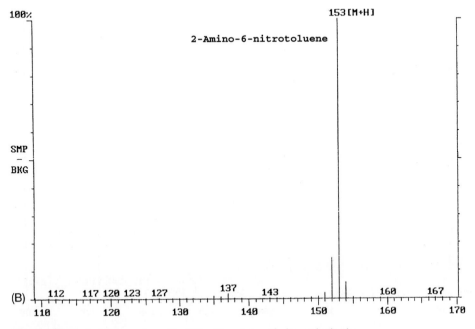

Fig. 4-167 Typical CI spectra with different amino and nitro substitutions
(A) Diaminotoluene
(B) Aminonitrotoluene

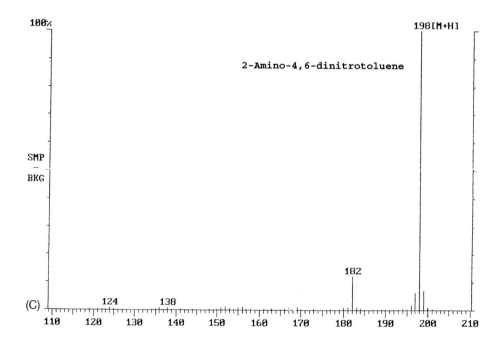

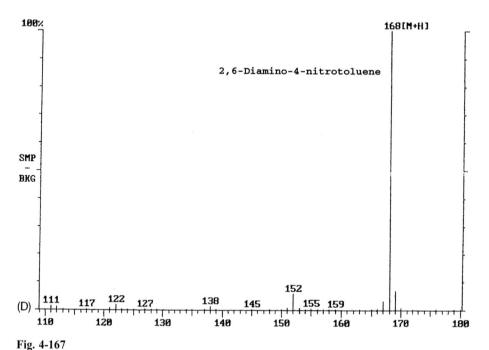

Fig. 4-167
(C) Aminodinitrotoluene
(D) Diaminonitrotoluene

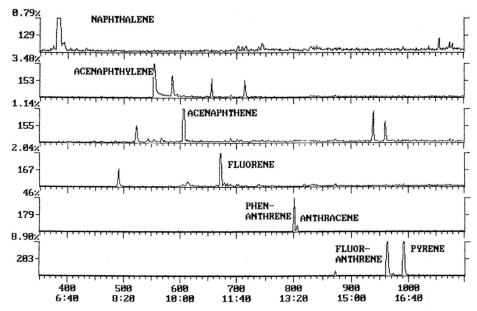

Fig. 4-168 Analysis of a TNT soil sample for PAH components (CI mode)

4.29 SFE Extraction and Determination of Nitroaromatics

Key words: military waste, nitroaromatics, chloronitroaromatics, SFE, soil, choice of column, ECD

SFE is suitable for the clean-up of soil samples for the analysis for nitroaromatics as it allows short extraction times and can be automated. In systematic investigations and the checking of decontamination procedures a large number of samples are obtained for which an analysis result is expected within a short time.

The extractability of nitroaromatics is demonstrated in a fortification experiment of standard substances into standard soils. Table 4-23 summarises the results for the extraction of the soils with ethyl acetate (EA) and with SFE (2% acetone as modifier).

The extracts of the ethyl acetate extraction (duration ca. 2.5 h) were purified on a silica gel column before the GC/MS determination. SFE extracts, on the other hand, could be analysed directly after a 30 min extraction. For the two differently fortified samples the recoveries based on the standard soil were of the same order of magnitude. Lower yields of volatile components in the extraction of nitroaromatics indicate losses through aerosol formation in spite of the low temperature of the cold trap used.

Table 4-23 Recoveries of nitrobenzenes (NB) and nitrotoluenes (NT) from various matrices for cold extraction with ethyl acetate (EA) and SFE (2% acetone as modifier)

Compound	Fortified [mg/kg]	Lufa2.1 EA WF [%]	Lufa2.1 SFE WF [%]	Lufa2.3 EA WF [%]	Lufa2.3 SFE WF [%]
NB	100	68.4	18.2	94.6	47.6
4-NT	100	84.1	46.0	110.0	80.7
1,3-DNB	102	51.9	67.9	76.6	73.5
2,6-DNT	101	72.6	69.5	94.4	81.3
2,4-DNT	102	57.1	69.2	76.9	77.6
3,4-DNT	102	52.2	69.0	67.3	73.2
NB	200	63.9	51.5	95.0	61.4
4-NT	200	74.7	72.5	97.4	76.8
1.3-DNB	204	76.0	80.1	81.9	75.8
2,6-DNT	202	76.6	79.7	91.7	77.5
2,4-DNT	204	65.3	81.1	84.6	77.5
3,4-DNT	204	62.7	80.3	76.3	74.9

Like the nitrobenzenes and nitrotoluenes, standard soils were likewise fortified with nitrophenols. Nitrophenols cannot be recovered by cold extraction in the lower ppm range while the SFE results are generally higher.

In the analysis of nitroaromatics and their amine metabolites there is a critical pair. The isomers 2,6-and 2,4-diaminotoluene have the same mass spectra so it is necessary to resolve these compounds by gas chromatography. Methylsilicone phases lead to co-elution of the compounds. However, good separation is possible using a fluoropropyl phase (e.g. Restek Rtx-200), which is also unusually thermally stable up to 360 °C (Fig. 4-165).

Detection with GC/MS systems has particular advantages for the analysis of nitroaromatics and their metabolites compared with ECD or NPD. As can be seen from the chromatogram shown in Fig. 4-165, the components 11 (2,4-dinitrotoluene) and 12 (2-amino-4-nitrotoluene) co-elute in the separation. These compounds can be differentiated in the GC/MS system through their characteristic mass traces. The partial co-elution of the components 18 (4-amino-2,6-trinitrotoluene), 19 (2,3,4-trinitrotoluene) and 21 (2,6-diamino-4-nitrotoluene) can also be resolved using the different mass spectra. Besides this specificity in detection the response of the detector plays a critical role. The use of ECD for residue analysis of mono- and dinitrated compounds has limitations, while the response of the mass spectrometer is independent of the degree of nitration. Furthermore, as important degradation products the amines are determined at the same detection level using the mass spectrometer. The selectivity can be increased still further by using chemical ionisation.

Analysis Conditions

SFE extraction:	Suprex PrepMaster with AccuTrap	
	Sample quantity:	5 ml
	Fluid:	CO_2
	Modifier:	2% methanol
	Pressure:	400 bar
	Temperature:	80 °C
	Static:	15 min
	Dynamic:	50 ml, 2.5 ml/min
	Restrictor:	variable, 150 °C
	Cold trap:	+5 °C
	Elution medium:	methanol
GC/MS system:	Finnigan ITS 40	
Injector:	PTV cold injection system	
	Start:	60 °C,
	Ramp:	20 °C/s
	Final temp.:	320 °C for 10 min
Column:	SGE HT5, 25 m × 0.22 mm × 0.1 µm	
	Start:	60 °C,
	Ramp:	12 °C/min
	Final temp.:	330 °C for 7.5 min
Mass spectrometer:	Scan mode:	EI, full scan, 100–300 u
	Scan rate:	1 s/scan

Results

The SFE extraction of a soil sample from a typical disused military site gives an extract which can be analysed directly by GC/MS without further processing. The total ion current chromatogram shows a broad matrix peak (Fig. 4-169) with intense peaks of multiply overlapping individual components. In the evaluation two peaks can be found which are identified as chlorinated nitroaromatics on the basis of a spectroscopic comparison with the NIST library (Figs. 4-170 and 4-171). This is checked by comparison of the mass spectra and retention times with those of commercial standards.

References: Holger Bittdorf, Kernforschungszentrum Karlsruhe.
Peter Lönz, TERRACHEM Essen GmbH, Essen.

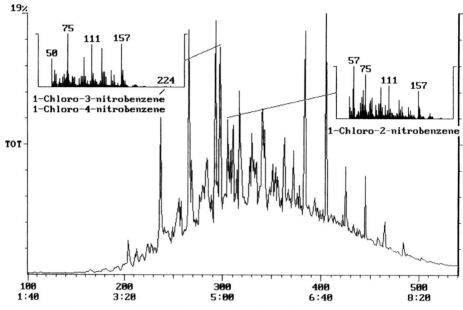

Fig. 4-169 Chromatogram of the SFE extract of a soil sample from a disused military site

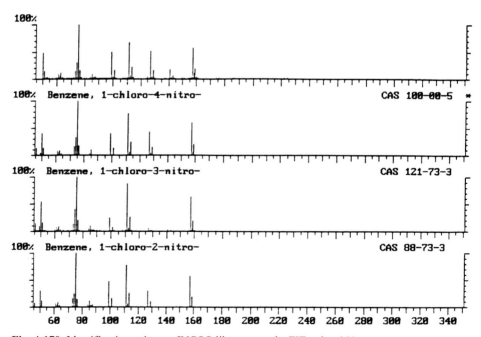

Fig. 4-170 Identification using an INCOS library search: FIT value 951

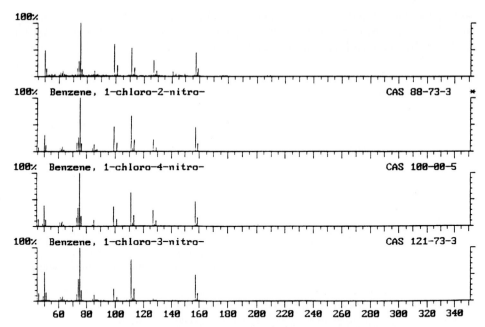

Fig. 4-171 Identification using an INCOS library search: FIT value 959

4.30 Detection of Drugs in Hair

Key words: SFE, opiates, pharmaceuticals, cocaine, methadone, fortifying, MS/MS

Analysis of hair allows information to be obtained concerning a period extending further back than that which can be examined using blood and urine samples. Testing of hair can be used to detect repeated and chronic drug misuse. Narcotics and other addictive drugs and, for example, medicines can be detected unambiguously using the method described and this satisfies the current legal requirements. With hair analysis the chronic misuse of these substances can be traced back over weeks and months (hair growth amounts to ca. 1 cm/month). Amphetamines, cannabis, cocaine, methadone and opiates, for example, can be detected. As this is a stepwise investigation, the development of drug use over a period can be visualised. Monitoring drug dependence in the methadone programme is one application of this type of analysis, as is doping detection in competitive sport. Hair analysis cannot be used to detect an acute drug problem, as a blood or urine test can. The quantity examined should be at least 50 mg. Sometimes a very small quantity such as the quantity of hair found daily in a shaver can be sufficient.

SFE is used for extraction, whereby it is assumed that all the active substances which can be determined by GC can be extracted with supercritical CO_2. SFE has the advantage of a higher extraction rate, as, unlike other procedures, it takes only a few minutes. In addition the extraction unit can be coupled directly to the GC/MS allowing the method to be automated and ever smaller quantities to be analysed.

Analysis Conditions

Sample: A bundle of hair of pencil thickness is used, which is fixed in such a way that movement of the individual hairs over one another is inhibited. Sections of 1 cm length are ground and extracted using SFE.

Extraction: 20 mg hair powder are treated with 20 µl ethyl acetate in the extraction cartridge (volume 150 µl).
Oven temp.: 60 °C
Static extraction: 15 min, 200–300 bar
Dynamic extraction: 5 min, 200–300 bar
Extract isolation: transfer of the eluate to a sample bottle with ca. 1 ml ethyl acetate; derivatisation with pentafluoropropionic anhydride (PFPA).

GC/MS/MS system: Finnigan TSQ 700
Injector: Gerstel KAS cold injection system, splitless injection
Column: J&W DB-5, 30 m × 0.32 mm × 0.25 µm
Helium, 1 ml/min
Program: Start: 60 °C, 2 min
Program 1: 25 °C/min to 230 °C
Isotherm: 230 °C, 5 min
Program 2: 25 °C/min to 280 °C
Final temp.: 280 °C, 5 min

Mass spectrometer: Scan mode: EI, MS/MS (product ion scan mode, Fig 4-172)
CID gas: argon, 2 mbar; collision offset –25 eV

Results

In the method described the SFE was not coupled directly to the GC/MS (on-line SFE-GC/MS), but the extraction, derivatisation and analysis were carried out separately. The use of different pressures (100, 200, 300 bar) shows the different extraction possibilities. At 200 bar the extract produced showed the best signal/noise ratio for heroin (Fig. 4-173). Using derivatisation morphine is detected as morphine-2-PFP and monoacetyl-morphine (MAM) as MAM-PFP together with underivatised heroin.

As it is known that the more lipophilic substances, such as heroin, cocaine and tetra-hydrocannabinol carboxylic acid (THC), can be deposited directly in hair and do not need to be derivatised for chromatography, SFE can in fact be coupled directly with GC/MS/MS.

Reference: Dr. H. Sachs, Institut für Rechtsmedizin, Munich.

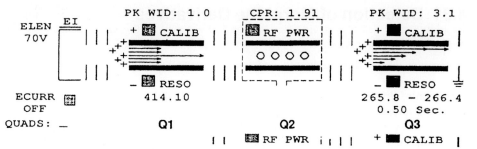

Fig. 4-172 Principle of MS/MS in the product ion scan mode
(1) The component mixture in the sample reaches the ion source of the MS and is ionised there.
(2) Substance-specific ions are filtered out of the ion mixture in the first quadrupole section (Q1, MS1)
(3) The ions are focused into a collision chamber (Q2) by a lens system. Collisions with argon effect fragmentations forming product ions (CID, collision induced dissociation). A high frequency field (RF) inhibits scattering effects without separating masses.
(4) The resulting mixture of ions (product ion spectrum) is separated by the second quadrupole section (Q3, MS2).

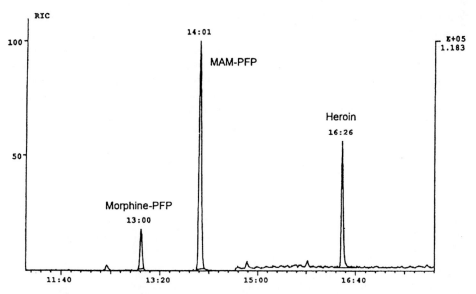

Fig. 4-173 Chromatogram of a hair sample extracted by SFE (200 bar), taken in product ion scan mode

4.31 Detection of Morphine Derivatives

Key words: drug screening, heroin, codeine, morphine, hydrolysis, derivatisation

Detection of drug taking by the investigation of blood and urine samples is one of the tasks of a forensic toxicology laboratory. The routine analysis of drug screening can be carried out by TLC, HPLC or by immunological methods. Positive results require confirmation. Determination using GC/MS is recognised as a reference method. The full scan method is preferred over the SIM procedure for differentiating between different drugs in a comprehensive screening procedure because of the higher specificity and universality. The decision limits (Table 4-24) are laid down by the US Ministry of Health and the US Ministry of Defense.

Heroin (3,6-diacetylmorphine) is usually not determined directly. The unambiguous detection of heroin consumption is carried out by determining 6-monoacetylmorphine (6-MAM) which is formed from heroin as a metabolite (Fig. 4-174). If morphine is difficult or impossible to detect when clean-up is carried out without hydrolysis, then the latter should be used. It should be noted that nonspecific hydrolases can also effect the degradation of 6-MAM to morphine. If the substitute drugs methadone and dihydrocodeine are present, these are also determined using the procedure described. As blood from corpses also has to be processed during routine operations to some extent, a comparatively time-consuming extraction including a re-extraction is carried out. For the gas chromatography of morphine derivatives derivatisation of the extract by silylation, acetylation, or pentafluoropropionylation, for example, is basically necessary. Mass spectrometric detection using negative chemical ionisation is possible by using fluorine derivatives. In the procedure described silylation with MSTFA (N-methyl-N-trimethylsilyl-trifluoroacetamide) and detection in the EI mode has been chosen.

Table 4-24 Decision limits in drug screening (data in ng/ml)

Active substance	Screening		Confirmation	
	HHS	DOD	HHS	DOD
THC	100*	50	15	15
BZE	300	150	150	100
Opiates	300	300	300	300
AMPs	1000	500	500	500
PCP	25	25	25	25
BARBs	k.A.	200	k.A.	200
LSD	k.A.	0.5	k.A.	0.2

* HHS suggestion 50 ng/ml

THC 11-nor-Δ-9-tetrahydrocannabinol-9-carboxylic acid PCP   Phencyclidine
BZE benzoylecgonin   BARB barbiturates
AMP amphetamines   LSD   lysergic acid diethylamide

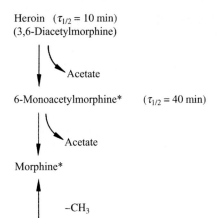

Heroin ($\tau_{1/2}$ = 10 min)
(3,6-Diacetylmorphine)

6-Monoacetylmorphine*   ($\tau_{1/2}$ = 40 min)

Morphine*

Codeine*
(3-Methylmorphine)

Fig. 4-174 Opiate metabolism (* determination by GC/MS)

Analysis Conditions

Clean-up for serum (blood):
    1 ml sample, addition of the internal standard (e.g. 100 ng morphine-d_3 or codeine-d_3), dilution to 7 ml with phosphate buffer (pH 6).
    SPE extraction using Bond-Elut-Certify (130 mg, Varian No. 1211–3050) after conditioning with 2 ml methanol and 2 ml phosphate buffer (pH 6).
    Elution with 2 × 1 ml chloroform/isopropanol/25% ammonia (70:30:4).
    Derivatisation of the extract residue with 50 µl MSTFA (80 °C, 30 min).

Clean-up for urine:
    Hydrolysis of 2 ml sample with 20 µl enzyme solution (β-glucuronidase) (60 min, 60 °C),
    adjustment of sample to pH 8–9 with 0.1 N sodium carbonate solution,
    activation of C_{18}-cartridge with methanol and conditioning with 0.1 N sodium carbonate solution,
    application of sample and washing of cartridge with 0.1 N sodium carbonate solution,
    elution of C_{18}-cartridge with 1 ml acetone/chloroform (50:50),
    evaporation of eluate to dryness, treatment of the residue with 50 µl MSTFA (80 °C, 30 min).

Gas chromatograph: Finnigan GCQ GC
split/splitless injector, 275 °C
splitless injection, split open at 0.1 min

Column: J&W DB-1, 30 m × 0.25 mm × 0.25 µm
Helium 40 cm/s

Program: Start: 100 °C
Ramp: 10 °C/min
Final temp.: 310 °C

Mass spectrometer: Finnigan GCQ
Scan mode: EI
Scan range: 70–440 u
Scan rate: 0.5 s/scan

Results

The chromatogram and mass spectra after clean-up of a typical serum sample are shown in Fig. 4-175. The mass traces for codeine, morphine-d_3 and morphine and the total ion current are shown. For morphine a concentration of 60 ng/l is calculated with reference to morphine-d_3 using a calibration curve. The resulting mass spectra allow identification, for example on comparison with relevant toxicologically orientated spectral libraries. The clean-up and analysis methods can be used for other basic drugs, such as amphetamine derivatives, methadone and cocaine and their metabolites.

A list of morphine derivatives and synthetic opiates which have been found by the procedure described after clean-up of serum samples is shown in Table 4-25. For the corresponding trimethylsilyl derivatives the specific search masses are given which allow identification in combination with the retention index.

Table 4-25 Selective masses of the opiates as TMS derivatives (after Weller, Wolf)

Opiate	Ret.- Index	m/z values
Levorphanol-TMS	2188	329, 314
Pentazocine-TMS	2262	357, 342
Levallorphan-TMS	2318	355, 328
Dihydrocodeine-TMS	2365	373, 315
Codeine-TMS	2445	371, 343
Hydrocodone-TMS	2447	371, 356
Dihydromorphine-TMS	2459	431, 416
Hydromorphone-TMS	2499	357
Oxycodone-TMS	2503	459, 444
Morphine-bis-TMS	2513	429, 414
Norcodeine-bis-TMS	2524	429
Monoacetylmorphine-TMS	2563	399, 340
Nalorphine-bis-TMS	2656	455, 440

Note: All TMS derivatives show an intense signal for the trimethylsilyl fragment at m/z 73, which does not appear when the scan is begun at m/z 100 and is unnecessary for substance confirmation. The retention indices were determined on DB-1, 30 m × 0.25 mm × 0.25 μm with an n-alkane mixture of up to C_{32}.

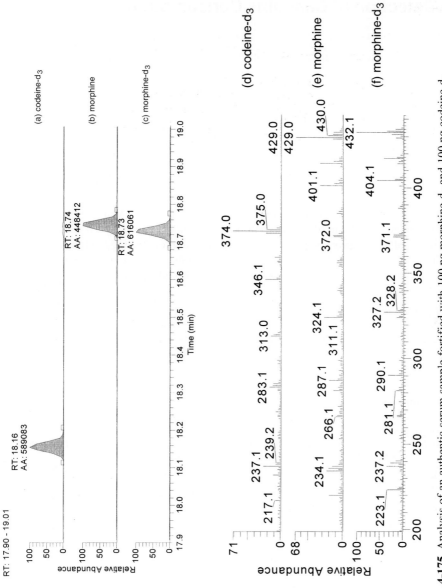

Fig. 4-175 Analysis of an authentic serum sample fortified with 100 ng morphine-d$_3$ and 100 ng codeine-d$_3$
Above: (a) Mass trace for codeine-d$_3$ (m/z 374). (b) Mass trace for morphine (m/z 429)
(c) Mass trace for morphine-d$_3$ (m/z 432)
Below: (d) Spectrum of codeine-d$_3$ (e) Spectrum of morphine
(f) Spectrum of morphine-d$_3$

4.32 Detection of Cannabis Consumption

Key words: THC, metabolite, SPE, derivatisation, quantitation, internal standard

Because motorists and those in the workplace are now tested for alcohol and drugs, more samples are being produced. The use of a rapid and reliable analysis procedure is therefore necessary for routine operations. The clean-up of serum and urine samples by solid phase extraction, combined derivatisation and elution and GC/MS detection fulfils these requirements.

To detect cannabis consumption THC (tetrahydrocannabinol) and the main metabolite 11-nor-Δ-9-tetrahydrocannabinol-9-carboxylic acid (9-carboxy-THC) are determined. The sample preparation method described here involves alkaline hydrolysis of THC bonded to glucuronide and clean-up by special solid phase extraction. Instead of the SPE cartridges usually used, a small SPE filter disc is employed, which is treated directly with the derivatisation agent after applying the sample, washing, and drying. This allows the derivatisation of the 9-carboxy-THC to take place at the same time as the elution and thus dispenses with several steps and the consumption of extra solvent. The 9-carboxy-THC is detected as the TMS derivative. Detection is confirmed from the mass spectrum. For quantitation the internal standard d_9-9-carboxy-THC is used and the mass chromatogram is evaluated. Alternatively MS/MS can be used for detection. Through the increased selectivity limits of detection in the lowest ng/ml range can be achieved for serum samples.

Analysis Conditions

Sample preparation:
Hydrolysis: 5 ml urine are treated with 200 µl 10 N KOH and incubated for 15 min at 60 °C.
Clean-up: after cooling of the hydrolysate, 700 µl conc. acetic acid are added, the sample passed through a ToxiLab Spec extraction disc (SPE), the disc washed with 1 ml 20 % acetic acid, and dried (15 min, vacuum). The SPE disc is placed in a sealable vessel, 75 µl BSTFA are added for derivatisation and elution and the disc left for 10 min at 90 °C. The sample is injected into the GC/MS system without further steps.

GC/MS/MS system: Finnigan Witness System
Column: SGE HT8, 25 m × 0.25 mm × 0.25 µm
Program: Start: 60 °C, 1 min
Ramp: 25 °C/min
Final temp.: 330 °C

Mass spectrometer: Scan mode: EI, full scan, 200–500 u
Scan rate: 0.6 s/scan

Calibration: for calibration d_9-9-carboxy-THC is added at a concentration of 50 ng/ml and 9-carboxy-THC in concentrations of 2.5, 5, 15, 50, 200 and 500 ng/ml as the internal standards to 5 ml THC-free urine.

Results

The extracts from the disc clean-up (SPE) are very clean, and give a high recovery and good quantitative precision. 9-Carboxy-THC can be detected in urine samples down to the threshold value of 15 ng/ml recommended by the NIDA (US National Institute on Drug Abuse) by examining the whole mass spectrum (Fig. 4-176). In the calibration a linear response in the range of 5 ng/ml to 500 ng/ml is achieved. For this procedure involving urine the calibration graph is particularly precise.

In the analysis of serum samples for the nonmetabolised active substance THC there is a much stronger chemical background in the chromatogram, which does not allow the limits of quantitation given for urine to be achieved so easily. The effect of the matrix can be effectively counteracted by detection of THC in the MS/MS mode and effective limits of quantitation of below 1 ng/ml serum can be achieved. Fig. 4-177 shows the product ion spectrum of THC-TMS arising from the precursor ion m/z 386. THC-d_3 is used as the internal standard. The GC peak is successfully integrated on the mass chromatogram of the base peak m/z 371 and that of the internal standard m/z 374.

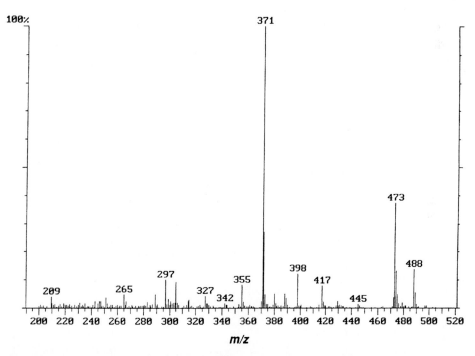

Fig. 4-176 Mass spectrum of the 9-carboxy-THC-TMS derivative at the NIDA threshold value of 15 ng/ml

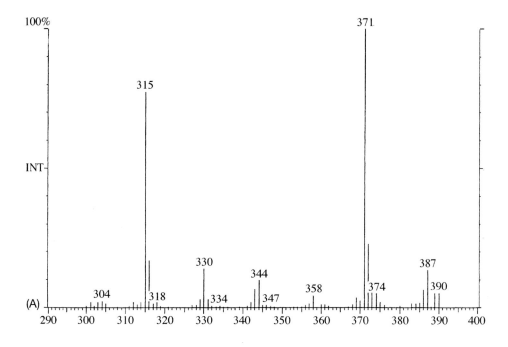
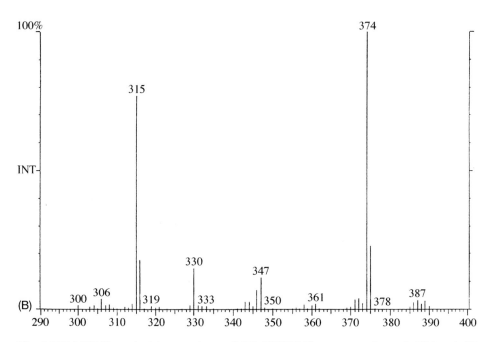

Fig. 4-177 MS/MS product ion spectrum of (A) THC-TMS, precursor ion m/z 386 and (B) THC-d$_3$-TMS, precursor ion m/z 389 (after Weller, Wolf 1995, Finnigan GCQ)

4.33 Determination of Phencyclidine

Key words: phencyclidine, PCP, drug analysis, SPE

Phencyclidine (PCP) is used illegally as an intoxicating substance. Originally it was used as an anaesthetic in surgery. Because of its wide misuse PCP has been included in the drug control programme in the USA. Laboratories which operate according to the guidelines of NIDA (US National Institute on Drug Abuse) have to identify and quantify PCP in urine down to a concentration of 25 ng/ml. PCP is detected in urine using solid phase extraction (SPE) and GC/MS detection. The identity is confirmed from the mass spectrum. For quantitation difluoro-PCP is used as the internal standard.

Analysis conditions

Sample preparation: 5 ml urine are extracted using SPE (Bondelut), the eluate is evaporated to dryness in a stream of nitrogen, taken up in 50 µl ethyl acetate and transferred to microvials.

GC/MS system: Finnigan ITS 40
equipped with autosampler A200S for injection from 200 µl microvials
Column: SGE HT8, 25 m × 0.25 mm × 0.25 µm
Program: Start: 60 °C, 1 min
Ramp: 25 °C/min
Final temp.: 330 °C

Mass spectrometer: Scan mode: EI, full scan, 50–400 u
Scan rate: 1 s/scan

Calibration: PCP-free urine is treated with the internal standard difluoro-PCP at a concentration of 50 ng/ml. For calibration of the procedure six different PCP concentrations of between 0.5 and 500 ng/ml are used.

Results

The method described for the determination of PCP shows good sensitivity. At the threshold limit of 25 ng/ml required by the NIDA the PCP peak can be detected with a signal/noise ratio of 100:1 (Fig. 4-178). At this dilution level PCP can be identified and confirmed unambiguously using the mass spectrum by comparison with spectral libraries. The mass spectrum (Fig. 4-179) shows the molecular ion and a characteristic dominant fragment at m/z 200, which is used for quantitation of the PCP peak. The calibration of the PCP determination is linear over a very wide range from 0.5 ng/ml to 500 ng/ml. The regression gives a correlation factor of 0.999. Consequently even concentrations below the required threshold value can be determined reliably with the procedure described.

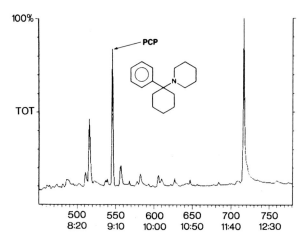

Fig. 4-178 Chromatogram of a Phencyclidine determination at the threshold value of 25 ng/ml urine

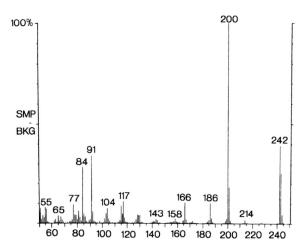

Fig. 4-179 Mass spectrum of Phencyclidine (concentration 25 ng/ml)

4.34 Analysis of Steroid Hormones using MS/MS

Key words: anabolics, derivatisation, GC/MS, MS/MS, signal/noise ratio, selectivity

Besides the area of pharmacology, the analysis of steroid hormones is important in doping analysis in sport. The steroids used as anabolics are derived from the androgenics. The use of the body's own steroids, such as testosterone, or compounds derived from them, such as methandienone, is forbidden. They are detected by analysing a urine sample and detection must be unambiguous. One difficulty with the analysis arises from the usually very low concentrations of individual active substances together with a high concentration of matrix in the extracts. Many laboratories only recognise a finding as positive when a full scan mass spectrum of the substance can be produced.

Methandienone, a widely used anabolic, is metabolised in the body and excreted as 6-β-hydroxymethandienone. The results presented here clarify the effect of the GC/MS/MS technique on the specificity of trace detection of the methandienone metabolite compared to GC/MS analysis. A standard solution of the metabolite at a dilution of 1 ng/µl and the extract of a fortified urine sample (10 ng/ml) with a concentration of 500 pg/µl are used for the measurement. For the gas chromatography the standard and the urine extract were converted into the tri-TMS derivatives.

Analysis Conditions

GC/MS system:	Finnigan GCQ *Tandem*	
Column:	HP 1, 25 m × 0.2 mm × 0.1 µm	
Program:	Start:	90 °C,
	Ramp:	15 °C/min
	Final temp.:	320 °C
Mass spectrometer:	Scan mode:	EI, full scan, 50–600 u
	Scan rate:	0.7 s/scan
	Ion source:	175 °C

Results

The mass spectra of 6-β-hydroxymethandienone are shown in Fig. 4-180. The EI-GC/MS spectrum shows a dominant base ion at m/z 517 arising from the molecular ion at m/z 532 by loss of the methyl group. A further fragmentation is insignificant. The product ion spectrum of the EI fragment m/z 517 is shown in Fig. 4-180 b. The intensity of the precursor ion has almost completely disappeared in favour of the characteristic product ions. The product ion signals m/z 229 and 337 have high intensity.

The EI-GC/MS analysis of the urine extract at a concentration of 10 ng/ml is shown in Fig. 4-181a together with the mass chromatograms of the molecular ion m/z 532 and the base peak m/z 517. At the expected retention time of the metabolites of 15:33 min no signal can be seen on any of the mass traces. Both masses are affected by chemical noise in the sample. It is not possible to obtain the mass spectrum of the metabolites at

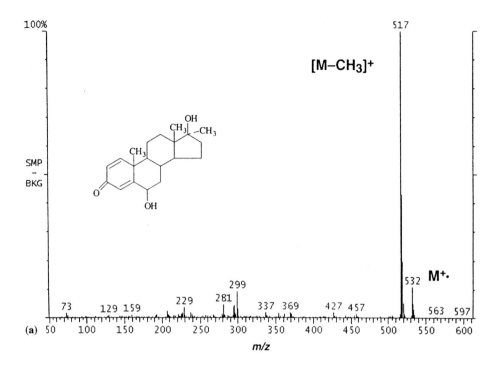
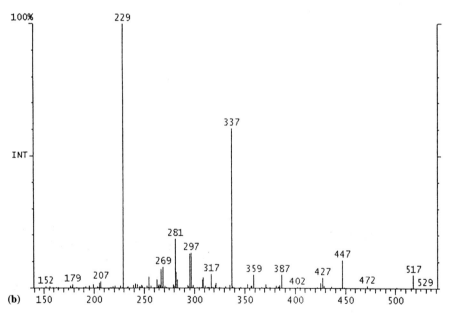

Fig. 4-180 Mass spectra of 6β-hydroxymethandienone as the tri-TMS derivative (M 532)
(a) EI-GC/MS
(b) EI-GC/MS/MS

4.34 Analysis of Steroid Hormones using MS/MS 511

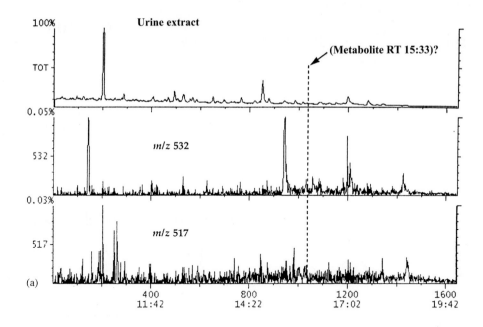

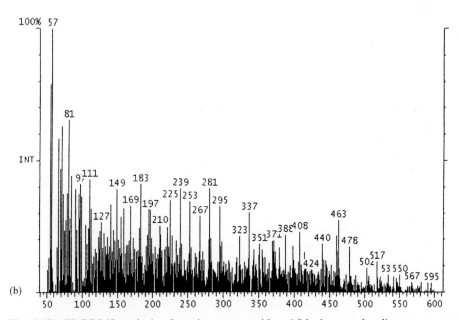

Fig. 4-181 EI-GC/MS analysis of a urine extract with a 6β-hydroxymethandienone concentration of 10 ng/ml
(a) Mass chromatograms m/z 532, 517
(b) Spectrum at the expected retention time 15:33 min

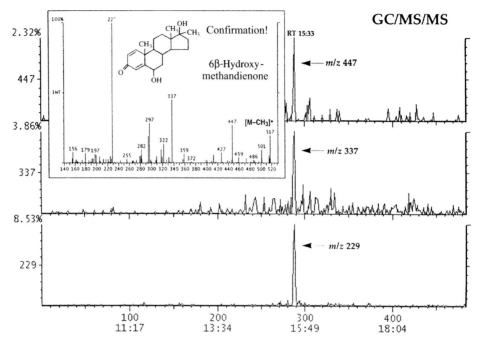

Fig. 4-182 EI-GC/MS/MS analysis of the urine extract from Fig. 4-181 with the product ion spectrum of 6 β-hydroxymethandienone

this point (Fig. 4-181 b). In the upper mass range also many unspecified matrix signals are present which completely mask the signal of the 6-β-hydroxymethandienone.

The effect of the co-eluting matrix can be eliminated using EI-GC/MS/MS analysis. The MS/MS analysis of the urine extract shows a clear picture (Fig. 4-182). Through the isolation of the precursor ion m/z 517 and the generation of the product ion spectrum, the appearance of the 6-β-hydroxymethandienone in the urine is unequivocal. The mass chromatograms of the product ions m/z 229, 337 and 447 show a common peak maximum at the expected retention time. For the intense product ion m/z 229 a peak with a high signal/noise ratio is detected, which is completely isolated from the matrix by the selectivity of the MS/MS technique. The product ion spectrum from the urine sample matches the standard very well and unquestionably confirms the detection of anabolics (Fig. 4-180 b).

4.35 Determination of Prostaglandins using MS/MS

Key words: prostaglandins, urine, derivatisation, GC/MS, MS/MS, co-elution, quantitation

Prostaglandins are cyclic derivatives of arachidonic acid (5,8,11,14-eicosatetraenic acid). They occur widely in organisms in very low concentrations and exhibit many types of mediator activity similar to that of hormones. For example, blood pressure, stomach secretion, aching muscles, general pain and fevers are affected by prostaglandins. Many human prostaglandins are synthesised in the kidneys. Synthetic derivatives are also known. The analysis of the arachidonic acid metabolism and investigations of the pharmacological spectrum of activity of prostaglandins has been of great scientific interest in recent years. While prostaglandins are degraded very rapidly in an organism, the concentration in urine can be used to check prostaglandin production in the kidneys.

For the analysis of body fluids GC/MS and GC/MS/MS are used as a reference and addition to immunological procedures. The classical procedures are not specific enough so the values determined can be too high as a consequence of the accompanying matrix (Schweer 1986). The concentrations recorded for healthy children lie in the range of 100 pg/ml urine, and in the case of children with Zellweger syndrome the values can rise by more than 100 fold (8-iso-$PGF_{2\alpha}$).

For sample preparation (typically 5 ml urine) SPE is used as described here with subsequent PFB derivatisation and additional purification of the extracts by thin layer chromatography. For quantitative determinations reversed phase HPLC can be used for further purification of the extracts. This may however be omitted when MS/MS detection is used without loss of quality (Tsikas 1998). The keto and OH groups are then derivatised through formation of methoximes and TMS ethers (BSTFA).

In spite of the time-consuming sample preparation, with one stage GC/MS analysis co-elutions cannot be reliably excluded in spite of selectivity by single mass registration (SIM, MID). Only by the selective GC/MS/MS procedure described here (Schweer 1988, Tsikas 1998) can prostaglandins be determined reliably for diagnostic purposes. The possibilities for MS/MS are shown graphically. A direct comparison between triple quadrupole and ion trap instruments is shown.

Analysis Conditions

(1) (Tsikas 1998)

GC/MS system:	Finnigan TSQ 45 (triple stage quadrupole GC/MS/MS mass spectrometer)
Gas chromatograph:	Finnigan GC 9611
Injector:	splitless injection, 280 °C, 2 min splitless
Column:	SPB 1701: 30 m × 0.25 mm × 0.25 µm
	Helium, constant pressure 70 kPa
Program:	Start: 80 °C, 2 min
	Program 1: 250 °C, 25 °C/min
	Program 2: 280 °C, 2 °C/min
	Final temp: 280 °C, 5 min

MS interface:	280 °C	
Mass spectrometer:	Ionisation:	NCI, argon/methane (0.2/65 Pa)
	Ion source:	140 °C
	Scan mode:	MS/MS product ion scan
	Collision gas:	argon/methane (0.2/65 Pa)
	Collision energy:	18 eV

(2) (Gummersbach)

GC/MS system:	Finnigan Polaris (ion trap GC/MS/MS mass spectrometer)	
Gas chromatograph:	CE Instruments Trace GC 2000	
Injector:	splitless injection, 260 °C, 2 min splitless	
Column:	J&W 1301: 30 m × 0.32 mm × 0.25 µm	
	Helium, constant flow ca. 30 cm/s	
Program:	Start:	80 °C, 2 min
	Program 1:	230 °C, 25 °C/min
	Program 2:	280 °C, 8 °C/min
	Final temp.:	280 °C, 22 min
MS interface:	280 °C	
Mass spectrometer:	Ionisation:	NCI, methane
	Ion source:	150 °C
	Scan mode:	MS/MS product ion scan
	Collision gas:	helium
	Collision energy:	1.4 V

Results

The GC-NCI-MS/MS procedure presented allows rapid and specific determination of prostaglandins in human urine. The retention times obtained with the most widely used GC columns are shown in Table 4-26. Table 4-27 shows the most important ions in the one-stage MS spectrum of the compounds arriving for analysis. In the case of $PGF_{2\alpha}$-PFB-TMS derivatives the fragment obtained after cleavage of the PFB group gives rise to the base peak. This ion is used as the precursor ion for the MS/MS analysis used in the selective determination. The resulting MS/MS spectra are summarised in Table 4-28.

To carry out the analyses only triple quadrupole systems are used in the literature cited. The use of the recently developed clearly better value ion trap GC/MS/MS systems gives results of comparable selectivity at the required level of sensitivity (Figs. 4-183 and 4-184). The spectra obtained show the fragmentation pattern used for identification in the same way and with comparable intensities (Fig. 4-185). Similarly their use for analysis of other precipitated prostanoid components in the urine of sick children is also shown (Fig. 4-186) and thus makes the alternative use of ion trap systems worth considering, for example, for the diagnoses of Zellweger syndrome in children, as well as for research tasks.

Reference: J. Gummersbach, ThermoQuest Applications Laboratory, Egelsbach, Germany

Table 4-26 Gas chromatographic retention times of PFB-TMS derivatives of the prostanoids investigated using two different capillary columns (after Tsikas 1998)

F_2-Prostaglandin derivative	Retention time (min)/rel. ret. time DB5-MS	Retention time (min)/rel. ret. time SPB-1701
9β,11α-PGF$_{2\alpha}$-PFB-TMS	22.88/1.0000	22.05/1.0000
^{2}H$_4$-8-iso-PGF$_{2\alpha}$-PFB-TMS	22.90/1.0009	22.19/1.0063
8-iso-PGF$_{2\alpha}$-PFB-TMS	22.97/1.0039	22.25/1.0091
9α-11β-PGF$_{2\alpha}$-PFB-TMS	23.48/1.0262	22.62/1.0258
^{2}H$_4$-PGF$_{2\alpha}$-PFB-TMS	23.70/1.0389	23.05/1.0454
PGF$_{2\alpha}$-PFB-TMS	23.77/1.0389	23.11/1.0481

Table 4-27 GC-NICI-MS Mass Spectra of PGF$_{2\alpha}$-PFB-TMS Derivatives, I = PGF$_{2\alpha}$-PFB-TMS, II = 3,3',4,4'-^{2}H$_4$- PGF$_{2\alpha}$-PFB-TMS (ions [M-PFB]$^-$ = [P]$^-$, after Schweer 1988)

Ion assignment	I	II	% Int.
[P]$^-$	569	573	100
[P-(CH$_3$)$_2$Si=CH$_2$]$^-$	481	485	8
[P-TMSOH]$^-$	479	483	10
[P-TMSOH-(CH$_3$)$_2$Si=CH$_2$]$^-$	407	411	1
[P-2xTMSOH-(CH$_3$)$_2$Si=CH$_2$]$^-$	317	321	1
[P-3xTMSOH]$^-$	299	303	1
[C$_6$F$_6$CHO]$^-$	196	196	1
[C$_6$F$_6$CH$_2$]$^-$	181	181	2
[C$_6$F$_4$CH$_2$O]$^-$	178	178	1

Table 4-28 Major mass fragments (intensity >5% is given in parentheses) in the GC-NCI-MS/MS mass spectra of the PFB-TMS derivatives of the prostaglandins (precursor ions [M-PFB]$^-$ = [P]$^-$, after Tsikas 1998)

Ion assignment	8-iso-PGF$_{2\alpha}$	^{2}H$_4$-8-iso-PGF$_{2\alpha}$	PGF$_{2\alpha}$	^{2}H$_4$-PGF$_{2\alpha}$
[P]$^-$	569 (26)	573 (15)	569 (10)	573 (17)
[P-TMSOH]$^-$	479 (17)	483 (15)	479 (12)	483 (30)
[P-2xTMSOH]$^-$	389 (32)	393 (37)	389 (32)	393 (40)
[P-2xTMSOH-(CH$_3$)$_2$Si=CH$_2$]$^-$	317 (23)	321 (33)	317 (46)	321 (28)
[P-3xTMSOH]$^-$	299 (100)	303 (100)	299 (100)	303 (100)
[P-2xTMSOH-(CH$_3$)$_2$Si=CH$_2$-CO$_2$]$^-$	273 (33)	273 (60)	277 (50)	277 (37)
[P-3xTMSOH-CO$_2$]$^-$	255 (76)	255 (62)	259 (50)	259 (33)

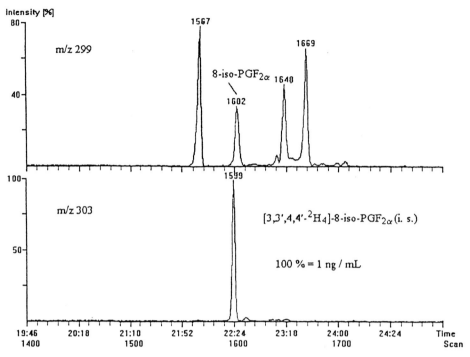

Fig. 4-183 Partial chromatogram from the TSQ GC-NCI-MS/MS analysis of urine samples from a healthy child (TLC fraction) containing the PFB derivatives of 8-iso-PGF$_{2\alpha}$ (m/z 299) and 3,3',4,4'-^{2}H$_4$-PGF$_{2\alpha}$ (m/z 303), (cap. column SPB-1701, after Tsikas 1998)

4.35 Determination of Prostaglandins using MS/MS

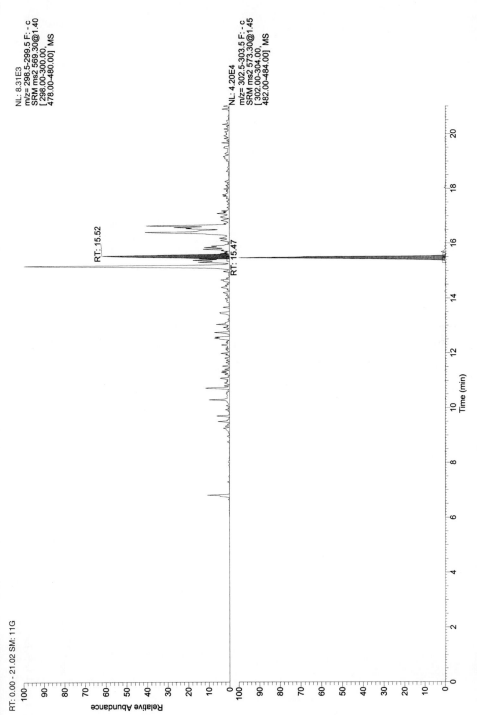

Fig. 4-184 Partial chromatogram from the polaris GC-NCI-MS/MS analysis of urine samples from a healthy child (TLC fraction) containing the PFB derivatives of 8-iso-PGF$_{2\alpha}$ (m/z 299, upper trace) and 3,3′,4,4′-^{2}H$_4$-PGF$_{2\alpha}$ (m/z 303, lower trace), (cap. column J&W 1301, after Gummersbach 1999)

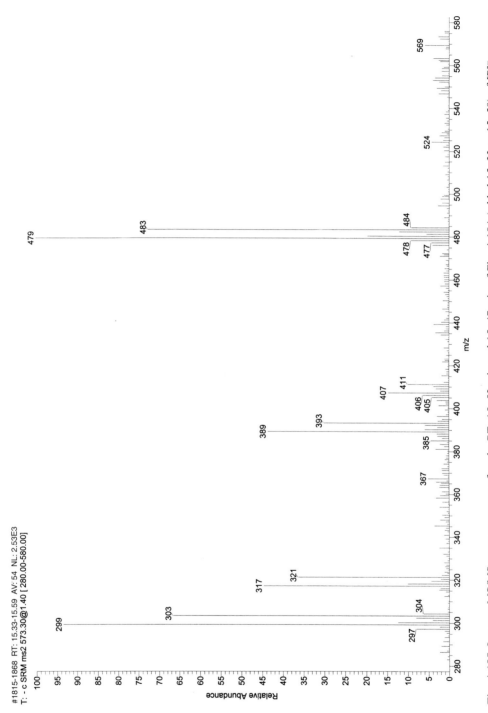

Fig. 4-185 Ion trap MS/MS mass spectrum of peaks RT: 15:52 min and 15:47 min of Fig. 4-184 (added 15:33 to 15:59) of [P]⁻ precursor ions m/z 569 (PGF$_{2\alpha}$-PFB-TMS) and m/z 573 (3,3',4,4'-^{2}H$_4$-PGF$_{2\alpha}$-PFB-TMS)

4.35 Determination of Prostaglandins using MS/MS

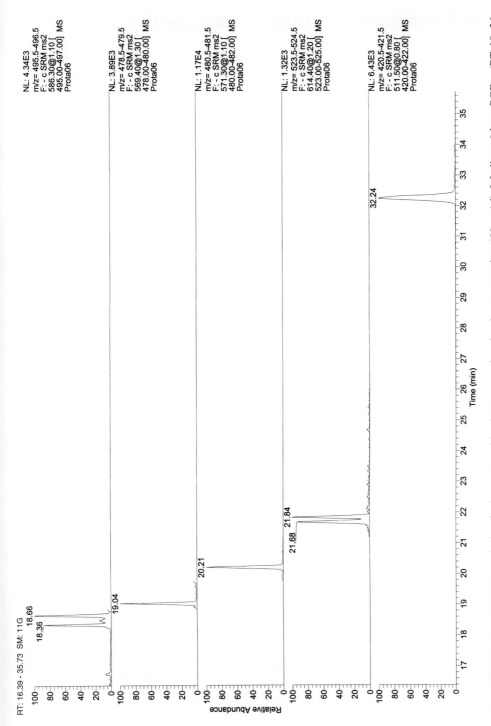

Fig. 4-186 Ion trap MS/MS analysis of other protanoid substances at low level concentration (50 pg/ul) 2,3-dinor-6-keto-PGF$_{1\alpha}$ (RT 18:36, 18:66 two isomers, precursor m/z 586.3), 8-iso-PGF$_{2\alpha}$ (RT 19:04, precursor m/z 569.4), PGF$_{1\alpha}$ (RT 20:21, precursor m/z 571.3), 6-keto-PGF$_{2\alpha}$ (RT 21:68, 21:84 two isomers, precursor m/z 614.4), 11-dehydro-TxB$_2$ (RT 32:24, precursor m/z 511.5)

4.36 Amphetamines – Differentiation by CI

Key words: amphetamines, TMS derivatives, HFBA derivatives, quantitation, chemical ionisation

Amphetamines act as stimulants. A large number of active substances in this class are used for therapeutic purposes. Amphetamines are also involved in doping and the consumption of illegal drugs (e. g. Ecstasy). The analysis conditions for the determination of amphetamines are given in detail in Section 4.39. Amphetamines are detected in serum or urine after basic C_{18}-SPE extraction as TMS derivatives (MSTFA) or heptafluorobutyric acid derivatives (HFBA).

Results

Figure 4-187 shows the analysis of a serum sample for amphetamines using full scan data acquisition in the EI mode. The extract was converted into the TMS derivative with MSTFA. The mass chromatogram of the ions m/z 116+130 with a total ion current shows the elution of amphetamine-TMS (M 207), Prophylhexedrin-TMS (M 227) as the internal standard, and MDMA-TMS (methylenedioxymethamphetamine-TMS, M 265).

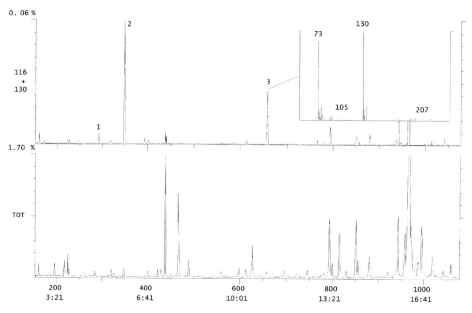

Fig. 4-187 Analysis of the serum sample from an Ecstasy case (ITD 800, after Weller, Wolf, Institut für Rechtsmedizin, Hanover)
(1) Amphetamine-TMS
(2) Propylhexedrin-TMS (ISTD)
(3) MDMA-TMS

4.36 Amphetamines – Differentiation by CI

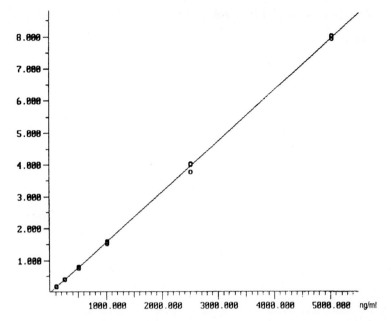

Fig. 4-188 Calibration function for amphetamine-HFB from 100 to 5000 ng/ml

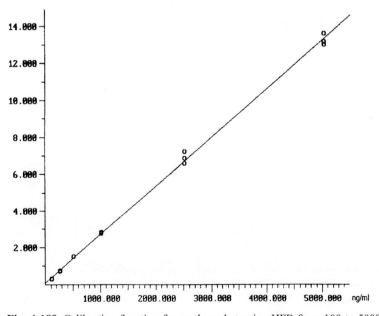

Fig. 4-189 Calibration function for methamphetamine-HFB from 100 to 5000 ng/ml

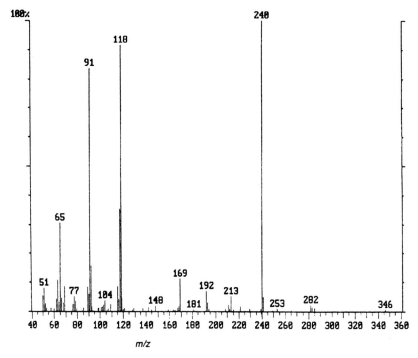

Fig. 4-190 Mass spectrum of amphetamine-HFB at 100 ng/ml

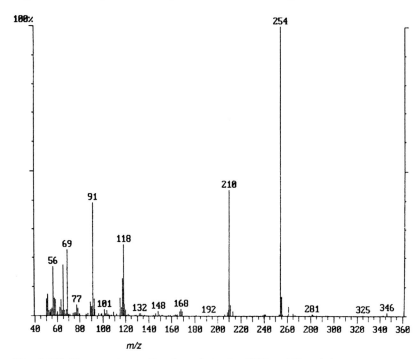

Fig. 4-191 Mass spectrum of methamphetamine-HFB at 100 ng/ml

4.36 Amphetamines – Differentiation by CI

The ease of chromatographing TMS derivatives and the high selectivity of the individual mass traces is evident. The molecular ions of the TMS derivatives do not appear in the EI spectra; only fragment ions are detected.

To quantify the amphetamines in urine, calibrations are carried out in the concentration range of 100–5000 ng/ml which gives good linearity under the measuring conditions chosen (Figs. 4-188 to 4-191). For the detection of methamphetamine it should be noted that amphetamine can be formed as a GC artefact depending on the injector temperature and the residence time. Even nonderivatised amphetamines can be separated well by GC, but, because of their basic character, require a clean, less active chromatography column.

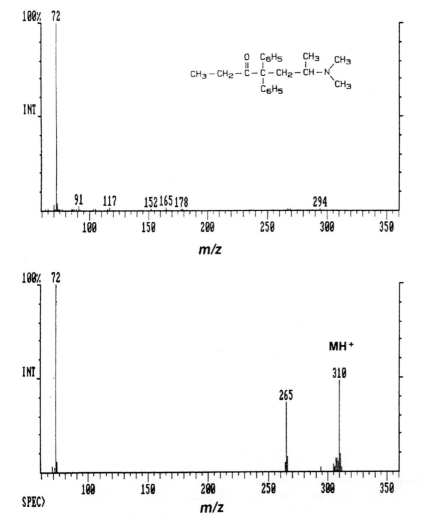

Fig. 4-192 Methadone (M 309)
Above: EI spectrum, below: CI spectrum (isobutane)

In spite of the soft ionisation, the product of α-cleavage (m/z 44, 58, 72) remains visible in the CI spectra of amphetamines. The protonated molecular ion is formed at high intensity with isobutane or ammonia. The methadone spectrum only shows the cleavage product m/z 72 in the EI mode; only under CI conditions can molecular information be obtained (Fig. 4-192). Mass spectrometric differentiation between amphetamines is possible using chemical ionisation with isobutane. Monoethylaminophenylpropanone (M 177) and monoethylaminopyhenylpropanol (M 179) give the same EI spectra. Only the detection of the (quasi)molecular ions with m/z 178 and m/z 180 gives an unambiguous identification (Figs. 4-193 and 4-194).

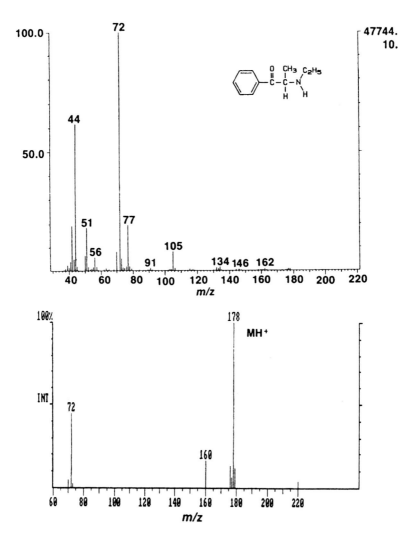

Fig. 4-193 Monoethylaminophenylpropanone (M 177)
Above: EI spectrum, below: CI spectrum (isobutane)

4.36 Amphetamines – Differentiation by CI

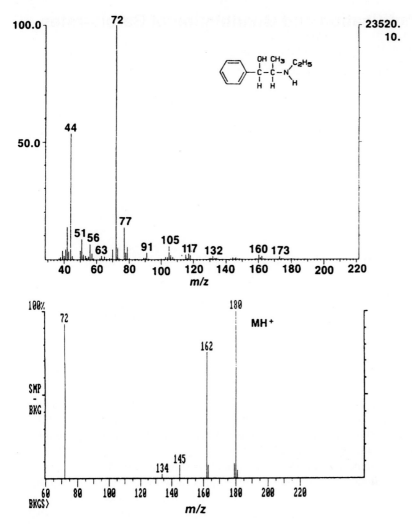

Fig. 4-194 Monoethylaminophenylpropanol (M 179)
Above: EI spectrum, below: CI spectrum (isobutane)

4.37 Identification and Quantitation of Barbiturates

Key words: barbiturates, chemical ionisation, confirmation, INCOS library search

Barbiturates have a calming effect on the central nervous system and over longer periods lead to tolerance resulting in the necessity of larger doses. The euphoric effect at higher doses is the cause of the frequently observed misuse. Routine analyses of barbiturates are based on immunological methods using Secobarbital antibodies. The sensitivity for the various active substances varies, however. Some active substances are confirmed by GC/MS with full scan data acquisition in the EI and CI modes. Clean-up, for example of urine samples, is carried out by solid phase extraction (SPE). Capillary columns with phenylmethylsilicone phases are used for chromatography.

Analysis Conditions

GC/MS system: Finnigan ITS 40
    Scan mode: EI, full scan, 50–400 u
          CI, full scan, 70–400 u, reagent gas isobutane
    Scan rate: 1 s/scan

Results

For the analysis a urine sample is fortified with Amobarbital, Butabarbital, Butalbital, Pentobarbital, Phenobarbital and Secobarbital in concentrations of 10, 25, 50, 100, 250 and 500 ng/ml and 3 ml are cleaned up using SPE. Hexobarbital is added as the internal standard at a concentration of 100 ng/ml.

The chromatographic separation of the barbitals is shown in Fig. 4-195 for the lower concentration of 10 ng/ml. The mass chromatograms show the co-elution of Butabarbital and Butalbital, which can be integrated separately using the selective mass traces. The results of the INCOS library search using the toxicology library (TX) suggests Butalbital with the best FIT value (Fig. 4-196). The difference spectrum (Fig. 4-197) is used for a new search and gives suggestions with very similar peak patterns which cannot be differentiated (Fig. 4-198). The differentiation and confirmation of the compounds was achieved by mass spectrometry using chemical ionisation with isobutane as the reagent gas. Figure 4-199 shows the CI spectrum of the co-eluting peak with the $(M+H)^+$ ions of Butabarbital m/z 213 and Butalbital m/z 225. The chromatogram of the CI analysis is shown in Fig. 4-200. The isomeric Amobarbital and Pentobarbital do not elute close together here. On the basis of their EI spectra these compounds can only be differentiated from the retention times (Fig. 4-201). Here again mass spectrometric confirmation is only possible with the CI spectra shown in Fig. 4-202. Quantitation of the barbitals is carried out using Hexobarbital as the internal standard. The calibration function for Phenobarbital is shown in Fig. 4-203.

4.37 Identification and Quantitation of Barbiturates 527

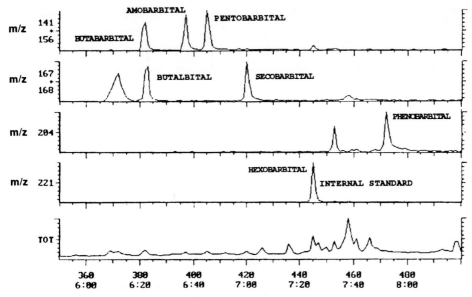

Fig. 4-195 Chromatogram of the barbiturates (10 ng/ml) in the EI mode

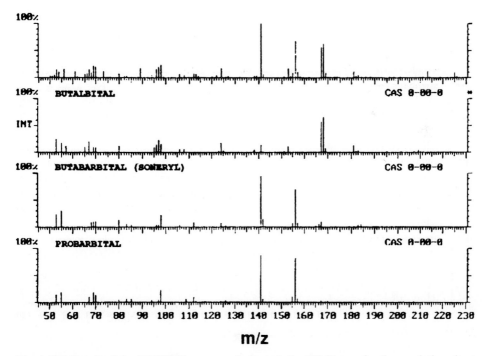

Fig. 4-196 Result of the INCOS library search through the TX library for the co-elution situation of Butabarbital and Butalbital

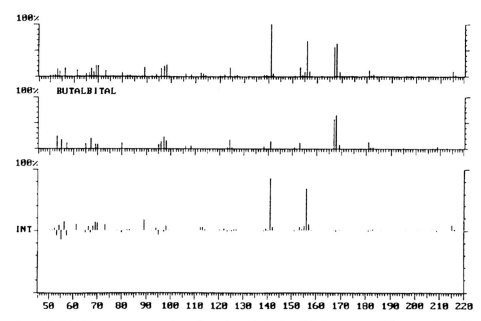

Fig. 4-197 Difference spectrum by subtraction of the Butalbital spectrum (1st suggestion with FIT 925 from Fig. 4-196) from the spectrum measured

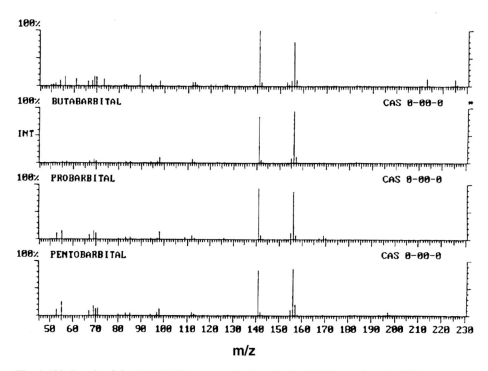

Fig. 4-198 Result of the INCOS library search through the TX library for the difference spectrum from Fig. 4-197

4.37 Identification and Quantitation of Barbiturates

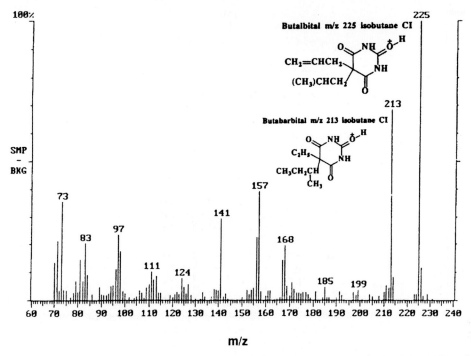

Fig. 4-199 CI spectrum for the co-elution situation from Fig. 4-195: Butabarbital (M 212) and Butalbital (M 224) are confirmed through their (M+H)$^+$ ions

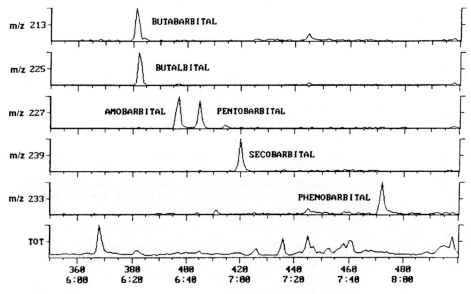

Fig. 4-200 Chromatogram of the barbiturates in the CI mode (isobutane)

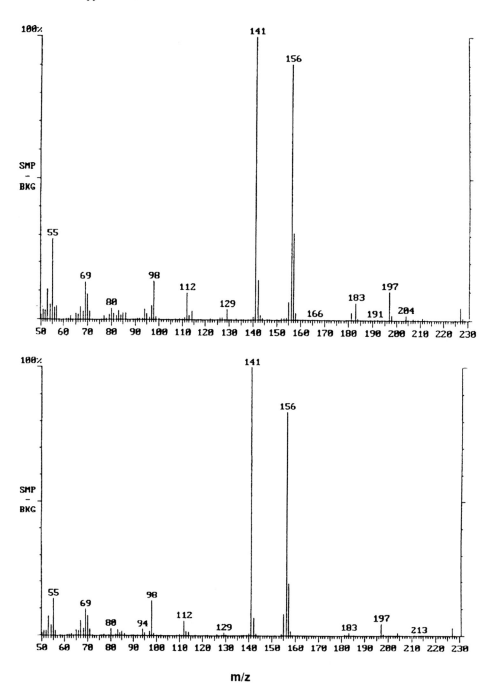

Fig. 4-201 EI spectra of the isomers Amobarbital (above) and Pentobarbital (below)

4.37 Identification and Quantitation of Barbiturates

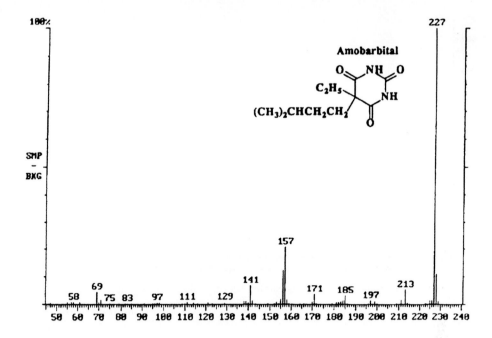

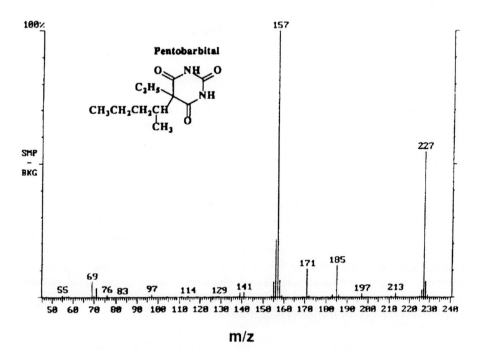

Fig. 4-202 CI spectra of the isomers Amobarbital (above) and Pentobarbital (below)

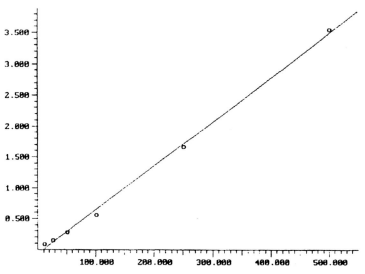

Fig. 4-203 Calibration function for Phenobarbital between 10 and 500 ng/ml

4.38 Detection of Clenbuterol by CI

Key words: Clenbuterol, Salbutamol, chemical ionisation

β-Sympathomimetics are used both in human and animal medicine for treating diseases of the respiratory passages. In treatment of animals with higher concentrations an increase in muscle growth and an improvement in the meat/fat ratio is achieved. The use of these compounds for fattening purposes is forbidden in some countries. Clenbuterol has become known as a fattening and doping agent. Another active substance in this group is Salbutamol.

HPLC, EIA and GC methods are used for analysis. GC/MS procedures also allow the positive detection of the substance as a TMS derivative even in the trace region. On-column injection has proved to be particularly effective. However, because of the high matrix concentrations, the use of the hot split injector for routine analysis is preferred. Electron impact ionisation is unsuitable for residue analysis as intense fragments are formed exclusively in the lower mass region which is occupied by the matrix. Clenbuterol can be detected with greater sensitivity using GC/MS through positive chemical ionisation (protonation by methane) (Fig. 4-204).

4.38 Detection of Clenbuterol by CI

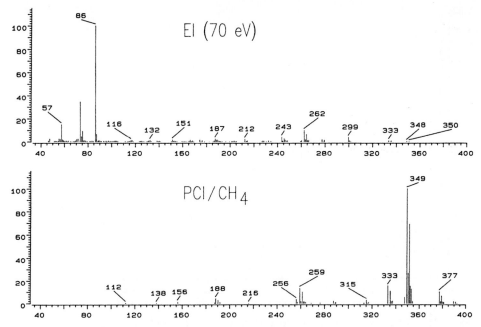

Fig. 4-204 Mass spectra of the Clenbuterol-TMS derivative (after Fürst 1994)

Analysis Conditions

Derivatisation: the extracts are dried in a stream of N_2 and treated with 200 µl HMDS (70 °C, 45 min). For injection, the excess HMDS can be blown out. The residue is taken up in acetone/5 % HMDS or dichloromethane.

GC/MS system: Finnigan ITS 40
Split injector: 270 °C isotherm
Splitless injection
Column: J&W DB-5, 30 m × 0.25 mm × 0.25 µm
Program: Start: 80 °C, 1 min
Program 1: 20 °C/min to 160 °C
Program 2: 10 °C/min to 250 °C
Final temp.: 250 °C, 1 min

Mass spectrometer: Scan mode: CI, full scan, 120–400 u, reagent gas methane
Scan rate: 1 s/scan

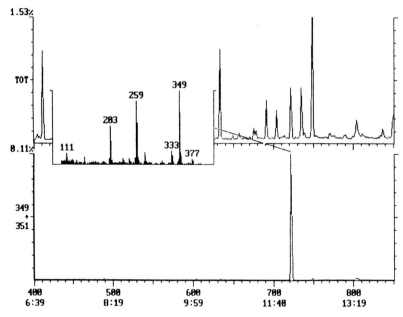

Fig. 4-205 PCI chromatogram of an extract fortified with 2 ng/µl Clenbuterol
Above: total ion current, below: selective mass chromatogram m/z 349, 351

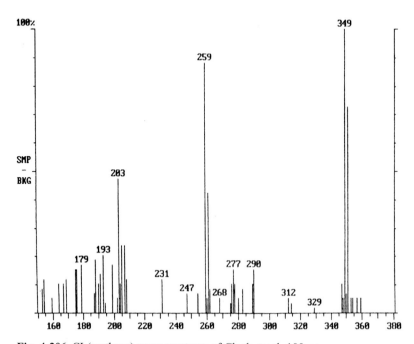

Fig. 4-206 CI (methane) mass spectrum of Clenbuterol, 100 pg

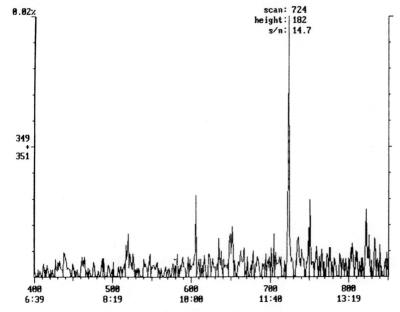

Fig. 4-207 Mass chromatogram of Clenbuterol at 50 pg/µl

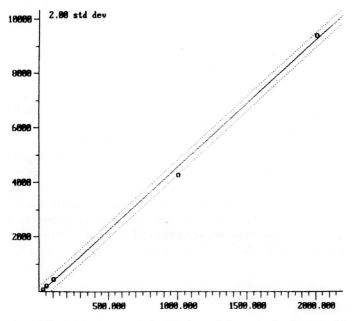

Fig. 4-208 Calibration graph for Clenbuterol determination between 25 pg and 2 ng

Results

Figure 4-205 shows the analysis of an extract to which 2 ng/µl Clenbuterol has been added. With a total ion current only a few signals are present in the chromatogram, which can be detected using chemical ionisation. The selective masses of Clenbuterol-TMS m/z 349, 351 represent the only active substance peak in the mass chromatogram. The mass spectrum in the relevant trace region of the 100 pg injected shows the unambiguous fragments of Clenbuterol-TMS (Fig. 4-206). Chemical ionisation allows the selective detection of the active substance with an ion trap mass spectrometer down to the lower pg region (Fig. 4-207). The calibration shows good linearity over the region 25 pg to 2 ng shown (Fig. 4-208). Clenbuterol-d_9 can be used as the internal standard.

4.39 Systematic Toxicological-chemical Analysis

Key words: diagnosis, emergency situations, general unknown analysis, detoxification, drug monitoring, quantitation, automatic evaluation, screening

Toxicological-chemical analysis can be very important for diagnosis and efficient treatment of acutely or chronically poisoned patients. In many cases, e.g. in emergency situations, poisoning is the result of the intake of mixtures. Therefore special requirements are made on an analysis system so that in one step as many active substances as possible can be identified and quantified by only one procedure. GC/MS is now the method of choice for the screening and subsequent unambiguous identification and quantitation of sufficiently volatile organic pharmaceutical substances and poisons (Figs. 4-209 and 4-210).

For the systematic toxicological analysis of general unknowns, an analysis procedure is described which allows the simultaneous detection and determination of a large number of pharmaceutical substances. Serum (or blood when no serum can be obtained, e.g. in the case of corpses) is analysed. Only the quantitative determination of serum or blood allows a good estimation of the degree of severity of poisoning, of the success of detoxification or therapy (drug monitoring) or the degree of damage to the body. In addition most active substances in serum are present in a nonmetabolised form, at least to some extent, so hydrolysis and derivatisation can be dispensed with. This cannot be circumvented for the determination of more hydrophilic metabolites from urine. Consequently sample preparation is quicker, as is the procedure as a whole. Besides isolation in sufficient yield, the separation of biogenic substances as far as possible is a requirement for the full utilisation of the dynamic range of the ion trap GC/MS system used.

4.39 Systematic Toxicological-chemical Analysis 537

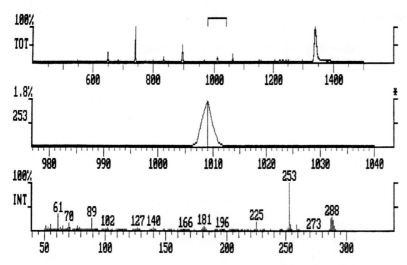

Fig. 4-209 Chromatogram of a serum sample with enlarged section (centre) and substance spectrum (below)

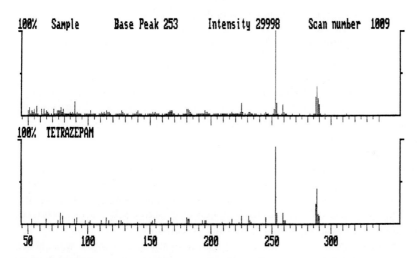

Fig. 4-210 Result of the INCOS search through the toxicology library TX with the suggestion of Tetrazepam (FIT value 967, spectrum from Fig. 4-207)

Analysis Conditions

Sample preparation: adjust the pH of 1 ml serum to 9 with NaHCO$_3$,
add the internal standard Etoloxamine (5 µl $\cong$ 100 ng Etoloxamine),
extract with 5 ml n-hexane/ethyl acetate 7:3 (v/v), 1% amyl alcohol,
2 min centrifuge,
re-extract the organic phase with 2 ml 0.5 M H$_2$SO$_4$, centrifuge and discard the organic phase,
adjust the pH of the aqueous phase to 9 with 0.2 ml 8 N NaOH and NaHCO$_3$,
extract with 2 ml CHCl$_3$, evaporate the organic phase to dryness and dissolve in 20 µl ethyl acetate, inject 1 µl.
Time required ca. 20–25 min

Gas chromatograph: Varian 3400 with split injector and split injection
Column: HP-1 15 m × 0.25 mm × 0.25 µm
Program:
Start: 80 °C, 1 min
Ramp: 10 °C/min to 280 °C
Final temp.: 280 °C, 4 min

Mass spectrometer: Finnigan ion trap detector ITD 800
GC coupling: 270 °C
Scan mode: EI, full scan, 50–450 u
Scan rate: 1 s/scan

Calibration and automatic evaluation: search for active substances in the total ion current chromatogram using stored retention times and mass spectra
Retention time window: 20 s
FIT threshold: 700
S/N ratio: min. 3
Internal standard: Etoloxamine
Duration of the evaluation: 15 s

Results

At the current state of development of the procedure the 77 active substances listed in Table 4-29 are determined in an automatic evaluation program (AutoQuan). For most substances calibration is available which was drawn up at five concentrations using a blind serum (drug-free serum) (Fig. 4-211). Metabolites of active substances are not calibrated. The compounds marked with a * are only detected in cases of poisoning. The other active substances are also detected in therapeutic doses. Etoloxamine was chosen as the internal standard because it is very stable and readily chromatographed. It is no longer used as a pharmaceutical substance. The ease with which an active substance can be determined using the GC/MS procedure described depends on the recovery, the relative response and the concentration in the serum. For this reason the isolation procedure is set up to obtain basic substances, as many neutral and acidic pharmaceutical substances (barbiturates, Phenytoin) are present at such a high concentrations within the range of therapeutic activity that they are readily detectable by GC/MS in spite of the low recovery from sample clean-up. A more complete recovery in the case of these compounds often leads to overloading of the column and thus impairs quantitation. In Table 4-30 examples are given of the experimentally determined recovery and relative response together with the average therapeutic concentration for 10 active substances

4.39 Systematic Toxicological-chemical Analysis

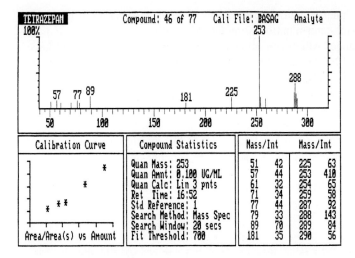

Fig. 4-211 Example of an entry from the calibration file with all data for the qualitative identification using retention time and mass spectrum as well as quantitative determination

Table 4-29 List of active substances which can be detected and quantified by the screening procedure (situation 1994/95)

Amitriptyline	Doxepin	Phenacetin
Amitriptyline oxide*	Doxylamine	Phenazone
Biperiden	Etoloxamine – ISTD	Phenobarbital
Bornaprin**	Haloperidol*	Phenytoin
Bromazepam*	Hydrocodone	Primidone
Carbamazepine	Imipramine	Primidone-M**
Carbamazepine artefact 2**	Levomepromazine	Promazine
Carbamazepine-M**	Lidocaine	Promethazine
Quinidine	Medazepam	Promethazine-M(HO)**
Quinidine acetate**	Metamizol-Desalkyl**	Propranolol*
Quinine**	Methadone	Propranolol artefact**
Quinine acetate**	Methaqualone	Propyphenazone
Chlordiazepoxide	Methylphenobarbital	Propyphenazone-(HO-Phenyl)**
Chlormezanone	Midazolam**	Propyphenazone-(HO-Propyl)**
Chlorpromazine	Moclobemid	Propyphenazone-(Isopropenyl)**
Chlorprothixene	Naloxone*	Propyphenazone-(nor-di-HO)**
Clomipramine	Nicotine**	Propyphenazone-(nor-HO-Phenyl)**
Clozapine	Nitrazepam*	Remoxiprid*
Cocaine	Nordazepam	Tetrazepam
Codeine	Nortilidine**	Theophylline**
Caffeine	Nortrimipramine**	Tilidine**
Cotinine	Nortriptyline	Tramadol
Desmethylclomipramine**	Paracetamol*	Trimipramine
Desmethylmedazepam**	Pentobarbital	Trihexyphenidyl
Diazepam	Perazine	Zotepine
Diphenylhydramine	Pethidine	

* active substances can only be determined in case of poisoning
** only qualitative detection

from different classes of pharmaceuticals. From this table the interaction of these three parameters for a successful determination using GC/MS is clear.

In principle all substances which can be extracted in sufficient quantity and have a retention index between 1200 and 3100 (cholesterol) can be determined using the procedure described. Besides basic pharmaceutical substances, neutral and acidic compounds, such as barbiturates, can be determined because of their high therapeutic dose, in spite of low recovery rates. From the times given for sample preparation, measurement and evaluation it can be concluded that the qualitative and quantitative analysis of a serum sample can be carried out in less than an hour. The development of the investigation programme by incorporating further substances is being carried out.

Reference: Dr. U. Demme, M. Thoben, Klinikum der Friedrich-Schiller-Universität Jena, Institut für Rechtsmedizin, Jena

Table 4-30 Determination of the product from recovery, response and concentration for ten selected active substances

Active substance	RI	Scan	m/z	a	c_{thp} [µg/mL]	RR	$RR \cdot a \cdot c$ [µg/mL]
Diphenylhydramine	1870	701	58	0.91	0.05	1.58	0.072
Phenobarbital	1965	753	204	0.02	30	0.03	0.018
Methaqualone	2160	870	235	0.60	2	0.07	0.084
Amitriptyline	2205	904	58	0.91	0.1	1.82	0.166
Trimipramine	2225	910	58	0.81	0.12	0.46	0.045
Promethazine	2270	937	72	0.66	0.25	0.27	0.045
Biperiden	2280	960	98	0.91	0.07	0.59	0.038
Clomipramine	2455	1012	58	0.70	0.1	0.34	0.024
Chlorprothixene	2510	1060	58	0.66	0.16	0.48	0.051
Levomepromazine	2540	1076	58	0.85	0.09	0.98	0.075

RI   Retention index
m/z Base ion
a Recovery on isolation from biological material
c_{thp} Mean therapeutic concentration
RR Relative response (sensitivity of recorder)

4.40 Clofibric Acid in Aquatic Systems

Key words: Clofibric acid, pharmaceuticals, contamination, SPE, pentafluorobenzyl esters, SIM technique

In 1991 in the course of the investigation of a series of samples of Berlin (Germany) ground water for phenoxyalkylcarboxylic acid herbicides, an unknown chlorine-containing carboxylic acid was discovered, the structure of which was unambiguously found to be that of 2-(4-chlorophenoxy)-2-methylpropionic acid. This compound is used to lower lipid levels in human beings under the name of Clofibric acid. To find out how this substance got into the ground water the investigation procedure was changed to improve the sensitivity of detection. Using the methods described here tap water samples from different districts of Berlin were systematically investigated for residues of Clofibric acid. The isomeric compound 2-(4-chlorophenoxy)butyric acid (4-CPB) was added to a water sample as a surrogate standard. Clean-up involved solid phase extraction with a special RP-C_{18}-adsorbent. 2,4-Dichlorobenzoic acid (2,4-DCB) was added as the internal standard for quantitation. After derivatisation with pentafluorobenzyl bromide the Clofibric acid is detected as the pentafluorobenzyl ester by GC/MS in the SIM mode and determined quantitatively. It is identified from the retention time and the intensity ratio of three characteristic ions. Because two standards are added, the results are very reliable even at the lowest concentrations.

Analysis conditions

Sample preparation:	adjustment of the pH of 1 l water to 1–2 addition of 4-CPB standard (methanolic solution) and adjustment of the concentration to 200 ng/l; 1 g RP-C_{18}-SPE cartridge (washed with 10 ml acetone, 10 ml methanol, conditioned with 10 ml water pH <2), flow rate ca. 8 ml/min, drying with N_2, elution with 2.5 ml methanol, solvent removal with N_2 (Fig. 4-212).
Derivatisation:	addition of the internal standard 2,4-DCB (200 ng in methanol), transfer to autosampler vessels and evaporation to dryness with N_2, treatment with 200 µl pentafluorobenzyl bromide solution (100 µl pentafluorobenzyl bromide in 4.9 ml toluene) and 2 µl triethylamine, closure of the vessel, 1 h reaction time at 90 °C, excess reagent driven out with N_2, residue taken up in 100 µl toluene.
GC/MS system:	HP 5890 with MSD HP 5970
Injection temp.:	210 °C
Column:	HP-5, 25 m × 0.2 mm × 0.33 µm
Program:	Start: 100 °C, 1 min Program 1: 30 °C/min to 150 °C, 1 min Program 2: 3 °C/min to 205 °C Program 3: 10 °C/min to 260 °C, 23 min
Mass spectrometer:	Scan mode: EI, SIM mode (Table 4-31) Scan rate: 1 s/scan (dwell times per ion, see Table 4-31)

Results

Recoveries of 90 to 100% were achieved for Clofibric acid, the surrogate standard 4-CPB, Mecoprop and Dichlorprop with the RP-C_{18}-material Europrep from Eurochrom (Fig. 4-210). Interference in the analysis can be caused by humic substances. With such samples negative results do not necessarily mean that no Clofibric acid is present. The use of the additional internal standard demonstrates the reliability of the GC/MS analysis. The peak area ratio of the internal and the surrogate standards is a measure of the quality of the analysis.

The GC/MS detection procedure involves investigating five compounds in three time windows in a typical target compound analysis (Table 4-31). In the first time window Clofibric acid and 2,4-DCB are determined with six ions, in the second Mecoprop and 4-CPB with four ions and in the third window Dichlorprop with three ions. The results are plotted as an MID chromatogram to give a better overview (addition of SIM Signals). The absence of an ion here is a definite sign of the absence of the target substance (Fig. 4-213 a, b). For confirmation, besides the exact agreement of the retention times, the relative ion intensities are also checked. In the samples these may not differ by more than 20% from the calibration standards measured under the same conditions.

Table 4-31 SIM programming and retention times

No.	Substance	Retention time [min]	Time window [min]	SIM masses [m/z]	dwell time* [ms]
1	Clofibric acid	24.54	24.00–25.19	128, 130, 173	150
	2,4-DCB	24.78		370, 372, 394	
2	Mecoprop	25.64	25.20–26.39	128, 142	200
	4-CPB	25.84		169, 394	
3	Dichlorprop	27.07	26.40–27.50	162, 400, 402	300

* Dwell time = duration of measurement per ion at 1 s-scan

The limit of detection is 1 ng/l for Clofibric acid, and the limit of quantitation 10 ng/l. The method is also suitable for the analysis of highly contaminated surface water. Even with very contaminated samples other substances present do not appear to affect the results, for example the analysis of a run-off from a sewage works (Figs. 4-214 and 4-215). This selectivity is traced back to the particular derivatisation which gives derivatives with high molecular weights. Polar, nonderivatised components do not dissolve in the toluene. Possible impairment of the chromatography by highly contaminated samples can be controlled using the internal standard 2,4-DCB.

The presence of Clofibric acid in water has not only been found in Berlin. The substance has been found in random samples taken outside Berlin in rivers and drinking water. It has not been found in drinking water from areas without contact with the filtrate from the river banks. It is assumed that Clofibric acid from therapeutic use gets

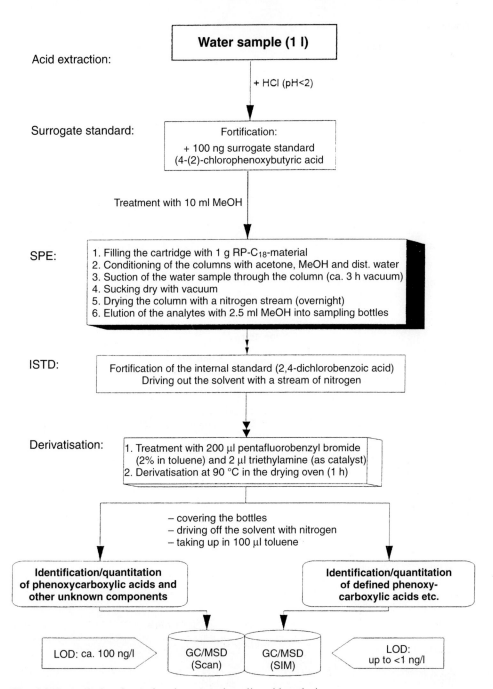

Fig. 4-212 Analysis scheme for phenoxycarboxylic acid analysis

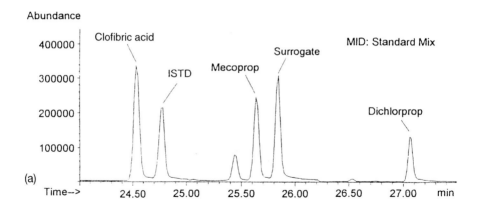

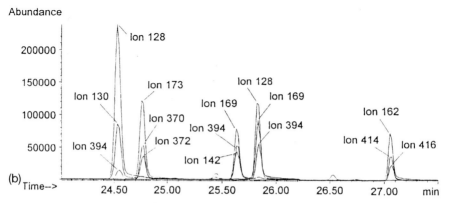

Fig. 4-213 Chromatogram of PFB esters in a standard mixture
(a) MID plot as a sum of SIM signals (2 ng/component)
(b) Individual ion traces of the PFB esters

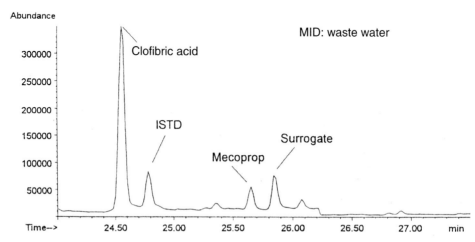

Fig. 4-214 Analysis of a waste water sample from a sewage works containing both Clofibric acid (1100 ng/l) and Mecoprop (175 ng/l)

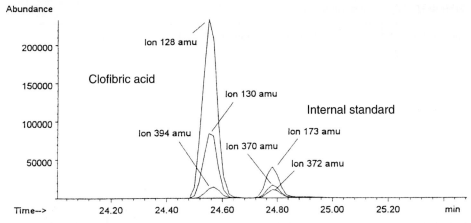

Fig. 4-215 Checking of the individual ion traces of Clofibric acid and the internal standard from Fig. 4-214

into surface water via municipal waste water. According to the working group this is the first known case of a medicine used for humans reaching aquatic systems in this way.

Reference: Prof. Dr. H.-J. Stan, Institut für Lebensmittelchemie, Technical University, Berlin

4.41 Polycyclic Musks in Waste Water

Key words: musk perfumes, surface water, waste water, solid phase extraction

Synthetic musk perfumes are added to washing and cleaning agents as well as fabric conditioners and many bodycare products to improve their odours. Because of the ecotoxicological problems caused by their poor degradability and lipophilic properties, there are reservations about their widespread use. In the examination of water, sewage sludge, sediments and fish for nitro-musk compounds (e.g. musk-xylene, 1-t-butyl-3,5-dimethyl-2,4,6-trinitrobenzene), three further compounds were found in the GC/MS chromatograms. These were the polycyclic nitro-musk compounds HHCB, AHTN and ADBI (Table 4-32). These substances are widely used in the cosmetics and perfumes industry and have already been detected in river water and various species of fish.

Table 4-32 Polycyclic musk perfumes

Specification	CAS-No.	Empirical formula	M	Structure
HHCB 1,3,4,6,7,8-Hexahydro-4,6,6,7,8,8-hexamethylcyclo-penta-(g)-2-benzopyran	1222-05-5	$C_{18}H_{26}O$	258	
AHTN 7-Acetyl-1,1,3,4,4,6-hexa-methyltetralin	1506-02-1	$C_{18}H_{26}O$	258	
ADBI 4-Acetyl-1,1-dimethyl-6-tert.butylindane	13171-00-1	$C_{17}H_{24}O$	244	

Analysis Conditions

Sample preparation:	2 l surface water are enriched using SPE with 2 g Bakerbond PolarPlus C_{18}-cartridges and eluted with 10 ml acetone; nitrogen is blown through the eluate and the residue taken up in n-hexane. Waste water samples are extracted directly with n-hexane.
GC/MS system:	Finnigan ITS 40
Injector:	PTV cold injection system, splitless injection
Program:	Start: 60 °C Program : 300 °C/min Final temp.: 300 °C, 15 min
Column:	J&W DB-5MS, 30 m × 0.32 mm × 0.25 µm and SGE HT8, 25 m × 0.22 mm × 0.25 µm
Program:	Start: 60 °C, 1 min Program: 10 °C/min Final temp.: 300 °C, 6 min
Mass spectrometer:	Scan mode: EI, full scan, 40–400 u Scan rate: 1 s/scan

Results

The EI mass spectra of the compounds taken in the full scan mode are suitable for unambiguous identification because of their typical fragmentation patterns. Intense fragment signals are seen at m/z 43, 258, 243 and 229. The mass spectrum of HHCB also contains the fragment m/z 213, unlike AHTN (Figs. 4-216 to 4-218). HHCB and AHTN are only partially separated on the DB-5MS column. All three substances are separated well on the slightly polar phase of the HT8 column. Figure 4-219 shows the chromatogram of a waste water sample with the mass chromatograms of the selective ions under the total ion current. In random samples significant measurable concentrations in the ppb range were found both in surface water and in the inflow and outflow of municipal sewage works. Very high concentrations were found in the muscle tissue of fish from these types of water (Fig. 4-220), indicating the low degradability and bioaccumulation of these compounds.

Reference: H.-D. Eschke, Chemisches und Biologisches Laboratorium des Ruhrverbandes, Essen.

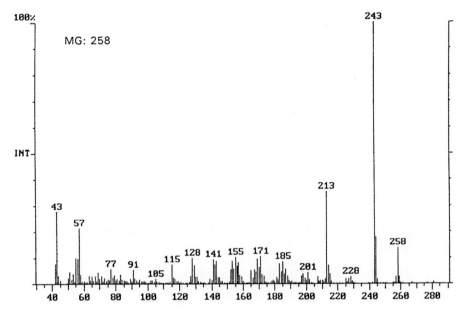

Fig. 4-216 Mass spectrum of HHCB

548   4 Applications

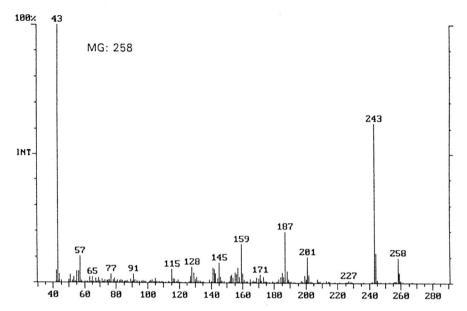

Fig. 4-217 Mass spectrum of AHTN

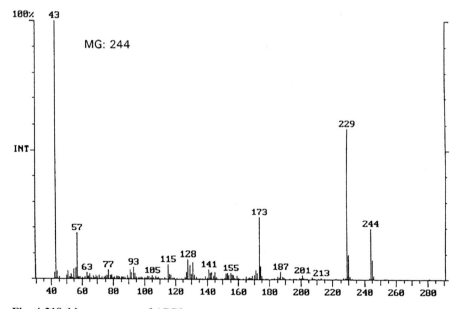

Fig. 4-218 Mass spectrum of ADBI

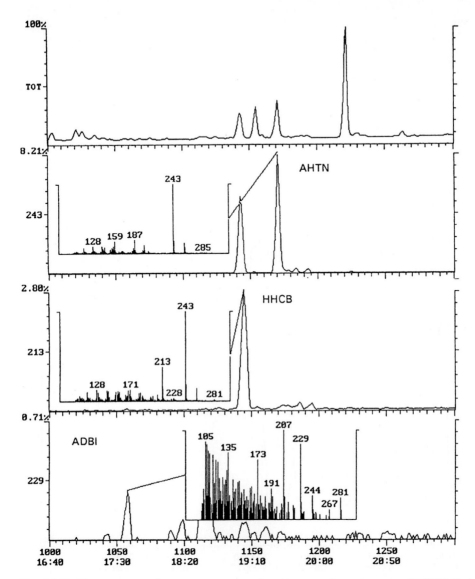

Fig. 4-219 Chromatogram of a waste water sample with mass chromatograms of HHCB (m/z 213/243), AHTN (m/z 243) and ADBI (m/z 229)

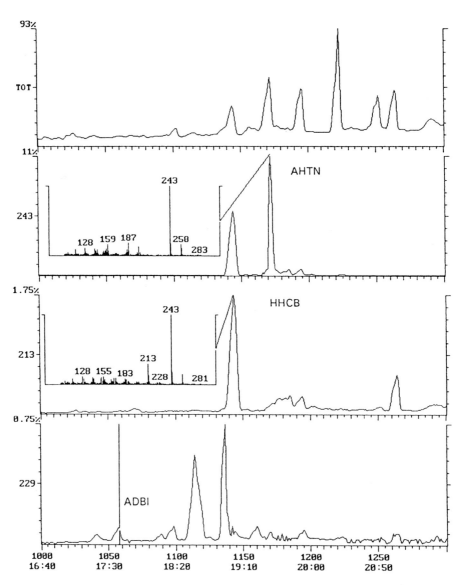

Fig. 4-220 Chromatogram of a fish sample (muscle tissue, eel) with the mass chromatograms of HHCB (m/z 213/243), AHTN (m/z 243) and ADBI (m/z 229)

References for Chapter 4

Section 4.1

Madden, Al, Analysis of Air Samples for the Polar and Non-Polar VOCs Using a Modified Method TO-14, Tekmar Application Report Vol 5.3, Cincinatti 1994.
Schnute, B., McMillan, J., TO-14 Air Analysis Using the Finnigan MAT Magnum Air System, Application Report No. 230, 1993.

Section 4.4

Analysis of Volatile Priority Pollutants in Water by Static Headspace Gas Chromatography/Mass Spectrometry (GC/MS), Application Note AN 1000, ThermoQuest GC and GC/MS Division, Manchester, UK.
Belouschek, P., Brand. H., Lönz, P., Bestimmung von chlorierten Kohlenwasserstoffen mit kombinierter Headspace- und GC/MS-Technik, Vorn Wasser 79 (1992) 3–8.
Rinne, D., Direkter Vergleich der Headspace-GC-Technik zum Extraktionsverfahren von leichtflüchtigen Halogen-Kohlenwasserstoffen, Gewässerschutz, Wasser, Abwasser 88 (1986) 291–325.

Section 4.6

Schlett, C., Pfeifer, B., Gaschromatographische Bestimmung von Vinylchlorid nach Purge-and-Trap-Anreicherung und massenspektrometrischer Detektion, Vom Wasser 81 (1993), 1–6.

Section 4.12

McMillan, J., Analysis of 2-Methylisoborneol and Geosmin Using Purge and Trap on the Finnigan MAT MAGNUM GC/MS, Finnigan MAT Environmental Analysis Application Report No. 232, San Jose 1994.
Preti, G., Gittelman, T. S., et al., Letting the Nose Lead the Way – Malodorous Components in Drinking Water, Anal. Chem. 65 (1993) 699A-702A.

Section 4.13

Schlett, C., Pfeiffer, B., Bestimmung substituierter Phenole unterhalb des Geruchsschwellenwertes, Vom Wasser 79 (1992), 65–74.

Section 4.14

Cairns, Th., Luke, M. A., Chiu, K. S., Navarro, D., Siegmund, E. G., Multiresidue Pesticide Analysis by Ion-Trap Technology: A Clean-up Approach for Mass Spectral Analysis, Rap. Comm. Mass Spectrom. 7 (1993), 1070–1076.
Worthing, C. R., Walker, S., The Pesticide Manual: A World Compendium, 8th Ed., The British Crop Protection Council, Lavenham Press: Lavenham UK 1987.

Section 4.15

Edwards, J., Fannin, S. T., Klein, D., Steinmetz, G., The Analysis of Pesticide Residue Compounds in Biological Matrices by GC-MS, TR 9137 Technical Report, Finnigan Corp., Austin, TX, 7/1998.

Matter, L., Bestimmung von Propham und Chlorpropham in Kartoffeln mittels hochauflösender Kapillargas-Chromatographie und Ion-Trap-Detection, GIT Fachz. Labor 11 (1989) 1116.

Section 4.16

Abke, W., Korpien, H., Post, B., Belastung des Grundwassers im Abstrom von Gleisanlagen durch Herbizide, Vom Wasser 81 (1993), 257–273.

Section 4.17

Brodesser, J., Schöler, H. F., Vom Wasser 1987, 69, 61–71.

Nick, K., Schöler, H. F., Bestimmung von Nitrophenolen mittels GC und GC/MS nach einer Derivatisierung mit Diazomethan, GIT Fachz. Labor 1993, 5, 393–397.

Peldszus, S., Dissertation, Universität Bonn 1991.

Section 4.20

DIN 38407–14, standard German procedure for the testing of water, waste water and sludge, groups of substances which can be determined together (Group F), determination of phenoxyalkylcarboxylic acids using gas chromatographic and mass spectrometric detection following solid-liquid extraction and derivatisation (F14), October 1994.

DIN 38407–20, standard German procedure for the testing of water, waste water and sludge, groups of substances which can be determined together (Group F), determination of Bentazon, Bromoxynil and eight selected phenoxyalkylcarboxylic acids using gas chromatographic and mass spectrometric detection following solid-liquid extraction and derivatisation with diazomethane (F20), draft, situation: March 1996.

Howard, S. F., Yip, G., „Diazomethane methylation of a mixture of chlorophenoxy acids and dinitrophenols", J. Assoc. Off. Anal. Chem. 54, 970–974 (1971).

Hübschmann, H.-J., Kriegsmann, F., "Determination of phenoxyalkylcarboxylic acids in water samples using automatic target compound analysis", GIT, 1997 (1).

Section 4.21

Schenker, D., Matter. L., Zur Bestimmung von Pentachlorphenol in Arbeitshandschuhen mittels Kapillar-Gaschromatographie und Ion-Trap-Detection, GIT Fachz. Labor 9 (1990) 1084.

Section 4.22

Gerbino, T. C., Garbarino, G., Spelta, D., Petit Bon, P., Comparative Evaluation of HRGC-MS-Ion Trap Mass Detector and Other Chromatographic Techniques in Determination of Polynuclear Aromatic Hydrocarbons in Environmental Samples, In: 16th Int. Symp. Cap. Chrom. Riva del Garda, P. Sandra (Ed), Huethig 1994.

Speer, K., Bestimmung von PAK in Lebensmitteln, In: Matter, L., Lebensmittel- und Umweltanalytik mit der Kapillar-GC, VCH: Weinheim 1994, 75–116.

Section 4.23

Götze, H.-J., Hundertmark, C., Moderne Methoden der Dieselruß- und PAK-Analytik, GIT Fachz. Lab. 7 (1995) 648–653.
Hahne, F., Schumann, H., Tillmanns, U., Bestimmung von polyzyklischen Aromaten in Stadtluft, Finnigan MAT Application Report Nr. 84, 1991.

Section 4.24

Hübschmann, H.-J., Niebel, J., Landrock, A., Richter, H., Merten, H., GC-MS Identification of Nitroaromatic Compounds with a New Selective and Highly Sensitive Detection Method for Ion Trap GC-MS Systems, In: Sandra, P., Devos, G. (Eds): 16th Int. Symp. Cap. Chrom., Riva del Garda, Sept. 27–30, 1994, Hüthig: Heidelberg 1994, 814–820.
Landrock, A., Richter, H., Merten, H., Water CI, a new selective and highly-sensitive detection method for ion trap mass spectrometers, Fresenius J. Anal. Chem. 351 (1995) 536–543.

Section 4.25

Ballschmitter, K., Bacher, R., Fischer, R., et al., The Analysis of Halogenated Biphenyls (HalxB), Dibenzodioxins (HalxD) and Dibenzofurans (HalxDF) by HRGC and Mass Spectrometric Detection (GC/MS). In: Sandra, P., 13th Int. Symp. Cap. Chromatogr., Riva del Garda, May 1991, Huethig: Heidelberg 1991, 501–525.
BDE – Bundesverband der deutschen Entsorgungswirtschaft e.V. (Hrsg.), Abfallanalytik. Merz Verlag: Bonn 1993.
Beck, H., Mathar, W., Analysenverfahren zur Bestimmung von ausgewählten PCB-Einzelkomponenten in Lebensmitteln, Bundesgesundheitsblatt 28 (1985), 1.
DIN 51527 Teil 1: Prüfung von Mineralöl-Erzeugnissen, Bestimmung des Gehaltes an polychlorierten Biphenylen (PCB), Beuth Verlag: Berlin.
Fürst, P., GC/MS-Bestimmung von Rückständen und Kontaminanten. In: Matter L., Lebensmittel und Umweltanalytik mit der Kapillar-GC, VCH: Weinheim 1994.
Hillery, B. R., Girard, J. E., Schantz, M. M., Wise, S. A., Quantitative Analysis of Selected PCB Congeners in Marine Matrix Reference Materials Using a Novel Cyanobiphenyl Stationary Phase, J. High Resol. Chromatogr. 18 (1995) 89–96.
Schantz, M. M., Koster, B. J., et al., Certification of Polychlorinated Biphenyl Congeners and Chlorinated Pesticides in a Whale Blubber Standard Reference Material, Anal. Chem. 67 (1995) 901–910.
Schantz, M. M., Parris, R. M., Kurz, J., Ballschmitter, K., Wise, S. A., Comparison of Methods for the Gas-Chromatographic Determination of PCB Congeners and Chlorinated Pesticides in Marine Reference Materials, Fres. J. Anal. Chem. 346 (1993), 766–778.
Steinwandter, H., Fresenius J. Anal. Chem. 331 (1988), 499.

Section 4.27

Bachhausen, P., 2. GC/MS-Symposium der Axel Semrau GmbH, Abstracts, Sprockhövel: Februar 1994.
Ball, M., Papke, 0., Lis, A., Weiterführende Untersuchung zur Bildung von polybromierten Dioxinen und Furanen bei der thermischen Belastung flammgeschützter Kunststoffe und Textilien, Texte 45/92, Umweltbundesamt: Berlin 1992.
Ballschmitter, K., Bacher, R., et al., The Analysis of Halogenated Biphenyls, Dibenzodioxins and Dibenzofurans by HRGC and Mass-Spectrometric-Detection (GCMS), In: 14th Int. Symp. Cap. Chrom. Riva del Garda, P. Sandra (Ed), Huethig 1991, 501–525.

Chang, R. R., Jarman, W. M., Hennings, J. A., Sample Cleanup by Solid Phase Extraction for the Ultratrace Determination of Polychlorinated Dibenzo-*p*-dioxins and Dibenzofurans in Biological Samples, Anal. Chem. 65 (1993) 2427–2430.

Erste Verordnung zur Änderung der Chemikalien-Verbotsverordnung vom 6. Juli 1994, Art. 1 (11), Bundesgesetzblatt, Teil I, 42 (1994), 1493–1495.

Fürst, P., GC/MS-Bestimmung von Dioxinen in Lebensmittel- und Umweltproben, GDCh-Fortbildungsprogramm Chemie, Seminar 719/94, Frankfurt/Main, 18. Nov. 1994.

Fürst, P., GC/MS-Bestimmung von Rückständen und Kontaminaten, in: L. Matter (Hrsg), Lebensmittel- und Umweltanalytik mit der Kapillar-GC, VCH: Weinheim 1994, 29–74.

Lamparski, L. L., Shao, J., Nestrick, T. J., et al., Validation of MAT 95 Instrument Operation for the Measurement of Various Chlorinated Dibenzo-p-dioxins and Dibenzofurans, Dioxin '91, Abstracts, North Carolina, Sept. 23th–27th 1991.

Tong, H. Y., Monson, S. J., Gross, M. L., Monobromopolychlorodibenzo-p-dioxins and Dibenzofurans in Municipal Waste Incinerator Flyash, Anal. Chem. 63 (1991), 2697–2705.

Section 4.28

Deutsche Forschungsgemeinschaft, Maximale Arbeitsplatzkonzentrationen und biologische Arbeitsstofftoleranzwerte, VCH: Weinheim 1990.

Kuitunen, M. L., Hartonen, K., Riekkola, M. L., Analysis of Chemical Warfare Agents from Soil Samples Using Off-Line Supercritical Fluid Extraction and Capillary Gas Chromatography. In: P. Sandra (Ed), 13th Int. Symp. Cap. Chrom., Riva del Garda May 1991, Huethig: Heidelberg 1991, 479–488.

Niebel, J., Cleanup und Detektion von Nitro- und Aminoaromaten mit dem GC-MSD/ECD/FID, Diplomarbeit, FH Technik Mannheim, 1992.

Schimmelpfeng, L. (Hrsg), Altlasten, Deponietechnik, Kompostierung, Teil 1: Erfassung, Untersuchung von Altlasten, Rüstungsaltlasten, Academia Verlag: Sankt Augustin 1993.

Wöhrle, D., Die neue Chemie-Waffen-Konvention, Nachr. Chem. Tech. Lab. 41 (1993) 291–296.

Yinon, J., Zitrin, S., Modern Methods and Applications of Explosives, John Wiley & Sons: Chichester 1993.

Section 4.30

Martz, R., Donelly, B., et al., The Use of Hair Analysis to Document a Cocaine Overdose following a Sustained Survival Period Before Death, J. Anal. Toxicol. 15 (1991), 279–281.

Möller, M. R., Fey, P., Wennig, R., Simultaneous Determination of Drugs of Abuse (Opiates, Cocaine and Amphetamine) in Human Hair by GC/MS and its Application to the Methadon Treatment Program, in: P. Saukko (Ed), Forensic Science International, Vol. 63, Special Issue: Hair Analysis as a Diagnostic Tool for Drugs of Abuse Investigation, Elsevier: Amsterdam 1993, 185–206.

Sachs, H., Raff, I., Comparison of Quantitative Results of Drugs in Human Hair by GC/MS, in: P. Saukko (Ed), Forensic Science International, Vol. 63, Special Issue: Hair Analysis as a Diagnostic Tool for Drugs of Abuse Investigation, Elsevier: Amsterdam 1993, 207–216.

Sachs, H., Uhl, M., Opiat-Nachweis in Haar-Extrakten mit Hilfe von GC/MS/MS und Supercritical Fluid Extraktion (SFE), T + K 59 (1992), 114–120.

Schwinn, W., Drogennachweis in Haaren, Der Kriminalist 11 (1992) 491–495.

Traldi, P., Favretto, D., Tagliaro, F., Ion Trap Mass Spectrometry, a New Tool in the Investigation of Drugs of Abuse in Hair, Forensic Science International 63 (1991) 239–252.

Section 4.31

Cowan, D. A., Woffendin, G., Drug Testing in Human Sports, Spectra 11 (1988) 4–9.
Donike, M., Derivatisierung für chromatographische Untersuchungen, GDCh-Kurs 306/92, Institut für Biochemie der Deutschen Sporthochschule Köln, 1992.
Jain, N. C., The HHS (formerly NIDA) Program for Drugs of Abuse Testing – The Views of a Certified Lab, Symposium Aktuelle Aspekte des Drogennachweises, Mosbach 15. April 1993.
Matthiesen, U., GC-MS in der klinischen Chemie, LaBo 1 (1988) 7–12.
Maurer, H. H., GC/MS contra Immunoassay?, Symposium Aktuelle Aspekte des Drogennachweises, Mosbach 15. Apr.1993, Abstracts.
Maurer, H. H., Identifizierung unbekannter Giftstoffe und ihrer Metaboliten in biologischem Material, GIT Suppl. 1 (1990) 3–10.
Pfleger, K., Maurer, H. H., Weber, A., Mass Spectra and GC Data of Drugs, Poisons, Pesticides, Pollutants and Their Metabolites, Parts 1,2,3, 2. erw. Aufl., VCH: Weinheim 1992.
Stellungnahme der Arbeitsgruppe „Klinisch-Toxikologische Analytik" der Deutschen Gesellschaft für Klinische Chemie zu den notwendigen Drogensuchtests insbesondere bei der Methadon-Substitutionsbehandlung von i.v.-Heroinabhängigen, DG Klinische Chemie, Mitteilungen 24 (1993) 212–214.
Uhrich, M., Tillmanns, U., Identifizierung und Quantifizierung von Codein und Morphin, GIT Fachz. Lab. 10 (1990) 1265–1267.
Weller, J.P., Wolf, M., Nachweis von Morphinderivaten im Blut mittels Ion-Trap-Detektor, Finnigan MAT Application Report No. 76, 1990.
Weller, J. P., Wolf, M., Szidat, S., Enhanced selectivity in the determination of Δ^9-tetrahydrocannabinol and two major metabolites in serum using ion trap GC-MS-MS, J. Anal. Toxicol. 24 (2000) 1–6.

Section 4.32

Johnson, K., Uhrich, M., Tillmanns, U., Identifizierung und Quantifizierung von 11-Nor-9-Carboxy-9-Delta-Tetrahydrocannabinol, GIT Spezial-Chromatographie 1 (1991) 16–18.
Uhrich, M. D., On-Disc Derivatisation and Full Scan GC/MS Analysis of 9-Carboxy-THC Create Competitive Advantages for Laboratories, Finnigan MAT Application Data Sheet No. 53, Firmenschrift, 1992.
Weller, J. P., Wolf, M., Med. Hochschule Hannover, Institut für Rechtsmedizin 1995, unveröffentlichte Ergebnisse.

Section 4.33

Berberich, D., Uhrich, M., Tillmanns, U., Drogenanalytik mit GC/MS – Identifizierung und Quantifizierung von Phencyclidin. GIT Fachz. Lab. 34 (1990) 629–632.
Werner, M., Hertzman, M., Pauley, C. J., Gas-Liquid Chromatography of Phencyclidine in Serum with Nitrogen-Phosphorus Detection, Clin. Chem. 32 (1986) 1921–1924.

Section 4.34

Cowan, D. A., Woffendin, G., Drug Testing in Human Sports, Spectra 11 (1988) 4–9.
Woffendin, G., Analysis of Steroids by GC/MS and GC/MS/MS. Finnigan MAT Application Report 238, Firmenschrift, 1995.

Section 4.35

Schweer, H., Seybarth, H. W., Meese, C. O., Fürst, O., Negative Chemical Ionization Gas Chromatography/Mass Spectrometry and Gas Chromatography/Tandem Mass Spectrometry of Prostanoid Pentafluorobenzyl Ester/Methoxime/Trimethylsilyl Ether Derivatives, Biomed. Environm. Mass Spectrom. 15 (1988) 143–151.

Schweer, H., Meese, C. O., Watzer, B., Seybarth, H. W., Determination of Prostaglandin E_1 and its Main Plasma Metabolites 15-Keto-prostaglandin E_0 and Prostaglandin E_0 by Gas Chromatography/Negative Ion Chemical Ionization Triple-stage Quadrupole Mass Spectrometry, Biolog. Mass Spectrom. 23 (1994) 165–170.

Schweer, H., Seyberth, H. W., Schubert, R., Determination of Prostaglandin E_2 Prostaglandin $F_{2\alpha}$ and 6-Oxo-prostaglandin $F_{1\alpha}$ in Urine by Gas Chromatography/Mass Spectrometry and Gas Chromatography/Tandem Mass Spectrometry: A Comparison, Biomed. Environm. Mass Spectrom., 13 (1986) 611–619.

Strife, R. J., Kelley, P. E., Weber-Grabau, M., Tandem Mass Spectrometry of Prostaglandins: A Comparison of an Ion-Trap and a Reversed Geometry Sector Instrument, Rap. Comm. Mass Spectrom. 2/6 (1998) 105–109.

Tsikas, D., Schwedhelm, E., Fauler, J., Gutzki, F.-M., Mayatepek, E., Frölich, J. C., Specific and rapid quantification of 8-iso-prostaglandin $F_{2\alpha}$ in urine of healthy humans and patients with Zellweger syndrome by gas chromatography – tandem mass spectrometry, J. Chrom B, 716 (1998) 7–17.

Section 4.36

Cowan, D. A., Woffendin, G., Drug Testing in Human Sports, Spectra 11 (1988), 4–9.

Uhrich, M., Chemical Ionization with the INCOS 50: The Key to the Accurate and Quantitative Analysis of Drugs, Finnigan MAT Application Report Nr. 213, 1988.

Section 4.37

Stephenson, J., Barbiturates: Identification and Quantitation with the ITS 40, Finnigan MAT Application Data Sheet No. 52, 1990.

Section 4.38

Courselle, P., Schelfaut, M., Sandra, P., et al., The Analysis of the β-Agonists Clenbuterol and Salbutamol by Capillary GC-MS. Considerations on Sample Introduction. In: P. Sandra (Ed), 13th Int. Symp. Cap. Chrom., Riva del Garda, May 1991, Huethig: Heidelberg 1991, 388–393.

Fürst P., GC/MS-Bestimmung von Rückständen und Kontaminanten, in: Matter L., Lebensmittel und Umweltanalytik mit der Kapillar-GC, VCH: Weinheim 1994, 29–73.

Tillmanns, U., Nachweis von Clenbuterol mit dem ITS 40, Finnigan MAT Application Report No. 74, 1990.

Section 4.39

Pfleger, K., Maurer, H. H., Weber, A., Mass Spectral and GC Data, Part 1, VCH: Weinheim 1992.

Demme, U., Müller, U., Zur Ausbeutebestimmung toxikologisch-chemischer Analysenverfahren, Toxichem + Krimitech 57 (1990), 121.

Deutsche Forschungsgemeinschaft, Orientierende Angaben zu therapeutischen und toxikologischen Konzentrationen von Arzneimitteln und Giften in Blut, Serum oder Urin, VCH: Weinheim 1990.

Meyer, L. V., Hauck, G., Der Nachweis gebräuchlicher Antihistaminika nach therapeutischer Dosierung, Beiträge gerichtl. Medizin 34 (1976), 129.

Section 4.40

Butz, S., Heberer, Th., Start, H.-J., Analysis of phenoxyalkanoic acids and other acidic herbicides at the low ppt-level in water applying solid phase extraction with RP-C18 material, J. Chromatogr. (1994), im Druck.

Heberer, Th., Butz, S., Stan, H.-J., Detection of 30 acidic herbicides and related compounds as their pentafluorobenzylic esters using GC/MSD, J. AOAC (1994), im Druck.

Stan, H.-J., Heberer, Th., Linkerhägner, M., Vorkommen von Clofibrinsäure im aquatischen System – Führt die therapeutische Anwendung zu einer Belastung von Oberflächen-, Grund- und Trinkwasser?, Vom Wasser 83 (1994), 57–68.

Section 4.41

Eschke, H.-D., Traud, J., Dibowski, H.-J., Analytik und Befunde künstlicher Nitromoschus-Substanzen in Oberflächen- und Abwässern sowie Fischen aus dem Einzugsgebiet der Ruhr, Vom Wasser 83 (1994), 373–383.

Eschke, H.-D., Traud, J., Dibowski, H.-J., Untersuchungen zum Vorkommen polycyclischer Moschus-Duftstoffe in verschiedenen Umweltkompartimenten, Z. Umweltchem. Ökotox. 6 (1994), 183–189.

Rimkus, G., Wolf, M., Analysis and Bioaccumulation of Nitro Musks in Aquatic and Marine Biota, in: P. Sandra, G. Devos (Eds): 16th Int. Symp. Cap. Chrom., Riva del Garda, Sept. 27th–30th 1994, Hüthig: Heidelberg 1994, 433–445.

5 Glossary

AC	Alternating current
AGC	Automatic gain control, variable ionisation time in the ion trap scan function. A build-up of ions formed in parallel in the ion trap is facilitated by the AGC until the maximum storage capacity has been filled. This results in the extraordinarily high sensitivity of the ion trap analyser for full scan data acquisition in the EI mode.
Analytical Scan	The part of the ion trap scan function which produces the mass spectrum. The pre-scan takes place first, which regulates the variable ionisation time for the analytical scan.
Reagent ion capture	Type of CI reaction (PCI/NCI). Ionisation of the analyte is achieved through addition of the reagent ion.
ARC	Automatic reaction control, variable ionisation and reaction time for chemical ionisation using the ion trap scan function. A build-up of CI ions formed in parallel in the ion trap is facilitated by the ARC until the maximum capacity or the maximum reaction time is reached. This results in the extraordinarily high sensitivity of the ion trap analyser for full scan data acquisition in the CI mode.
Bake out	Purification step for the adsorption trap in the purge and trap technique.
Base peak	The most intense signal (100%) in a mass spectrum.
Benzyl cleavage	A fragmentation reaction of alkylaromatics forming the benzylic carbenium ion, which appears as the tropylium ion (m/z 91) in many spectra of aromatics.
Limit of quantitation (LOQ)	Unlike the limit of detection, the limit of quantitation is confirmed by the calibration function. The value gives the lower limiting concentration which differs significantly from a blank value and it can be determined unambiguously and quantitatively with a given precision. The LOQ value is therefore dependent on the largest statistical error which can be still tolerated in the results.
Blank value	Many analysis methods require the determination of blank values in order to be able to correct measured values for samples by a nonspecific component. A differentiation is made between reagent blank samples and sample blank samples.
BSTFA	Bis(trimethylsilyl)trifluoroacetamide, silylating agent for derivatisation.
BTEX	Abbreviation for the analysis of the aromatic compounds benzene, toluene, ethylbenzene and xylene isomers (also BTX).

BTV	Breakthrough volume of an adsorption trap.
BTX	→ BTEX
CAD	Collision-activated decomposition, → C I D
Carbosieve	Carbon molecular sieve used as an adsorption material for air analysis, → VOCARB (Supelco)
Carboxen	Carbon molecular sieve used as an adsorption material for air analysis, → VOCARB (Supelco)
Carry-over	Taking the analyte into the next analysis.
CAS No.	The unambiguous registration number of a chemical compound or substance assigned by the Chemical Abstracts Service. The CAS No. is usually given in the substance entries of mass spectral libraries.
Centroid	The calculated centre of gravity of a mass peak. Centroids can also be calculated precisely in → LRMS and give the mass numbers used for annotating line spectra. The decimal places determined in the calculation are often associated with the resolving power of an LRMS instrument, but should strictly be separated from mass details given in high resolution MS (→ HRMS). In spite of centroid calculations the resolving power of an LRMS instrument should still be taken into consideration on unit mass resolution with a peak width of 1000 mu.
CI	Chemical ionisation, ionisation of substances brought about by the presence of a reagent gas. Through the chemical reaction of the reagent ion with the analyte, soft ionisation takes place and usually gives the (quasi)-molecular ion. The selectivity of the reaction and the fragmentation are controlled by the choice of reagent gas. Both positive and negative ions can be formed by chemical ionisation.
CID	Collision-induced dissociation; leads in the MS/MS technique to formation of the product ion spectrum of a selected ion.
Clean-up	General sample preparation procedure involving removal of the matrix and concentration of solutions.
Cold trapping	A technique for focusing volatile compounds at the beginning of the chromatography column by cooling the section of the column, e.g. with liquid CO_2, below the boiling point of the solvent (→ Cryofocusing).
Dalton	Unit of relative atomic or molecular mass (Da). Since 1961 this has been given relative to 1/12 of the mass of the ^{12}C isotope; frequently used in biochemistry for describing the masses of macromolecules.
DAT	Diaminotoluene isomers
DC	Direct current
Desorb preheat	Purge and trap technique step for heating up the adsorption trap. The desorption and transfer of the analyte into the GC/MS system is effected by switching a 6-way valve after the preselected preheat temperature (5 °C below the desorption temperature) has been reached.

DNB	Dinitrobenzene isomers
DOD	US Department of Defense
Pressure	The SI unit is given in Pascals

$1 \text{ Pa} = 1 \text{ N/m}^2$
$10^5 \text{ Pa} = 10^5 \text{ N/m}^2 = 1 \text{ bar}$

Pressure is still often given in traditional units. In MS vacuum techniques pressures are given in [Torr] or [mTorr] and gas pressures of GC supplies frequently in [kPa], [bar] or [psi].

Conversion table

	Pa	bar	Torr	psi	at	atm
Pa	1	$1 \cdot 10^{-5}$	$7.5 \cdot 10^{-3}$	$1.45 \cdot 10^{-4}$	$1.02 \cdot 10^{-5}$	$9.87 \cdot 10^{-6}$
bar	$\mathbf{1 \cdot 10^5}$	1	750	14.514	1.02	0.987
Torr	133	$1.33 \cdot 10^{-3}$	1	$1.94 \cdot 10^{-2}$	$1.36 \cdot 10^{-3}$	$1.32 \cdot 10^{-3}$
psi	$6.89 \cdot 10^3$	$6.89 \cdot 10^{-2}$	51.67	1	$7.03 \cdot 10^{-2}$	$6.80 \cdot 10^{-2}$
at	$9.81 \cdot 10^4$	0.981	736	14.224	1	0.968
atm	$1.0133 \cdot 10^5$	1.0133	760	14.706	1.033	1

at technical atmosphere 1 kp/cm^2
atm physical atmosphere 1.033 kp/cm^2
psi pound per square inch

Dry purge	Purge and trap analysis step whereby moisture is removed from the trap by the carrier gas before desorption.
Duty cycle	Degree of effective analysis time of, for example, a mass spectrometer; here it gives the dwell time per ion relative to the time required for a scan.
Dwell time	Effective acquisition time for an ion using the SIM technique.
ECD	Electron capture detector, gives a mass-flow-dependent signal.
Eddy diffusion	Multipath effect in chromatography, cause of peak broadening through diffusion processes.
Effective plates	The number of effective theoretical plates in a column, taking the dead volume into consideration ($\rightarrow$ also HETP).
EI	Electron impact ionisation. With modern GC/MS instruments ionisation usually takes place at an energy of 70 eV. Positive ions are formed predominantly.
ELCD	Electrolytic conductivity detector (also Hall detector), gives a mass-flow-dependent signal.
Electron capture	The capture of thermal electrons by electronegative compounds. This process forms the basis of the ECD (electron capture detector) and is also made use of in negative chemical ionisation (NCI, ECD-MS).

Electron impact ionisation	→ EI
Elution temperature	The temperature of the GC oven at which an analyte reaches the detector.
Energy	The SI unit is the Joule [J] $1 J = 1 Nm = 1 Ws$ In MS energy is usually given in [eV] and is calculated from the elemental charge and the acceleration voltage. Conversion factors: $1 eV = 23.0$ kcal $= 96.14$ kJ $(1$ cal $= 4.18$ J$)$
Decision limit	The smallest detectable signal from the substance that can be clearly differentiated from a blank signal (assessed in the signal domain, detection criterion). The decision limit is calculated as the upper limit of the distribution range of the blank value.
EPA	US Environmental Protection Agency
ESI	Electrospray ionisation, an LC/MS coupling technique.
ESTD	External standard, quantitation by external standardisation. The signal height for a known concentration of the analyte is used for the calibration procedure. The calibration analyses are carried out separately (externally) from the analysis of the sample.
External ionisation	Process for the production of substance ions in a storage mass spectrometer, e.g. in ion trap GC/MS systems. Ionisation does not take place inside the ion trap analyser, but in the ion source situated in front of the analyser, into which the end of the chromatography column is led directly. This arrangement ensures that a mass analysis is unaffected by the operating parameters of the GC and also allows the use of negative chemical ionisation (NCI) as well as several MS/MS scan techniques.
FAME	Fatty acid methyl ester
FC-43	Perfluorotributylamine (PFTBA), a widely used reference substance for calibration of the mass scale, M 671.
FID	Flame ionisation detector, gives a mass-flow-dependent signal.
Forward search	Strategy in library searching of mass spectra in which only the intensities of the signals in the unknown spectrum are compared with the library data (unlike the reverse search). Signals in the reference spectra, which are not present in the unknown spectrum, are not taken into consideration.
FPD	Flamephotometric detector, gives a mass-flow-dependent signal.
Fragment ion	An ion formed by the decomposition of a (quasi)molecular ion.
Frit sparger	U tube for the purge and trap analysis of water samples with built-in frit for fine dispersion of the purge gas.

5 Glossary

Fritless sparger	U tube without a frit for the purge and trap analysis of water samples, for solid samples, or moderately foaming water samples.
Full scan	Complete mass scan, acquisition of the complete mass spectrum.
GCB	Graphitised carbon black, → VOCARB (Supelco)
Glass cap cross	→ Werkhoff splitter
Balanced pressure injection	Headspace injection technique whereby the equilibrated sample vial is depressurised to its maximum against the column pre-pressure.
GPC	Gel permeation chromatography, gel chromatography used for sample preparation e.g. to remove lipids from fatty foodstuffs.
Hall detector	→ ELCD
Headspace	Static headspace GC
Headspace sweep	Technique in purge and trap analysis for treating foaming samples whereby the purge gas is passed only over the surface of the sample (headspace) instead of through it.
HETP	Height equivalent to a theoretical plate. The dependence of the HETP value on the carrier gas velocity determined from the van Deemter curve is used to optimise separation.
HFBA	Hetptafluorobutyric anhydride, derivatisation agent for preparing hetptafluorobutyrates. It is frequently used for introducing halogens into compounds to increase the response in ECD and NCI.
HHS	US Human Health Services
High resolution MS	The separation of C, H, N, and O multiplets in MS (resolution >10 000). The empirical formula of an ion is usually obtained through fine mass determination as part of the structure elucidation. In GC/MS the extremely selective detection of specific ions is possible (e.g. in dioxin analysis).
Hot needle	GC injection technique in which a drop of sample is drawn up into the body of the syringe and, before injection, the needle is allowed to heat up in the injector.
HpCDD	Heptachlorodibenzodioxin isomers
HRGC	High resolution gas chromatography, GC with fused silica capillaries. The term is used to contrast the currently widely used capillary technique with that using packed 1/8″ columns.
HRMS	High resolution mass spectrometry. Here HRMS describes the use of a double focusing magnetic sector mass spectrometer with a resolution of greater than 10 000 (as opposed to LRMS with quadrupole and ion trap instruments) (→ LRMS, → high resolution).
HSGC	Headspace gas chromatography
HxCDD	Hexachlorodibenzodioxin isomers

ICR-MS	Ion cyclotron resonance mass spectrometer, an ion storage MS with very high resolution from which spectra are obtained from the turnaround frequency of the ions by Fourier transformation.
INCOS	Integrated computer systems, it refers to the search process for mass spectra in Finnigan GC/MS systems.
Ion storage MS	Mass spectrometer equipped with internal or external ion production which first collects the ions. The collection and analysis of the mass spectra take place discontinuously: ion trap (LRMS) and ICR mass spectrometer (HRMS).
Ion injection	Transfer of the substance ions formed in an external ion source to the analyser of a storage mass spectrometer, e. g. the ion trap analyser, and subsequent mass analysis, → external ionisation.
Isomers	Two substances which have the same molecular formula, but which differ chemically and physically; they have different structural formulae or they differ in the spatial arrangements of the atoms (stereoisomers). Frequently isomers give mass spectra which cannot be differentiated, but, because of different interactions with the stationary phase, they can be separated by GC.
Isotope	Although they have the same nuclear charge (number of protons) most elements exist as atoms which have nuclides with varying numbers of neutrons. These are known as isotopes. They belong to the same chemical element but have different masses. While in chemical synthesis the natural isotope distribution is not taken into consideration (use of the average molecular weight), in mass spectrometry the distribution of the isotopes over the different masses is visible and the isotope pattern is assessed (calculation of the molecular weight on the basis of the most common isotope).
Isotope pattern	Characteristic intensity pattern (arrangement of lines) in a mass spectrum, based on the different frequency of occurrence of the isotopes of an element. From the isotope pattern in a spectrum conclusions can be drawn on the number of atoms of this element. Important isotope patterns in organic analysis are shown by Cl, Br, Si, C and S. Organometallic compounds show isotope patterns which are extremely rich in lines. Molecular ions show the isotope patterns of all the elements contained in the compound.
ISTD	Internal standard. A substance not contained in the sample is added to it in a defined quantity and used as an internal reference for the quantitation. The relative signal heights at known concentrations of the analyte relative to the internal standard at a constant concentration are used to calibrate the procedure.
Jet separator	Interface construction for MS coupling of wide bore and packed chromatography columns. The quantity of the lighter carrier gas is reduced by dispersion; loss of analyte cannot be prevented.
Capacity factor	The k' value of a column describes the molar ratio of a substance in the stationary phase to that in the mobile phase from the relationship of the net retention time to the dead time.
Continuous injection	Process for the production of defined atmospheres for the calibration of thermodesorption tubes.

Kovats Index	The most well known index system for describing the retention behaviour of substances. The retention values are based on a standard mixture of alkanes: Alkane index = number of C-atoms × 100.
Cryofocusing	Capillary GC sample injection technique for volatile compounds in headspace, purge and trap or thermodesorption systems. Instead of an evaporation injector, on-line coupling to the sample injection is set up in such a way that a defined region of the column is cooled by liquid N_2 or liquid CO_2 to focus the analyte. Once a sample has been injected, this region is heated up.
Charge exchange	Type of CI reaction (PCI). Ionisation is effected by electron transfer from the substance molecule to the reagent ion with formation of a positive molecular ion.
Charge transfer	Type of CI reaction (NCI). Ionisation is effected by electron transfer from the reagent ion to the substance molecule with formation of a negative molecular ion.
Line spectrum	Representation of a mass spectrum as a series of vertical lines (intensities) across the mass scale (m/z values).
LRMS	Low resolution mass spectrometry. For LRMS, quadrupole or ion trap mass spectrometers with nominal mass resolution (unit mass resolution) are used, as opposed to HRMS where double focusing magnetic instruments are used (→ HRMS, high resolution).
MAGIC 60	A term referring to the ca. 60 analytes of the combined volatile halogenated hydrocarbon/BTEX determination (VOC), which are analysed together in EPA methods e. g. by purge and trap GC/MS.
MAM	Monoacetylmorphine
Mass unit	Mass units are given in the literature in different ways. In the USA a mass unit is frequently given as an amu (atomic mass unit), and the mmu (millimass unit) is used for 1/1000 amu. The SI unit for the atomic mass m is given in kg. The additional unit used in chemistry is the atomic mass unit defined as $1\ u = 1.660 \times 10^{-27}$ kg. In mass spectroscopy the Dalton is also used as a mass unit (1 Da is 1/12 of the mass of the carbon isotope ^{12}C → Dalton).
Mass filter	A quadrupole analyser works as a mass filter. A mass spectrum is run by filtering out individual m/z values from the large number of those ions formed in the ion source and accelerated in the direction of the analyser. By cyclic changes in the control (scan) the whole mass range is covered (full scan). Switching to certain values of individual masses (SIM, selected ion monitoring) only filters out preselected ions.
Mass scale	A mass scale always implies the m/z scale (mass/charge) because in EI/CI ionisation singly charged ions are formed, apart from a few exceptions (e. g. polyaromatic hydrocarbons).
Maximising masses peakfinder	A routine method of working out GC/MS analyses. The change in each in individual ion intensity with time is analysed. If several ions have a peak

	maximum at the same time, the elution of individual substances is recognised and noted even in the case of co-elution. The peakfinder is used for automatic analysis of complex chromatograms.
McLafferty rearrangement	The loss of neutral particles from a molecule by an H-migration in a 6-membered-ring transition state (named after Prof. McLafferty, Cornell University). The structural prerequisites for the rearrangement are as follows: – a double bond as C=C, C=O or C=N, – a chain of three σ-bonds ending in a double bond, – an H-atom in a γ-position which can interact with the double bond. The McLafferty rearrangement is a typical fragmentation reaction of aldehydes, organic acids and methyl esters (FAMES, fatty acid methyl esters).
MCM	Moisture control module, a device for removing moisture in the desorption step of the purge and trap procedure by means of condensation.
MCS	Moisture control system, → MCM
MDL	Minimum detection limit, → limit of detection
Metastable ions	Ions which lose neutral particles during the time of flight through the analyser give information on fragmentation pathways and can only be detected using magnetic sector instruments.
MHE	Multiple headspace extraction, a quantitation procedure used in static headspace involving multiple extraction and measurement from a single sample.
MID	Multiple ion detection, the recording of several individual masses (unlike full scan) → SIM
Mixture search	One mode of the PBM search procedure whereby a mixture is analysed by forming and searching difference spectra.
MNT	Mononitrotoluene isomers
Modifier	The addition of organic solvents to the extraction agents in SFE. The modifier can be added directly to the sample (and is then effective only during the static extraction step) or can be added continuously using a second pump. Usually up to 10% of the modifier is added.
Molecular weight	The sum of the atomic weights of all the atoms present in a molecule. The term molecular weight is generally used although actually masses are involved. The molecular weight is defined as how much heavier a molecule is than 1/12 of the carbon isotope ^{12}C. The average molecular weight is calculated taking the natural isotopic distribution of the elements into account (stoichiometry). Molecular weights in MS: the calculation is carried out exclusively using the atomic masses of the most frequently occurring isotope, → nominal mass.
Molecular ion	The nonfragmented positive or negative ion with the mass (m/z) of the nominal molecular weight.
MQL	Minimum quantitation limit, → limit of quantitation (LOQ).

M-Series	n-Alkyl-bis(trifluoromethyl)phosphine sulfides, a homologous series used to construct a retention index system. The compounds can be used with FID, ECD, ELCD, NPD, FPD, PID and MS.
MS/MS	Measuring technique in mass spectrometry whereby collision-induced decomposition gives the product ion spectra of selected ions. It can be carried out with magnetic sector, triple, quadrupole and ion trap instruments.
MS/MS^n	The multiple MS/MS technique whereby granddaughter spectra are produced by collision-induced decomposition of selected product ions.
MSTFA	N-Methyl-N-trimethylsilyltrifluoroacetamide, a derivatisation agent for silylation.
m/z	The mass to charge ratio based on the atomic mass unit m [u] (1 u = $1.66 \cdot 10^{-27}$ kg) and the elemental charge e_0 [C] (1 e_0 = $1.6 \cdot 10^{-19}$ C). As there are only whole multiples of e_0, according to a IUPAC recommendation m/n · e_0 is given as m/z, e.g. Cl^+ with m = 34.96989 u and z = 1.
Limit of Detection	The lowest concentration at which a substance can still be detected unambiguously (assessed in the substance domain). The value is obtained from the decision limit (smallest detectable signal) using the calibration function or the distribution range of the blank value.
Nafion Drier	Instrument for drying gas streams using a semipermeable membrane.
NB	Nitrobenzene
NBS	The National Bureau of Standards, now → NIST (US)
NCI	Negative chemical ionisation, production and detection of negative ions in chemical ionisation → CI.
Needle sparger	Glass vessel used in purge and trap analysis of solids or foaming samples. The purge gas is passed into the sample via a needle perforated on the side or by means of headspace sweep.
Neutral loss scan	MS/MS experiment for analysing for precursor ions which undergo a common loss of neutral particles of the same mass. Analytes with functional groups in common are detected.
NIH	US National Institute of Health
NIST	National Institute of Standards and Technology, part of the US Department of Commerce.
Nominal mass	The value of the molecular weight calculated from the atomic masses of the most frequently occurring isotopes of the elements used in mass spectrometry (e.g. 12 for C, 1 for H, 35 for Cl, 79 for Br etc). According to this convention CH_4 and CD_4 have the same nominal mass!
Nominal mass resolution	The mass spectrometric resolution for the separation of mass signals with a uniform peak width of 1000 mu (1 mass unit) in a quadrupole or ion trap analyser (→ LRMS). The resulting mass numbers are generally given as whole numbers or to one decimal place (unlike high resolution).
NPD	Nitrogen phosphorus detector, gives a mass-flow-dependent signal.

NT	Nitrotoluene isomers
OCDD	Octachlorodibenzodioxin
On-column	Sample injection technique for GC whereby a diluted liquid extract is injected directly (without evaporation) on to a pre-column or on to the column itself.
PAHs	Polyaromatic hydrocarbons, usually the group of polycyclic hydrocarbons from naphthalene to coronene.
Precursor ion	Selected precursor ion for collision-induced decomposition in the MS/MS acquisition technique.
Precursor ion scan	MS/MS experiment for the analysis of precursor ions which lose one particular fragment. It is used to detect substances with related structures which give common fragments.
Precursor-product ion scan	Acquisition of product ion spectra in the MS/MS technique.
Partition coefficient	The partition coefficient k used in the headspace technique is given by the partition of a compound between the liquid and gaseous phases c_{liq}/c_{gas}.
PAT	Purge and trap, dynamic headspace procedure.
PBBs	Polybrominated biphenyls
PBBEs	Polybrominated biphenyl ethers
PBM	Probability-based match, comparison procedure for mass spectra developed by Prof. McLafferty.
PCBs	Polychlorinated biphenyls
PCI	Positive chemical ionisation, production and detection of positive ions during chemical ionisation, → CI.
PCXEs	Polychlorinated xanthenes
PCXOs	Polychlorinated xanthones
PEEK	Polyether ether ketone, hard plasticiser-free polymer with the general structure (with $x = 2$, $y = 1$):

$$\left[\left(\!\!\bigcirc\!\!\right)\!-\!O\right]_x\!\!\left[\left(\!\!\bigcirc\!\!\right)\!-\!\overset{O}{\underset{\|}{C}}\right]_y$$

PEEK is used as a sealing and tubing material and for screw joints in SFE, SFC, and HPLC and in high vacuum areas of MS, thermally stable up to 340 °C.

PFBA	Pentafluoropropionic anhydride, derivatisation agent. It is also frequently used for introducing halogen to increase the response for ECD and NCI.
PFK	Perfluorokerosene, used to calibrate the mass axis, widely used with magnetic sector instruments, → FC43.

PFTBA	Perfluorotributylamine, → FC43.
Phase ratio	In the headspace technique the phase ratio V_{gas}/V_{liq} gives the degree of filling of a headspace bottle. In capillary GC the phase ratio is the ratio of the internal volume of a column (volume of the gaseous mobile phase) to the volume of the stationary phase. High performance columns are characterised by high phase ratios. A table showing different combinations of internal diameters and film thicknesses can be used to optimise the choice of column with regard to the analysis time, resolution and capacity.
PID	Photoionisation detector, gives a concentration-dependent signal.
PLOT	Porous layer open tubular GC column, in which the stationary phase consists of a solid adsorption material. Instead of being filled completely (packed column), the material is applied in a thin layer to the column wall (e. g. Porapak).
PPINICI	Pulsed positive ion negative ion chemical ionisation, alternating data acquisition of positive and negative ions during chemical ionisation (patented by Finnigan, San Jose, CA).
Pre-purge	A preliminary step in purge and trap analysis. The atmospheric oxygen is removed before the sample is heated, i. e. at room temperature, by the purge gas to avoid side reactions.
Pre-scan	The step before the analytical scan in the ion trap scan function. In pre-scan the variable ionisation time is adjusted in order to run the mass spectrum.
Pre-search	Part of library searching of mass spectra in which a small group of candidates is selected from the whole number for detailed comparison.
Press fit	Glass connector for fused silica capillaries. The cross cut column end is simply pushed into the conical opening and a seal is achieved with the external polyimide coating.
Primary reaction	Conversion of the reagent gas used for CI into reagent ions by electron impact ionisation, → CI.
Proton abstraction	Type of CI reaction (NCI). Ionisation is effected by transfer of a proton from the analyte (abstraction) to the reagent ion, e. g. from analytes with phenolic OH groups.
Protonation	CI reaction type (PCI) involving ionisation by proton transfer to the substance molecule. Protonating reagent gases include methane, methanol, water, isobutane and ammonia.
PTV	Programmable temperature vaporiser, used for sample injection into GC instruments, cold injection system for splitless and split injection, solvent split technique and cryo-enrichment.
Pure search	A mode of the PBM search procedure, which only uses the forward search capability of the library search, → forward search.
Purge and trap	Dynamic headspace procedure.

Quasimolecular ion	An ion to which the molecular mass is assigned, which is formed by, for example, chemical ionisation as $(M+H)^+$, $(M+NH_4)^+$, $(M-H)^+$, $(M-H)^-$.
QUISTOR	Quadrupole ion storage device, developed by Prof. Paul, University of Bonn, Germany. The name was given to the arrangement of electrodes currently known as the ion trap analyser. The name reflects the relationship to the quadrupole filter. Prof. Paul received the Nobel Prize for physics in 1989 for his work on QUISTOR developed at the beginning of the 1950s.
Reagent gas cluster	The spectrum of ions which is formed from the reagent gas through chemical ionisation.
Recovery	This can be determined in an analysis from the ratio of the surrogate standard to an internal standard.
Response	Specific height of the detector signal, usually calculated as the ratio of the peak area or height to quantity of substance.
Retention gap	Pre-column or guard column with a larger internal diameter (typically 0.53 mm) for on-column injection on to capillary columns. The retention gap absorbs the entire volume of solvent. As it is not coated with a stationary phase, there is no retention of analytes. After evaporation of the solvent, the analyte is focused at the beginning of the main column.
Retention volume	This is the carrier gas volume required to elute a component.
Reverse search	Library search of mass spectra whereby only the mass intensities from the library data are compared with the unknown spectrum. Other mass signals which are present in the unknown spectrum are disregarded (opposite of forward search). This comparison procedure serves to expose mixed spectra or to search for a substance in a GC/MS chromatogram.
RF	Radio frequency, high frequency.
RI	Retention index, → Kovats.
RIC	Reconstructed ion chromatogram. The total ion current (TIC) is calculated from the sum of the intensities of all the mass signals in a mass spectrum. This value is generally stored with the mass spectrum and shown as an RIC or TIC. Plotting the TIC values along the scan axis (number of the mass spectrum and retention time) gives the conventional chromatogram diagram. (Older magnetic sector instruments had their own total ion current detectors for this purpose). For this reason the shape of the chromatogram is currently dependent on the mass range scanned.
RT	Retention time
Sandwich technique	→ solvent flush.
Scan	Running through the masses of the MS analyser while a mass spectrum is being taken, → full scan.
µ scan	The shortest scan unit for ion trap mass spectrometers. Depending on the preselected scan rate of the chromatogram, several µ scans are added together.

Scan function	Time control of the ion trap or quadrupole mass analyser by changing the applied voltage. It is represented as a diagram of voltage U [V] against time t [ms].
Skewing	Reversal of the relative intensities in a mass spectrum by changing the substance concentration during the scan. Skewing occurs with beam instruments during slow scans in the rising and falling slopes of steep GC peaks.
Magnetic sector MS	Single or double focusing mass spectrometer with magnetic (and electric) analyser for separation of the flight paths of the ions and focusing them on to the detector; used for high resolution mass spectrometry, → HRMS, high resolution.
Secondary reaction	The chemical reaction of the analytes with the reagent ions during CI to give stable products, → CI.
SFC	Supercritical fluid chromatography, uses a supercritical fluid as the mobile phase. The importance of SFC is continually increasing because of on-line coupling with MS and as an alternative procedure to HPLC. The use of fused silica capillary columns involves a less complicated direct coupling to MS for SFC compared with HPLC.
SFE	Supercritical fluid extraction. Pure CO_2 is mostly used, although modifiers, such as methanol or ethyl acetate, can be added for optimal extraction yields.
SIM	Selected ion monitoring, recording of individual masses (as opposed to → full scan, → MID).
SIM descriptor	Control file for an SIM analysis. Contains the selected specific masses of the analytes (m/z values), the individual dwell times and the retention times for switching the detection to other analytes.
SIS	Selective ion storage, describes the SIM measuring technique for ion trap MS (→ SIM, → waveform ion isolation).
SISCOM	Search for identical and similar compounds, search procedure for mass spectra developed by Henneberg, Max Planck Institute for Coal Research, Mülheim, Germany.
S/N	Signal to noise ratio of, for example, a GC peak. It is calculated from the noise width in the peak region and the height of the peak is measured from the middle of the noise width to the peak maximum.
Solvent effect	Focusing the analyte to a narrow band at the beginning of a capillary column by means of the condensation of the solvent and directed evaporation.
Solvent flush	GC injection technique whereby a liquid (solvent, a derivatisation agent) is first drawn up into a syringe, followed by some air to act as a barrier, and finally the sample (also known as the sandwich technique).
Solvent split	Injection technique using the PTV injector so that larger quantities of solvent can be applied (>2 μl up to the LC/GC coupling).
α-Cleavage	Basic MS fragmentation reaction. α-Cleavage involves breaking the bond between the α- and β-carbon atoms and occurs after ionisation of a mole-

cule by loss of one nonbonding electron from a heteroatom (e.g. amines, ethers, ketones) or the formation of an allylic or benzylic carbenium ion (e.g. alkenes, alkylaromatics).

SPE	Solid phase extraction
SPI	Septum equipped programmable injector, special form of the cold injection system which can be used exclusively for total sample transfer.
SPME	Solid phase microextraction
Spray and trap	Extraction procedure for foaming aqueous liquids. The liquid sample is sprayed into the purge gas stream.
SRM	Selected reaction monitoring, MS/MS scan technique, whereby a product ion spectrum is produced by collision-induced decomposition of a precursor ion, in which only one or several specific fragments are detected. In MS/MS instruments the SRM technique offers the highest possible specificity with the highest possible sensitivity and is used like the SIM technique for multistep MS.
Beam instruments	Mass spectrometers in which beams of ions are formed continuously from an ion source, pass through a lens system (ion optics), and are resolved in the analyser (mass analysis): magnetic sector and quadrupole mass spectrometers.
SUMMA canister	Passivated stainless steel canister for air analysis (e.g. EPA method TO-14/15) for collection of samples and standardisation. The inner surface is deactivated by a patented procedure involving a Cr/Ni oxide layer.
Surrogate standard	This is an internal standard which is added to a sample before clean-up. The ratio of the surrogate standard to other internal standards added in the course of the clean-up is used to calculate the recovery.
Tandem-in-space	MS/MS analysis with beam mass spectrometers. Ion formation, selection, collision-induced decomposition, and acquisition of the product ion spectrum take place continuously in separate compartments of a mass spectrometer, e.g. triple-quadrupole MS.
Tandem-in-time	MS/MS analysis with storage mass spectrometers. Ion formation, selection, collision-induced decomposition and acquisition of the product ion spectrum take place in the same location of the mass spectrometer but at different times, e.g. ion trap MS, ICR-MS.
Tandem MS	A term describing the MS/MS technique using triple-quadrupole mass spectrometers. Two scanning quadrupoles with a collision cell situated between them work as a tandem.
TCA	Target compound analysis, multicomponent analysis, multimethod. Compounds grouped together in the list for analysis are searched for using automatic routines in GC/MS analysis by spectrum comparison in a retention time window, identified and then quantified if they are found.
TCDD	Tetrachlorodibenzodioxin isomers
TCDF	Tetrachlorodibenzofuran isomers

Tenax	Nonpolar synthetic polymer based on 2,6-diphenyl-p-phenylene oxide used as an adsorption material for the concentration of air samples or in purge and trap analysis.
Theoretical plates	→ HETP
TCMS	Trimethylchlorosilane, trimethylsilyl chloride, silylation agent (catalyst).
TMS Derivative	Trimethylsilyl derivative of a compound. The TMS group appears as a fragment in the mass spectrum at m/z 73.
TMSH	Trimethylsulfonium hydroxide, derivatisation agent for esterification and methylation, e. g. of free fatty acids, phenylureas etc. TMSH was used successfully to save on extra steps for the derivatisation in the insert of the PTV cold injection system.
TNT	Trinitrotoluene isomers
Product ion spectrum	The mass spectrum produced by the collision-induced decomposition of a selected ion in MS/MS analysis. The starting ion is known as the precursor ion.
TIC	Total ion current. Sum of all the intensities in a mass spectrum, → RIC.
Total sample transfer	GC injection technique whereby the entire injection volume reaches the column with or without evaporation (splitless injection).
Dead time	The time taken for a substance which is not retarded (e. g. air, or methane for silicone phases) to pass through a chromatography column. The carrier gas velocity is calculated from the dead time and the length of the column, → HETP.
Number of theoretical plates	This describes the separating capacity of a column.
Triple-quadrupole	Arrangement of three consecutively situated quadrupoles for MS/MS analysis. The middle quadrupole is designed as a collision cell (CID), the first (Q_1) and the third (Q_3) for the scanning operation, → tandem MS.
Tropylium ion	The characteristic ion in the spectrum of alkylaromatics, $C_7H_7^+$ m/z 91, is formed by benzyl cleavage of the longest alkyl chain. The structure of the seven-membered aromatic ring (after internal rearrangement) is responsible for the high stability of the ion.
TRT	Temperature rise time, heating up period in pyrolysis until the required pyrolysis temperature is reached. This is a quality feature of the pyrolyser.
TSP	Thermospray LC/MS coupling.
UAR	Unknown analytical response. This refers to GC peaks in multicomponent analysis (TCA) which are not mentioned in the list of results.
van Deemter plot	→ HETP
VOC	Volatile organic carbon, volatile organic hydrocarbons, a group of analytes which are determined together using an EPA method and which contain both volatile halogenated hydrocarbons and BTEX.

VOCARB	Nonpolar adsorbent filling for concentration of air samples, and in purge and trap analysis, multilayer filling based on graphitised carbon black (Carboxen) and carbon molecular sieves (Carbosieve), Supelco.
Waveform ion isolation	Ion storage technique for the ion trap analyser. By using resonance frequencies the ion trap can exclude several m/z values from storage (e. g. the matrix) and collect selectively determined preselected analyte ions ($\rightarrow$ SIM, $\rightarrow$ SIS).
Werkhoff splitter	Adjustable flow divider for sample injection on to two capillary columns, or a split situated after the column, also known as a glass cap cross divider.
Wisswesser line notation	Coding chemical structures by an alphanumerical system, e.g. for Lindane, L6TJ AG BG CG DG EG FG *GAMMA, contained in the Wiley library.
Thermal conductivity detector	This gives a concentration-dependent signal.
XAD	Synthetic polymer (resin) used as an adsorption material in air or purge and trap analysis.
Standard addition method	Quantitation procedure involving repeated defined addition of the analyte to the sample, e.g. in headspace analysis; one vial is left without standard addition.

Subject Index

Acceleration voltage   131
Acetates   273
Acetylation   389
Activated charcoal   52
Addition products   151
Adduct formation   153, 154
Adduct ions   162
Adsorption   52
Adsorption material   54
Adsorption trap   35
Adsorption tube   54
Air analysis   52, 333
Air samples   333
Air sampling   470
Alkenes   260
Alkyl halides   155
Alkyl phosphates   309
Alkylaromatics   262
Alkylphenols   392
Amines   259
Anabolics   509
Analyser
– ion trap   147
– Nier/Johnson   5
– Quadrupole   5
Analysis documentation   416
Analysis error   321
Analysis time   77, 89
Antioxidant   322
Aromas   27
Aromatic rubble   262
Automatic evaluation   250
Automation   83
AutoQuan   538
Axial diffusion   114

Background spectrum   189
Background subtraction   138
Baffle   195
Baking out phase   43
Band broadening   74
Barbiturates   526
Base line   175
Beam instruments   141

Bench top/PBM   243
Benzyl cleavage   260
Biphenyls   454
Blank sample   311
Blank value   315
Breakthrough temperatures   88
Breakthrough volume   54, 151
Broad band excitation   183
Bromophenols   389
BTEX   262 ff
– limit of detection   341
– limits of quantitation   343
– recoveries   343

Caffeine   15
Calibration   56
Calibration file   379
Calibration function   319, 314 ff
Calibration level   315
Calibration standards   187
Cannabis   504
Capacity factor   106
Capacity ratio   116 f
Capillary columns   88 ff
– carrier gas flow   99
– choice of   88
– column length   98
– film thickness   97
– internal diameter   89
– packed columns   198
– polarity   88
– sample capacity   89
– special GC/MS recommendation   94
– stationary phases   92
Carbamates   16, 285 ff
Carbenium ions   260
Carbonyl function   261
Carbopak   41
Carbosieve   41
Carbotrap   53
Carbowax phases   322
Carcinogenicity   483
Carrier effect   317

Carrier gas   192
- flow rates   71
- flow   192
- helium   192
- hydrogen   192
- supply tubes   323
- velocity   115
Cartridges   10
CAS registry numbers   236
Centroid   8137
Charge exchange   152
Charge transfer   153
Chemical background   222
Chemical ionisation   144 ff
- changing between EI and CI   162
- high pressure CI   162
- low pressure CI   162
- negative chemical ionisation   153
- positive chemical ionisation   151
- primary reactions   148
- reagent gas systems   158 ff
- secondary reaction   148
- signal/noise ratio (S/N)   417
Chemical ionisation   5, 144
Chemical noise   169
Chemical warfare agents   306
Chemiluminescence   127
Chlorophenols   389
Chromatograms   213 ff
Chromatography   103 ff
- capacity factor   106
- chromatographic resolution   107 ff
- model   103
- parameters   103
- separation process   104
Clean-up   251
Closed loop stripping technique   385
Cocaine   497
Co-elution   216
Cold extraction   494
Cold injection systems   72 ff
Cold trap   58
Collision activated decomposition (CAD)   180
Collision induced dissociation (CID)   180
Collision stabilisation   156
Collision-activated decomposition (CAD)   180
Column bleed   89
Column length   98
Commercial products   231
Component of a mixture   244
Compression ratio   193
Concentration   334
Confidence interval   314
Confirmation   215 f
Control samples   482

Conversion dynode   148
Correlation factor   507
Cryo-enrichment   80
Cryofocusing   86
Curie point pyrolysis   63

Data formats   235
Data points   310
Dead time   106
Decision limit   311
Decomposition   285
Deconvolution   247
Degree of chlorination or bromination   254
Derivatisation   80
- BSTFA   513
- HFBA   520
- HMDS   533
- methylation   289
- MSTFA   520
- TMAH   442
- TMSH   80
Desorption   13
Desorption phase   39
Detergent   36
Deuterated standards   317
Deuteration   317
Diazomethane   41
Difference spectrum   528
Diffusion pump   195
Diffusion pump oil   322
Diffusion rate   193
Dinitrophenol herbicides   418 f
- specific ions   426
Dioxin analysis   138, 482
- most intense molecular ions   478
- quality assurance measures   482
Dioxins   473
Direct coupling   198
Direct injection   83
Discrimination   69
Division of the flow   128
Drinking water   35
Drug monitoring   536
Drug screening   500
Drugs   298, 497
Duty cycle   165
Dwell time   166
Dynamic region   459

ECD-MS   154
Eddy diffusion   105
Effective plates   117
Electrolytical conductivity detector (ELCD)   125, 140

Electron affinity   153
Electron beam   141
Electron capture   154
Electron capture detector (ECD)   10, 121 ff
– conversion rates   122
Electron impact ionisation (EI)   140 ff
Electronic pressure control (EPC)   99
Elimination   261
Elution   10, 103
– peak   105
– temperatures   97
– volume   10
Empirical formula   131
Enrichment   11
EPA (US Environmental Protection Agency)   58, 355
EPA methods   15
– 524.2   355
– 8240   355
– 8260   355
– TO   14   333
Errors
– injection 317
– volume   317
Ethers   260
Exact mass   253
Excess energy   140
Exchange of substances   113
Expansion volume   74
Explosives   301 ff

FC   43
FID   117
Field analysis   369
Film thickness   97
FIT value   242
Flameproofing agents   311
Flow optimisation   114
Flow rate   82, 113
Focusing   44
Foil pyrolysis   61 ff
Fomblin   195
Forepump   193
– oil   322
Fortifying   497
Forward search   242
FPD   127
Fragmentation   143 f
Fragments   220
Fragrances   35, 233
Full scan   164 ff
– safeguarding   167
Gas
– chromatography   67 ff
– phase acidity   154

– purification   323
– sampling   80

GC/MS interface   196
Gel permeation column (GPC)   463
General unknown analysis   536
Glass cap cross divider   128

Half width   105, 108
Hall detector 125
Headspace   26, 349
– dynamic   35 ff
– equilibration time   31, 32
– injection techniques   33
– multiple headspace MHE   28
– sample matrix   31
– sample volume   30
– static   27
– sweep   36
Heating rate   76
Height equivalent to a theoretical plate   117
Helium   71
Heptafluorobutyric anhydride   157
Herbicides   415
Herbicides   437
High resolution   131 ff
– exact mass   131
– interference   138
– nominal mass   131
High temperature phase   90
High volumes samples   470
Hit list   245
Homologous compounds   244
Humic substances   389
Hydride abstraction   152
Hydrocarbon matrix   12
Hydrocarbon, spectra of   235
Hydrogen   71
Hydrogenator   118
Hydrolysis   426
Hydroxide ions   154
Hydroxybenzonitrile herbicides   430
– specific ions   432

Impurities   322, 325
INCOS   236 ff
– spectral comparison   242
Individual mass registration (SIM, MID)   166
Initial bands   86
Injection   68
– cold injection system   72
– hot sample injection   68
– PTV cold injection system   72

- PTV solvent split injection   77
- PTV split injection   77
- PTV total sample transfer   74, 75
- rate   78
- split injection of   70
- total sample transfer   70
Injection techniques   68 f
- cold needle   69
- filled needle   69
- hot needle   69
- solvent flush/cold needle   69
- solvent flush/hot needle   68
Injection temperature   76
Injection times   82
Injection volumes   81
Injector system   82
Interactions with the stationary phase   111
Interfering signals   322
Internal diameter   81, 89
Internal standard   349
- choice of   317
Interpretation of mass spectra   251
Ion series spectrum   249
Ion source   141
Ion trap SIM technique   169
Ion yield   141
Ionisation energy   140
Ionisation potential   141
Ionisation procedures   140 ff
IsoMass- PC program   254
Isomerism   262
Isomers   250, 454
Isotope frequencies   253
Isotope intensities   185
Isotope patterns   253 ff
Isotope ratio   230
Isotopes   4
Isotopic labelling   297

Jet separator   200

Ketones   260
Key fragments   252
Kovat's index   117

Labile component   72
Lens potential   185
Levels of contamination   482
Libraries of mass spectra   230 ff
- CI library   230
- quality index   232
- substructure libraries   230
Library comparison   454

Library search   236
- Biemann   236
- INCOS/NIST   236
- PBM   243
- SISCOM   247
Light phase rotation perforator   422
Limit of detection   312
Line spectrum   145
Linearity   357
Liquid/liquid extraction   422
Local linearisation   315
Local normalisation   241
Localised charge   140
Loop injection   334
Loss of alkenes   261
Loss of analyte   26
Loss of CO   261
Loss of HCN   261
Low voltage ionisation   140
LRMS   473, 477

Magic 60   262, 359
Magnetic sector instruments   181
Magnetic sector mass spectrometer   476
Main search   241
MAK value list   483
Makeup gas   197
Mass calibration   185
- drift   198
- stability   190
Mass correlations   312
Mass scale   143
Mass selectivity   12
Mass spectrometer   130
- double focusing   132
- high resolution   5
- ion cyclotron   5
- magnetic sector   131
- quadrupole   133
- time of flight   5
Mass spectrometry   130 ff
Mass spectrum   143 ff
- skewing   230
Mass stability   138
Mass
- axis   213
- chromatograms to   15
- defect   187
- filter   164
- range   136
MassLib   247
Matrix   20
- tea   394
- air   454
- automotive paint   60

– blood   49
– cell cultures   49
– combustion gases   51
– compost   473
– cooking oils   48
– crude oil   181
– drinking water   389
– drinks   49
– exhaust gases
– fermentation units   49
– filtrate from the river banks   542
– fish   545
– floor covering   66
– fly ash   21
– gases from landfill sites   51
– ground water   483
– hair   497
– human milk   140
– hydrocarbons   273
– indoor air   54
– leather goods   442
– milk   463
– outdoor air   54
– plastic sheeting   48
– polymers   49
– polystyrene   31
– run off from a sewage works   542
– sealants   470
– sediments   545
– seepage water   345
– serum   504
– sewage sludge   14, 545
– soil air   51
– soil   14
– surface water   366, 538
– suspended particulates   454
– traffic emissions   444
– untreated water   389
– urine   504
– waste water   36
– water from the river Rhine   48
– water   341
Matrix effects   320
Maximising behaviour   218
Maximising masses peakfinder   213
Maximum sample capacity   112
McLafferty rearrangement   261
Megabore columns   89
Metabolites   181
Methadone programme   497
Methaniser   118
Methoximes   513
Methyl esters   273
Microbore capillary   89
MID   166, 254
Migration solutions   377

Military waste   444, 483
Minimum number to search   241
Mixed spectra   244
Mixture correction   248
Mixture search   244
Mobile GC/MS analyses   36
Modifier   23
Moisture   26
Molecular ion   140
Molecular sieve   53
Morphine derivatives   500
MS/MS   177 ff
MS/MS product ion spectrum
– 6 β-hydroxymethandienone   510
– THC   506
M-series   226
Multicomponent process   36
Multimethod   16
Multiple split technique   59
Multipoint calibrations   316
Musk   545

Nafion drier   334
Narrow bore   89
NBS   5
Needle sparger   356
Net retention time   106
Neutral loss scan   181
Neutral particle   181
NIST/EPA/NIH mass spectral library   231
Nitro compounds   301
Nitroaromatics   446, 493, 483
– analysis of, by ECD and PID   485
– chloronitroaromatics   493
– metabolites   484, 485
– SFE extraction   493
Nitrogen rule   252
Nitrophenol herbicides   421
– molecular masses of methyl ethers   424
Nitrosamines   233
Noise filter   241
Noise width   173
Nominal mass   136
Normal bore capillaries   192
NPD   119
Number of theoretical plates   117

Odour contamination   380
Off-line techniques   7
O-FID   119
On-column injection   83 ff
One point calibrations   315
On-line techniques   7
Open split coupling   196 ff

580 Index

Open split interface 335
Opiates 502
– selective masses 502
Organoarsenic compounds 306
Organochlorine pesticides 279
Ortho effect 301
OV 140
Oven temperature 74

Packed columns 199
Partition coefficient 117
Partition equilibrium 114
Partition ratio 117
Partition series 104
Passive collection 369
PBM 243 ff
– spectral comparison 244
PCDDs 478
PCDFs 478
Peak area 30
Peak matching 189
Peak shape 68
Peak slope 223
Peak symmetry 113
Peak width 106
Peak width at half height 108
PEEK material 18
Pentafluorobenzoyl chloride 157
Pentafluorobenzyl bromide 541
Pentafluoropropionylation 500
Perfluorobenzoyl chloride 157
Perfluorokerosene (PFK) 187
– spectrum 188
Perfluoropolyethers 196
Perfluorotributylamine 187
Perfume oils 15
Personal air samplers 470
Pesticide library 234, 394
Pesticides 394
Pfleger/Maurer/Weber toxicology library 233
Pharmaceuticals 497
Phase ratio 30, 98, 117
– of a capillary column 98
Phenols 273 ff, 454
Phenoxyalkylcarboxylic acids 291, 437
Phenyl acetates 392
– retention times 391
– specific masses 392
Phenylureas 289
Phosphoric acid esters 286 ff
Photons 182
Phthalate plasticisers 322
PID 123
Plant protection agents 279 ff
Plasticisers 10

Plates
– effective 116
– theoretical 99
Plausibility 229
PLOT column 363
Point to point calibration 315
Polarity 9
Pollution measurements 56
Polyaromatic hydrocarbons (PAHs)
  273, 444, 454
– carcinogenicity 445
– internal standards 446, 461
– molecular ions 444
– PAHs according to EPA 461
– parallel detection with nitroaromatics 488
– quasimolecular ions 461
– structural formulae 445
  see also: Polycondensed aromatics,
  Polycyclic aromatic hydrocarbons
Polybrominated biphenyl ethers (PBBE)
  309
Polybrominated biphenyls (PBB) 309
Polychlorinated biphenyls (PCB) 293 ff
– in milk and milk products 463
– processing of 13
Polymerisation additives 377
Poly(phenyl ethers) 195
Polyurethane foam cartridges 470
Porapak 53
Precursor ion 180
Pre-search 244
Preservatives 233
Pressure balanced injection 33
Pressure equivalence 15
Pressure surge 71
Probability values 244 f
Product ion spectrum 178
Prostaglandins 513
Proton abstraction 154
Proton affinities 149
Protonation 151
Pumping capacity 99
Pure search 244
Purge & trap 35
– back pressure control 38
– baking out phase 36
– contamination 357
– desorb-preheat step 39
– desorption phase 39
– dry purge phase 39
– glass apparatus 38
– limits of detection 359
– percentage recovery 37
– poly(ethylene glycol) 355
– preparation of water samples 355
– purge efficiency 37

– purge phase   36
– traps   40
Purge gas pressure, control of   38
Purity   250, 242
Pyrethroids   16
Pyrolysis   60, 596

Quantitation   310
– deuterated standards   444
– external standard calibration   316
– internal standard calibration   316
– standard addition procedure   320
Quantitation   318, 349
– limit of   312
Quasimolecular ion   146
QUISTOR   5

Reagent gas pressure   147
Reagent gas systems   158
– ammonia   160
– argon   152
– benzene   152
– carbon monoxide   152
– isobutane   160
– methane   158
– methanol   159
– nitric oxide   152
– nitrogen   152
– water   159
Reagent ion capture   154
Rearrangement reactions   259
Reconstructed ion chromatogram   214
Recovery   15
Redundancy filter   241
Re-extraction   500
Reference table   187
– for FC   43   187
– for PFK   189, 192
Refocusing   71
Regression
– analysis   250
– coefficient   315
Reproducibility   27
Residual monomers   377
Residual solvents   48
Residues   74
Resolution   110
– chromatographic   107 ff, 117
– conditions   132
– dispersion term   112
– factors affecting   110
– in mass spectrometry   131 ff
– retardation term   111
– selectivity term   110

Resolving power   89
Response factors   23
Restriction   23
Restrictor   17
Retention   10
– behaviour of structural isomers   229
Retention gap   83
Retention index   225
– calculation from empirically determined values   228
– determination from the empirical formula   229
Retention time   105 f, 213
Retention volume   116 f
Retro-Diels-Alder   261
Reverse search   242
RFIT value   242
Rotary vane vacuum pump   196

Salting out process   31
Sample collection   54
Sample
– throughput   26
– capacity   89
– loop   33
– preparation   7 ff
– quantity   23
Sampling   321, 320
Sampling intervals   341
Santovac   195
Satellites   253
Scan   214
– function   163
– rate   165, 182
– starting mass   214
– techniques   180
– time   246
Scanning conditions   231
Screening   473
Screening analysis   380
Search depth   244
Secular frequency   183
Selectivity   91
Sensitivity   46
Separation factor   117
Sewage sludge   14
SFE-Solver   23
Shark fin   89
Signal domain   311
Signal/noise ratio (S/N)   74
Significance weighting   241
Silanised glass wool   77
Silica gel   40
Silicone phases   322
Silicones   254

Silylation   500
SIM   166
– data acquisition   168
– gain in sensitivity   169
– ion statistics   254
– qualifiers   167
– set-up   168
– technique   166
– tuning the analyser   169
Similarity search   248
SISCOM   247 ff
Skewing   230
Software filters   311
Soil   349
Soiling of ion volumes and lenses   137
Solid phase extraction (SPE)   10 ff
– cartridges   11
– columns   9
– discs   12
– microextraction (SPME)   12 ff
– phases   9
Solid samples   65
Solvent effect   71
Solvent peak   71
Solvent split injection   77
Soxhlet   454
Soxhlet extraction   15
– characteristic ions from amines   259
– characteristic ions from ethers   260
– characteristic ions from ketones   260
– cleavage   259
– time required   50
Spectrum qualities   231
SPI injector   76
Spiro molecular pump   195
Split injection   70
Split ratio   70
Splitless injection   70
Spray and trap   36
Standard addition procedure   320
Standard atmospheres   56
Standard
– internal   317
– deuterated   317
– external   71
– isotopically labelled   180
– deviation   32
– conditions   230
Standardisation   318
Steroid library   235
Steroids   235
STIRS   243
Stock solution   317
Storage mass spectrometer   162
Substance domain   312
Substance exchange   115

Substance identification   218 ff
Substitution reactions   154
Subtraction of mass spectra   220
Suction capacity   193
SUMMA canister   333
Supercritical fluid extraction (SFE)   10
– aerosol formation   17, 493
– extraction agent   16
– extraction temperature   70
– modifier   17
– on-line coupling   23
– recovery   22
– restrictor   18
– sensitivity   24
– solubility   16
Supercritical fluids   16
Surrogate standard   317
Suspended particulates   454
Syringe injection   33

Tailing   198
Tandem mass spectrometry (MS/MS)   177 ff
– 10 % trough definition   132
– collision energy   182
– collision gas   182
– quantitation   513
– tandem-in-space   181
– tandem-in-time   181
Tangents   109
Target compound analysis (TCA)   476
Tenax   40 f
Terpene spectra, collection of   233
Thermal decomposition   83
Thermal extraction   65
Thermodesorption   51
– calibration   56
– desorption temperatures   58
– interfering components   58
– quantitation   465
– sample collection   54 ff
– transfer rates   471
Thick film columns   97
Thin film columns   97
Time of flight   144
TMS ethers   513
TNT   483
Total ion current   214 ff
Total sample transfer   70 ff
– PTV   74, 75
Toxicity equivalents   473, 474
Toxicological-chemical analysis   536
Toxicology library   233
Transfer line   24
Transmission   137
Trennzahl factor   117

Triazine herbicides 415
Triazines 283
Trimethylaniline hydroxide (TMAH) 442
Triple-quadrupole technique 6
Trivial names 231
Tropylium ion 260
Tuning 183
– calibration of the frequency scale 183
– SIM 230
Turbo molecular pump 193
Types of ions in mass spectrometry 144
TX library 527

Unit mass resolution 133 154

Vacuum systems 192 ff
Van Deemter curve 97, 99
Viscosity 99
VOCARB 41
VOCs 349

Volatile halogenated hydrocarbons 262 ff
– limits of quantitation 343
Volatile priority pollutants 349

Waste oil 12
Water 39, 349
– affinity for 51
– content 23
– loss of 151
– MCS 39
– Nafion drier 58
– removal of 39
Waveform 182
Wide bore capillaries 192
Wiley library 231

XAD resins 52

Zero point 311

Index of Chemical Substances

Acenaphthene 90, 170, 444, 461, 493
Acenaphthene-d_{10} 170, 446, 461
Acenaphthylene 90, 444, 461, 493
Acephate 170
Acetaldehyde 379
Acetic acid 379
Acetone 8 f, 31, 124, 155, 437
Acetylene 124
Acetylsalicylic acid 145
Acridine 276
Acrylonitrile 8 f, 379
Adamsite (DM) 308
ADBI 546
AHTN 546
Alachlor 170, 400 ff, 404
Aldicarb 170
Aldrin 170, 282, 404
Alkanes 149, 226
Alcohols 149, 261
Aldehydes 260
Alkenes 260
Alkylamines 149
Alkylaromatics 260, 266
n-Alkylbis(trifluoromethyl)phosphine sulfide 227
Alkylphenols 389, 392, 456 f
Allethrin 170
Ametrin 405
Amines 94, 149 f, 259 f, 262
4-Aminobiphenyl 483 f, 489
2-Amino-4,6-dinitrotoluene 483 f, 489
4-Amino-2,6-dinitrotoluene 483 f, 489
6-Amino-2,4-dinitrotoluene 483 f, 489
Amitryptyline 539
Amitryptylin oxide 299, 539
Ammonia 160
Amobarbital 526 f
Amphetamine 14, 298, 500 ff, 520 ff
Amphetamine-TMS 520
Anabolics 509 f
Aniline 141, 150, 262
Anthracene 90, 170, 240, 275, 444, 461, 493
Anthracene-d_{10} 275
Argon 124, 152

Atrazine 170, 228, 283, 405, 413, 416 ff
Azinphos-methyl 170, 405
Azulene 122, 170

Barban 170
Barbiturates 155, 500 ff, 526 ff
Benazoline-methyl ester 170
Bendiocarb 170, 285, 405, 413
Benalaxyl 405
Bentazon 438, 447
Benz(a)anthracene 22, 170, 444, 461
Benzene 32, 40 ff, 57, 124, 141, 155, 170, 239 f, 246, 266 f, 269, 319, 337 ff, 343 f, 346 f, 351 f, 359 f, 371 f, 377, 458
Benzene-d_6 170
Benzo(b)chrysene 170
Benzo(a)coronene 170
Benzo(b)fluoranthene 90, 170, 444 f, 448, 461
Benzo(k)fluoranthene 90, 170, 444 f, 461
Benzo(b)fluoranthracene 90
Benzo(k)fluoranthracene 90
Benzofluorene 170
Benzo(g,h,i)perylene 22, 90, 170, 444, 461
Benzo(g,h,i)perylene-d_{12} 446, 461
Benzo(a)pyrene 22, 90, 170, 276, 444 ff, 461
Benzo(e,a)pyrene 170, 448
Benzo(e)pyrene 170, 448, 461
Benzoylecgonin 500 f
Benzyl chloride 337
Binapacryl 425 f
Biperidene 529
Biphenyl 170, 293 ff, 458, 461 ff
Bornaprin 539
Bromacil 170, 416
Bromacil N-methyl derivative 170
Bromfem 431
Bromine 253, 256 f
Bromobenzene 32, 48, 54, 319, 359 f
2-Bromobutane 124
Bromochloromethane 336, 359 f
Bromodichloromethane 54, 343, 346 f, 351 f, 359 f

Bromofluorobenzene 319, 336
4-Bromofluorobenzene 360
Bromoform 42, 48. 54, 319, 343, 346 f, 351 f, 359 f
Bromophenols 389 ff
α-Bromostyrene 378 ff
Bromkal 309
Bromophos 170
Bromophos-ethyl 170, 227
Bromoxynil 422, 430 ff, 438 f
Bromoxynil-methyl ether 171, 424, 438 f
Bromoxynil-octyl ester 430 f
BTEX 41, 266, 341 ff, 345 ff, 369, 374, 378
Butabarbital 526 f
Butachlor 400
1,3-Butadiene 379
Butalbital 526
Butane 141
Butanol 30 f
n-Butanol 31
n-Butyl acetate 30, 124
Butyl acrylate 379
Butylaldehyde 379
n-Butylamine 124
n-Butylbenzene 359 f
sec-Butylbenzene 359 f
tert-Butylbenzene 359 f

Caesium iodide 187
Cannabis 504 ff
Captafol 171
Captan 171
Carbamates 285
Carbamazepin 539
Carbamazepin artefact 539
Carbaryl 171
Carbendazim 171
Carbenium 260
Carbetamid 171
Carbofuran 171
Carbon 253
Carbon dioxide 17, 141, 152, 355
Carbon disulfide 351 ff
Carbon tetrachloride 42, 48, 54, 220, 224, 319, 337, 343, 346 f, 353, 360 f
Carbonyl compounds 150, 261
Carbophenothion 227
9-Carboxy-THC 504 f
Carbowax 14
Charcoal 40
Chloral hydrate 368 f
Chlorbromuron 171
Chlorbufam 171
Chlordane 171, 404 ff

Chlordecon 399
Chlordiazepoxide 339
Chlorfenprop-methyl 171
Chlorfenvinphos 171, 216
Chloridazon 171
Chlormezanon 539
Chloroacetophenone 308
Chlorobenzylidenemalnonitrile 308
Chlorobenzene 42, 48, 54, 319, 337, 343, 351 f, 359 f, 371 f
Chlorobenzene-d5 336
1-Chloro-2-bromopropane 360
2-Chlorobutane 124
Chloroethyl vinyl ether
2-Chloroethyl vinyl ether 42
Chloroform 8 f, 42, 48, 54, 57, 122, 267, 319, 337, 343, 347, 351 ff, 359 f, 368
3-Chloro-4-methylaniline 171
2-Chloro-5-methylphenol 392
4-Chloro-2-methylphenol 392
4-Chloro-3-methylphenol 392
Chloroneb 171
Chloronitroaromatics 493 ff
1-Chloro-2-nitrobenzene 496
1-Chloro-3-nitrobenzene 496
1-Chloro-4-nitrobenzene 496
Chlorophenols 389 ff
Chloroxuron 171
Chloroxynil 430 ff
Chloroxynil-methyl ether 432 f
o-Chlorophenol 276
2-Chlorophenol 391
3-Chlorophenol 391
4-Chlorophenol 391
2-Chlorotoluene 359 f
4-Chlorotoluene 351 ff, 359 f
Chlorpromazin 539
1-Chloropropane 141
3-Chloropropane 337
Chloropropham 171
Chlorprothixene 539
Chlorpyrifos 171
Chlorthal-dimethyl 171
Chlorthiamide 171
Chlortoluron 171
Chrysene 22, 90, 170, 444, 446, 461
Chrysene-d_{12} 170, 461
Cinerin 171
Clenbuterol 233, 532 ff
Clenbuterol-TMS 233, 533
Clofibric acid 541 ff
Clomipramin 539
Cocaine 300, 497, 539
Codeine 299, 499 f, 539
Codeine-TMS 502

Caffeine 539
Coronene 170, 276
Cotinine 539
p-Cresol 277
Cumene 54
Cyanazine 171, 405, 416
Cyclohexane 30
Cypermethrin 171

2,4-D 437 ff
Dalapon 171
Dazomet 171
2,4-DB 438
2,4-DB methyl ester 171
4,4'-DDD 281, 404
DDE 138, 281, 404
DDT 138, 171, 281, 404
Decachlorobiphenyl 296
Demeton-S-methyl 171, 404, 413
Desethylatrazine 405, 413, 416
Desmethylclomipramin 539
Desmethylmedazepam 539
Desmetryn 171, 405, 412
Diacetone alcohol 324
Dialifos 171
Di-allate 171
2,3-Diaminotoluene 483 f, 489
2,4-Diaminotoluene 304, 483 f, 489
2,6-Diaminotoluene 304, 483 f, 489
Diazepam 539
Diazinon 171, 227
Dibenz(a,h)anthracene 90, 444, 461
Dibenzanthracene 170, 444
Dibenzodioxin 170, 474
Dibenzofuran 170, 448 f, 474
Dibenzopyrenes 170
Dibromochloroethane 267, 269, 343, 346 f, 351 f
Dibromochloromethane 54, 319, 359 f
2,6-Dibromo-4-chlorophenol 278
1,2-Dibromo-3-chloropropane 222, 359
1,2-Dibromoethane 337, 359
Dibromomethane 359
2,4-Dibromophenol 391 f
2,6-Dibromophenol 391 f
Di-tert-butylphenol 456
Dibutyl phthalate 326
Dicamba 438
Dicamba methyl ester 171
Dichlobenil 171
Dichlofenthion 171
Dichlofluanid 171
1,2-Dichlorobenzene 319, 337 ff, 351 f, 359 f
1,3-Dichlorobenzene 32, 337 ff, 340, 360 f

1,4-Dichlorobenzene 337 f, 360 f
1,2-Dichlorobenzene-d4 360
1,4-Dichlorobenzene-d4 446, 461
Dichlorobiphenyl 294
Dichlorodifluoromethane 336 f, 360 f
Dichloroethane 351 ff, 360 f
1,1-Dichloroethane 42, 338, 342, 351 ff, 360 f
1,2-Dichloroethane 42, 48, 54, 337, 343, 351 ff, 360 f
1,2-Dichloroethane-d4 360
1,1 Dichloroethylene 42, 337, 343, 351 ff, 360 f
cis-1,2-Dichloroethylene 337, 351, 258, 360 f
trans-1,2-Dichloroethylene 351, 360
Dichloromethane 8 f, 48, 57, 122, 267, 319, 337, 343, 351 ff, 360 f
Dichlorophenol 392 ff
2,3-Dichlorophenol 278
2,4-Dichlorophenol 391
2,6-Dichlorophenol 391
3,4-Dichlorophenol 391
3,5-Dichlorophenol 391
2,4-Dichlorophenyl acetate 278
Dichloropropene 42
Dichloropropane 54
1,2-Dichloropropane 337, 343, 351, 360 f
1,3-Dichloropropane 360 f
2,2-Dichloropropane 360 f
Dichloropropene 269
1,1-Dichloropropene 376
1,2-Dichloropropene 376
1,3-Dichloropropene 220, 222, 376
cis-1,3-Dichloropropene 337, 351 f
trans-1,3-Dichloropropene 337, 351 f
Dichlorprop 438 ff, 542 ff
Dichlorprop isooctyl ester 171
Dichlorprop methyl ester 171, 438 f
1,1-Dichloropropylene 360 f
4,6-Dichlororesorcinol 391 f
Dichlorvos 171, 405, 413
Dicofol 171
Dieldrin 171, 282, 404
Diethylbenzene 170
1,4-Difluorobenzene 336
Dihydroanthracene 170
Dihydrocodeine-TMS 502
Dihydromorphine-TMS 502
Dimethyirimol methyl ether 171
Dimethoate 171, 227, 405, 413
Dimethylbenz(a)anthracene 170
Dimethylbenzene 170
Dimethylbiphenyl isomers 457
N,N-Dimethylformamide 124
Dimethylnaphthalene 170

1,3-Dimethylnaphthalene   274
1,6-Dimethylnaphthalene   275
2,6-Dimethylnaphthalene   461
2,6-Dimethyl-4-nitrophenol methyl ether   422 f
Dimethylphenanthrene   170
Dimethylphenol   391 ff
Dinitrobenzenes   305, 483 f, 489
2,2'-Dinitrobiphenyl   483 f, 489
1,3-Dinitronaphthalene   483 f, 489
4,6-Dinitro-o-cresol methyl ether   424
2,4-Dinitrophenol   424 f
Dinitrophenol methyl ether   424 f
2,3-Dinitrotoluene   483 f, 489
2,4-Dinitrotoluene   184, 483 f, 489
2,6-Dinitrotoluene   184304, 483 f, 489
3,4-Dinitrotoluene   184, 483 f, 489
3,5-Dinitrotoluene   303
Dinobuton   425 f
Dinoseb   424 ff
Dinoseb acetate   425
Dinoseb methyl ether   425
Dinoterb   424 ff
Dinoterb acetate   425
Dinoterb methyl ether   172, 424 ff
Dioctyl phthalate   326
Diol   8
Dioxacarb   172
Dioxan   30, 149
Dioxins   see: Polychlorinated dioxins
Diphenamid   172
Diphenylanthracene   170
Dipropetryn   416
2,4-D isooctyl ester   171
Disulfoton   172
Ditalimphos   227
Diuron   172, 290
Divinylbenzene isomers   379
2,4-D methyl ester   171
DNOC   422, 425
DNOC methyl ether   172
Dodine   172
Doxepin   539
Doxylamine   539
Dympilate   405, 413

Endosulfan   172, 401 f, 404
α-Endosulfan   402
Endrin   172, 282, 404
Ephedrin   298
Epichlorohydrin   379
Ethane   124, 376
Ethers   260
Ethylene   124

Ethiofencarb   172
Ethion   396 ff, 402
Ethirimol   172
Ethirimol methyl ether   172
Ethopropos   405, 413
Ethyl acetate   8 f, 20, 30, 494
Ethyl acrylate   379
Ethylbenzene   32, 40, 42, 48, 54, 57, 170, 266, 271, 319, 337, 341 ff, 346 f, 351 f, 360 f, 377
Ethyl chloride   42, 337, 359 f
Ethylene oxide   150, 379
Ethylmethacrylate   379
Ethylnaphthalene   274
Ethylphenol   391 f
4-Ethyltoluene   337
Ethylvinylbenzene isomers   379
Etoloxamine-ISTD   539
Etrimfos   172, 405

FC 43   see: Perfluorotributylamine
Fenarimol   172, 405
Fenitrothion   172
Fenoprop isooctyl ester   172
Fenoprop methyl ester   172, 438 f
Fenthion   227
Fenuron   172
Flamprop isopropyl   172
Flamprop methyl   172
Fluoranthene   22, 90, 170, 444, 454, 461, 493
Fluorobenzene   319, 360 f
Fluorene   90, 170, 444, 461, 493
Fluorine   253
Fluoroxypyr   438 f
Fomblin   187
Formaldehyde   14
Formothion   172
Freons   336 f
Furans   see: Polychlorinated furans

Geosmin   385 ff

Haloperidol   539
Haloxifop   438
HDN   416
Helium   124, 130, 141, 152
Heptachlor   172, 399, 404
Heptachlorobiphenyl   138, 295, 468
Heptane   8 f
n-Heptane   54
1-Heptene   54
Heptenephos   405, 413

Heroin   300, 499 f
Hexabromobiphenyl (HBB)   157, 309
Hexachlorobenzene (HCB)   280
Hexachlorobiphenyl   175 f, 295, 468
Hexachlorobutadiene   319, 337, 351 f, 360
Hexane   8 f, 437
n-Hexane   30, 124
Hexaphene   170
Hexazinon   284, 405, 416
Hexogen (RDX)   302
HHCB   546
HpCDD   478
HpCDF   478
HxCDD   478
HxCDF   478
Hydrazines   150
Hydrocarbons   220, 224, 262 f, 279 f, 324, 345 ff, 374, 378, 444, 454
Hydrocodone   502, 539
Hydrogen   124, 130, 134, 253
3-Hydroxy-5α-androst-16-ene   232
Hydroxybenzonitrile   430 ff
6-β-Hydroxyemthanedienone   509 f
Hydroxytetrachlorodibenzofuran   138

Iminoethers   150
Imipramine   539
Indeno(1,2,3-cd)pyrene   22, 90, 170, 444, 461
Iodine   253
2-Iodobutane   124
Iodofenphos   172
Ionol   327
Ioxynil   422, 430 f, 488 f
Ioxynil isooctyl ether   172
Ioxynil methyl ether   172, 423 f, 438 f
Isobutane   160, 298
Isobutyraldehyde   124
2,4-D Iso-octyl ester   171
Isopropanol   30
Isopropylbenzene   360 f
4-Isopropyltoluene   319, 360 f
Isoproturon   172

Jasmolin   172

Ketones   260

Lenacil   172
Lenacil N-methyl derivative   172
Levallorphan-TMS   502

Levomepromazin   539
Levorphanol-TMS   502
Lidocain   539
Lindane   172, 280, 404
Linuron   172, 290
LSD   500 ff

Malathion   172, 287
MAM   499
MCPA   438 f
MCPA methyl ester   292
MCPB   438 f
MCPB isooctyl ester   172
MCPB methyl ester   172, 292
MDMA-TMS   520
Mecoprop   430 ff, 438 f, 542
Mecoprop isooctyl ester   172
Mecoprop methyl ester   172, 432
Medazepam   539
Metamitron   172
Metamizol-Desalkyl   539
Metazachlor   416
Methabenzthiazuron   172
Methacrylonitrile   379
Methadone   497, 523, 539
Methamphetamine HFB   521
Methane   124, 158, 376
Methandienone
Methanol   8 f, 20, 159, 355, 395, 437
Methaqualon   539
Methazole   172
Methidathion   172, 405
Methiocarb   172
Methomyl   172
Methoprotryn   235
Metoxychlor   404
Methyl acrylate   379
Methyl anthracene   170
Methyl bromide   42, 336 f, 359 f
Methyl chloride   42, 336 f, 359 f
Methylcholanthrene   170
Methylchrysene   170
Methyl-Diuron   290
2,4-D Methyl ester   292
Methylene chloride
Methylethylbenzene   170
2-Methyl-1-ethylbenzene   272
Methylfluoranthene   170
Methylisoborneol   385 f
Methyl isobutyl ketone   30
Methyl isothiocyanate   124
Methyl Linuron   291
Methyl Monuron   289
3-Methylmorphine   501

Methylnaphthalene   170, 461
2-Methyl-3-nitrophenol methyl ether   424
3-Methyl-4-nitrophenol methyl ether   424
4-Methyl-2-nitrophenol methyl ether   424
4-Methyl-3-nitrophenol methyl ether   424
5-Methyl-2-nitrophenol methyl ether   424
Methylphenanthrene   170
1-Methylphenanthrene   461
Methylphenobarbital   539
1-Methyl-2-propylbenzene   273
Methylstyrene   379
Methyl-tert-butyl ether, MTBE   351 ff
Methyl vinyl ketone   379
Metobromuron   172
Metolachlor   416
Metoxuron   172
Metribuzin   172
Mevinphos   172, 405, 413
Midazolam   539
Mirex   283, 399
Moclobemid   539
Monoacetylmorphine   501 f
Monoacetylmorphine-PFP   502
Monoacetylmorphine-TMS   502
Monochlorobiphenyl   293
Moncrotophos   172
Monoethylaminophenylpropanol   525
Monoethylaminophenylpropanone   524
Monolinuron   172
Monuron   289
Morphine   299, 500 ff
Morphine-2-PFP   499
Morphine-bis-TMS   502
Morpholine   149

Nalorphin-bis-TMS   502
Naloxone   539
Naphthalene   90, 170, 274, 319, 346, 351 f, 360 f, 444, 461, 493
Naphthalene, 1-(2-propenyl)-   448
Naphthalene-d_8 170, 444, 461
1,4-Naphthoquinone   122
Napropamide   172
Nicotine   172, 539
Nitrazepam   539
Nitroaromatics   483 ff, 493 ff
Nitrobenzene   122, 141, 306, 494
Nitrobiphenyl   483 f, 489
Nitrofen   173
Nitrogen   124, 130, 134, 141, 253, 405
2-Nitronaphthalene   483 f, 489
Nitropenta (PETN)   303
Nitrophenols   421 ff
Nitrotoluene   494

2-Nitrotoluene   301 f, 483 f, 489
3-Nitrotoluene   305, 483 f, 489
4-Nitrotoluene   301 f, 483 f, 489
Nonachlor   404
Nonachlorobiphenyl   138, 296
Nonane   404
Norcodeine-bis-TMS   502
Nordazepam   539
Nortidine   539
Nortrimipamin   539
Nortryptyline   539
Nuarimol   173

OCDD   478
OCDF   478
Octachlorobiphenyl   296, 468
Octachlorostyrene   280
1,7-Octadiene   379
1-Octene   379
Oleic acid amide   324
Omethoat   173
Oxadiazon   173
Oxides   149
Oxycodon-TMS   502
Oxygen   134, 141, 253

Paracetamol   539
Paraoxon(-ethyl)   287
Paraoxon-methyl   287
Parathion   173
Parathion-ethyl   228, 288
Parathion-methyl   173, 288
PCXE   473 ff
PCXO   473 ff
Penconazole   405
Pendimethalin   173, 416
Pentachlorobiphenyls   138, 175, 295, 453
Pentachlorobiphenylene   138
Pentachloromethyoxybenzene   443
Pentachlorophenol, PCP   442 ff, 500 ff
Pentane   141
Pentazocine-TMS   502
Pentobarbital   530, 539
Perazine   539
Perfluorokerosene, PFK   187, 192
Perfluorotributylamine   187
Perfluoroheptylazine   187
Perfluorotrinonyltriazine   187
Permethrin   173
Perylene   170, 461
Perylene-d_{12} 170, 446, 461
Pethidine   539
Phenacetin   539

Phenanthrene   22, 90, 170, 240, 246, 444, 461, 493
Phenanthrene-d_{10} 170, 446, 461
Phenazone   539
Phencyclidine   507 f
Phenmedipham   173
Phenobarbital   526 f
Phenols   124, 273, 277, 389 ff, 456 f
Phenoxalkylcaboxilic acids   437 ff
Phenylanthracene   9, 170
Phenyl urea   14, 289 f
Phenylnaphthalene   170
Phenytoin   539
Phosalone   173, 288
Phosphoro compounds   150
Phosphorus   253, 405
Phosphorus acid esters, PAE   286
β-Pinene   234
Piperazine   149
Piperidine   149 f
Pirimicarb   173, 285, 405
Pirimiphos-ethyl   173
Pirimiphos-methyl   173, 405
Polyacrylate   14
Polychlorinated Biphenyls, PCBs   22, 175, 293 f, 316, 463 f, 470 ff
Polychlorinated dioxins, PCDDs   21, 297 ff, 473 ff, 478
Polychlorinated furans, PCDFs   297 f, 473 f
Polydimethyl siloxane, PDMS   14
Poly(ethylene glycol)   94, 355
Polynuclear aromatics   273
Poly(perfluoropropylene oxide)   187
Polypropylene   11
Polysilioxanes   92 f
Primidone   539
Primidon-M   539
Promazine   539
Promecarb   285
Promethazine   539
Promethazine-M(HQ)   539
Propachlor   173
Propanes   141
Propanil   173
Propanolol   539
Propazine   416, 419
Propene   124, 376
Propham   173
Propoxur   173, 405, 413
Propylamines   141
Propylene oxide   379
Propylhexedrin-TMS   520
Propyphenazon   539
Propyl phthalate   326
Prostaglandins   513 ff

Pyrazophos   405
Pyrene   20, 90, 170444, 455, 461, 493
Pyrethin   173
Pyridine   150 f
Pyrrolidine   149

Quinalphos   216
Quinidine   539
Quinidine acetate   539
Quintozene   173

Remoxiprid   539
Resmethrin   173
Rubicene   170

Salbutamol   532 ff
Sarin   307
Sulfur   253, 258
Sulfur hexafluoride   122
Sebutylazine   416
Secobarbital   526 f
Sephadex   9
Silica   13, 40, 385
Silicon   253, 257
Silicones   323, 327
Siloxanes   253, 257
Simazine   173, 284, 413, 416, 419 f
Soman   307
Squalene   325
Styrene   31, 57, 272, 337, 351 f, 360 f, 370
Sulfides   149
Sulfur   325

2,4,5-T   438 f
Tabun   307
TCDD   1, 156, 478
TCDF   478
Tebutan   405, 413
Tecnazene   173
Tenax   40, 373, 385, 470
Terbacil   173, 405
Terbacil N-methyl derivative   173
Terbufos   396 ff
Terbutryn   405
Terbutylazine   284, 405, 413, 416, 419
Tetrachlorobenzyltoluene   138
Tetrachlorobiphenyl   294
2,3,7,8-Tetrachlorodibenzodioxin   297
2,3,7,8-Tetrachlorodibenzofuran   297
Tetrachloroethane   30, 42, 48, 337, 343, 351 f, 360 f, 376

Tetrachloroethylene 48, 54, 57, 269, 337, 343, 346 f, 351, 360 f, 371
Tetrachloromethane 224
Tetrachloromethoxybiphenyl 138
Tetrachlorvinphos 173
Tetrahydrocannabinol 504 ff
Tetrahydrocannabinol-carboxylic acid ,THC 500 ff
Tetrahydrofuran 48, 149
Tetrahydropyran 149
Tetrazepam 537 ff
THC-d$_3$-TMS 506
THC-TMS 506
Tetrasul 173
Thiabendazole 173
Thiofanox 173
Thiometon 173
Thiopene 351 f
Thiophanat-methyl 173
Thiram 173
Tilidine 539
TNT see: trinitrotoluenes
Toluene 20, 30, 32, 40, 42, 48, 54, 57, 124, 141, 170, 269, 318 f, 337, 343 f, 346 f, 360 f, 371, 375 f
Toluene-d$_8$ 170, 318, 346 f, 351 f, 360
Tramadol 539
Triadimefon 228
Triadimenol 405
Tri-allate 173
2,4,6-Triaminotoluene 483 ff, 489
Triazines 187, 283, 415 ff
Tribromophenol 392
Tributyl phosphate 325
Trichlopyr 438
Trichlorobenzene 175, 351 f, 450
1,2,3-Trichlorobenzene 351 f, 360 f
1,2,4-Trichlorobenzene 32337, 351 f, 360 f, 451
Trichlorobiphenyl 294
Trichlorodifluoromethane 336 f
Trichloroethane 30, 343
1,1,1-Trichloroethane 48, 54, 57, 319, 337, 343, 346 f, 351 f, 360
1,1,2-Trichloroethane 337, 351 f, 360
Trichloroethylene 42, 48, 54, 57, 268, 319, 337, 343, 346 f, 351 f, 360 f, 371

Trichlorofluoromethane 42, 319, 360 f
Trichloropropane 42, 360
Trichlorfon 173
Trichloronat 227
Trichlorophenol 390 ff
Tricresyl phosphate 328
Tridemorph 173
Trietazine 173
Triethylamine 48
1,1,2-Trifluoro-1,2,2-trichloroethane 268
Trifluorobromomethane 343
Trifluorochloromethane 343
Trifluralin 173
Trihexylphenidyl 539
Trimethylbenzenes 170, 319, 337, 360 f
Trimethylhydroxides 442
2,3,5-Trimethylnaphthalene 461
Trinitrobenzenes 490
Trinitrotoluenes, TNT 303, 483 f, 489, 493
1,3,5-Trioxan 48
Triphenylphosphine 327
Triphenylphosphine oxide 328

Ugilec 175

Vamidothion 173
Vinclozolin 173, 416
Vinyl acetate 379
Vinyl chloride 42, 54, 268, 336 f, 355 f, 360 f, 368 ff, 379
Vinylidene chloride 379

Water 8 f, 20, 159, 343 f, 362, 385 f, 415 f, 422 f

Xanthene 473 f
Xanthone 473 f
Xylenes 30, 32, 48, 54, 57, 124, 155, 266, 271, 319, 337, 341 ff, 346 f, 360 f, 371, 545

Zotepin 539